The ESD Control Program Handbook

The ESD Control Program Handbook

Jeremy M Smallwood
Electrostatic Solutions Ltd
Southampton, UK

This edition first published 2020

Registered Offices
John Wiley & Sons, Inc., 111 River Street, Hoboken, NJ 07030, USA
John Wiley & Sons Ltd, The Atrium, Southern Gate, Chichester, West Sussex, PO19 8SQ, UK

Editorial Office
The Atrium, Southern Gate, Chichester, West Sussex, PO19 8SQ, UK

For details of our global editorial offices, customer services, and more information about Wiley products visit us at www.wiley.com.

Wiley also publishes its books in a variety of electronic formats and by print-on-demand. Some content that appears in standard print versions of this book may not be available in other formats.

Library of Congress Cataloging-in-Publication Data

Names: Smallwood, J. M. (Jeremy M.), author.
Title: The ESD control program handbook / Jeremy M Smallwood, Electrostatic Solutions Ltd, Southampton, UK.
Description: First edition. | Hoboken, NJ : John Wiley & Sons, Inc., [2020] | Series: IEEE | Includes bibliographical references and index.
Identifiers: LCCN 2020006384 (print) | LCCN 2020006385 (ebook) | ISBN 9781118311035 (hardback) | ISBN 9781118694572 (adobe pdf) | ISBN 9781118694558 (epub)
Subjects: LCSH: Electronic apparatus and appliances–Protection. | Electric discharges. | Electrostatics.
Classification: LCC TK7870 .S5265 2020 (print) | LCC TK7870 (ebook) | DDC 621.381/044–dc23
LC record available at https://lccn.loc.gov/2020006384
LC ebook record available at https://lccn.loc.gov/2020006385

Cover Design: Wiley
Cover Image: © pinging/Getty Images

Set in 9.5/12.5pt STIXTwoText by SPi Global, Chennai, India
Printed and bound in Singapore by Markono Print Media Pte Ltd

10 9 8 7 6 5 4 3 2 1

To Jan, who has often put up with my grumpy non-communication while I've been writing this book. To Alia, now making her own life journey. To Caroline, who tragically died so young.
To the subject of electrostatics and ESD, that has kept me occupied, perplexed and challenged for many hours.
To the many people who have attended my courses and asked so many awkward questions that helped me understand while trying to explain. To my fellow ESD practitioners whose opinions and expertise give us many interesting, and sometimes heated, debates – and who mainly agree the answer to most electrostatic questions is – "it depends."

Contents

Introduction

Electrostatic discharge (ESD) can damage or destroy many types of modern electronic components, or modules or assemblies containing electrostatic discharge–susceptible (ESDS) components.

In electronics manufacturing, sensitivity to ESD became a general concern after the adoption of metal-oxide-semiconductor (MOS) transistor technology exacerbated by decreasing internal physical size of semiconductor component features and the rise of integrated circuits (ICs). The first Electrical Overstress/Electrostatic Discharge Symposium in the USA was organized in 1979 (Reliability Analysis Center 1979). The 1980 Symposium (Reliability Analysis Center 1980) shows papers on topics as diverse as theory and practice, device failure analysis studies, failure mechanisms and modeling, design of device protective networks, and implementing ESD controls, facility evaluation and effective training. Standards and technical handbooks also emerged around this time. The standards gave requirements for an ESD control program, while the technical handbooks gave technical data and tutorial material useful for educating the user and developing the ESD control program. ElectroStatic Attraction (ESA) of contaminant particles is a problem for manufacturers of semiconductor devices and displays. For operating electronic systems, ESD provides a source of electromagnetic interference (EMI) that can result in system crash, malfunction or data corruption.

Thus, the issues of ESD in electronic components and systems give two areas of interest. Issues of ESD control during electronic component, assembly and system manufacture are largely concerned in preventing damage in the unpowered non-operational state and ensuring that product reaches the customer in good condition without compromise to appearance or reliability. This area can itself be further subdivided into

- Electrostatic and ESD issues affecting product yield and quality during semiconductor wafer scale fabrication
- ESD issues affecting product yield and quality during component, assembly and system manufacture and assembly, sometimes known as "factory issues"
- Design of semiconductor devices to withstand ESD up to target levels

ESD interference and damage during working electronic system operation is generally viewed as part of ElectroMagnetic Compatibility (EMC) and the responsibility of a different community. In some areas (e.g. Europe) electronic products are subject to ESD immunity test as part of their evaluation for fitness to be placed on the market (and qualification for CE marking) (Williams 2007).

Nevertheless, there is some overlap and often confusion between these areas. EMI caused by ESD in the manufacturing process can cause interference to product testers and lead to rejection of product and hence reduction of yield. In ESD test in EMC, ESD applied to exposed circuit connectors can lead to component or system hardware failure and may lead to requirements for ESD robustness of components that connect to the outside world.

This book is largely concerned with the development and maintenance of an ESD control program for protection of ESD susceptible components and assemblies during electronic system and assembly manufacture. This is the so-called "factory issues" area of ESD control. It is intended that the book can be used as a handbook or practical guide for personnel working in ESD control in electronics or other companies that handle unprotected ESD susceptible parts. At the same time, sufficient background information and technical explanation is given to enable the user to understand the principles and practice of effective ESD control.

Personnel working in this field can have a wide variety of technical background and do not necessarily have strong electronics or electrical understanding. Many will not have had opportunity to attend courses on ESD control other than basic ESD awareness. It is surprising that at the time of writing very few University electronics related courses offer any modules on ESD control. Conversely there are as yet few industry courses available that deal with the subject in any depth other than basic ESD awareness. Worldwide, there are still only a very few courses and qualifications available to those who wish to obtain a good grounding in the field.

So, I have attempted to present the subject with a minimum of theory to make it accessible to those who do not have a strong relevant theoretical background. This is balanced by sufficient description of background theory for understanding the material presented, with references and a bibliography of further reading for those who wish to go into the subject in greater depth. The intention is to reveal and clarify the principles behind an area often considered a mysterious "black art." In many ways, I have tried to write the book I would have liked to have found when I started learning about ESD control in the electronics industry.

A widespread current approach to development of an ESD control program is to comply with the requirements of an ESD control standard such as ANSI/ESD S20.20 or IEC 61340-5-1. It is often thought that this is sufficient to ensure that product ESD damage is brought under control. While this can be successful, if applied with insufficient knowledge it can lead to a program that is not well optimized or fails to address all the ESD threats (Smallwood et al. 2014; Lin et al. 2014). With knowledge and understanding an optimized and effective ESD control program may be achieved and maintained, often with lower costs. Nevertheless, compliance with an ESD control standard is advantageous and can help demonstrate, especially to customers, the seriousness with which ESD control is treated in the facility. Development of an ESD control program in compliance with the most widely used and respected ESD control standards 61340-5-1 and S20.20 is therefore discussed in some length. Properly specified ESD control programs compliant with these standards are held to be adequate to protect ESD sensitive devices with withstand voltages down to 100 V Human Body Model (HBM), while also addressing basic ESD risks due to charged metal objects and charged devices.

ESD susceptible components become ever more sensitive to ESD damage as time goes on, due to component technology developments. The need for development of ESD control programs through knowledge and understanding rather than rote application of standard techniques will grow as a greater number of more ESD sensitive components are handled in electronic manufacturing, assembly, and maintenance processes in the future. In parallel with the development of ESD control techniques and standards, a massive research effort has supported on-chip ESD protection networks aimed at reducing device ESD susceptibility, with target withstand voltages of 2 kV HBM, 200 V Machine Model (MM) and 500 V Charged Device Model (CDM) (Industry Council 2011). In the early 2000s The Industry Council on ESD Target Levels was formed with members from IC manufacturing and electronics assembly companies, and independent consultants in the industry. In the face of increasing difficulty in achieving the target ESD withstand levels, and the belief that modern electronic manufacturing companies have ESD control programs routinely achieving the standard protection levels, they recommended reduction of on-chip target protection levels to 1 kV HBM, 30 V MM and 250 V CDM (Industry Council 2011, 2010a,b). At the same time, many discrete components and ICs exist that for various reasons do not have on-chip ESD protection or otherwise have lower ESD withstand voltage than these levels. It seems likely that this will be the first of many reductions in target level driven by technology changes and the assumption that industry can handle lower ESD withstand components.

While this book is primarily intended to support the industry factory practitioner, I hope that this book will encourage and enable Universities and Further Education organizations to offer courses and modules on ESD control for personnel who wish to make a career in electronics production and related fields.

This book does not attempt to address electrostatics and ESD control in semiconductor wafers and device manufacture, or device design for ESD protection. The former may be still as yet inadequately covered by the very few books available on the subject but is better discussed in a book more focused on this technology area such as Welker (2006). The topic of device design is best covered by specialist books such as Amerasekera and Duvvury (2002) and Wang (2002).

ESD immunity of operating electronic systems is left to be treated in other books as part of EMC issues, except for some discussion confined to areas of overlap with the ESD factory issues topic. This field is more concerned with the design of electronic systems for immunity to ESD than it is with ESD control (Ind. Co. White Paper 3, Johnson and Graham 1993; Montrose 2000; Williams 2007).

While electrostatic control is used in other industries such as explosives and flammable materials handling (the latter known in Europe as "ATEX"), these are typically governed by other standards or regulations. They are only mentioned in this book to draw attention to possible confusion areas and help avoid mistakes, for example in equipment specification and sourcing.

While this book could be read "cover to cover" it is probably more likely that the reader will "dip into" specific chapters as the need arises to learn about different topics while working in ESD control. The book has been written with this in mind. For those who wish to go deeper into the subject, lists of references are provided with each chapter.

Every specialist subject has its own set of specialist terms or uses specific terms in specific ways. Chapter 1 defines and introduces the reader to the commonly used terminology in ESD control. Whilst this chapter forms a general introduction to the key concepts and terminology in the field, it is also likely to be used to revise or clarify the meaning of terms during reading of other chapters. This is why the definitions and terminology has been provided together in one chapter rather than being defined and explained as required during the remainder of the book. Chapter 2 then explains in more detail the principles that underlie ESD control work.

Chapter 3 discusses ESD susceptible devices, and how ESD susceptibility of a component is measured. The range of ESD susceptibility of components, and current trends in ESD susceptibility are reviewed. The topic of failure analysis as it is applied to ESD failed components is outlined. Some case studies of ESD failures from the literature are briefly described.

Chapter 4 describes the "seven habits of a highly effective ESD program." This is a way of explaining the essential activities of an effective ESD control program, that the author has used in ESD training work for many years. If these activities are effectively and habitually implemented, it is likely that an ESD control program will be, and remain, effective. If any one of them is neglected, it is likely the effectiveness of the ESD control program will eventually suffer.

Most basic ESD control techniques and standards mainly address manual handling of ESD susceptible devices, components, and assemblies. Chapter 5 extends the discussion to ESD control in automated systems, processes and handling, which form a major part of modern electronic manufacture.

Chapter 6 explains the approach and requirements given by the IEC 61340-5-1 and ANSI/ESD S20.20 ESD control standards at the time of writing. These standards are continually updated as time goes on, and so the reader is advised check for current versions available at the time of reading.

Chapter 7 gives an overview of the equipment and furniture commonly used in ESD control and commonly specified for use in an electrostatic discharge protected area (EPA) to control common ESD risks. The chapter explains how these often work together as part of a system and must be specified with that in mind.

ESD protective packaging is one of the most misunderstood areas of ESD control. ESD packaging is now available in an extraordinary range of forms from bags to boxes and bubble wrap to tape and reel packaging for automated processes. The principles and practice of ESD protective packaging are explained in Chapter 8. This is a deep and constantly developing subject in itself, and this chapter can barely do more than give an introduction to it.

The thorny question of how to evaluate an ESD control program is addressed in Chapter 9 with a goal of compliance with a standard as well as effective control of ESD risks and possible customer perceptions.

Whilst evaluating an existing ESD control program provides challenges, developing an ESD control program from scratch provides others. Chapter 10 gives an approach to this.

ESD control product qualification and compliance verification is an essential part of an ESD control program. Standard test methods have been developed and specified to go with compliance with ESD control standards. These are explained in Chapter 11. The ESD control program may also need to use control measures and equipment that are not currently specified in the standards. Some examples of test methods that may be used with these are also given in this chapter.

ESD Training has long been recognized as essential in maintaining effective ESD control. Chapter 12 discusses this in more detail. It describes some demonstrations and techniques which the author has used to help trainees understand static electricity, ESD and static control in practice.

Finally, Chapter 13 attempts to look at where ESD control may go in the near future.

References

Amerasekera, A. and Duvvury, C. (2002). *ESD in Silicon Integrated Circuits*, 2e. Wiley. ISBN: 0 470 49871 8.

Industry Council on ESD Target Levels (2010a) White paper 2: A case for lowering component level CDM ESD specifications and requirements. Rev. 2.0. http://www.esdindustrycouncil.org/ic/en/documents/6-white-paper-2-a-case-for-lowering-component-level-cdm-esd-specifications-and-requirements [Accessed: 10th May 2017]

Industry Council on ESD Target Levels (2010b) White paper 3: System Level ESD Part I: Common Misconceptions and Recommended Basic Approaches. Rev. 1.0 http://www.esdindustrycouncil.org/ic/en/documents/7-white-paper-3-system-level-esd-part-i-common-misconceptions-and-recommended-basic-approaches [Accessed: 10th May 2017]

Industry Council on ESD Target Levels (2011) White paper 1: A case for lowering component level HBM/MM ESD specifications and requirements. Rev. 3.0. Available from: http://www.esdindustrycouncil.org/ic/en/documents/37-white-paper-1-a-case-for-lowering-component-level-hbm-mm-esd-specifications-and-requirements-pdf [Accessed: 10th May 2017]

Johnson, H. and Graham, M. (1993). *High Speed Digital Design – A Handbook of Black Magic*. Prentice Hall. ISBN: 0 13 395724 1.

Lin N, Liang Y, Wang P. (2014) Evolution of ESD process capability in future electronics industry. In: *15th Int. Conf. Elec. Packaging Tech.*

Montrose, M. (2000). *Printed Circuit Board Design Techniques for EMC Compliance*, 2e. Wiley Interscience/IEEE Press. ISBN: 0 7803 5376 5.

Reliability Analysis Center (1979). *Electrical Overstress/Electrostatic Discharge Symposium Proceedings. EOS-1*. Griffiss AFB, NY: Reliability Analysis Center.

Reliability Analysis Center (1980). *Electrical Overstress/Electrostatic Discharge Symposium Proceedings. EOS-2*. Griffiss AFB, NY: Reliability Analysis Center.

Smallwood J., Taminnen P., Viheriaekoski T. (2014) Paper 1B.1. Optimizing investment in ESD Control. In: *Proc. EOS/ESD Symp. EOS-36*.

Wang, A.Z.H. (2002). *On-Chip ESD Protection for Integrated Circuits*. Klewer Academic Press.

Welker, R.W., Nagarajan, R., and Newberg, C. (2006). *Contamination and ESD Control in High-Technology Manufacturing*. Wiley-Interscience/IEEE Press. ISBN-10: 0 471 41452 2 ISBN-13: 978 0 471 41452 0.

Williams, T. (2007). *EMC for product designers*, 4e. Newnes. ISBN-13: 978-0750681704 ISBN-10: 0750681705.

Further Reading

Danglemeyer, T. (1999). *ESD Program Management*, 2e. Springer. ISBN: 0-412-13671-6.

ESD Association (2014) ANSI/ESD S20:20-2014. *ESD Association Standard for the Development of an Electrostatic Discharge Control Program for – Protection of Electrical and Electronic Parts, Assemblies and Equipment (excluding Electrically Initiated Explosive Devices).* Rome, NY, EOS/ESD Association Inc.

ESD Association. (2016) ESD Association Electrostatic Discharge (ESD) Technology roadmap, revised 2016. Available from: https://www.esda.org/assets/Uploads/docs/2016ESDATechnologyRoadmap.pdf [Accessed: 10th May 2017].

International Electrotechnical Commission (2016) IEC 61340-5-1: 2016. *Electrostatics – Part 5-1: Protection of electronic devices from electrostatic phenomena - General requirements.* Geneva, IEC.

Foreword

I was quite flattered when Dr. Jeremy M Smallwood asked me to write this foreword for his book. I view this as a significant honor. Dr. Smallwood (Jeremy) and I have worked together on standards for electrostatics since the mid-1990s after the IEC formed Technical Committee 101 – *Electrostatics*. Both of us had considerable time invested in the standards process on our sides of the Atlantic: Jeremy with BSI and myself with the ESD Association (ESDA). In the early 1990s, several things occurred: The IEC formed Technical Committee-TC101 – *Electrostatics*. Most of the original members were from the CENELEC committee that produced the electrostatics document CECC00015 along with delegates from other non-European countries; and the ESDA became a recognized American National Standards (ANSI) development body, thus able to officially represent the United States to the IEC. I was appointed the first lead delegate from the United States National Committee, and Jeremy was appointed delegate from the UK National Committee. I had the pleasure of hosting an early IEC TC101 working group meeting in Austin, Texas, in 1996, following our annual EOS/ESD Symposium held that year in Orlando, Florida. That was the first time I met Jeremy. While the early years of deliberation on standards for electrostatics were contentious at times, the committee eventually came together to form a cohesive group and has prepared important standards, recognized around the world. Jeremy had considerable input and influence during the formative years of TC101, including his long-term appointment as the TC101 Chairman. Currently, he is the lead delegate from the UK and participates actively on many working groups.

As a past president of the ESDA, I was delighted to present Jeremy with the ESD Association's Industry Pioneer Award in 2010 in recognition of his contribution to the science of electrostatics, and in particular, for his role in standards development.

There are many good books on the fundamentals of electrostatics by well-known authors and practitioners such as Professor Niels Jonassen, University of Denmark and Professor A.D. Moore (University of Michigan). There are other books that deal with specific issues in electrostatics and these can be found in the Bibliography of this book. However, this is one of the few that covers all the aspects of the modern standardization process for electrostatic control in the manufacturing of electronics and other materials sensitive to electrostatic influences. This book is designed to assist a novice to the world of electrostatics as well as the expert practitioner. I believe the book could be a useful reference to anyone who has to deal with electrostatics in any field and in particular, electronics manufacturing operations. The principles and standards discussed herein may be applied to any manufacturing area and process. This book would also be useful as a text for a college level course dealing with electronic design when the subject matter turns to reliability and sensitivity of electronic parts and assemblies.

Dr. Smallwood uses his considerable experience in standards development and practical experience to guide the reader through the maze and tangle often associated with application of standards to a manufacturing process. Since electrostatics is a natural phenomenon and around us all the time, one would think standardization is an improbable task (maybe leaning toward impossible). The practical approach taken by Jeremy in Chapters 9 and 10 helps sort out the implementation and program management processes.

Chapter 12 on Training should be extremely useful to anyone that has to set-up, run and maintain an ESD control program. Jeremy has provided some very useful "tips" on training considerations, garnered from his years of experience in providing basic to advanced classes on the "art and science" of electrostatics.

Another important section is Chapter 6 which covers standards very well. Since the early 1990s, the understanding of how to deal with the "mysteries" of electrostatics in industry has increased dramatically, with the result being able to develop very useful and practical standards, standard practices, advisories and operating guides. While we must appreciate and understand that static electricity cannot be prevented, we now know how to live with it in the manufacturing world and generally can provide mitigation techniques to resolve most issues before serious problems occur. That being said, fires and explosions, product damage and loss of life occur every year due to static electricity happening when least expected or when proper procedures are not followed. Simply having a bad connection to ground (earth) in the wrong place at the wrong time can (and does) have catastrophic results more often than most will realize.

The first step in resolving or preventing electrostatic issues is to develop an understanding of the phenomenon. Chapters 1–5 provide an excellent introduction to the basic considerations involved in electrostatics and will be a great starting point for the novice in this subject. Even the experienced practitioner will find these chapters useful for review and provide a reminder of what they have "forgotten." Overall the book is written for the non-technical person to be able to understand the electrostatic phenomenon but the "expert" in the field will also find it useful. The extensive bibliography provides a great resource for anyone needing details on any of the related subjects.

I am confident that you will find this book as entertaining, enlightening and useful as I have. Happy reading and stay safe.

David E. Swenson, President, Affinity Static Control Consulting, LLC
deswenson@affinity-esd.com
2609 Quanah Drive, Round Rock, Texas, USA 78681
Past President, ESD Association – 1998, 1999, 2008, and 2009
ESD Association Outstanding Contributions Award, 2002
Joel P. Weidendorf Memorial Award – for contribution to ESDA Standards Development – 2004
Edward G. Weggeland Memorial Award – for contributions to the operation of the ESD Association – 2014

Preface

Although the subject of ESD has been a concern to the electronics industry since the 1970s, there have been relatively few books written on the subject from the point of view of the person who has to put together, or evaluate, maintain, and update an effective ESD control program. In my work as a consultant and trainer I have met many people in this position. Some of these, when I met them, had recently had responsibility for the ESD program thrust upon them with little or no experience of the subject. Others had some knowledge but were confused by the array of facts and myths they had learned about the subject. Still others had developed true expertise over a considerable experience in the subject and developing and running their own company ESD Programs. Of course, I have learned from them all – as a trainer and consultant I find that I learn most from trying to explain my subject to others. Doing so often challenges my own understanding and causes me to think things through with greater clarity.

In the process, I started trying to simplify the presentation of the principles of ESD prevention. With a nod to Steven Covey's "Seven Habits of a Highly Effective People" the "Seven Habits of a Highly Effective ESD Program" was born. Why "Habits"? Because if these things are done habitually when handling ESD susceptible components and assemblies, we are well on the way to having an effective ESD control program.

In the mid 1990s, then working for ERA Technology Ltd., I started participating in development of British Standards at BSI in Chiswick, London. Through this work I soon found myself participating also in international standards through BSI's participation in International Electrotechnical Commission (IEC) Technical Committee101 "Electrostatics" which was formed in the mid 1990s. Standards work was an eye-opener and I found it simultaneously highly stimulating and very frustrating. Stimulating, because I found myself talking with experts from around the world whose knowledge of their field could be exceptionally deep, and their experience widely varied and practical. At BSI we could argue for hours about technical issues, and how best to write a standard test method, that could be understood and reproduced by anyone with reasonable technical ability and expected to produce results which would agree with others who had done the same. In international standards, we would first have to agree on a test method acceptable to the participating National Committees. There could be several to choose from already in use amongst the standards from the participating countries. Naturally each National Committee expert would favor a particular approach, especially if it was already adopted within their country. Mostly, of course, these would produce different results due to different conditions and methodologies.

We would discuss these at length amongst a group of international experts whose cultures and experiences could be very different, and English often not their mother tongue. The resulting methodology would have to be acceptable in the very different climates, conditions, and working practices in Japan, UK, USA, Canada, Scandinavia, France, Germany, Italy or wherever the participating experts were from. In most cases the end product would have to be translatable into the expert's mother language for publication. We found that some common ways of writing in English can be difficult to translate or may be unclear in other languages. English phrases or words can even mean different things to an American or a UK English speaker – leading to long discussions about the best wording for a single sentence!

In the early 2000s the ESD Association standards were becoming widely accepted in the electronics industry, and their use spreading from North America to other areas of the world. Encouraged by ESDA standardization experts working with IEC TC101, the decision to rewrite IEC 61340-5-1 with an unofficial harmonization with ANSI/ESDA S20.20 was a landmark decision by the IEC TC101 Working Group 5 revising this standard. Subsequent further harmonization has simplified the task of the ESD Coordinator, especially in multinational companies.

As time goes on, the components commonly handled in electronics facilities are becoming more susceptibility to ESD damage. The variety of facilities and processes in which they are handled, stored or transported grows greater. This means that there is an increasing necessity for the person developing, implementing and maintaining an ESD control program to understand, analyze and specify effective protection against ESD risks.

Part way through writing this book, I realized I was trying to write the book I would have liked to find when I first got into the subject of ESD control in the electronics industry. This book aims to help the reader understand the principles and practice of ESD control to the point where they can make the decisions needed to develop an effective and optimized ESD control program compliant with the current ESD control standards. To do this one needs to understand the purpose of ESD control equipment and materials, and how to specify and test that it does the job intended. If the reader wishes to improve their knowledge further, the references and bibliography given should give them a good starting point. Perhaps most importantly, I hope this book will help the reader find that an initially mysterious set of practices is actually based on sound engineering principles that they can learn to apply with confidence.

Acknowledgments

I would like to acknowledge my debt to all the experts in the field of ESD control and standardization who through discussions and published work have contributed to the current state of my understanding. I am also indebted to all my clients and course attendees who have challenged me to clarify, explain and justify effective ESD control techniques applied in many different situations.

I would especially like to thank David E Swenson for his comments on the text and for contributing photographs and other material, and for writing the Foreword, as well as for many enlightening discussions over the years. Dave performed the extraordinary feat of reading and commenting on almost all the draft Chapters at least once. This helped enormously in picking up my mistakes and typographic errors, adding or clarifying important points and generally improving my work.

Several other friends and colleagues have also very kindly read and commented on Chapters of this book and encouraged me in this work. Special thanks are due to Rainer Pfeifle, Charvakka Duvvury and Christian Hinz who each reviewed and commented on various Chapters in detail. Bob Willis also contributed comments, and Charles Cawthorne kindly provided me some photographs from his own ESD training materials. Lloyd Lawrenson kindly allowed me to use Kaisertech facilities for some of my photography. I'm indebted to Lisa Pimpinella of the ESD Association for arranging permission for me to include figures from the 2016 ESD Association Electrostatic Discharge (ESD) Technology Roadmap.

Last, but definitely not least, I would like to thank my wife Jan for her good-natured tolerance of my absent mindedness and lack of communication when engrossed in my work, and my daughter Alia for helping improve some of my photographs in preparation for publication in this book.

1

Definitions and Terminology

As with any specialist subject there are many terms that are the "jargon" of the subject that can be confusing to the newcomer. There are also terms that have specific meanings in the context of these standards but may have different meanings in common parlance. The intention here is not to give strict and rigorous academic definitions, but to assist the newcomer to the field to understand the following chapters.

Sometimes a range of meanings is common in different industries. For example, the terms *conductive*, *static dissipative*, *insulative* or *insulating*, and *antistatic* can mean many different things to different people from different industry areas or in the context of different standards or electrostatic discharge (ESD) control product types. In most cases, only the meaning common in ESD control, and in particular in the context of the IEC 61340-5-1 and ANSI/ESD S20.20 and related standards, is emphasized here.

The task of supervising an ESD control program is often given to personnel from many technical and educational backgrounds. For this reason, the minimum of prior technical knowledge is assumed in this book.

Despite this, some of the terms used in this document are defined with basic mathematical relationships given where appropriate. This is because simple mathematics often helps to clarify the subject and, in some cases, may be essential to helping the user understand how to specify aspects of an ESD control program. In many cases, these aspects, and their practical importance and application, are discussed further in Chapter 2.

1.1 Scientific Notation and SI Unit Prefixes

In electrostatics and ESD work, we often have to deal with very large or very small numbers. For example, the resistance of a material may be measured to be 10 000 000 000 Ω. For convenience and clarity, we use scientific notation and SI unit prefixes as shorthand for numbers (http://physics.nist.gov/cuu/Units/prefixes.html).

In scientific notation, the number is rewritten in the form $a \times 10^b$, where a lies between 1 and 10, and b is the number of decimal places a must be shifted to get the full number. This is probably most easily understood by examples of resistance and capacitance (Table 1.1). Sometimes, when the number a is simply 1, it is omitted.

The ESD Control Program Handbook, First Edition. Jeremy M Smallwood.

Table 1.1 Examples of use of scientific notation and SI prefixes.

Value	Scientific notation	SI prefix
150 Ω	1.5×10^2 Ω	150 Ω
22 000 Ω	2.2×10^4 Ω	22 kΩ
35 000 000 Ω	3.5×10^7 Ω	35 MΩ
1 000 000 000 Ω	1.0×10^9 Ω or 10^9 Ω	1 GΩ
1 000 000 000 000 Ω	1.0×10^{12} Ω or 10^{12} Ω	1 TΩ
0.000 022 F	2.2×10^{-5} F	22 μF
0.000 000 001 F	1.0×10^{-9} F or 10^{-9} Ω	1 nF
0.000 000 000 15 F	1.5×10^{-10} F	150 pF
0.000 000 000 001 F	1.0×10^{-12} F	1 pF

1.2 Charge, Electrostatic Fields, and Voltage

1.2.1 Charge

The *charge* is a property of elementary particles – electrons and protons – that make up the atoms of all materials. All materials are made up of positively charged atomic nuclei and negative electrons (Cross 1987). The charge on a proton in the nucleus is labeled positive, and the charge on an electron is labeled negative. The charge effects of a proton are equal and opposite to those of an electron, so if a proton and electron are together in an atom, their effects cancel exactly. In this case, the atom is neutral. Atoms of different elements have many protons and electrons, depending on the particular element. For example, a hydrogen atom has 1 proton and 1 electron, and a carbon atom has 12 protons in the nucleus and 12 electrons surrounding it. An object or material is made of huge numbers of atoms and so extraordinary numbers of electrical charges.

When talking of static electricity, it is often said that charge is "generated" in certain circumstances. That is not so – all that happens is that a small number of negative charges become separated from their positive companions in a material and end up in a different place. For every negative charge appearing somewhere, there must be a positive charge appearing somewhere else. An object is described as charged if it has a net imbalance of the number of positive and negative charges that it contains. The electrical effects of the charges are no longer balanced, and a net static electrical charge exists at that location. It is this net charge imbalance that we are referring to when we talk about the charge on an object or material.

The unit of charge is the coulomb (C). In practice, the coulomb is a rather large amount of charge and microcoulombs (μC, 10^{-6}C), nanocoulomb (nC, 10^{-9}C), or even picocoulombs (pC, 10^{-12}C) are more usual. A single electron or proton has a charge of 1.6×10^{-19}C. So, an object having even 1nC of net charge has a large number, 6.2×10^9, of unneutralized electrons or protons.

1.2.2 Ions

Ions are very small particles having a small electrostatic charge. They are naturally present in the air but may also be generated deliberately or accidentally around objects at high voltage.

Charges are naturally present in atoms. Protons in the atomic nucleus have positive charge, and electrons in the atom have negative charge. A negative ion is formed when a particle gains one or more electrons. A positive ion is formed when a particle loses one or more electrons. Ions may consist of free electrons, single atoms, many atoms, or molecules (Wikipedia 2018). Sometimes these ions may become attached to larger particles.

1.2.3 Dissipation and Neutralization of Electrostatic Charge

Where there is an imbalance of charges present, there will usually be voltage differences.

Charges exert forces on each other and create an electrostatic field in which various effects occur. Like charges repel each other, and unlike charges attract each other.

If like charges have built up in a region, they repel each other, and if they are able to move, they will move apart and gradually spread out and dissipate. Unlike charges will be attracted and move together.

When opposite polarity charges are sufficiently close together, their effects cancel, and the charge is said to be neutralized.

1.2.4 Voltage (Potential)

Electric *potential* is defined in terms of the work done in moving a charge from one place to another in an electric field (Cross 1987). If a charge Q is moved a distance s against a uniform electric field E, the potential difference V between the start and end positions is

$$V = QEs$$

The energy taken to move a charge between two points is the same, no matter what route is taken between the points. Potential difference is measured in volts (V) and is often referred to as *voltage*. The unit volt (V) is equivalent to joules per coulomb. Voltage is a measure of the potential energy at a point and is perhaps analogous to pressure in a fluid system or height in a gravitational system.

Engineers often talk about the potential of (for example) a conductor (see Section 1.7.3 for a discussion of conductors and insulators), as a synonym to voltage. This is not strictly correct as potential is strictly the work done in bringing a charge from infinity to the place of measurement (Jonassen 1998).

A voltage or potential difference at a place of measurement must always be referred to another place. In practice, the potential difference is usually quoted with reference to the potential of the earth (also referred to as *ground*; see Section 1.5), which is defined for convenience as zero volts. If this other place is not specifically stated, it is usually ground (the earth).

All points in space surrounding a charge have a voltage (potential) – typically this voltage will be different from its neighboring points. For a conducting surface, if it is not initially an equipotential, voltage differences cause charge (current) to flow until the voltage around the surface is eventually equal. So, an electrically conducting surface in equilibrium is an equipotential surface.

1.2.5 Electric or Electrostatic Field

Any charge has a region of influence around it, in which various electrostatic effects are noticed – this region is the electric (electrostatic) field due to the charge. Charge is the fundamental source of static electricity, and the electrostatic field shows the effect of the charge in the world around the charge source. In this field, we find that

- Like polarity charges are repelled.
- Opposite polarity charges are attracted.
- Conductors (e.g. metals) redistribute their charges and experience a change of potential (voltage) in response to the field.
- Particles of many materials may be attracted or repelled within the field.

Static electricity phenomena are due to these basic effects.

Dust particles, and small objects, are attracted or repelled by a field, especially if they are themselves charged (e.g. ionized particles in the air). The force F experienced by a charge q in an electrostatic field E is (Cross 1987)

$$F = qE$$

If equal positive and negative charge are sufficiently close together, from a distance their electric field effects cancel, and no external field is noticed. The charges are said to be *neutralized*.

Electrostatic fields and potentials around an object are not easy to visualize. One way of doing so is by use of field and equipotential lines. A field line represents the path a small charge would take, if it were free to move under the influence of the force due to the electrostatic field alone. Field lines always leave a conductor at a right angle (90°) to the surface.

In Figure 1.1 a charged spherical conducting object has a potential V. Each point in the surrounding space can also be assigned a potential, according to the work required to move a unit charge to that position. If all the positions of equal potential are marked, an equipotential line (or in three dimensions a surface) is marked out. A system of equipotential surfaces could be marked, forming contours of potential showing the presence of the peak in potential rather like the contours on a map showing the presence of a hill. Equipotential lines are always at a right angle (90°) to the field lines.

Equipotential lines are like contour lines on a map of an area of the earth's surface. Height is a form of potential energy. If a ball is released on a smooth hillside, it will roll down the hill perpendicular to the contour lines. Similarly, if a same polarity charge (e.g. a positive charge, next to a high positive potential) is present in the electrostatic field, it will move away from the peak in potential, in a path perpendicular to the equipotential lines. These paths form lines of electrostatic field. The intensity of the field is given by how close together the field and equipotential lines are.

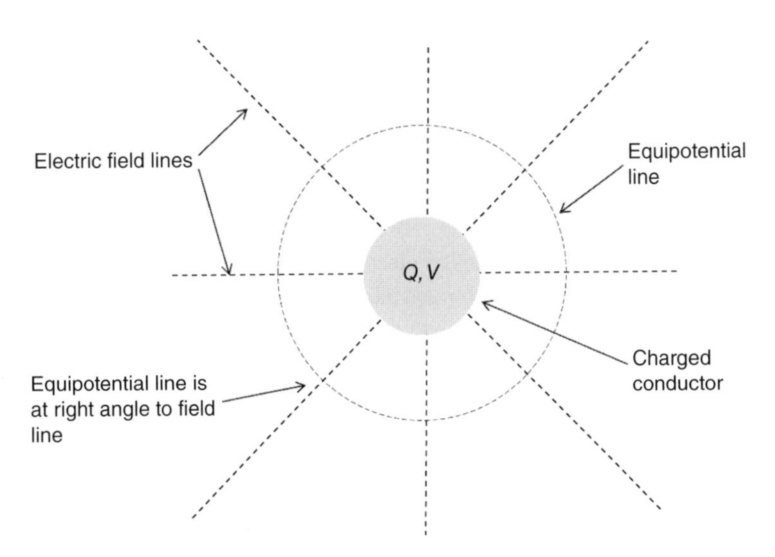

Figure 1.1 Field lines and equipotential around a charged sphere.

The electric field E (vector, as it has magnitude and direction) is the gradient of voltage V over a distance s. Electric field, therefore, has the units volts per meter (V/m).

$$E = \frac{-dV}{ds}$$

In Figure 1.1 if the charged sphere is very small, it is effectively a point of charge. The electrostatic field around the charge Q at a distance r from this point is proportional to the charge present, according to Coulomb's law (Cross 1987)

$$E \propto \frac{Q}{r^2}$$

From this equation, in this case the field strength decreases rapidly with the distance from the charge, with $1/r^2$. This is also indicated by the spreading of the field lines with distance from the charge. Field lines can be considered to begin and end on electrostatic charges, and so a high density of field lines at a surface implies a high charge density as well as high electrostatic field.

For other shapes of charge patterns, the equipotentials will not in general be spherical, and field lines will in general be curved rather than straight lines. Field lines are always perpendicular to the equipotentials and are always perpendicular to conducting surfaces as these are also equipotentials.

1.2.6 Gauss's Law

In Figure 1.1, eight field lines cut the equipotential line. Each of these lines in principle originates on a charge. So, the field lines cutting a surface is related to the net charge within it. Gauss's law generalizes this to state that the component of electric field perpendicular to a surface is proportional to the charge enclosed by the surface. For further information, the reader should refer to more academic texts such as Cross (1987).

1.2.7 Electrostatic Attraction (ESA)

A charge in an electrostatic field experiences a force, as described in Section 1.2.2. So, a charged particle or object will also experience a force according to the charge it carries. This causes charged particles and objects to be attracted or repelled by other objects, some of which may be product or items that are required to be kept clean. This effect is known as ESA.

A lesser known phenomenon that contributes to electrostatic attraction or repulsion is dielectrophoresis (Cross 1987). In this case, uncharged particles can be attracted or repulsed in a divergent or convergent electrostatic field due to differences in the permittivity of the particle and the material in which it is immersed.

1.2.8 Permittivity

Coulomb's law shows that the field due to a point charge is proportional to the charge and inversely proportional to the distance from it squared (Cross 1987).

$$E \propto \frac{Q}{r^2}$$

Permittivity (dielectric constant), ε, was defined to give a convenient constant of proportionality in this relation.

$$E = \frac{1}{4\pi\varepsilon}\frac{Q}{r^2}$$

For air, the permittivity is very close to the permittivity of free space ε_0 (vacuum) $\varepsilon = \varepsilon_0 = 8.8 \times 10^{-12}$ Cm^{-1}. Other materials have different permittivity and affect field strengths correspondingly. In general, a dielectric material has a permittivity greater than air. This is conveniently expressed as a relative permittivity ε_r, and the permittivity is given by

$$\varepsilon = \varepsilon_r \varepsilon_0$$

Polymers often have relative permittivity in the range 2–3 and many other materials in the range 2–10. Materials such as ceramics can have far higher permittivity.

1.3 Electric Current

Moving charges form electrical currents. One coulomb of charge has passed if 1 ampere has flowed for one second.

$$Q = It$$

or for a varying current

$$Q = \int_0^t I\,dt$$

and so

$$I = \frac{dQ}{dt}$$

1.4 Electrostatic Discharge (ESD)

IEC 61340-1:2012 defines an electrostatic discharge as "transfer of charge by direct contact or by breakdown from a material or object at a different electrical potential to its immediate surroundings." IEC 61340-5-1:2016a gives a slightly different definition of "Rapid transfer of charge between bodies that are at different electrostatic potentials."

There are various types of electrostatic discharge that are important in different fields. In ESD in electronics handling, the main types of concern are

- Spark discharges between conducting objects or materials
- Brush discharges between a conducting object and an insulating material
- Corona discharges from sharp conducting objects and materials

Electrostatic discharges are discussed further in Chapter 2.

1.4.1 ESD Models

ESD from different sources produces very different discharge current waveforms. These can be modeled and simulated by simple electronic circuits. Three standard ESD source circuit models, human-body model (HBM), machine model (MM), and charged device model (CDM), have been developed and standardized for testing ESD susceptibility of electronic components. This is discussed further in Chapter 3.

1.4.2 Electromagnetic Interference (EMI)

An ESD event can produce very large and fast-changing currents and voltages. These produce fast-changing electromagnetic fields with strong and fast-changing magnetic and electric field components and a broad frequency spectrum, sometimes extending to over GHz frequencies. This can be radiated and conducted to be picked up by nearby electronic circuits and can cause temporary malfunction. This is known as *electromagnetic interference*.

1.5 Earthing, Grounding, and Equipotential Bonding

Electrostatic discharges occur because of voltage differences between the objects between which the discharge occurs. If there were no voltage difference, then no ESD could occur.

So, one way to prevent ESD from occurring is to eliminate voltage differences between objects. If the two objects are conductors, connecting them electrically ensures that they are eventually at the same voltage. This must be so, as if any voltage difference were to arise, charge (current) would flow due to the voltage difference, until there is no voltage difference. The practice of connecting conductors together to eliminate voltage differences is known as *equipotential bonding*.

If two conductors at two different voltages are brought into contact, an electrostatic discharge will occur as part of the voltage equalization process. If one of the conductors is susceptible to ESD damage, it could risk being damaged as a result. So, ESD-susceptible parts must only make contact with other conductors, including grounded conductors, in circumstances designed to protect against damage.

In many practical cases, one of the conductors concerned may already be electrically connected to an electrical earth or can be conveniently connected to earth. The earth is often defined as our 0 V reference point in electricity power distribution, electrostatics, and ESD control. So, it is often convenient and is common practice to electrically connect all conductors to earth. *Earth* is also known as ground, and *earthing* is also known as grounding.

The terms *earthing* and *grounding* can have different meanings and requirements in different contexts or industries. An electrical engineer may require an earth resistance less than an ohm. A plant engineer may earth bond two items of plant, requiring a resistance less than 10 Ω. An electromagnetic compatibility (EMC) engineer may require an extremely low impedance to be maintained from direct current (DC) to hundreds of MHz or even GHz. To an ESD control practitioner, a resistance to ground $<10^9\ \Omega$ at dc may be sufficient.

In practice in ESD control, there are various types of ground that can be used. In the ESD standards IEC 61340-5-1:2016a and ANSI/ESD S20.20-2014, the term *grounding* is used to mean any of the following:

- Connection to electrical earth (the safety earth wire of a mains electrical system)
- Connection to a functional earth (e.g. an earth rod driven into the ground)
- Connection to an equipotential bonding system

1.6 Power and Energy

Energy is the ability to do work. Physics recognizes many types of energy – heat, light, gravitational, mechanical, and of course electrical.

Mechanical energy expended is the product of force and the distance moved. If a force qE is applied to move a charge q over a distance s between points A and B, the work done, W_{AB}, is

$$W_{AB} = qEs$$

Energy (work) expended, W, is also the product of power P and the time duration t that the power is applied.

$$W = Pt$$

The electrical power expended is the product of voltage V and current flowing I.

$$P = VI$$

So, the electrical energy expended is

$$W = VIt$$

1.7 Resistance, Resistivity, and Conductivity

1.7.1 Resistance

Electrical resistance is the ratio between the dc voltage applied to a circuit or material and the current flowing through it, given by Ohm's law.

$$R = \frac{V}{I}$$

1.7.2 Resistivity and Conductivity

1.7.2.1 Surface Resistivity and Surface Resistance

Surface resistivity is defined as a material surface property. It is based on the theoretical resistance of a square of material surface with sides of unit length, with a voltage applied to two opposing sides of the square (Figure 1.2). In theory, the current flows across the surface of the square. For a material of surface resistivity ρ_s with linear electrodes of width w placed parallel on the surface a distance d apart, the surface resistance R_s measured between the electrodes is

$$R_s = \frac{\rho_s \mathrm{d}}{w}$$

where $d = w$, which reduces to $\rho_s = R_s$.

The unit of surface resistivity is ohms (Ω). In some industries, it is quoted as ohms per square (Ω/sq). This reflects the property that the value of the surface resistance measured with a square electrode pattern ($d = w$) is the same, no matter what the dimension of the side of the square is.

In practice, standards exist for measuring surface resistivity using concentric ring electrodes (IEC 62631-3-2 (International Electrotechnical Commission 2015), IEC 16340-2-3, ANSI/ESD STM 11.11 (EOS/ESD Association Inc. (2015a)). This is further discussed in Chapter 11.

Surface resistance is a resistance measured between two electrodes on the surface of a specimen. The electrodes may be of any convenient form. Sometimes this measurement is made using electrodes designed so that conversion from surface resistance to surface

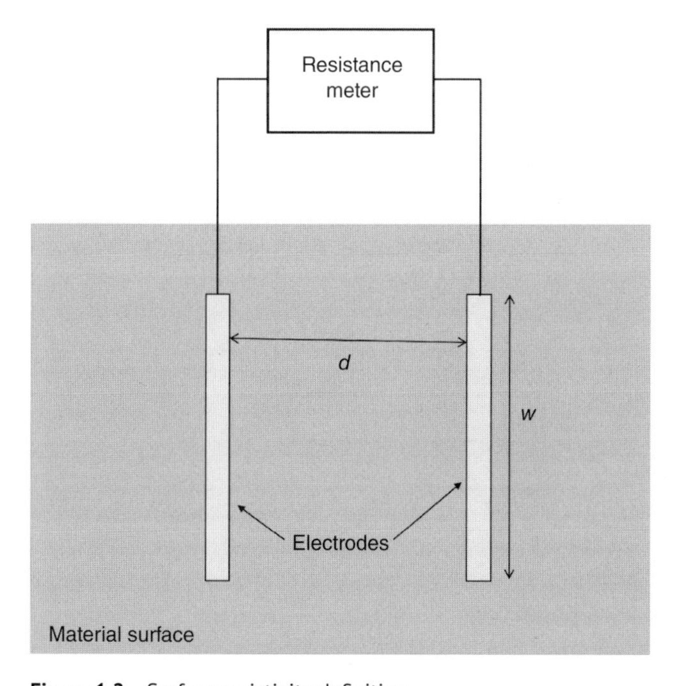

Figure 1.2 Surface resistivity definition.

resistivity is a simple calculation. In ESD control practice, conversion to surface resistivity is often not needed, and the surface resistance result obtained with defined standard electrodes is used directly.

1.7.2.2 Volume Resistance, Volume Resistivity, and Conductivity

Volume resistivity is a bulk material property based on the resistance of a cube of material with sides of unit length, with a voltage applied to two opposing faces of the cube (Figure 1.3).

The volume resistance R_v measured through a material of volume resistivity ρ_v using electrodes of area A is given by

$$R_v = \frac{\rho_v t}{A}$$

where $t = A = 1$, or $t/A = 1$, which reduces to $\rho_v = R_v$.

The unit of volume resistivity is ohm meter (Ωm). The volume resistivity of a material is often simply referred to as its *resistivity*.

In practice, standards exist for measuring volume resistivity using concentric ring electrodes (IEC 62631-3-1 (International Electrotechnical Commission 2016c), IEC 61340-2-3 (International Electrotechnical Commission 2016b), ANSI/ESD STM 11.12 (EOS/ESD Association Inc. 2015b))

Volume resistance, R_v, is a resistance measured between opposing faces of a material. The electrodes may be of any convenient form. Often this measurement is made using electrodes

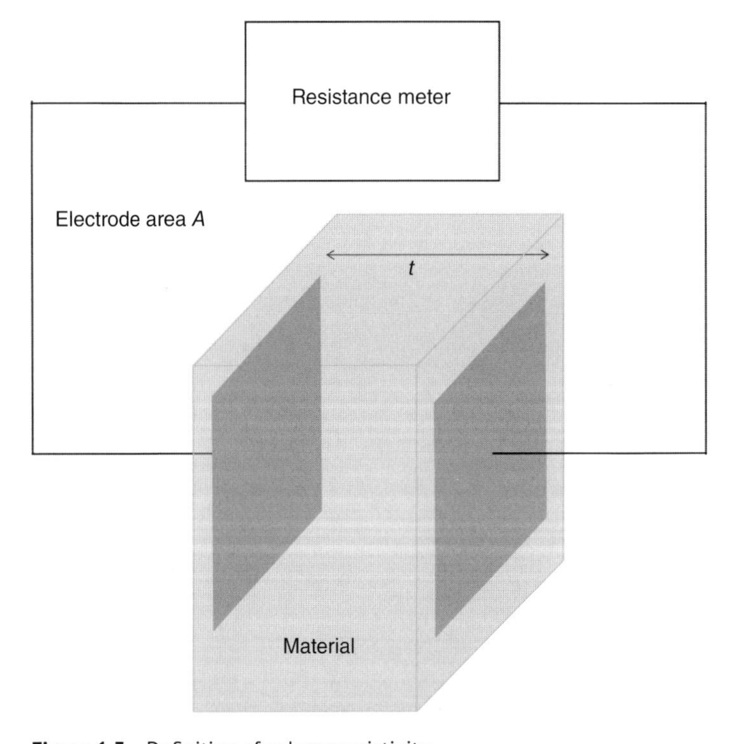

Figure 1.3 Definition of volume resistivity.

designed for volume resistivity measurement so that conversion from volume resistance to volume resistivity is a simple calculation. In ESD work, conversion to volume resistivity is often not needed. The volume resistance obtained with defined standard electrodes is used directly, saving the effort of calculation. Examples of surface and volume resistance measurement methods are given in Chapter 11.

The conductivity, σ, of the material is simply the inverse of its resistivity.

$$\sigma = \frac{1}{\rho_v}$$

The units of conductivity are siemens per meter (Sm^{-1}).

The resistivity of materials can vary by many orders of magnitude from 10^{-8} Ωm (e.g. copper) to more than 10^{15} Ωm (e.g. mica, quartz, polytetrafluoroethylene, polyethylene).

1.7.3 Insulators, Conductors, Conductive, Dissipative, and Antistatic Materials

There is no fundamental definition of insulators and conductors in electrostatics. In reality, there is a continuum of material resistivity from highly conducting (low resistance) to highly insulating (very high resistance). Different industry areas may have differing views on the resistance level at which a material is considered to have insulating properties.

For our purposes, a conductor is a material that allows charge to move around on the surface or in the bulk of the material and can thereby be used to transport charge from one place to another. An insulator (nonconductor) is a material that does not allow the charge to move in this way.

One problem in practice is that a material that is considered "insulating" in one application may be considered significantly conducting in electrostatics. So, for some years I have offered the following pragmatic definitions for use in practical electrostatics and ESD control:

- A conductor is a material that allows charge to move away quickly enough to avoid significant electrostatic charge build up.
- An insulator is any material that is not a conductor, in other words, a material that does not allow charge to move quickly enough to avoid charge build up.

Conductors are easily maintained at a low voltage by connecting them to earth (ground). However, an insulator in electrostatic terms cannot be maintained at a low voltage by installing a ground connection. The charge on the material simply does not move to the ground connection quickly enough to be conducted away in the desired timescale.

Materials or equipment are often defined as conductors or insulators based on either their measured resistance or a charge decay time. This is discussed further in Chapter 2.

Table 1.2 shows how the terms *insulating*, *dissipative*, *conductive*, and *antistatic* are widely used in ESD control. Take care when using these terms, because they may be defined differently in different contexts and may mean different things to different people. When defined in the standards, the precise definition can change as the standards evolve into new editions.

The situation becomes worse if usage of these terms in other industries and for specific products is considered (Table 1.3). In general, these words should be considered unreliable in meaning unless specified by standards as part of an ESD control system.

Table 1.2 Example of how meanings of *conductive*, *static dissipative*, *insulative*, and *antistatic* can vary with context in ESD control in electronic manufacturing.

Term	Application	General use	Meaning under 61340-5-1:2016a	Meaning under S20.20-2014
Conductive	General	Resistance $<10^6\ \Omega$	Not defined	Not defined
	ESD control footwear		Not defined	Not defined
	ESD control flooring	$<10^6\ \Omega$	Not defined	Not defined
	ESD protective packaging		Surface resistance $<10^4\ \Omega$	Surface and volume resistance $<10^4\ \Omega$
Static dissipative	General	Resistance between 10^6 and $10^{11}\ \Omega$	Not defined	Not defined
	ESD control footwear		Not defined	Not defined
	ESD control flooring	$\geq 10^6\ \Omega$	Not defined	Not defined
	ESD protective packaging		Surface resistance $\geq 10^4$ and $\leq 10^{11}\ \Omega$	Surface and volume resistance $\geq 10^4$ and $<10^{11}\ \Omega$
Insulative	General	Resistance over $10^{11}\ \Omega$	Not defined	Not defined
	ESD control footwear		Not defined, but by implication $>10^8\ \Omega$	Not defined, but by implication resistance $>10^9\ \Omega$
	ESD control flooring		Not defined	Not defined
	ESD protective packaging		Surface resistance $\geq 10^{11}\ \Omega$	Surface and volume resistance $\geq 10^{11}\ \Omega$
Antistatic	General	Widely used to described materials used in static control; can mean almost anything	Not defined	The property of a material that inhibits triboelectric charging (ESD ADV1.0-2009)
	ESD control footwear	Note: Has defined meaning under ISO 20345 in process industry hazard work	Not defined	Not defined
	ESD control flooring		Not defined	Not defined
	ESD protective packaging		Not defined	Materials that have reduced amount of charge accumulation as compared with standard packaging materials

Table 1.3 Example of how meanings of *conductive*, *dissipative*, and *insulative* can vary with context in static control in other industries (IEC 60079-32-1:2013).

Object	Measurement	Conductive	Dissipative	Insulative
Material	Volume resistivity (Ωm)	$<10^5$	$\geq 10^5$ to 10^9	$\geq 10^9$
Clothes	Surface resistance (Ω)		$<2.5\times10^{10}$	$\geq 2.5\times10^{10}$
Footwear	Leakage resistance (Ω)	$<10^5$	$\geq 10^5$ to $<10^8$	$\geq 10^8$
Gloves	Leakage resistance (Ω)	$<10^5$	$\geq 10^5$ to $-<10^8$	$\geq 10^8$
Floor	Leakage resistance (Ω)	$<10^5$	$\geq 10^5$ to $<10^8$	$\geq 10^8$

1.7.4 Point-to-Point Resistance

In ESD control, it is convenient to make simple measurements to evaluate the surface properties of a material or item of equipment. One simple way of evaluating a surface is to place two electrodes on it and measure the resistance between them. The electrodes are often cylindrical in form. This is called a *point-to-point* resistance measurement. Standard test methods based on this approach are often used. Examples of point-to-point resistance test methods are given in Chapter 11.

1.7.5 Resistance to Ground

As explained earlier, in ESD control work, voltages on conductors are often eliminated or controlled by providing an electrical connection for the charge to pass to earth (ground). It is often required to know the resistance from an object or surface to ground to help understand the charge dissipation paths. This is known as *resistance to ground*. Examples of measurement methods for this are given in Chapter 11.

1.7.6 Combination of Resistances

In practice, the resistance of a ground path may be due in part to several components. If these are effectively in series (Figure 1.4), the effect is to add the resistance of all component contributors $R_1 \ldots R_n$ to get a total resistance R_{tot}.

$$R_{tot} = R_1 + R_2 \ldots . + R_n$$

If resistance of ground paths is in parallel (Figure 1.5), they are combined as

$$\frac{1}{R_{tot}} = \frac{1}{R_1} + \frac{1}{R_2} \ldots . + \frac{1}{R_n}$$

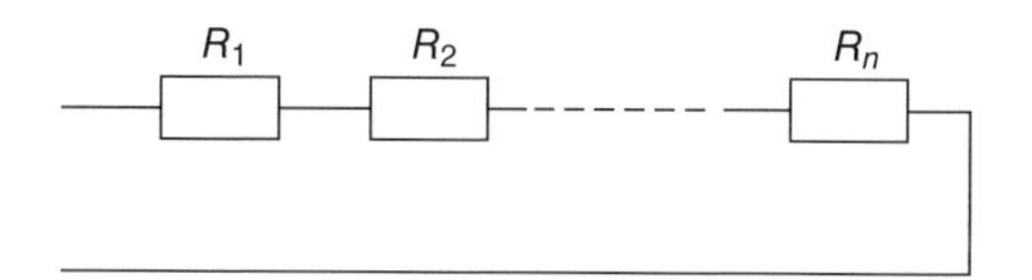

Figure 1.4 Resistances in series.

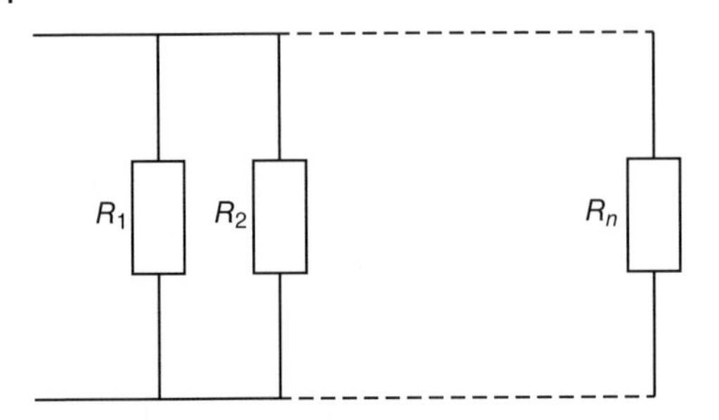

Figure 1.5 Resistances in parallel.

1.8 Capacitance

The voltage V on a conductor is related to the stored charge Q as

$$CV = Q$$

The variable C is the capacitance of the conductor. In electrostatics, any conductive object has capacitance; it is just the relationship between the stored charge and the object's voltage.

In practice, the capacitance of an object can vary with proximity of other conductors and materials (see Chapter 2).

A charged capacitor stores energy. The energy W stored in a capacitance C at voltage V is given by

$$W = 0.5CV^2$$

This can also be expressed as

$$W = 0.5\,QV$$

An object in free space (with nothing in the near vicinity) still has capacitance. For a spherical conductor of radius r in air or a vacuum, this capacitance C is

$$C = 4\pi\varepsilon_0\varepsilon_r r$$

In practice, the capacitance of an object may be due in part to the proximity to several objects. If these are effectively in parallel (Figure 1.6), the effect is to add the capacitance of all component contributors $C_1 \ldots C_n$ to get a total capacitance C_{tot} as

$$C_{tot} = C_1 + C_2 \ldots . + C_n$$

If capacitances between objects are in series (Figure 1.7), they are combined as

$$\frac{1}{C_{tot}} = \frac{1}{C_1} + \frac{1}{C_2} \ldots . + \frac{1}{C_n}$$

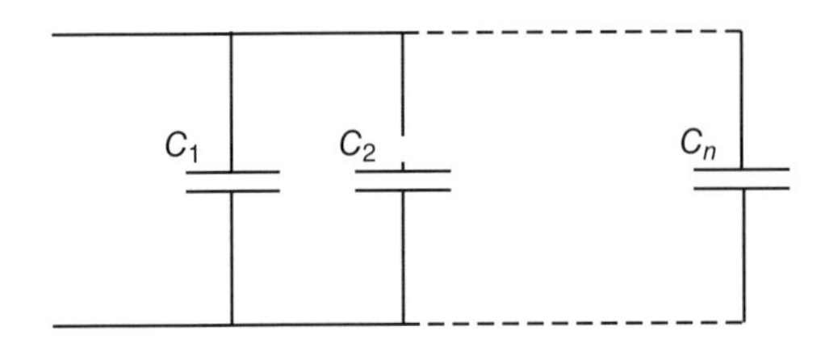

Figure 1.6 Capacitors in parallel.

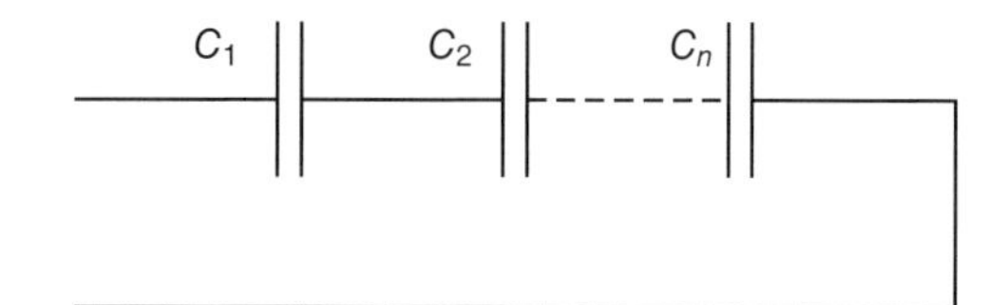

Figure 1.7 Capacitors in series.

1.9 Shielding

The term *shielding* is used in ESD control in a different way to other disciplines, especially EMC and radio frequency work. Shielding definitions and tests used in ESD control are often highly specific to the standards used. Typically, the term is used to describe the attenuation of electrostatic fields or electrostatic discharge energy applied to the outside of a protective package, measured at the inside of the package. This is discussed further in Chapter 8.

1.10 Dielectric Breakdown Strength

If a low voltage is applied across an insulating material, very little current will flow due to the high resistivity of the material. If, however, the voltage is increased, a level may eventually be reached where the current suddenly increases to a high value. Typically, this current flow may lead to formation and thermal heating of a small electrically conducting channel through the material. For a solid material, melting or damage of a small channel through the material may occur. This is dielectric breakdown of the material.

Typically, very high electrostatic field strengths are required for dielectric breakdown to occur. The breakdown strength of air is, for planar parallel electrodes, around 3 MV m^{-1} or about 3 kV mm^{-1}. For curved or sharp electrodes, it is much lower. The breakdown strength of most insulating solids is much higher than air. For polyethylene, it is about 20 MV m^{-1} (IEC 61340-1 (International Electrotechnical Commission 2012)).

1.11 Relative Humidity and Dew Point

The relative humidity (rh) or dew point of the atmosphere has a large influence on electrostatic phenomena (see Section 2.3.5). At any temperature, moisture-saturated air in equilibrium contains a maximum amount of moisture determined by the saturated vapor pressure of water at that temperature (Lawrence 2005). The saturated vapor pressure of water and hence the amount of moisture in saturated air increase strongly with increasing temperature. This saturated state is defined as 100% r.h. The relative humidity of air with lower than the saturated amount of water vapor present is given by

$$\textit{relative humidity} = \frac{\textit{vapor pressure of water present}}{\textit{saturated vapor pressure of water at the temperature}}$$

As the saturated vapor pressure increases strongly with temperature, if the amount of moisture present remains the same, increasing the air temperature will result in a reduction in relative humidity. Conversely, lowering the temperature will increase the humidity.

If the temperature is lowered sufficiently, the water vapor pressure eventually becomes equal to the saturated vapor pressure, and the air becomes saturated. Any further reduction in temperature may result in moisture condensing from the air on to surfaces in contact with it or in fog forming. This temperature is called the *dew point*.

References

Cross, J.A. (1987). *Electrostatics Principles, Problems and Applications.* Adam Hilger. ISBN: 0-85274-589-3.

EOS/ESD Association Inc. (2014) ANSI/ESD S20.20-2014. ESD Association Standard for the Development of an Electrostatic Discharge Control Program for – Protection of Electrical and Electronic Parts, Assemblies and Equipment (excluding Electrically Initiated Explosive Devices). Rome, NY, EOS/ESD Association Inc.

EOS/ESD Association Inc. (2015a) ANSI/ESD STM 11.11-2015. *ESD Association Standard for Protection of Electrostatic Discharge Susceptible Items – Surface Resistance Measurement of Static Dissipative Planar Materials.* Rome, NY, EOS/ESD Association Inc.

EOS/ESD Association Inc. (2015b) ANSI/ESD STM 11.12-2015. *ESD Association Standard for Protection of Electrostatic Discharge Susceptible Items – Volume Resistance Measurement of Static Dissipative Planar Materials*, Rome, NY, EOS/ESD Association Inc.

International Electrotechnical Commission. (2012) IEC/TR 61340-1: 2012. *Electrostatics – Part 1: Electrostatic phenomena — Principles and measurements.* Geneva, IEC.

International Electrotechnical Commission. (2013) PD/IEC TS 60079-32-1. *Explosive atmospheres Part 32-1. Electrostatic hazards, guidance.* Geneva, IEC.

International Electrotechnical Commission. (2015) IEC 62631-3-2. *Dielectric and resistive properties of solid insulating materials - Part 3-2: Determination of resistive properties (DC methods) - Surface resistance and surface resistivity.* Geneva, IEC.

International Electrotechnical Commission. (2016a) IEC 61340-5-1: 2016. *Electrostatics – Part 5-1: Protection of electronic devices from electrostatic phenomena - General requirements.* Geneva, IEC.

International Electrotechnical Commission. (2016b) IEC 61340-2-3:2016. *Electrostatics. Methods of test for determining the resistance and resistivity of solid planar materials used to avoid electrostatic charge accumulation. Section 3: Methods of test for determining the resistance and resistivity of solid planar materials used to avoid electrostatic charging.* Geneva, IEC.

International Electrotechnical Commission. (2016c) IEC 62631-3-1. *Dielectric and resistive properties of solid insulating materials - Part 3-1: Determination of resistive properties (DC methods) - Volume resistance and volume resistivity - General method.* Geneva, IEC.

Jonassen, N. (1998). *Electrostatics.* Chapman & Hall. ISBN: 0 412 12861 6.

Lawrence, M.G. (2005). The relationship between relative humidity and the dewpoint temperature in moist air. A simple conversion and applications. *Bull. Am. Meteorol. Soc.*: 225–233. https://doi.org/10.1175/BAMS-86-2-225 [Available from htt.s://journals.ametsoc .org/doi/pdf/10.1175/BAMS-86-2-225. Accessed 15th Aug. 2018.].

Wikipedia (2018) *Ion,* viewed 17 October 2018, [Available from https://en.wikipedia.org/wiki/ Ion]

2

The Principles of Static Electricity and Electrostatic Discharge (ESD) Control

2.1 Overview

ESD stands for electrostatic discharge or, according to some, electrostatic damage. This chapter provides the basis of how static electricity arises and can lead to ESD in the real world. It also provides the principles that underlie ESD control techniques and equipment design.

Electrostatic charge can build up in a variety of ways. The charged object has an electrostatic field that could conceivably lead to an ESD event in several ways:

- Direct breakdown of sensitive parts due to high electric field
- Generation of an ESD event directly subjecting the part to discharge currents
- Generation of an ESD event subjecting a part to induced transient electric or magnetic fields, or some other stress

At the root of any ESD event there is an object or surface that has a voltage that is different to its surroundings. Without this voltage difference, no electric field is present, and no ESD current can flow. Hence, the objective of ESD prevention measures has been to keep surface voltages and electric fields to a low level, below which damaging ESD cannot occur.

A review of the explanation of electrostatic charge build-up and ESD sources included here quickly reveals many ways in which ESD risks can be generated in the real world. These are summarized in brief in this chapter.

2.2 Contact Charge Generation (Triboelectrification)

The first thing to state is that charge is never generated, nor is it ever destroyed. The phenomenon that we often describe lazily as the "generation" of charge is more correctly "separation." Some practitioners speak of the charge being "liberated" or "set free." The charge is initially present in the atoms that make up all materials. There are positive charges as protons in the atomic nucleus, and there are negative charges as electrons around the nucleus. Normally these are present in equal numbers so that in an uncharged atom the number of positively charged protons equals the number of negatively charged electrons present. Static electricity arises when an imbalance is created and the local amounts of positive and negative charge become different.

The ESD Control Program Handbook, First Edition. Jeremy M Smallwood.

In practice, the amount of charge imbalance required to give strong electrostatic effects is surprisingly small. The limit of the amount of charge that can be built up on a surface is governed by the electrical breakdown field strength of air, around 3×10^6 V m^{-1}. The surface charge density required to give this field is only 2.64×10^{-5} Cm^{-2} (Cross 1987). This is equivalent to about 1.7×10^{14} electrons m^{-2}, or 8 atoms per million on the surface acquiring or losing an electron!

One common way in which static electrical charge imbalances can arise is when two materials make contact and then are separated. While in contact, electrons move from one material to the other at points of contact; this material gains a net negative charge, and the donor material gains a net positive charge. When the objects are separated, the negatively charged object can take its charge with it, leaving an equal positive charge on the other object. Although it is really charge separation that takes place, it is common to refer to the "generation" of static electrical charge.

2.2.1 The Polarity and Magnitude of Charging

The polarity of charge left on a material can be positive or negative and depends on a range of factors, especially on the other material with which it made contact. Materials may be arranged in a table according to the polarity of charge they take in contact with other materials, called the *triboelectric series* (see Table 2.1).

A material in the table (e.g. aluminum) can be expected to charge positively against another material below it in the table (e.g. polytetrafluoroethylene (PTFE)) and negatively against a material above it (e.g. wool). The amount of charge generated is a function of the separation of the materials on the table; aluminum and paper can be expected to charge relatively little against each other, but polyvinylchloride (PVC) and nylon can be expected to charge strongly against each other.

In practice, triboelectrification is a variable phenomenon and is highly dependent on surface conditions, contaminants, and humidity. Small amounts of surface contaminants can have a large effect on triboelectrification. One result is that the order of triboelectric series is not unique. Different experiments and samples of the same materials may produce different results especially if the experimental conditions are varied. While it could be assumed from the triboelectric series that contact between two surfaces of the same material would not generate charge, this is generally not what happens in practice.

2.3 Electrostatic Charge Build-Up and Dissipation

Any two materials in contact give charge separation that can lead to static electrical charge build-up. This may or may not lead to charge and voltage build-up, depending on the circumstances.

The key to this build-up is the balance between charge generation and charge dissipation (or neutralization). If charge is dissipated (or neutralized) more quickly than it is generated, no static electricity builds up, and no effects are noticed. If charge is generated more quickly than it is dissipated (or neutralized), then high voltages and static electricity effects are quickly built up.

Table 2.1 An example of a triboelectric series.

Acetate	Charge to positive polarity
Glass	↑
Mica	
Human hair	
Nylon	
Wool	
Lead	
Silk	
Aluminum	
Paper	
Cotton	
Steel	
Wood	
Epoxy-glass	
Copper	
Stainless steel	
Acetate rayon	
Polyester polyurethane	
Polyethylene	
Polypropylene	
PVC	
Silicon	↓
PTFE	Charges to negative polarity

2.3.1 A Simple Electrical Model of Electrostatic Charge Build-Up

Static electricity can be modeled as a charge generator, and a simple electrical model can be used to understand many practical situations (Figure 2.1).

The separation of charge is effectively a small electrical current represented by a current source I. The capacitor C represents the charge storage properties of the system and could be a material surface or a conducting object with a capacitance to earth. The resistance R represents charge dissipation processes (other than ESD) and can vary from less than 1 Ω to more than 10^{14} Ω for good insulators. (See Sections 1.7.3 and 2.3.4 for discussions on the meaning of insulators and conductors.)

If the current is constant (i.e. the effect of capacitance can be neglected), it's easy to see by Ohm's law that the voltage developed is highly dependent on the resistance R. If a charge generation rate of 1 nA (1 nCs^{-1}) is present, with a resistance of 10^9 Ω, a steady state voltage of 1 V is produced. If, however, the resistance was 10^{12} Ω, a voltage of 1 kV would be produced, and for a resistance of 10^{14} Ω, a voltage of 100 kV would theoretically be produced!

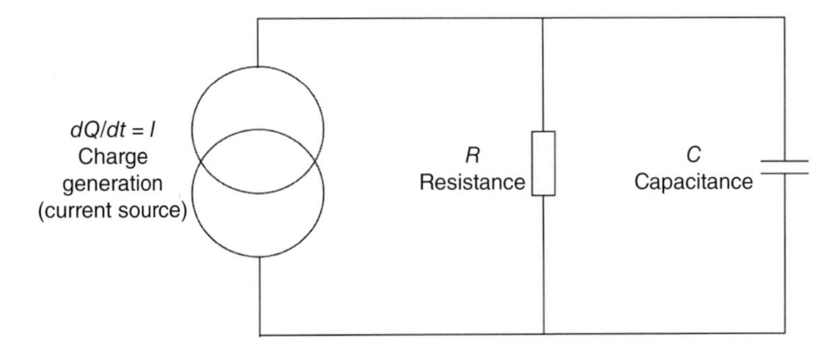

Figure 2.1 A simple electrical model of electrostatic charge build-up.

For a charge generation rate of 1 μA, a resistance of 10^{10} Ω would yield a voltage of 10 kV. In practice, electrostatic sources rarely generate charge at this rate or on a steady current basis unless there is steady movement involved (e.g. in a conveyor system).

The rate of electrostatic charge generation is affected by many factors. Some of the key factors are as follows:

- Relative position of the materials in the triboelectric series
- Rate of separation of contact area (high rates of movement)
- Condition of the surfaces that make contact
- Rubbing of the contacting surfaces
- Ambient humidity and temperature

The many factors involved in triboelectric charge separation make it a highly unpredictable phenomenon.

2.3.2 Capacitance Is Variable

The capacitance C represents charge storage. There is a simple relation between charge Q and voltage V, and capacitance is the ratio of charge and voltage.

$$CV = Q$$

$$C = \frac{Q}{V}$$

In practice, capacitance is usually a variable that depends on the materials and nearby objects and on the proximity to earth. Objects move around in daily life, and so their capacitance changes.

As an example, we can consider the human body. It is, in electrostatic terms, a conducting object, being mainly composed of water, which is a conducting material. Even if we neglect the nearby objects and earth, the human body can be approximated as a sphere that has a similar surface area. The "free space" capacitance of a sphere is given by $4\pi\varepsilon_0 r$, where r is the radius and ε_0 is the permittivity of free space, 8.8×10^{-12} Fm^{-1}. Typically, a 1 m radius sphere gives a useful approximation and has a "free space" capacitance around 110 pF.

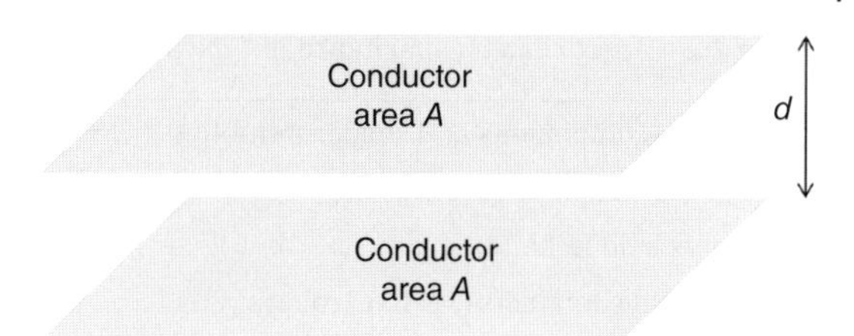

Figure 2.2 Parallel plate capacitor.

Nearby objects and earth increase this value. In fact, the human feet, on the earth, approximate two parallel plate capacitors in parallel with the free space capacitance. Each capacitor is made up of two electrodes: the sole of the foot and the earth. These are separated by a layer of material (the shoe sole, typically an insulating polymer of relative permittivity ε_r around 2.5). Each foot capacitance varies from moment to moment as the feet are lifted and replaced on the ground during walking. Each foot capacitance can be modeled as a parallel plate capacitor (Figure 2.2), with plates of area A separated by a distance d and capacitance C given by

$$C = \varepsilon_0 \varepsilon_r \frac{A}{d}$$

In general, if either the area A or the distance of separation d are changed, then the capacitance will change. If the charge is held constant, increasing the capacitance will decrease the body voltage, and reducing capacitance will increase body voltage. Reducing capacitance can be achieved by reducing the area (e.g. standing on tip-toe) or increasing the separation distance (e.g. raising the foot from the floor).

The previous equation shows that if the charge on conductor is unchanged and the capacitance changes, then the voltage of the conductor changes. For example, if a person's body capacitance changes between 50 and 150 pF while walking and the charge on their body is constant at 5 nC, their body voltage will vary between 100 V (at 50 pF) and 33 V (at 150 pF). If a printed circuit board (PCB) conductor has a capacitance of 20 pF and charge 5 nC when resting close to a large earthed machine part, its voltage will be 250 V. If its capacitance is reduced to 5 pF when far away from this machine part, its voltage will rise to 1000 V.

It can be useful to have an idea of the approximate capacitance of everyday objects, especially when estimating the possible effect of ESD to or from such items. Table 2.2 gives some examples (IEC 61340-1).

The variable ratio between charge and voltage behaves similarly for nonconductors.

When seated in a chair, the body generates charge on the clothes surfaces in contact with the chair. This forms a large area of charged material with a small distance between the charges (the two surfaces are in contact). The person's body voltage is low in this situation even though their clothing may be highly charged. On rising from the chair, the person can take much of the separated charge with them. The effective "capacitance" between the body and the chair is rapidly reduced (separation is rapidly increased), and a high body voltage quickly results if the charge cannot dissipate to ground. It is common to feel a shock on touching something metal after rising from a chair or car seat – voltages over 10 kV have been measured on people after getting out of a car seat (Pirici et al. 2003; Andersson et al. 2008).

Table 2.2 Approximate capacitance of typical everyday objects.

Electronic components and small assemblies	0.1–30 pF
Drinks can, small metal parts	10–20 pF
Tweezers held in hand	25 pF
Small metal containers (1–50 l), trolleys	10–100 pF
Larger metal containers (250–500 l)	50–300 pF
Human body	100–300 pF
Small signal MOSFET gate capacitance	100 pF
Power MOSFET gate-source capacitance	900–1200 pF
Car	800–1200 pF

MOSFET, Metal Oxide Silicon Field Effect Transistor

A corollary of this is that the voltage and field surrounding a charged object may be suppressed by the presence of a nearby conducting object. If the capacitance of the system is increased, the voltage is decreased.

As an example, a charged garment that fits snugly to the body has voltage suppressed due to the proximity of its surfaces to the body. Even if the garment is highly charged, the external field may be limited due to this. If the garment flaps open, the body and garment surfaces move apart. "Capacitance" is reduced, and a high voltage and electrostatic field appears outside the garment.

2.3.3 Charge Decay Time

The resistance and capacitance form a resistor-capacitor (RC) network that has a characteristic time constant τ.

$$\tau = RC$$

In a time τ, the voltage will decay to about 37% of its initial value.

In the example, if the charge current is suddenly halted at time $t = 0$ with initial voltage V_0, the voltage V on the capacitor reduces as

$$V = V_0 \exp \frac{-t}{\tau}$$

An electrostatic field meter monitoring the material surface would measure this exponential decay of voltage. The product of a material's resistivity ρ and permittivity $\varepsilon_0\ \varepsilon_r$ gives a physical time constant for the material.

$$\tau = \rho\varepsilon_0\varepsilon_r$$

This behavior has important practical implications. If we consider a situation in which the capacitance is fixed at 100 pF (the order of magnitude of capacitance of a person) and the charging current 100 nA, we can consider the effect of different resistances. With a resistance of 1 GΩ, the voltage generated is only 100 V, and on cessation of the current, the

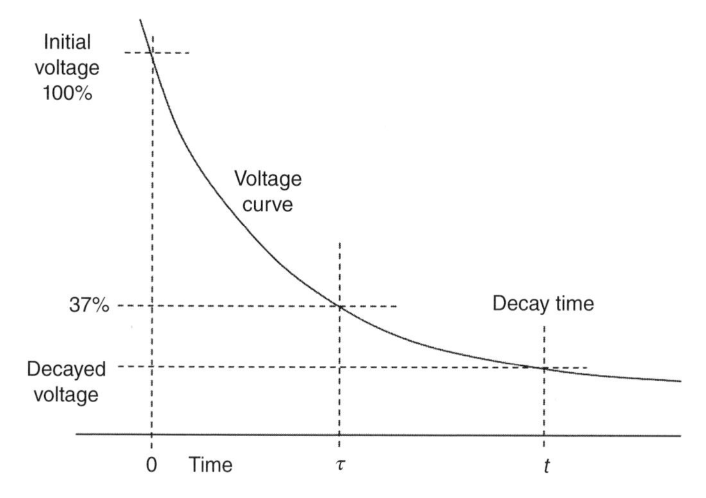

Figure 2.3 Charge or voltage decay curve.

voltage will fall to 37% of its initial value within $10^9 \times 10^{-10} = 0.1$ seconds. The effect of a short duration charging current of this magnitude is unlikely to be noticed.

If the resistance is increased 10 GΩ, not only is the voltage generated increased to 1 kV, but on cessation of the current, the voltage will take $10^{10} \times 10^{-10} = 1$ seconds to fall to 37% of its initial value. The presence of this voltage may or may not be noticeable or cause a problem, depending on the circumstance.

If the resistance is increased 100 GΩ, not only is the voltage generated increased to 10 kV, but on cessation of the current, the voltage will take 10 seconds to fall to 37% of its initial value. The presence of this voltage for such a long time could lead to the person experiencing shocks on touching something or discharging to cause some problem.

In ESD control, a different definition of charge decay time is usually used in standard measurements, and often the time for charge to reduce to one-tenth of its initial value is measured (Figure 2.3). This value is theoretically equal to 2.3τ.

In practice, the charge decay time is often measured from the starting voltage down to a certain threshold voltage, e.g. 100 V. Polymers may have time constants of many tens or hundreds of seconds, or even days under clean dry conditions.

In practice, the simple model does not always correspond well with material behavior. Measured charge decay curve may depart considerably from the ideal exponential, and the measured time "constant" varies with measurement conditions. Often with high resistance materials the decay time lengthens as the surface voltage drops and may become very long at low voltages.

2.3.4 Conductors and Insulators Revisited

In many engineering fields, conductors are often thought of as materials such as copper or aluminum that have very low resistance or resistivity (see Section 1.7), much less than 1 Ω. In ESD control, materials that have a much higher resistivity than this may be thought of as conductors. In practical electrostatic control, materials and equipment are often defined as conductors or insulators based on either a measured resistance or a charge decay time, or both. The model of Figure 2.1 can be used to explain this.

As charge generation rate (current I) in static electricity is often low, even a relatively high value of leakage resistance R (Figure 2.1) may pass the current to give low voltage, $V = IR$. In ESD control, a resistance of 1 MΩ (10^6 Ω) could be considered quite conductive and would reduce the electrostatic voltage in the previous example to 1 V. As an example, in a case where the charge generation currents normally experienced in practice are expected to be no more than 1 nA, calculations can be made on this basis. Alongside this, it may be wished to limit voltages to some level, e.g. 100 V. Given these constraints, the model and Ohm's law show that resistances up to $V/I = 10^2/10^{-9} = 10^{11}$ Ω would be acceptable.

In an application (e.g. electrostatic hazards avoidance in industrial processes) where higher charge generation is expected, the allowable resistance may be considerably smaller (IEC 60079-32-1).

A second way of looking at the matter is to decide how long a transient charge built up on a material or object may tolerably be allowed to remain without problems occurring. This may be evaluated in terms of the charge decay time. If a conductor has capacitance around 10 pF, resistance to ground of 10^{11} Ω will give a charge decay time of one second, and in the absence of charge generation a stored charge will reduce to only 5% of its initial value within three seconds. In manual assembly and handling processes, this will usually be fast enough to avoid problems. For materials, this decay time corresponds to a permittivity of 10^{-11} Fm^{-1} and resistivity of 10^{11} Ω. The permittivity of air is around 0.9×10^{-11} Fm^{-1}, and many plastics are around 2×10^{-11} Fm^{-1}. The presence of higher capacitance or material permittivity, or a requirement for faster charge decay, may lead to a lower maximum acceptable resistance.

2.3.5 The Effect of Relative Humidity

Water is an electrically conducting material. Moisture from the air forms a thin layer on the surface of many materials and can contribute to their apparent electrical conductivity. Some materials, especially natural materials such as paper, reduce by orders of magnitude in their resistivity as relative humidity increases from dry conditions.

As material surface resistance is increased under dry conditions, electrostatic charge build-up is often greatly enhanced. Some ESD control materials use additives to attract moisture to a polymer surface and provide static dissipative behavior. These materials may not work well at low humidity. As a rule of thumb, electrostatic charge build-up is generally increased for humidity less than about 30% rh.

The external atmospheric humidity varies daily with the climate and weather, in a range from below 10% rh (cold and dry winter conditions) to 100% rh (fog). The atmospheric relative humidity often has a large effect on material resistance, especially for materials that have resistance above about 1 MΩ. The effective resistance and charge decay times can be reduced over several orders of magnitude with increasing relative humidity for some materials.

Air relative humidity is a strong function of temperature and reduces as temperature increases for a given moisture content. Relative humidity is approximately halved by a 10 °C rise in temperature, if no moisture is added or removed. If, as in winter, cold air is brought

Table 2.3 The effect of humidity on typical electrostatic voltages (MIL HDBK 263).

Action	Voltage observed	
	@ 10–20% rh	@ 65–90% rh
Person walking across carpet	35 000	1 500
Person walking across vinyl floor	12 000	250
Person working at bench (not grounded)	6 000	100
Vinyl envelope	7 000	600
Polythene bag picked up from bench	20 000	1 200
Chair padded with polyurethane foam	18 000	15 000

indoors and heated, very low relative humidity can result. Hence, ESD problems can be seasonal and occur often in winter. Even in a room where the relative humidity is controlled, dry local microclimates can form where there are heat sources such as equipment, especially if air circulation is restricted.

A view of the effect of relative humidity on static electricity in daily life is indicated by the following typical voltages (Table 2.3) given by MIL HDBK 263 as observed at different ambient humidities. These are indicative and cannot be used to predict voltages occurring in real situations.

2.4 Conductors in Electrostatic Fields

2.4.1 Voltage on Conducting and Insulating Bodies and Surfaces

Like charges repel, and in a conductor where charges are free to move rapidly, charge will rapidly move to the outer surface to minimize their proximity to each other. After charge has redistributed, the voltage on all parts of the conductor is equal (equipotential). This must be so – current flows due to voltage differences, until the equipotential state is achieved.

For an insulating object, charge does not flow freely and so the voltage at each point on the surface is typically different from its neighbor. For intermediate materials, the time taken to achieve near equipotential surface is several times the time constant.

For objects that have high resistivity and a long time constant, charge will redistribute to equipotential if we wait long enough (and if the field source is not changing rapidly) – but in the meantime surface voltages can be different.

2.4.2 Electrostatic Field in Practical Situations

For a small point or spherical charge with the balancing charge a long way distant, the strength of the electric field falls off rapidly with distance r, as it is proportional to $1/r^2$. The field line spread out radially (Figure 2.4). In many practical situations the object presenting an electrostatic field source is too large to be considered a point source.

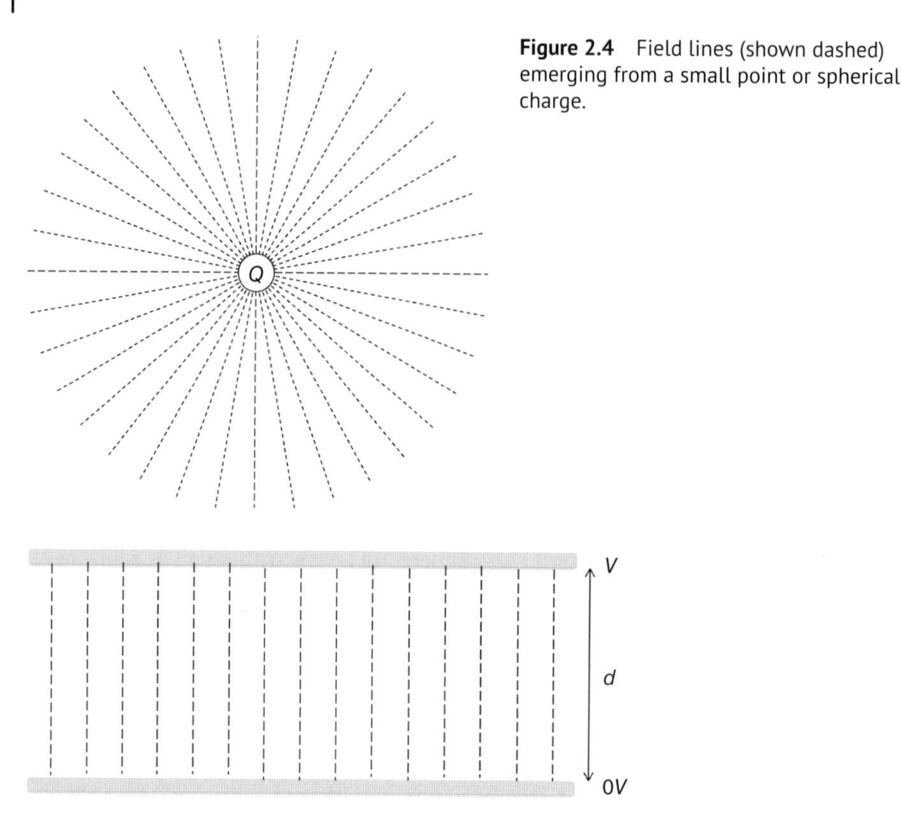

Figure 2.4 Field lines (shown dashed) emerging from a small point or spherical charge.

Figure 2.5 Electrostatic field between parallel plates.

For larger and different shaped objects, the field line pattern can be very different and the fall-off in field strength can be much less rapid. In practice, the field lines start and finish on conductors at different voltages in the region, and field lines may be more or less curved at any region in space between them.

The electrostatic field between two large flat parallel plates, well away from the plate edges, is uniform (Figure 2.5), and the field lines are parallel between the electrodes. Away from the plate edges, the field E is uniform and is easily calculated from the voltage difference between the plates V and the distance between them d.

$$E = \frac{V}{d}$$

If a conductor is placed in an electrostatic field, it has the effect of drawing the field to itself with field lines always emerging at right angles to the conductor surface. In response, charges on the conductor are redistributed until the voltage is the same all over the conductor surface. One of the consequences of this is that any instrument we use to measure an electrostatic field inevitably changes the field it is measuring. Figure 2.6 shows how this happens with the electrostatic field between a metal plate at voltage V and a grounded electrostatic field meter. The field meter actually sees a field higher than the V/d value. The same effect happens of course for any component, PCB or any other electrostatic discharge–sensitive (ESDS) device brought into the field. The density of the field lines at the

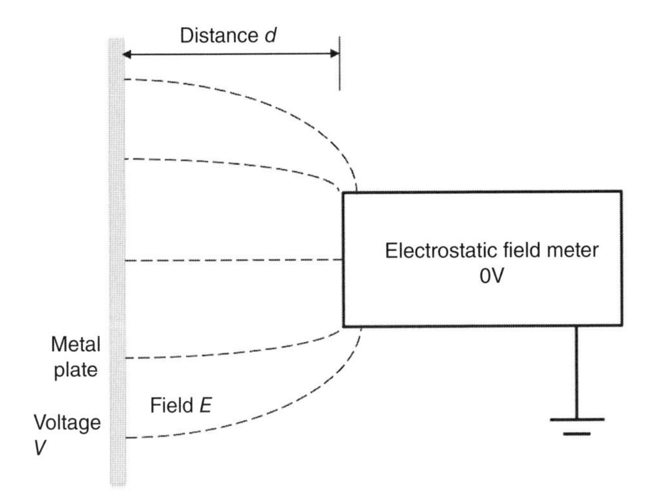

Figure 2.6 Electrostatic field between a field meter and metal plate at voltage *V*.

surface are related to the field strength and surface charge density induced at the surface. A high concentration of field lines indicates a high field strength.

Field lines tend to congregate at the tip or edge of an object, and the electrostatic field strength becomes more intense in these regions. Discharges tend to occur preferentially from high field strength regions at sharp edges on objects. This is used to an advantage, for example, in using sharp pins to produce intense fields and corona discharges as a source of ions in an ionizer for charge neutralization.

For a charged insulating surface, the situation is even more complicated. A charged insulator will normally have a highly variable charge density over its surface. The surface voltage is highly dependent on the surface charge density and presence of other materials nearby.

2.4.3 Faraday Cage

For a conducting object in an electric field, charge flows until all points on the surface are at the same voltage. This must be so, as any voltage difference would cause a current to flow in the conductor. If the object is hollow, then the field inside the object is zero, because an equipotential conductor surrounds it (Figure 2.7). An object placed within the hollow conductor would therefore be shielded from the effects of an external field. This hollow conductor is called a *Faraday cage*.

2.4.4 Induction: An Isolated Conductive Object Attains a Voltage When in an Electric Field

If we consider the behavior of a conductive object as a capacitor, we can quickly understand that the voltage on an isolated conductive object will change under the influence of a nearby charged object and its electric field.

In Figure 2.8 an earthed electrostatic field meter is monitoring the voltage V_m of a metal plate. There is an effective capacitance C_m between the metal object and the field meter. The metal object has no net charge and is initially at zero volts.

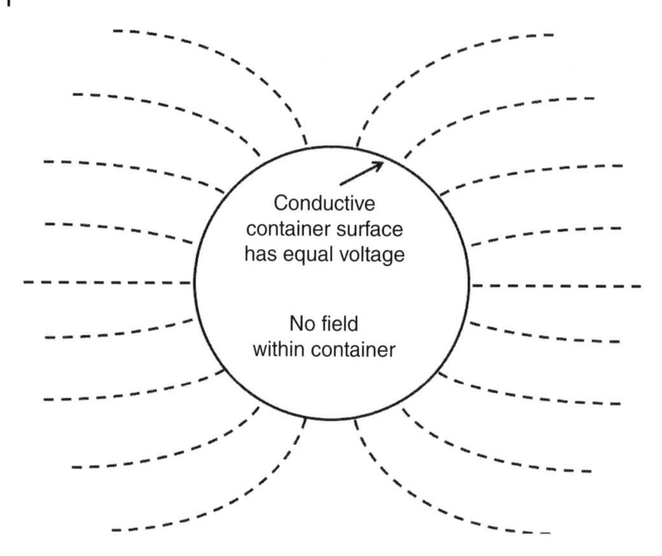

Figure 2.7 Faraday cage.

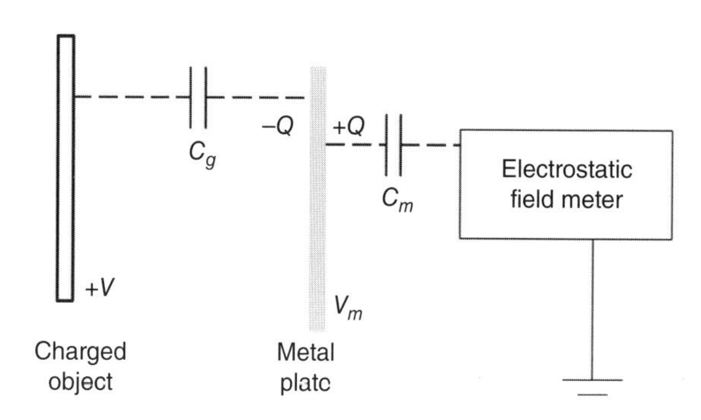

Figure 2.8 Voltage developed on a metal plate in an electric field.

A positively charged object is then brought near. As it approaches, it couples to an increasing amount of negative charge Q on the metal object attracted to the side nearest the positively charged object. The same amount of positive charge is repelled and appears on the side of the metal object nearest the field meter, coupled to an equivalent negative charge on the (grounded) field meter. The field meter sees a positive voltage on the metal object, as voltage on the metal object increases (the capacitor C_m is charged) by an amount.

$$V_m = \frac{Q}{C_m}$$

While the total amount of charge on the plate does not change, an amount $-Q$ is attracted toward the positively charged object to charge C_g, and an amount $+\ Q$ is repelled to charge C_m.

$$Q = C_m V_m = C_g(V - V_m)$$

$$V_m = \frac{C_g V}{(C_g + C_m)}$$

Note that the metal plate would have a capacitance and voltage with respect to ground, even if the field meter were not present. Although there is now a voltage on the plate, the net charge remains zero ($+Q$ $-Q$), and it is not charged!

This changing voltage on an object happens in practice if any conductive object passes through an electric field. If, for example, an integrated circuit passed into an electric field arising from a charged garment surface, it could acquire a voltage in this way. If it were subsequently grounded in this state, an ESD event would happen.

2.4.5 Induction Charging: An Object Can Become Charged by Grounding It

In Figure 2.8, we saw that a voltage was induced on the metal object when a charged object was brought nearby. In that situation, we can discharge the capacitor C_m by connecting a ground wire between the metal object and earth. The positive charges on the metal object flow to earth. The voltage V_m is then zero (Figure 2.9). Note that at the time of connection, an ESD occurs as the charge flows to earth. This is an important phenomenon in ESD control – an ESD occurs when two conductors at different voltages make contact, e.g. when grounding a conductor in the presence of an electrostatic field.

If the earth wire is then removed, the metal object remains at zero volts. However, it has a net negative charge Q, as the balancing positive charge Q has flowed away. The metal plate is now charged although the voltage on it is zero! If the charged field source object is then taken away, the voltage on the metal object rises to a negative voltage due to its negative charge.

$$V_m = \frac{-Q}{C_m}$$

This process is called *charging by induction*. It can happen in practice if an object, tool, device, or person becomes grounded temporarily when in an electrostatic field.

If a person or object can become charged and can act as a source of electrostatic field, then a nearby device can be subjected to that field. If the device is momentarily grounded, ESD occurs at that time, and it can become charged by induction. This can leave it in a charged state, at risk of ESD occurring on subsequent contact with another conductor at different voltage – grounded or not. If a grounded person moves to pick up a sensitive device within an electrostatic field, they may cause an ESD event when they touch the ESD-sensitive device. Practical demonstrations of these processes are given in Section 12.7.10.

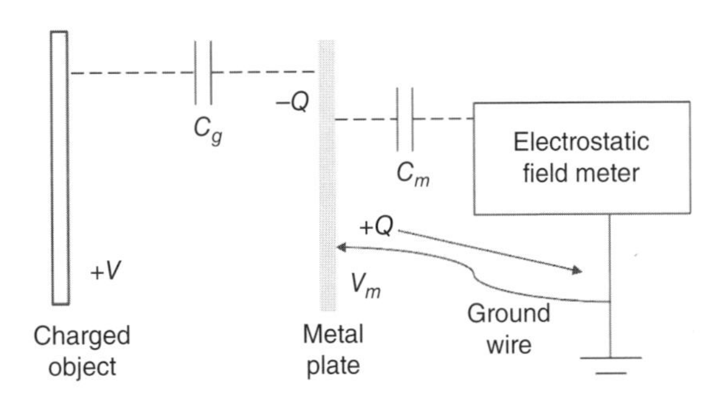

Figure 2.9 An earthed metal plate in an electric field becomes charged by grounding.

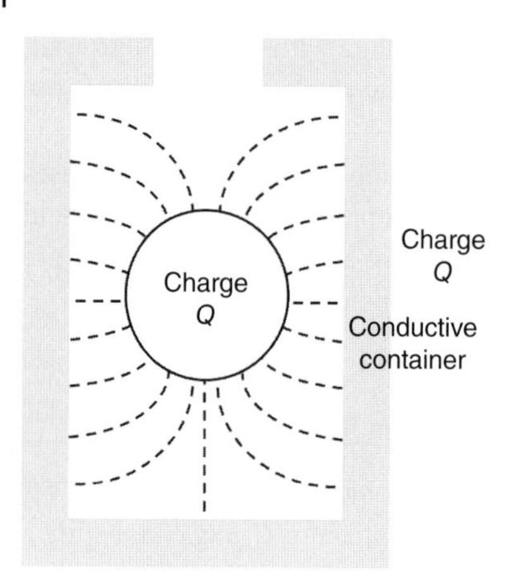

Figure 2.10 Faraday pail.

Induced voltage differences can also lead to breakdown over small gaps between nearby conductors in a field, if the voltage difference exceeds the gap breakdown voltage. This can also lead to ESD risks.

2.4.6 Faraday Pail and Shielding of Charges Within a Closed Object

If a charged object is placed within a closed hollow conducting object such as a box, then the field lines from the charge couple to the surrounding conductor (Figure 2.10). The net charge contained within the conducting box induces an equal net charge on the box.

This principle is used to measure electrostatic charge on items by placing the charged item in a container, which is known as a *Faraday pail.*

If the container is grounded, then the outside world is shielded from electrostatic fields arising from the charges. If it is not grounded, then it is itself a charged object and can be a source of ESD.

2.5 Electrostatic Discharges

Normally air is an excellent insulator. If, however, the electrostatic field strength exceeds about 3 MV m^{-1} (3 kV mm^{-1}), the insulating properties of air breaks down and ESD occurs. A large amount of stored charge can be rapidly dissipated by this event. The discharge may be sudden, as in sparks, or it may be gradual as in corona discharge.

An understanding of ESD is important in understanding the characteristics of ESD sources.

2.5.1 ESD (Sparks) Between Conducting Objects

The spark discharge occurs between conducting electrodes that initially have a high voltage difference between them. Large energies (μJ to >1 J) may be dissipated in very short, or long, times (ns to >ms) depending on discharge circuit (including the load characteristics). Peak currents are typically greater than about 0.1 A and can exceed 100 A. The discharge waveform is highly dependent on the source and "load" circuit characteristics and can have unidirectional or oscillatory waveforms (see Section 2.7).

The energy E stored in a capacitor C charged to voltage V is easily calculated using this simple formula

$$E = 0.5CV^2$$

In the absence of significant series resistance, it is often reasonable to assume that all this energy is transferred to the discharge.

The electrical breakdown field strength of about $3\,\mathrm{MV\,m^{-1}}$ is valid for normal air pressure and rather large distances (e.g. for a gap of 10 mm and large diameter or flat electrodes, the breakdown voltage would be about 30 kV). The relationship between breakdown field strength and air pressure is given by Paschen's law (Kuffel et al. 2001) and is nearly linear for larger gaps and uniform fields. At smaller gap distances d the breakdown voltage reaches a minimum (known as the *Paschen minimum*). As breakdown voltage V_b is also dependent on atmospheric pressure P, the Paschen curve is usually plotted as breakdown voltage against the product Pd (Figure 2.11). For air, according to Paschen's law, below about 350 V no breakdown occurs (minimum Pd 0.55 Torr cm, or 7 μm at 1 atm), and ESD can happen only with direct metal-to-metal contact. There is evidence that in practice discharges can occur through small gaps below the Paschen minimum voltage, possibly due to field emission (Wallash and Levitt 2003).

2.5.2 ESD from Insulating Surfaces

If a conductive electrode approaches a charged insulating surface, a "brush" discharge can occur. Several contributory discharges occur on the insulating surface, radiating from a central spark channel – the whole looks rather like an old-fashioned twig brush.

Brush discharges are less well documented than spark discharges. They typically have a lower peak discharge current than sparks (0.01–10 A) and unidirectional waveforms with fast rise and quasi-exponential decay (Figure 2.12) (Norberg et al. 1989; Norberg 1992;

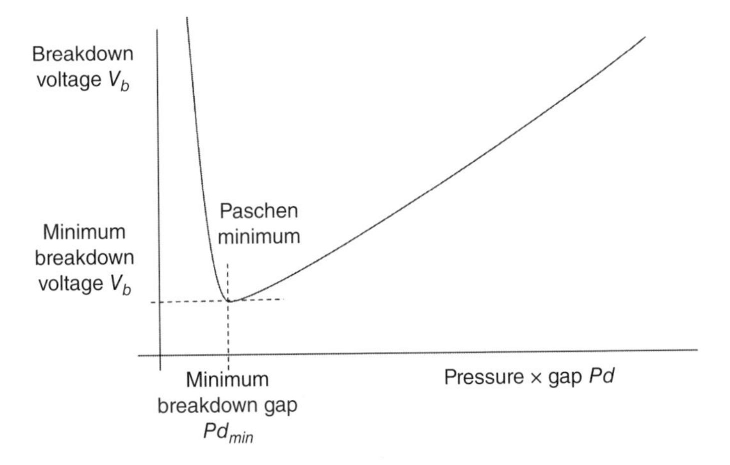

Figure 2.11 The relationship between breakdown voltage and spark gap Pd (Paschen curve).

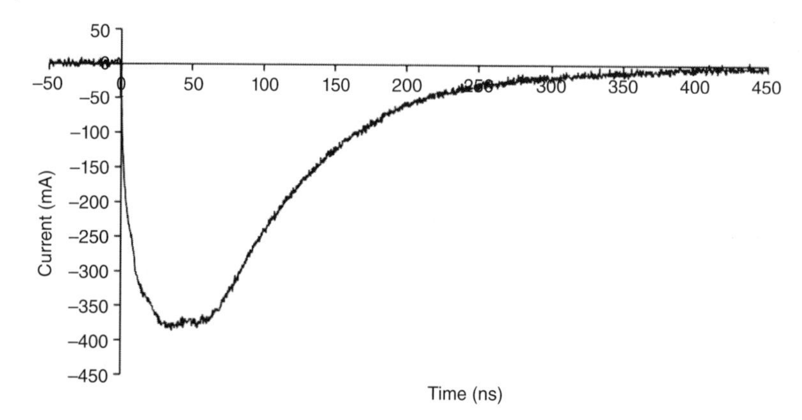

Figure 2.12 Discharge from negatively charged (>20 kV) insulating surface.

Norberg and Lundquist 1991; Smallwood 1999; Landers 1985). The power dissipation and energy of a brush discharge is not easy to calculate.

2.5.3 Corona Discharge

Very high electrostatic fields can occur at sharp edges or points on conductors in an electrostatic field. When this field reaches or exceeds a threshold, ions can be sprayed from the point or edge into the air, as a small continuous ion current. This effect is used in ionizers to create a source of ionized air for neutralizing electrostatic charges.

2.5.4 Other Types of Discharge

Where an insulating surface is backed by a conducting material, and high charge levels can be generated, a strong propagating brush discharge can occur. This type of discharge is not usually of concern in electronic component handling, but it can be of concern as an ignition source in industrial processes.

2.6 Common Electrostatic Discharge Sources

Any object that is at a different voltage from an ESDS device can be a source of ESD if the object can touch the device or come close enough for a discharge to jump a small air gap between them. The ESD that occurs may be more or less damaging or problematic according to its characteristics. Different ESD sources produce waveforms with very different characteristics in terms of parameters such as peak current, duration, energy and charge transferred to the device, and frequency spectrum. Even an apparently similar source can give widely different ESD waveforms under different circumstances. Some examples of real ESD waveforms are given next – these may or may not be representative of ESD produced from similar sources in other real situations, which may be highly variable.

2.6.1 ESD from the Human Body

The charged human body is an important source of ESD, both in device damage in manufacturing processes and in electromagnetic susceptibility of working systems. The body is a conductor in electrostatic terms and can have a variable capacitance up to about 500 pF, although considerably higher capacitance has been measured under some circumstances (Jonassen 1998; Barnum 1991). The capacitance of the human body is dependent on its proximity to other objects such as furniture and walls. When standing, the characteristics of footwear and the nature of the floor are important factors.

Although the body is a conductor, it has significant resistance, and this limits the current flow and causes ESD waveforms from the human body charged to higher voltages (more than a few kV) to have a characteristic unidirectional wave shape (Figure 2.13). The peak discharge current is typically in the range 0.1–10 A with duration of around 100–200 ns. Discharges from the human body at lower voltages can have highly variable waveform and current characteristics (Kelly et al. 1993; Bailey et al. 1991; Viheriäkoski et al. 2012). This can significantly affect related risks of ESD damage.

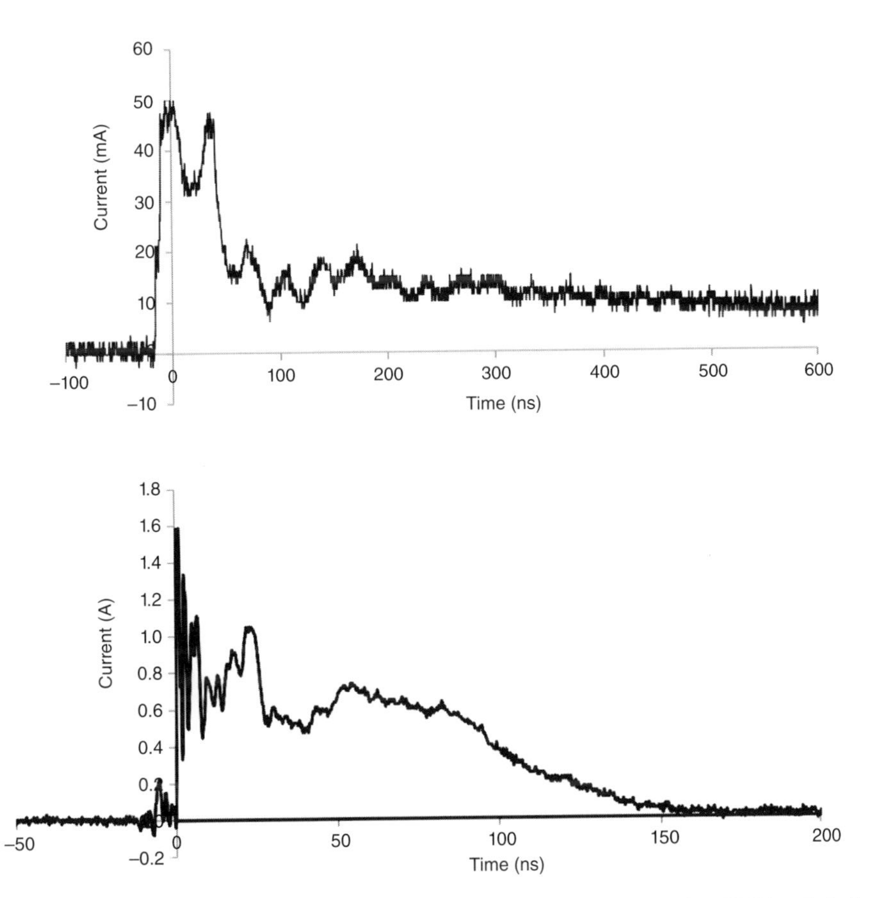

Figure 2.13 Example of waveform of discharge from the author charged to 500 V and discharging via skin of a finger (above) and small metal object (coin, below).

2.6.2 ESD from Charged Conductive Objects

When a highly conductive (e.g. metal) object is not grounded, it can gain a high voltage either through triboelectrification or through induction in an electrostatic field. If this conductor now touches another grounded conductor or device, an ESD event will occur.

The waveform of real-world ESD of this type can be highly variable depending on the characteristics of the source and discharge path. Typically, with low resistance source and discharge path materials, a high discharge current reaching tens of amps can occur. The waveform is often oscillatory, with the frequency determined mainly by capacitance and inductance of the source and discharge circuit. The waveform duration may be from a few nanoseconds to hundreds of nanoseconds.

If there is significant resistance in the discharge circuit, the peak ESD current and duration of the discharge are reduced. (For small ESD sources, the effective resistance of the discharge can be significant.) The number of oscillation cycles is also reduced. Eventually with sufficient circuit resistance, a single peak may occur. In practice, discharges from small metal items can look like charged device ESD (Figures 2.14 and 2.15).

If the resistance of the discharge circuit is sufficiently high, the peak ESD current is further reduced, and a unidirectional waveform with fast-rising edge but long decay may occur.

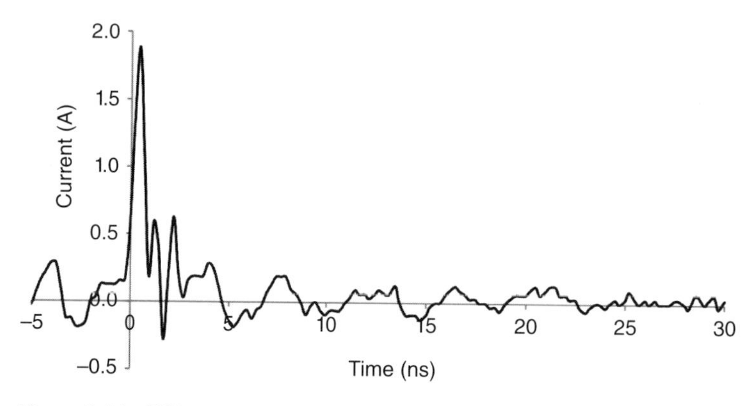

Figure 2.14 ESD waveform from screwdriver blade charged to +530 V. Charge transferred 0.03 nC.

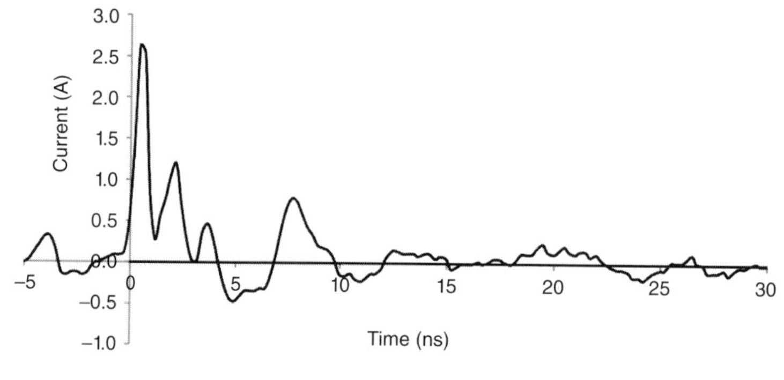

Figure 2.15 ESD waveform from a160 × 180 mm metal plate charged to 550 V. Charge transferred 2.5 nC.

2.6.3 Charged Device ESD

When a component touches a highly conductive object (e.g. metal) at a different voltage, a very short duration high discharge current ESD event occurs. The voltage difference may occur if the component is charged or the object is charged, or both. The same type of discharge will occur if either the component or the object is grounded.

The voltage on the device may arise from tribocharging or induced as a result of nearby electrostatic field sources. Often field-induced voltages can give the highest voltages arising on the device. Some examples of field-induced charged device ESD obtained in a laboratory experiment are given in Figure 2.16. In this experiment, the devices were slid down a charged PVC tube onto a 1.7 Ω target plate connected to a fast digital storage oscilloscope (500 MHz bandwidth, 2 Gs s^{-1} sample rate).

The fast high current peak typical of charged device ESD can be seen. The indicated peak current and rise and fall times of the waveform peaks are probably under-represented, as these waveforms are typically faster than the measurement system used here.

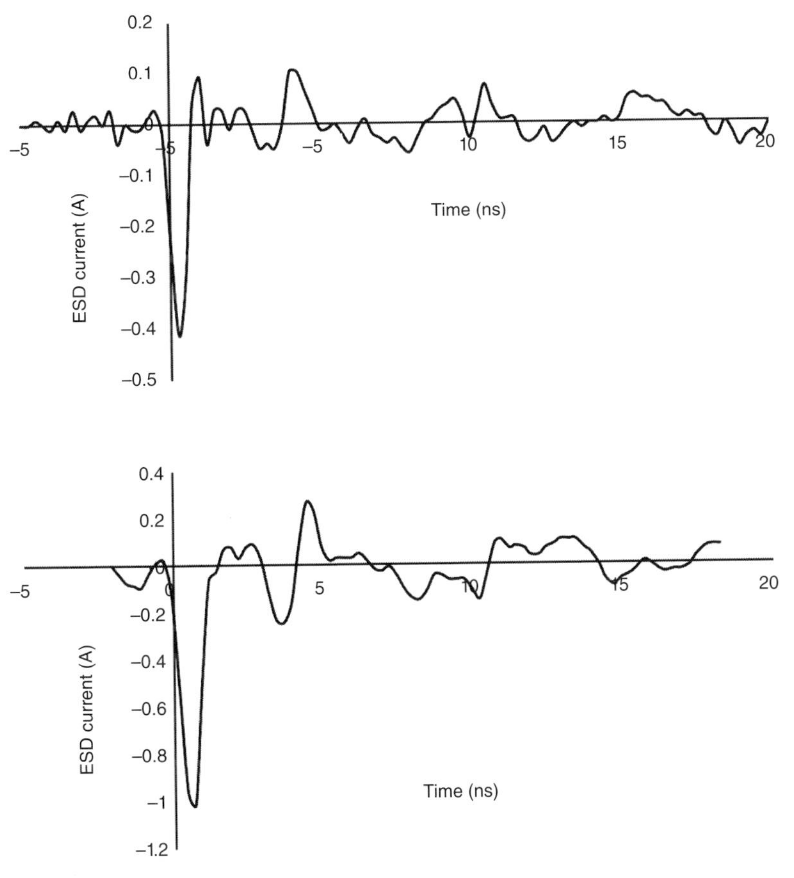

Figure 2.16 ESD waveforms from charged integrated circuits: (above) 32-pin plastic-leaded chip carrier and (below) 24-pin dual-inline package.

2.6.4 ESD from a Charged Board

PCBs often enter a production process highly charged. They can remain charged for long periods or can become charged as they are transported or go through a handing or assembly process. Voltages on the board up to 1000 V are not unusual, although the measured voltage will typically change with the proximity of the PCB to other objects. A PCB can also have high induced voltage if there is a highly charged insulator or another source of electrostatic field nearby.

If a conductor (e.g. track or component pin) on the charged board touches a highly conductive machine part (e.g. stop pin), a charged board ESD event can occur (Figure 2.17). The PCB can have high effective capacitance, so this type of discharge can be quite energetic.

2.6.5 ESD from a Charged Module

Many products, modules, or subassemblies have an insulating plastic housing containing a circuit board. The connections to this may be brought out to terminations at flying leads or a connector.

The housing can become highly charged, e.g. by rubbing or removal from packaging, inducing a high voltage on the PCB within the housing (Figure 2.18). If a connection is

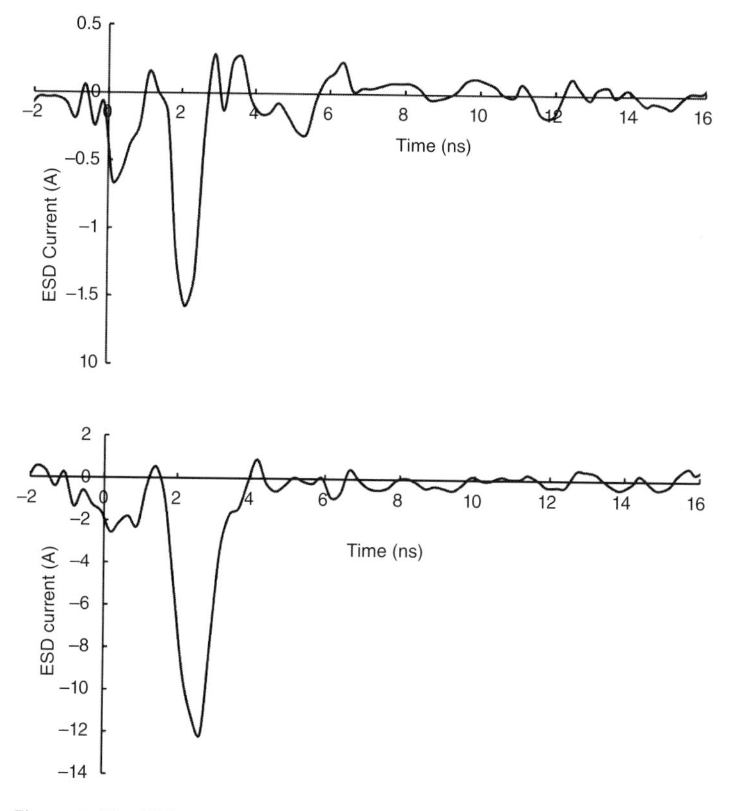

Figure 2.17 ESD waveform from a printed circuit board (above) charged to 1 kV (below) field induced charged by insulator 40 mm away.

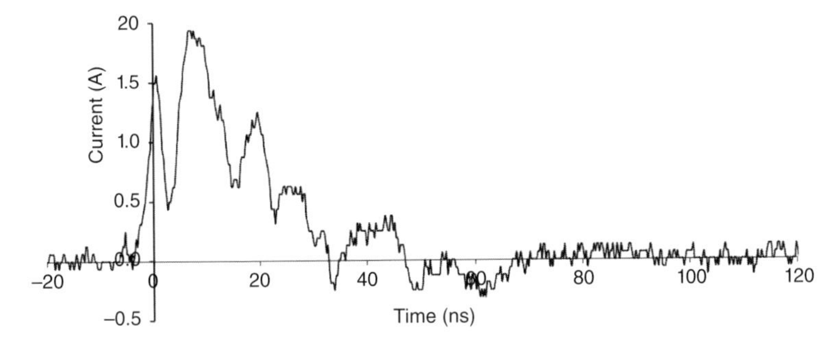

Figure 2.18 ESD waveform from a charged automotive module taken out of a polythene bag. Charge transferred 35 nC.

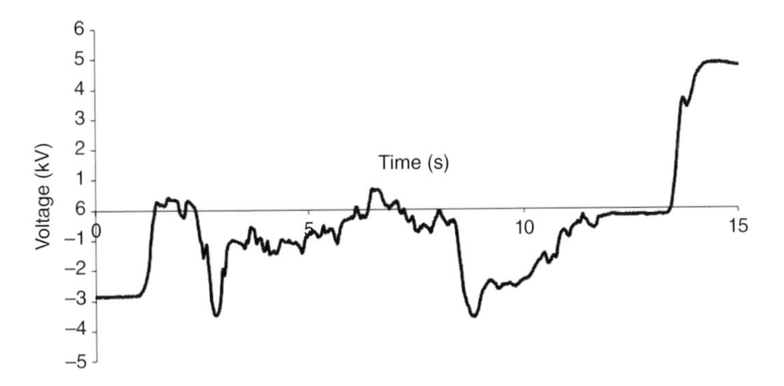

Figure 2.19 Voltage on an automotive cable core as polythene packaging is removed.

made to the module in this state, a discharge can occur at the termination at which contact is made.

2.6.6 ESD from Charged Cables

Cables and wiring looms can have significant capacitance between the wires in the cable and between the wires and ground. This can be of the order of 100 pFm^{-1}. Wires in the cable can become charged by various means such as by movement of the cable or by removal of the cable from packaging (Figure 2.19). If the cable is connected to equipment in this state, a charged cable ESD event can occur to the first terminal to make a connection (Figure 2.20).

2.7 Electronic Models of ESD

Many ESD sources can be simply modeled using a simple R-L-C circuit (Figure 2.21). The values of each component vary widely between different sources and help to explain the different types of waveforms observed.

At the heart of any ESD source is charge build-up and storage. This is represented by the capacitance in the model C. In many cases in real life, this charge storage may be on a conductor (e.g. metal item).

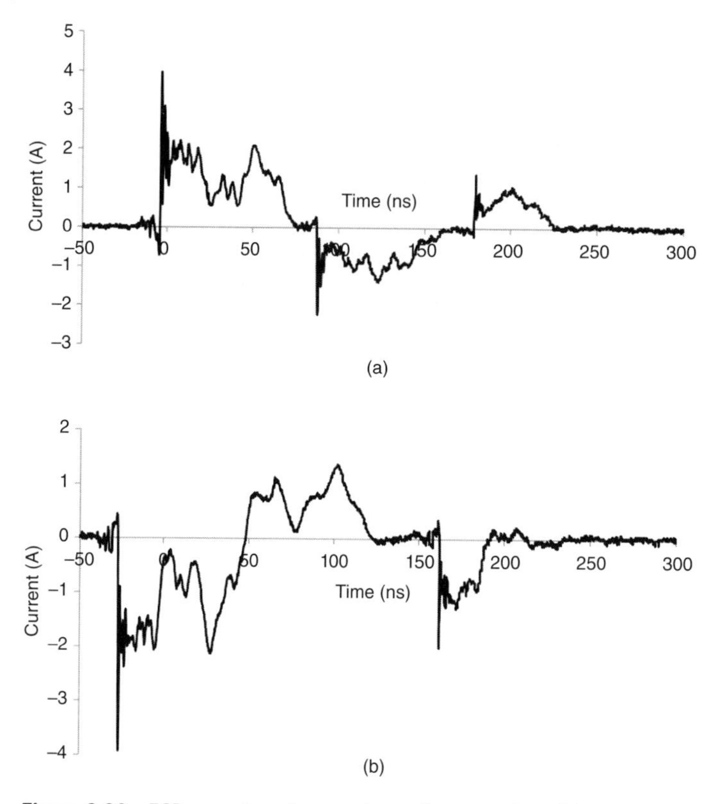

Figure 2.20 ESD waveform from a charged automotive wiring loom cable lying against an earthed metal plate. Positive (above) and negative (below) charging polarity.

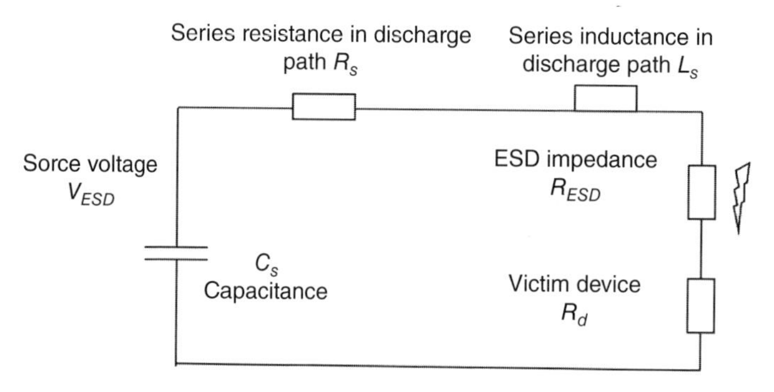

Figure 2.21 Electronic model of a simple ESD source.

The discharge is usually initiated by a breakdown of an air gap or some other insulating medium. At low voltages, it can also be initiated by contact or near-contact between two conductors. The discharge can itself have significant impedance R_{ESD} that can affect the waveforms produced and the energy delivered into the victim device. Often, however, this is negligible compared to the other impedances in the circuit, especially for larger ESD events.

After the discharge commences, the current flows through some circuit that includes some elements of resistance R_s and inductance L_s. These are normally due to the resistance and electrical properties of the materials in the current path.

In the case of ESD to a victim device, the device also has impedance, modeled in this simple circuit by a resistance R_d. In practice, a nonlinear impedance would be more typical of a semiconductor device. The impedance of the spark channel is highly variable and nonlinear.

For simplification, the total circuit resistance R is assumed to be linear and is the sum of the circuit resistances.

$$R = R_s + R_{ESD} + R_d$$

The discharge current I_{ESD} of this circuit has the form

$$I_{ESD} = A\exp(\alpha t) - \exp(\beta t)$$

For derivation of the equations for this and the following equations, the reader is referred to other texts (e.g. Agarwal and Lang 2005, https://en.wikipedia.org/wiki/RLC_circuit). This equation has two roots α, β given by

$$\alpha, \beta = -\left(\frac{R}{2L_s}\right) \pm \sqrt{\left(\frac{R^2}{4L_s^2} - \frac{1}{L_s C_s}\right)}$$

The waveform shape takes very different forms depending on the circuit component values. If the total circuit resistance is large and dominates the discharge path impedance, the waveform has a unidirectional shape, simulated in Figure 2.22 using model component values given for human-body model ESD (see Table 3.12). This occurs when

$$\frac{R^2}{4L_s^2} \gg \frac{1}{LC}$$

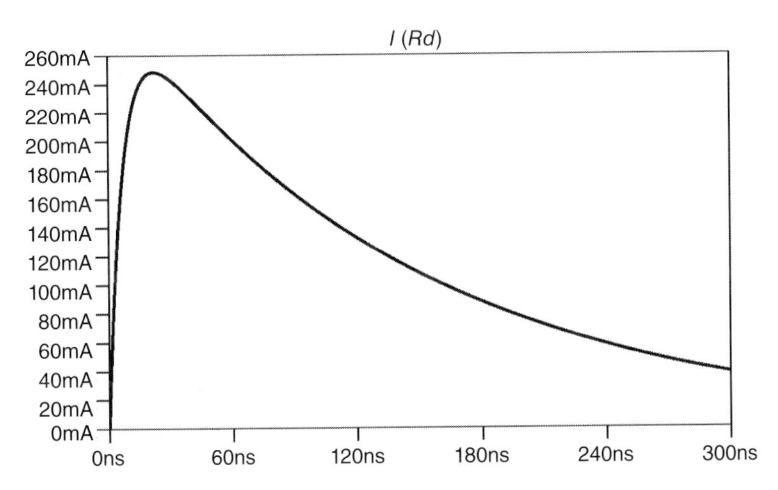

Figure 2.22 Simulated overdamped device current waveform IESD for dominant circuit resistance: $R_s = 1500\,\Omega$, $R_d = 10\,\Omega$, $R_{ESD} = 0\,\Omega$, $L_s = 10\,000$ nH, $C_s = 100$ pF, $V_{ESD} = 500$ V.

The discharge current rises rapidly to a peak I_p that, when inductance is small, approaches the value and polarity near that predicted by Ohms law.

$$I_p \approx V_{ESD}/R_S$$

Thereafter, the current drops nearly exponentially with decay time approaching $R_s C_{ESD}$.

At the other extreme, if the circuit resistance is insignificant compared to the inductive and capacitive impedance, the waveform is quite different. This occurs when

$$\frac{R^2}{4L_s^2} \ll \frac{1}{LC}$$

The waveform rises to a peak and then oscillates negative and positive about zero. The overall amplitude decays exponentially with time, simulated in Figure 2.23 using values given for machine model ESD (Table 3.12).

Between the two extremes, the waveform duration decreases and is minimum around the point of critical damped waveform, where the waveform changes between the two different shape types. This is simulated in Figure 2.24 using model values close to those given for the charged device model (Table 3.12). This occurs at the condition

$$\frac{R^2}{4L_s^2} = \frac{1}{LC}$$

Practical ESD sources often require the addition of more components (e.g. additional capacitors) to better represent additional charge storage (e.g. metal parts) and other features that may be present. These may contribute further current peaks or modify the shape of the waveform (Verhage et al. 1993).

The stored energy E_{ESD} in a conductive ESD source is given by

$$E_{ESD} = 0.5C_s V_{ESD}^2$$

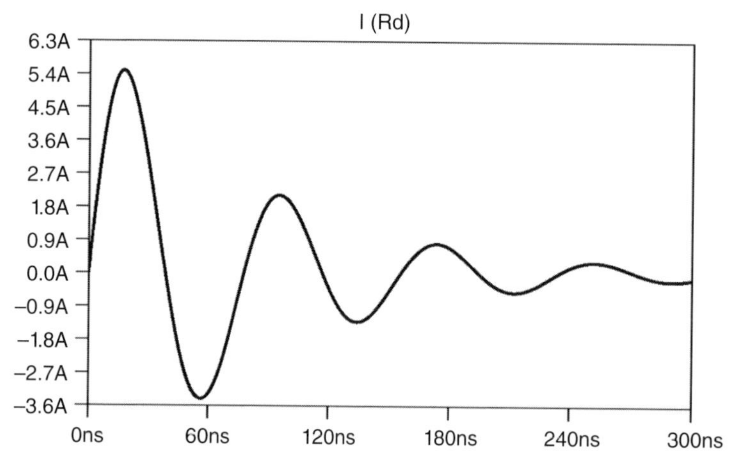

Figure 2.23 Simulated underdamped device ESD current waveform for the case of low circuit resistance (dominant inductive and capacitive impedance): $R_{ESD} = 10\,\Omega$, $R_d = 10\,\Omega$, $L_s = 750$ nH, $C_s = 200$ pF, $V_{ESD} = 500$ V.

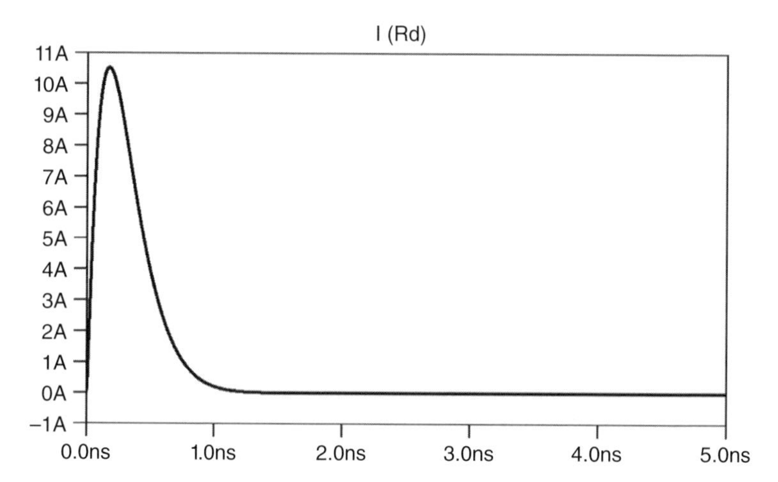

Figure 2.24 Simulated device ESD current waveform for near critical damping: $R_{ESD} = 20\,\Omega$, $R_d = 10\,\Omega$, $L_s = 2.5$ nH, $C_s = 10$ pF, $V_{ESD} = 500$ V.

All this energy is dissipated in the total circuit resistance R. Only a fraction of this is the energy dissipated in the device E_d.

$$E_d = \frac{E_{ESD} R_d}{R}$$

In a source such as the charged human body that has significant resistance, most of the stored energy is dissipated in the circuit (body) resistance, and only a small fraction is dissipated in the victim device. In contrast, a charged metal object is a low-resistance ESD source, and most of the stored energy can be dissipated in the victim device. This is one reason why some components may be damaged by a lower voltage with a metal ESD source compared to a charged person. In general, the likelihood of ESD damage to a component by ESD from a source will depend on the susceptibility of the device to ESD current, voltage, energy, or other parameter of the discharge. This is further discussed in Chapter 3.

2.8 Electrostatic Attraction (ESA)

Where there is an electrostatic field, charged particles in the vicinity will experience an attractive or repulsive force. A lesser known effect is that uncharged particles can be attracted or repelled in a convergent or divergent field – this is known as *dielectrophoresis* (Cross 1987).

The direction of the force depends on the polarity of the charged particles and the field. The force acts such that like polarity charges repel and unlike polarity attract. So, a positive charge will experience a force toward a more negative potential, and vice versa.

2.8.1 ESA and Particle Contamination

These effects can be important where cleanliness of the product is essential. In a clean room for wafer fabrication, an electrostatic field can cause charged dust particles that are present to be transported to a wafer within the field. Particle contamination can then cause loss of product yield (Welker et al. 2006).

Other processes in which product cleanliness is important can include

- Manufacture of flat-screen displays. Loss of even a small number of pixels due to contamination can result in rejection of the product.
- Packing of consumer products where dust or particle contamination can mar the appearance of the product before purchase.
- Assembly of optical systems where performance can be reduced by contamination.
- Assembly of medical systems where infection of the user may be a risk.

2.8.2 Neutralization of Surface Voltages by Air Ions

Clean air is naturally a good insulator with very few mobile charged particles present. A small number of ions are naturally generated when air molecules are split into positive and negative ions by the action of natural radioactivity or cosmic rays (Jonassen 1985). These ions will be repelled or attracted by surface charges due to the electrostatic field. The ions move in the direction of the electrostatic field.

The charge migration rate and direction are dependent on the ion charge and other factors, as well as the electrostatic field strength and direction at the point in space where the ion is located. In still air, the ion drift velocity v_d is related to electrostatic field E by the ion mobility μ.

$$v_d = \mu E$$

The mobility of the ion is dependent on the ion size. In air, charges bind to water, nitrogen, and other molecules or particles and form small or large ions. Small ions have mobility in the range $1-2\times 10^{-4}$ m^2 V^{-1} s^{-1} (Jonassen 1985). Large ions have mobility in the range 8×10^{-7} to 3×10^{-8} m^2 V^{-1} s^{-1}.

The number of air ions present can be increased using an ionizer. These produce air ions by various means such as corona discharge, radioactive, or X-ray ionization of the air. Radioactive and X-ray ionization sources provide both polarity ions by splitting air molecules into positive and negative ions.

Corona discharge sources use a high voltage applied to a sharp electrode (e.g. needle) to produce ions of one polarity. A nearly balanced ion source can be produced by this method by using an alternating current (AC) high voltage or two separate sources of opposite polarity.

A charged surface produces an electrostatic field surrounding it that repels like polarity ions and attracts opposite polarity ions. That is, a negatively charged surface repels negative ions and attracts positive ions. A positively charged surface attracts negative ions and repels positive ions. Opposite polarity ions will drift to the charged surface at a rate proportional to the field strength and in numbers proportional to the ion concentration. An opposite polarity charge on reaching the charged surface neutralizes an equal charge, reducing the net surface charge and electrostatic field. The ion drift represents a neutralizing current, limited by the ion concentration and field strength.

2.8.3 Ionizers

Ionizers are devices used to generate air ions for neutralization of surface charges on charged materials and objects (Jonassen 1985, 1986). These are available in various types based on passive, radioactive, electrical, and other principles of operation.

Passive ionizers rely on high electric fields developed around sharp points or edges on earthed conductors to generate air ions by corona discharge. These will always generate ions of the opposite polarity to the voltage producing the field at the point or edge. Unfortunately, corona discharges do not occur below a threshold field strength, and this means there is a threshold voltage, known as the *corona inception voltage*, below which ions are not produced, and neutralization does not occur. This threshold voltage may be several kilovolts (kV). This means that passive ionizers cannot be used to reduce voltages to below this threshold, and they are seldom useful in ESD control in electronics manufacture. They do find major applications in industrial processes such as plastic film manufacturing, printing, copiers, and other processes involving insulating materials.

Electrical ionizers also use the high electrostatic fields at points, usually needles, to generate air ions. In this case, however, a power supply is used to raise the needle to a voltage above the corona inception voltage to ensure that sufficient ions are produced. A balanced ion source with equal numbers of positive and negative ions is usually required. This can be produced either by using an AC driving voltage or by using two sets of needles, one at positive voltage and the other at negative voltage. This can produce an ion stream that is nearly balanced, but it is difficult to produce an exact balanced ion stream by this method.

Radioactive ionizers use a radioactive source to ionize the air by impact of radioactive particles with molecules in the air. Exactly balanced ion streams are produced by this means, as the molecule splits into two ions, one positive and one negative ion. The rate of production of ions is small and limited by the level of radioactivity of the ionizing material.

2.8.4 Rate of Charge Neutralization

Each ion that reaches a charged surface will add its charge to the surface. As this is usually the opposite polarity to the surface charge, the arriving ion will normally neutralize an equal and opposite polarity surface charge. The rate at which ions arrive at the surface depends on the electrostatic field strength, the concentration of ions in the air, and the mobility of the ions (Jonassen 1986). This can be thought of as an ion current flow. Different air ions can have different mobility depending on the size, polarity, and charge of the ions.

As the charged surface neutralization process continues, surface voltages and field strength reduce, and the drift of ions to the surface reduces with the reducing field strength. The neutralization process continues ever more slowly until the electrostatic fields and thus forces on the ions are insufficient to attract further charge from the air.

If the surface voltage is monitored, the voltage is seen to reduce at a reducing quasi-exponentially decaying rate. This is used in a charge plate monitor (CPM) instrument to measure the effectiveness of ionizers in neutralizing charges on surfaces. A typical voltage decay curve obtained is shown in Figure 2.25.

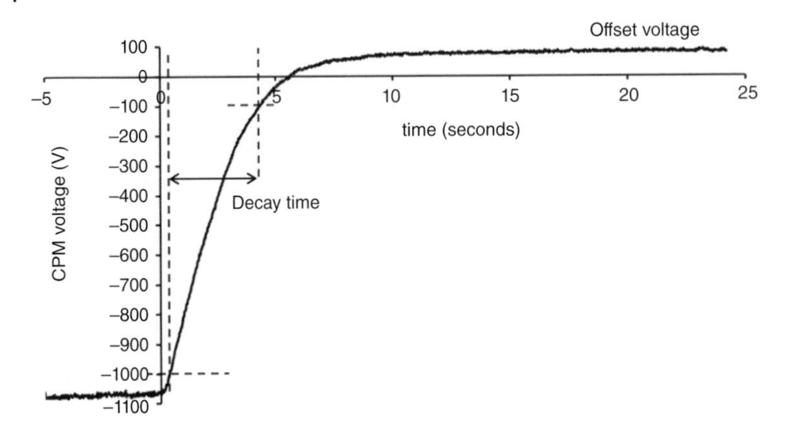

Figure 2.25 Electrostatic voltage on a CPM plate reducing during charge neutralization showing decay time and offset.

Charge neutralization by ionizers is typically quite a slow process and can take tens of seconds or longer, depending on the electrostatic field strength and ion density in the region surrounding the charged surface. This charge density, and the speed of charge neutralization, can be affected by many factors.

2.8.5 The Region of Effective Charge Neutralization Around an Ionizer

The ion density in the air typically reduces with the distance from the ionizer. This is in part because the ion cloud tends to spread out due to repulsion forces between like polarity ions. In addition, attraction between positive and negative charges in the cloud causes them to move toward each other and recombine to neutralize each other (Jonassen 1985).

As the ion cloud is drifting in air, movement of the air can have a significant effect on the location of the cloud and the local ion concentration. Ions can be blown toward or away from the region where they are wanted (e.g. by drafts from a fan or open window). Conversely, in many situations a fan can be used to blow ions to the location they are needed to improve charge neutralization effectiveness.

2.8.6 Ionizer Balance and Charging of a Surface by an Unbalanced Ionizer

An ion cloud is rarely accurately balanced in terms of ion charge polarity density. One consequence of this is that any surface within the ion cloud tends to reach a state of excess charge reflecting the ion density balance in the ion cloud. If, for example, the ion cloud has excess positive ions, then surfaces within the ion cloud will tend to become positively charged until electrostatic fields are sufficient to prevent further ion deposition. This effect will create residual voltages on insulating surfaces and isolated conductors within the ion cloud. With a well-balanced ionizer, this effect will be small, and the excess voltage achieved may be only a few volts or tens of volts. This "offset voltage" is an important parameter of an ionizer specification (Figure 2.25).

A poorly maintained ionizer can become seriously out of balance. This can happen in service due to erosion or contamination of the ionizer needles. If this occurs, then surfaces within the ion cloud may become charged to hundreds or even thousands of volts. Monitoring this effect over time and performing appropriate maintenance of an ionizer to prevent the imbalance from reaching an excessive level is an important aspect of ionizer maintenance (Simco Ion TN-003 2019).

2.9 Electromagnetic Interference (EMI)

ESDs can give waveforms that have very fast (nanosecond or subnanosecond) rise times, high peak current levels of tens of amps, and oscillatory waveforms. Discharge current rise rates can be of the order of 10^9 A s^{-1}, and electrostatic fields may collapse with rates of the order of 10^{12} V s^{-1}. The ESD source may radiate or conduct strong transient electromagnetic fields over a wide frequency band up to GHz frequencies.

These induce transients in nearby conductors, especially tracks or wires of significant length that can act as efficient antennas at the frequencies radiated in the discharge. This radiated or conducted interference can cause upset or malfunction to nearby equipment such as test equipment. Giant magnetoresistive (GMR) heads have been demonstrated to be damaged by this type of ESD transient (Wallash and Smith 1998).

EMI can also be an issue in highly automated facilities where automated test equipment (ATE) can test large numbers of production items. EMI may cause the ATE to register a product fail or to otherwise malfunction (Tamminen et al. 2015). This can represent a loss of production or output.

2.10 How to Avoid ESD Damage of Components

2.10.1 The Circumstances Leading to ESD Damage of a Component

ESD damage does not occur unless there is ESD. The circumstances leading to ESD are as follows:

- A conductor attains a high voltage through induction or triboelectrification. The conductor may be
  - A person
  - A metal or other conductive item such as a tool, machine part, or cable
  - An ESDS component
- The conductor touches, or comes sufficiently near to another conductor or ESDS device, for ESD to occur.
- The ESD current passes through a part of the ESDS device that may be damaged by ESD.
- The discharge current, charge, energy, or voltages developed on the ESDS device must exceed a threshold at which damage is likely to occur.

Most ESDS devices are not directly damaged by exposure to electrostatic fields. The risk is usually due to induced voltages on the device or a nearby conductor causing ESD when the ESDS device touches the conductor. For some high-impedance, voltage-sensitive devices, induced voltages can cause a risk of breakdown within the device.

2.10.2 Risk of ESD Damage

The occurrence of ESD to a susceptible component does not necessarily cause a failure. Indeed, it is likely that ESD happens before detectable damage occurs. So, ESD damage is perhaps best thought of as a risk with a probability that may depend on many factors.

- The likelihood that ESD occurs to an ESDS device.
- Most devices have many possible ESD entry points. Each of these will typically have a different ESD susceptibility.
- The likelihood that the ESD strength will exceed the ESD damage threshold of the device at the point of discharge.

Considering these conditions shows why ESD damage is a probabilistic phenomenon. Only a small fraction of ESD events may cause ESD damage, because

- The capacity of the ESD source to produce damage varies with the characteristics of the source.
- The magnitude of the source voltage and energy will vary tremendously with the conditions that cause it.
- The variation in discharge paths will cause great variation in characteristics of the discharge such as peak current, waveform, rise and fall times and duration, and power and energy deposited in the ESDS device.
- The susceptibility of the ESDS device to peak current, power, energy, waveform, and other parameters of the discharge through it varies between devices.
- The ESD current path through the ESDS component may be through parts that have different damage thresholds.
- The ESD event may not be sufficiently strong to exceed the damage threshold of its path through the component.

So, ESD may occur many times during handling or processing a component, without ESD damage occurring. The probability of ESD damage occurring may be one in hundreds or thousands of ESD events. Combine this with the fact that damage will be discovered often at a much later manufacturing stage, and may not be identified as due to ESD, it is small wonder that many people get the impression that ESD damage is something that does not happen to them!

Nevertheless, each time that a conductor contacts an ESDS component is an opportunity for ESD to occur. Given sufficient ESD opportunities and insufficient ESD control, ESD damage becomes inevitable.

2.10.3 The Principles of ESD Control

The principles of ESD control are remarkably simple. Each principle is aimed at reducing a particular ESD risk.

- ESDS are handled only within an electrostatic discharge–protected area (EPA; see Chapter 4) or under ESD-controlled conditions. This ensures that when ESDS devices are handled, the ESD risks are controlled to an acceptable level.

- Outside the EPA in unprotected areas (UPAs), ESD protective packaging is used to protect the ESDS device. The packaging is designed to prevent ESD sources outside the package from having a significant effect on the ESDS device within the package. It also provides a safe region within the package in which ESD risks are controlled.

Within the EPA, ESD risks are controlled by eliminating as far as is possible the sources of ESD likely to damage the ESDS.

- Conductors that may touch an ESDS, especially metal items and people, are grounded wherever possible. This ensures that as far as possible, electrostatic voltages on conductors are the same and near zero. This is to prevent them from becoming charged and a source of significant ESD to ESDS devices.
- Where conductors cannot be grounded and might contact an ESDS, the voltage difference between the conductor and the ESDS device must be reduced to a sufficiently low level to prevent significant ESD risk.
- Electrostatic fields that may occur near ESDS devices are eliminated or reduced to a low level. Insulators that may become charged and the source of electrostatic fields are, where possible, removed from the vicinity of ESDS devices. This is to reduce the risk of induced charging of isolated ESDS devices or other ungrounded conductors.
- Items that may contact an ESDS device are preferably made from materials that have appreciable electrical resistance. If ESD occurs, this helps to reduce discharge current to a safe level and absorb much of the discharge energy within the material rather than in the ESDS device.

These measures do not eliminate ESD but help reduce the numbers of ESD occurring, the magnitude of any ESD that occurs, and the likelihood of it damaging the ESDS. Reducing the number of times contact is made with an ESDS device can also help reduce the likelihood of ESD damage occurring. So, the simple measures of not handling the ESDS device any more than necessary and reducing to minimum contact between the ESDS device and other conductors that might be at different voltage can make a useful contribution to reducing ESD risk. If the materials are resistive rather than low resistance where they contact ESDS devices, then a further reduction of risk of ESD damage is made. This is due to reduction of peak current in the discharge and absorption of energy by the resistance of the material.

Suitably specified ESD protective packaging can reduce the risk of ESD damage occurring in the UPA to an insignificant level. The packaging must be specified to address the ESD risks appropriate to the ESDS component or item.

References

Agarwal, A. and Lang, J.H. (2005). *Foundations of Analog and Digital Electronic Circuits*. Morgan Kaufmann, ISBN 1-55860-735-8.

Andersson, B., Fast, L., Holdstock, P., and Pirici, D. (2008). Charging of a person exiting a car seat. Electrostatics 2007. *J. Phys. Conf. Ser.* 142: 012004.

Bailey, A.G., Smallwood, J.M., and Tomita, H. (1991). Electrical discharges from the human body. In: *Electrostatics –Inst. Phys. Conf. Se. 118 Sec. 2.* Inst.Phys.

Barnum, J.R. (1991). Sandia's severe human body electrostatic discharge tester (SSET). In: *Proc. EOS/ESD Symp. EOS13*, 29–30. Rome, NY: EOS/ESD Association Inc.

Cross, J.A. (1987). *Electrostatics Principles, Problems and Applications*. Adam Hilger. ISBN 0-85274-589-3.

Department of Defense. Military Handbook. (1994) *Electrostatic Discharge Control Handbook for protection of electrical and electronic parts, assemblies and equipment (excluding electrically initiated explosive devices) (metric). MIL HDBK-263B*. Washington DC, Department of Defense.

International Electrotechnical Commission. (2013) PD/IEC TS 60079-32-1. *Explosive atmospheres Part wp2-1. Electrostatic hazards, guidance*. Geneva, IEC.

Jonassen N. (1985) The physics of air ionization. In: *Proc. EOS/ESD Symp. EOS-7 Minneapolis USA*. Rome, NY, EOS/ESD Association Inc. pp. 59–66.

Jonassen, N. (1986). The physics of air ionization. In: *Proc. EOS/ESD Symp. EOS-8*, 35–40. Rome, NY: EOS/ESD Association Inc.

Jonassen, N. (1998). Human body capacitance – static or dynamic concept? In: *Proc. EOS/ESD Symp. EOS-20*, 111–117. Rome, NY: EOS/ESD Association Inc.

Kelly MA, Servais G E, Pfaffenbach T V. (1993) An Investigation of Human Body Electrostatic Discharge. *19th International Symposium for Testing & Failure Analysis Los Angeles, California, USA*. Russell Township, OH, ASM International.

Kuffel, E., Zaengl, W.S., and Kuffel, J. (2001). *High Voltage Engineering*. Newnes ISBN 0 7506 3634 3.

Landers, E.U. (1985). Distribution of charge and fieldstrength due to discharge from insulating surfaces. *J. Electrostat.* 17: 59–68.

Norberg, A. (1992). Modelling current pulse shape and energy in surface discharges. *IEEE Trans. Ind. Appl.* 28 (3): 498–503.

Norberg, A. and Lundquist, S. (1991). A distributed RC transmission line model for electrostatic discharges from insulator surfaces. In: *Inst. Phys. Conf. Se. 118*, 269–274. Electrostatics 1991.

Norberg, A., Szedenik, N., and Lundquist, S. (1989). On the pulse shape of discharge currents. *J. Electrostat.* 23: 79–88.

Pirici, D., Rivenc, J., Lebey, T. et al. (2003). A Physical model to explain electrostatic charging in an automotive environment: Correlation with experimental approach. In: *Proc. EOS/ESD Symp. EOS-25*, 161. Rome, NY: EOS/ESD Association Inc.

Simco Ion (2019) Emitter point maintenance. Technical note TN-003. Available from: https://technology-ionization.simco-ion.com/DesktopModules/Bring2mind/DMX/Download.aspx?command=core_download&entryid=112&language=en-US&PortalId=0&TabId=145 [Accessed 16 April 2019]

Smallwood, J.M. (1999). Simple passive transmission line probes for electrostatic discharge measurements. In: *Inst. Phys. Conf. Se. 163*, 363–366. Electrostatics 1999, Inst.Phys.

Tamminen P, Viheriäkoski T, Ukkonen L, Sydänheimo L (2015) ESD and Disturbance Cases in Electrostatic Protected Areas. In: *Proc EOS/ESD Symp. 5B.2,* Rome, NY, EOS/ESD Association Inc.

Verhage, K., Roussel, P.J., Groeseneken, G. et al. (1993). Analysis of HBM ESD Testers and specifications using a 4th order lumped element model. In: *Proc. EOS/ESD Symp*, 129–137. Rome, NY: EOS/ESD Association Inc.

Viheriäkoski T, Peltoniemi T, Tamminen T (2012) Paper 4A3. Low Level Human Body Model ESD. In: *Proc. EOS/ESD Symp. Tucson Ariz. USA*. Rome, NY, EOS/ESD Association Inc.

Wallash A, Levitt L. (2003) Electrical breakdown and ESD phenomena for devices with nanometer-to-micron gaps. In: *Proc. SPIE 4980, Reliability, Testing, and Characterization of MEMS/MOEMS II*. San Jose, CA, SPIE.

Wallash A, Smith D. (1998) Paper 4B.6. Electromagnetic Interference. (EMI) damage to giant magnetoresistive (GMR) recording heads. In: *Proc. EOS/ESD Symp.*, Rome, NY, EOS/ESD Association Inc. pp. 368–74.

Welker, R.W., Nagarajan, R., and Newberg, C.E. (2006). *Contamination and ESD Control in High-Technology Manufacturing*. Wiley ISBN-13: 978-0-471-41452-0.

Further Reading

EOS/ESD Association Inc. (2015) ANSI/ESD STM3.1-2015. *ESD Association Standard for the Protection of Electrostatic Discharge Susceptible Items – Ionization*. Rome, NY, EOS/ESD Association Inc.

International Electrotechnical Commission. (2017) IEC 61340-4-7:2017. *Electrostatics - Part 4-7: Standard test methods for specific applications – Ionization*. Geneva, IEC.

3

Electrostatic Discharge–Sensitive (ESDS) Devices

3.1 What Are ESDS Devices?

Electrostatic discharge susceptibility is defined in ESD ADV1.0-2009 as "the propensity to be damaged by electrostatic discharge." Various types of components can be susceptible to damage from electrostatic fields or ESD (Figure 3.1). A component that is susceptible to ESD damage is often called an *ESD-sensitive device*, often abbreviated to ESDS. Items susceptible to electrostatic discharge are defined in ESD ADV1.0-2009 as "electrical or electronic piece part, device, component, assembly or equipment item that has some level of electrostatic discharge susceptibility."

The list of ESDS technologies has grown, and is still growing, as new technologies have developed over time. New device technologies may be inherently more or less sensitive to ESD damage than earlier technologies. MIL HDBK 263 gave a list of types of ESDS components and outlined damage mechanisms. Some of these and more recent technologies are briefly reviewed in Section 3.4.

Electrical overstress (EOS) is damage to components due to their absolute maximum ratings being exceeded by some means. ESD is a form of EOS, with other EOS stress sources including lightning, electromagnetic pulse (EMP), and electrical transients that can occur at the board or system level during test or operation. EOS is increasingly becoming a serious concern and failure mode to electronic components (Amerasekera et al. 2002).

During an ESD event, electronic devices can be pushed outside their normal operating range. Very high currents may flow for short times and may pass through unintended routes in the component. Voltages may be applied, or may develop internally, that can greatly exceed the component design ratings.

ESDS devices and systems can fail completely, partially, or temporarily in various ways (MIL-HDBK-263B (Department of Defense 1994); Baumgartner n.d.). Complete or catastrophic (also sometimes called *hard*) failures occur when a component or system is permanently damaged by an electrostatic field or ESD. Some catastrophic failures may occur as the cumulative effect of several ESD events. ESD damage is often difficult to distinguish from other EOS sources in failure analysis, although the scale of damage to a component in EOS damage is often rather greater than ESD damage.

Unpowered components and devices tend to usually suffer catastrophic failures. It is possible that components can sometimes be damaged and be weakened or suffer parametric

The ESD Control Program Handbook, First Edition. Jeremy M Smallwood.

Figure 3.1 ESD-sensitive devices range from individual transistors or diodes to PCBs or modules.

change. They may still pass functional tests to fail at a later stage. This is called a *latent failure*.

It is not just individual components that may be susceptible to ESD. Any printed circuit board (PCB), module, or assembly containing ESDS is likely to be itself an ESDS device if not in some way protected against ESD. In some cases, the susceptibility to ESD may be limited to remaining contact points such as flying leads, connectors, or limited exposed conductors.

Fully assembled systems that include ESDS, protected within an enclosure or housing that acts as an effective barrier to ESD, are normally no longer considered an ESDS device. In many cases, these systems must be tested for ESD immunity during operation, using standard ESD immunity tests required by electromagnetic compatibility (EMC) regulations or requirements. These system ESD immunity tests are aimed at ensuring functionality of the system is not compromised during operation by ESD occurring in the uncontrolled operating environment or during consumer operations.

Working electronic systems tend to suffer partial or temporary (*soft*) failures due to ESD. The system may recover almost immediately, or it may suffer effects such as program crash or malfunction, spontaneous reset, or data corruption. Often effects such as program crash or malfunction may be completely recovered by restarting the system. Data corruption effects may lead to corrupted data remaining in the system.

Often the remaining connector terminations and other potential ESD entry points will have been designed for immunity to expected ESD threats. Sometimes, a risk of ESD damage can remain from strong ESD to components within the system connected to lines emerging on connectors or accessible via user interfaces such as keyboards or touch screens. Severe ESD to system parts such as cable discharge to connectors can sometimes result in physical, hard damage to the system. For this reason, some ESD protection may need to be built into the system.

3.2 Measuring ESD Susceptibility

3.2.1 Modeling Electrostatic Discharges

ESD can often be modeled using a simple electronic circuit (Figure 3.2). Additional components may, however, be added to tailor the waveform to better represent real-world waveforms (Wang 2002).

A capacitor C is charged to an ESD voltage $V_{esd.}$ The capacitor models the charge stored on a human body, a metal object, or the device itself in a real situation. On initiating the ESD event, the ESD current flows through a circuit resistance R, inductance L, and the load device or spark gap. Although the circuit shows a switch, in real ESD the current flow is usually initiated by breakdown of the insulating properties of the material through which the discharge occurs, often air.

The ESD waveform characteristics can be matched to real ESD events by careful choice of the capacitance, circuit resistance, and inductance. Occasionally, additional components are required for tailoring the waveform to a real application.

This type of model is used both in ESD immunity test in EMC and in semiconductor device ESD sensitivity tests. It has also been used in other industry areas, for example ignition tests of explosives and flammable atmospheres. Typical component values are shown in Table 3.1.

The simple model works well for many simple situations, particularly where the source is a conductive material and a simple resistance and inductance can approximate the discharge path. In practice, the ESD waveform is as much dependent on the properties of the "load" circuit, which can be highly nonlinear, as on the source. The victim device or spark channel may have significant impedance and may affect the resulting peak current, duration, and other ESD waveform properties, including whether the waveform is unidirectional or oscillatory (See Section 2.7). A device may also have a nonlinear rectifying action due to the presence of semiconductor junctions (Figure 3.3).

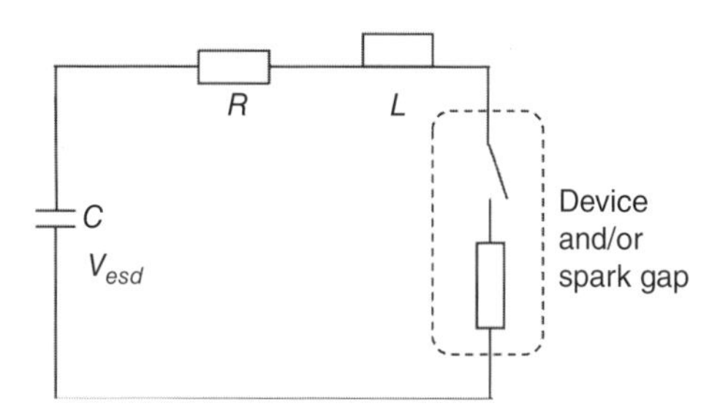

Figure 3.2 Simple electronic model of ESD circuit.

Table 3.1 Typical ESD model simulation component values for common sources.

Model of ESD source	Resistance R (Ω)	Capacitance C (pF)	Inductance L (nH)
Human body	300–1500	100–300	Stray
Large metal object	Stray	200	Stray
Charged device	<10	Capacitance of device under test (1–30 pF)	<10
Charged insulator surface (Norberg 1992)	Spark resistance 2–8 kΩ	8–11	

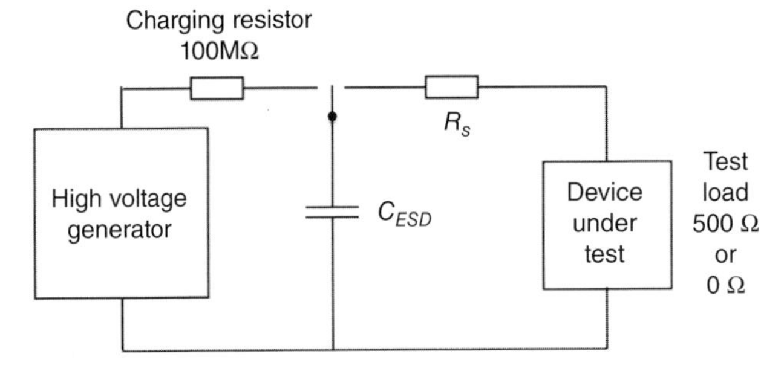

Figure 3.3 Typical ESD generator circuit.

3.2.2 Standard ESD Susceptibility Tests

The three main ESD sensitivity test models used in ESD susceptibility testing of electronic components are the human body model (HBM), machine model (MM), and charged device model (CDM). These simulate particular ESD sources known to give significant ESD damage in practice (Wang 2002; Amerasekeera 2002). The exact specification of the different models varies in detail with standards and implementations.

Components are normally characterized using the HBM test during device qualification. Components that may be assembled using automated handling techniques are normally also characterized using the CDM test. Some manufacturers, but unfortunately not all, publish the results in the component data sheet.

Component characterization using the MM test is, at the time of writing, falling out of practice. This is because the failure modes of MM ESD are similar to HBM, and the withstand voltages of these tests normally correlate to a reasonable degree. Manufacturers regard the MM test as giving little additional information over the HBM test (Duvvury et al. 2012).

Typical examples of widely used component ESD susceptibility test standards are ANSI/ESDA/JEDEC JS-001(EOS/ESD Association Inc., JEDEC 2017) and IEC60749-26 (HBM) (International Electrotechnical Commission 2013), IEC 60749-27 (MM) (International Electrotechnical Commission 2012), ANSI/ESDA/JEDEC JS-002 (CDM) (EOS/ESD Association Inc., JEDEC. (n.d.)), and IEC 60749-28 (CDM) (International Electrotechnical Commission 2017).

Table 3.2 ESD model component values to achieve specified waveforms in standards.

Model	Standard	R_s (Ω)	C_{ESD} (pF)
Human body model (HBM)	IEC 60749-26 ANSI/ESD/JEDEC JS-001	1500	100
Machine model (MM)	IEC 60749-27 ANSI/ESD/JEDEC STM 5.2		200
Human metal model (HMM)	IEC 61000-4-2 ANSI/ESD S5.6	330	150

These are defined in terms of waveform characteristics when discharged into a defined load, rather than model component values. The waveforms defined by these standards are given in Sections 3.2.4–3.2.8. Typical model component values used to create these waveforms are given in Table 3.2.

3.2.3 ESD Withstand Voltage

During device ESD susceptibility test, the voltage on the capacitor is increased in steps (stress levels), with a test performed at each stress level until damage occurs. The highest stress level at which damage does not occur is recorded as the ESD withstand voltage of the device. So, a 100 V HBM device was not damaged in the HBM test with 100 V capacitor charging voltage. It may be damaged by the next test voltage level up, depending on the voltage increments used.

3.2.4 HBM Component Susceptibility Test

The charged human body is the most common and damaging source of ESD in manual assembly. In daily life, people charge through normal movement, often up to several thousand volts (kV). Typically, a person will not feel an ESD event unless their body voltage exceeds about 2 kV (Brundrett 1976; Wilson 1972).

So, the primary measurement of device ESD sensitivity is by HBM ESD. In this test, a charged 100 pF capacitor is discharged into the device via a 1500 Ω resistor. The 100 pF capacitor simulates charged stored on the average human body, and the resistor simulates the resistance of the human body and skin.

The ANSI/ESD/JEDEC JS-001standard defines HBM current waveforms for device test purposes. The waveform parameters are defined in terms of peak current, rise time, and duration (Figure 3.4), and they give required values for qualification of the HBM test equipment measured with 0 and 500 Ω calibration loads (Table 3.3). With a device in the circuit, waveforms will typically be different from those obtained with the calibration loads.

Device evaluation typically is done using three samples of the device. Each sample is tested with negative and positive polarity ESD with a minimum pulse interval of 300 ms. Pulses are applied to each pin in turn, with each power pin in turn grounded. The test starts at the lowest voltage level and is increased progressively, providing the devices do not fail. Pins that are not grounded or under test are left floating. Pulses are also applied to all nonsupply pin pair combination.

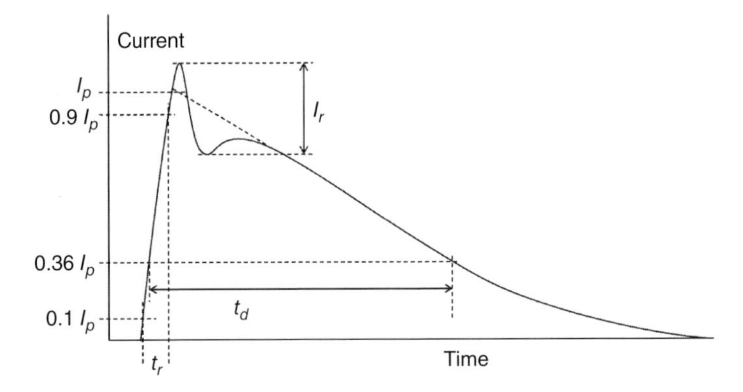

Figure 3.4 HBM waveform definition with 0 Ω calibration load.

Table 3.3 ANSI/ESDA/JEDEC JS001-2017 HBM waveform parameters with 0 Ω calibration load.

ESD voltage (V)	Peak current I_p (A)		Rise time (ns)		Decay time t_d (ns) 0 Ω load	Ringing current I_r (A)
	0 Ω load	500 Ω load	0 Ω load	500 Ω load		
250	0.15–0.18		2.0–10		130–170	15% of I_p
500	0.30–0.37		2.0–10			
1000	0.60–0.73	0.37–0.55	2.0–10	5.0–25		
2000	1.20–1.47		2.0–10			
4000	2.40–2.93		2.0–10	5.0–25		

After pulses have been applied to all pin combinations at the test level, the device is tested for failure. A failure is concluded when the device no longer meets the required parametric and functional parameters. Devices are classified after test, according to their failure voltage level (Table 3.4).

The peak current in an HBM discharge is mainly determined by the series 1.5 kΩ resistance. Stray inductance and circuit resistance have relatively little effect. In this model, the device has little effect on the waveform, providing it has low impedance compared to the series resistance. In this case, much of the energy stored in the capacitor is dissipated in the series resistor and not in the device.

The rather long rise time limit of 25 ns allowed equipment manufacturers to build testers (that typically have high stray capacitance and slow the rising edge) for testing high pin count devices.

3.2.5 System Level Human Body ESD Susceptibility Test

The ESD susceptibility of powered electronics systems has for many years been tested using a variant of a human body ESD source model that uses a 150 pF capacitor and 330 Ω series

Table 3.4 ANSI/ESDA/JEDEC JS001-2017 HBM device classification.

Classification	ESD voltage failure range
0Z	<50 V
0A	50 to <125 V
0B	125 to <250 V
1A	250 to <500 V
1B	500 to <1000 V
1C	1000 to <2000 V
2	2000 to <4000 V
3A	4000 to <8000 V
3B	≥8000 V

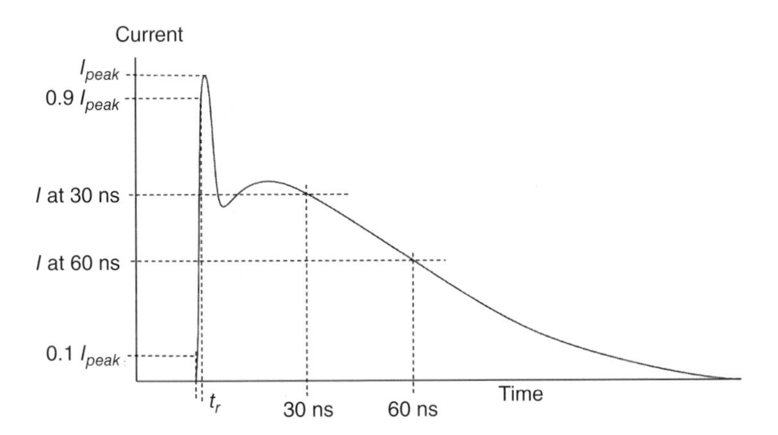

Figure 3.5 IEC 61000-4-2 system ESD test waveform.

resistor. A typical example is IEC 61000-4-2 (International Electrotechnical Commission 2008) (Figure 3.5). This is also known by some as the "*human metal model*."

In system ESD tests these are applied to the system under test in two ways – as an air discharge or a contact discharge applied from a handheld ESD gun. As the terminology implies, a contact discharge is applied with the gun tip in contact with a discharge point on the item under test. For an air discharge, the tip of the ESD gun is moved toward the discharge point until a discharge occurs or the gun's tip touches the system under test. The 61000-4-2 standard defines four levels for test (Table 3.5). Typical waveform parameters for contact discharge are given in Table 3.6.

Although the 61000-4-2 system test waveform is in some ways like the HBM test waveform, there are significant differences. The IEC 61000-4-2 150 pF/330 Ω model discharge is more severe than the 100 pF/1500 Ω model due to the higher stored energy and peak current for the same charging voltage. The rise time is also significantly faster (ON Semiconductor 2010).

Table 3.5 IEC 61000-4-2 test levels.

Level	Contact discharge test voltage (kV)	Air discharge test voltage (kV)
1	2	2
2	4	4
3	6	8
4	8	15
x	User defined	User defined

Table 3.6 61000-4-2 contact discharge waveform parameters.

Level	Source voltage (kV)	Peak current I_p (A) ±10%	Rise time (ns) ±25%	Current ±30%	
				@ 30 ns	@ 60 ns
1	2	7.5	0.8	4	2
2	4	15	0.8	8	4
3	6	22.5	0.8	12	6
4	8	30	0.8	16i	8

The rise time is measured between 10% and 90% of the value of the first current peak. The current is measured at 30 and 60 ns from the time at which the current reaches 10% of the first peak value.

Nevertheless, the 61 000-4-2 waveform has sometimes been used to test ESD withstand of ESDS devices where particularly severe test conditions are required, especially for components that have pins that are likely to emerge directly to system connectors. Recently the IEC 61000-4-2 waveform has been adopted for component tests as the human metal model (HMM) (ANSI/ESD S5.6 (EOS/ESD Association Inc. 2009)). This practice has arisen from demand that components that have pins that are likely to emerge directly to system connectors should be tested for susceptibility using the system test waveform (Ashton 2008; Industry Council 2010a,b).

Other system test models exist, for example ISO 10605 "Road vehicles – Test Methods for electrical disturbances from electrostatic discharge," which is intended for use in evaluating ESD susceptibility of electronics modules for vehicle use (International Organization for Standardization 2008). This uses combinations of 150 or 330 pF capacitances with 330 or 2000 Ω resistance and test voltages up to 25 kV air discharge or 15 kV contact discharge.

3.2.6 MM Component Susceptibility Test

A cart (trolley), tool, machine part, or other metallic or highly conductive object can become charged by the rolling action of wheels on the floor or by other contact with materials and by movement. The MM ESD simulates a discharge between a large conductive object and an ESD-sensitive device.

The IEC 60749-27 standard defines MM current waveforms for device test purposes. The waveform parameters are defined in terms of peak current and period (Figures 3.6 and 3.7), and they give required values for qualification of the MM test equipment measured with 0 and 500 Ω calibration loads (Table 3.7). With a device load, waveforms will typically be different from those obtained with the calibration loads as the device under test adds significant impedance and is likely to act as a nonlinear load.

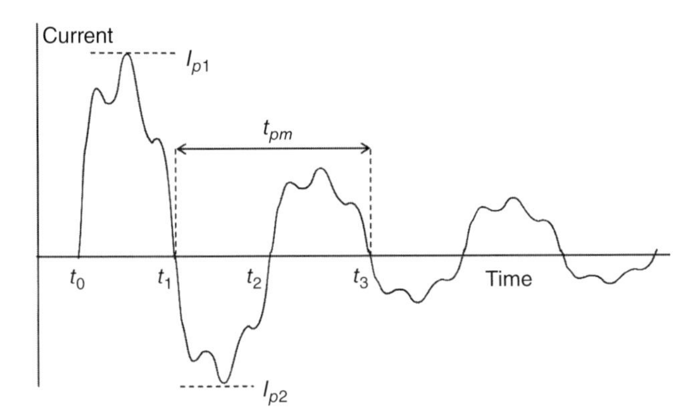

Figure 3.6 IEC 60749-27 MM waveform definition with 0 Ω calibration load.

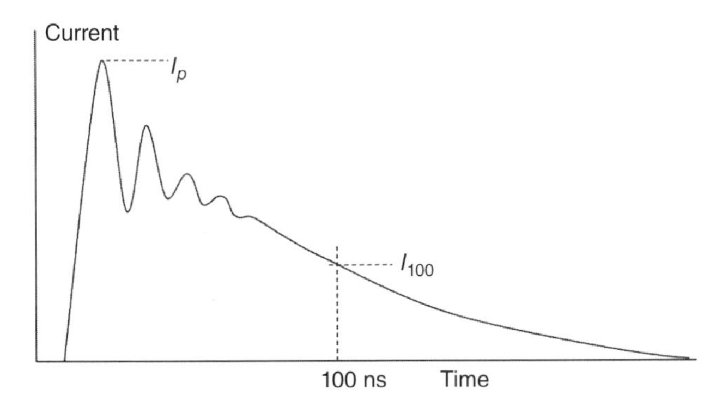

Figure 3.7 IEC 60749-27 MM waveform definition with 500 Ω calibration load.

Table 3.7 IEC 60749-27 MM waveform parameters with 0 and 500 Ω calibration loads.

	1st Peak current I_{p1}				
Voltage (V)	**0 Ω load**	**500 Ω load**	**2nd Peak current I_{p1} 0 Ω load**	**Waveform I_{100} current**	**Period t_{pm} (ns)**
100	1.7 ± 15%		67–90% of Ip1	0.29 ± 15%	63–91
200	3.5 ± 15%				
400	7.0 ± 15%	$<I_{100} \times 4.5$			
800	14.0 ± 15%				

A 200 pF capacitor simulates charge storage on the conductive object. There is no defined additional series resistance or inductance, but in practice stray inductance and resistance are always present in the discharge circuit. The stray inductance and circuit resistance, including the impedance of the device subjected to the ESD, determines the peak current. The fact that these are not well defined makes the MM ESD withstand test more prone to variation than HBM. As the circuit resistance is low, most of the stored energy is dissipated in the device.

As with HBM, device evaluation typically is done using three samples of the device. Each sample is tested with negative and positive polarity ESD with a minimum pulse interval of 300 ms. The test starts at the lowest voltage level and is increased progressively, providing the devices do not fail. Pulses are applied to each pin in turn, with each power pin in turn grounded. Pins that are not grounded or under test are left floating. Pulses are also applied to all nonsupply pin pair combination.

After pulses have been applied to all pin combinations at the test level, the device is tested for failure. A failure is concluded when the device no longer meets the required parametric and functional parameters. Devices are classified after test, according to their failure voltage level (Table 3.8).

The MM waveform is highly dependent on the load and can in practice be unidirectional or oscillatory, whereas the HBM waveform is largely defined by the series resistance and is always unidirectional. The peak current, for a given ESD voltage, is typically an order of magnitude greater for MM than for HBM.

3.2.7 CDM Component Susceptibility Test

In automated assembly and storage, ESDS devices can become charged by contact with packaging or other materials or by induction in an electric field (see Section 4.4.7). A device may become charged also due to accidental rubbing by an operator's garment or by induction due to the electric field of a nearby charged material, garment, or operator. CDM is of growing importance, especially in automated handling and assembly systems.

When the charged device is brought in contact with a grounded metal object, a very short, high-current ESD event occurs. This is simulated by the CDM ESD test. Many devices have been found to have high sensitivity to CDM ESD, which have very fast (nanosecond) high-current transient waveforms. Large-area, thin, multilayer devices with high overall device capacitance and small internal feature sizes tend to increase the CDM damage risk.

As the charge is stored on the device itself, the capacitor in the model is the capacitance of the device under test conditions. The device capacitance is dependent on the package and

Table 3.8 IEC 60479-27 MM device classification.

Classification	ESD voltage failure range
A	Fails at 200 V or less
B	Survives 200 but fails at 400 V
C	Survives 400 V

any air gaps or dielectrics between the device and the ground plane. The series resistance and inductance in the circuit are the resistance and inductance of the tester, of the spark, and within the device. These determine the peak current and waveform in the discharge.

Two versions of the CDM test exist (Brodbeck and Kagerer 1998). In the field-induced charged device model (FICDM) test, the device is placed on a metal plate that can be raised to the test voltage. A thin layer of insulator separates the device from the plate (Figure 3.8). After the plate has been raised to the test voltage, a metal "pogo pin" is touched to the device leg to initiate discharge. The FICDM simulates real-world events well but is time-consuming and expensive to perform.

The first and second peak current, rise time, full-width, and half-maximum height duration of the waveform are key parameters (Figure 3.9). The bandwidth of the measuring oscilloscope also affects the measurement results, and tables are given for use with 1 and 6 GHz oscilloscopes.

The system is calibrated using a small or a large verification module in the place of the device. These are metal discs, with the smaller giving a capacitance of 6.8 pF ±5% and the larger giving 55 pF ±5%. The parameters for small (Table 3.9) and large (Table 3.10) verification modules using a 1 GHz oscilloscope are given here as examples. The test condition

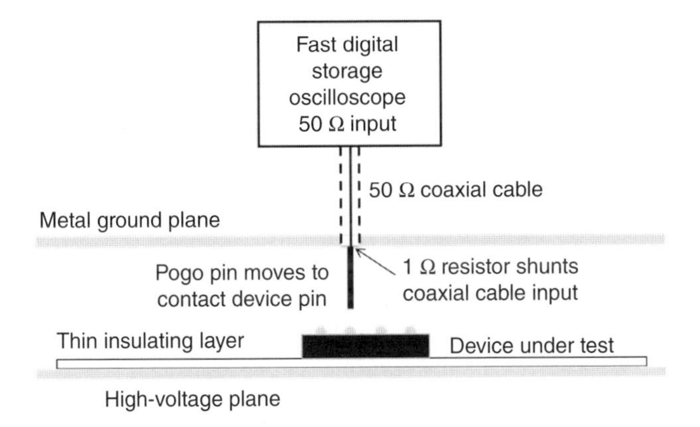

Figure 3.8 ANSI/ESDA/JEDEC JS-002 field-induced CDM test arrangement.

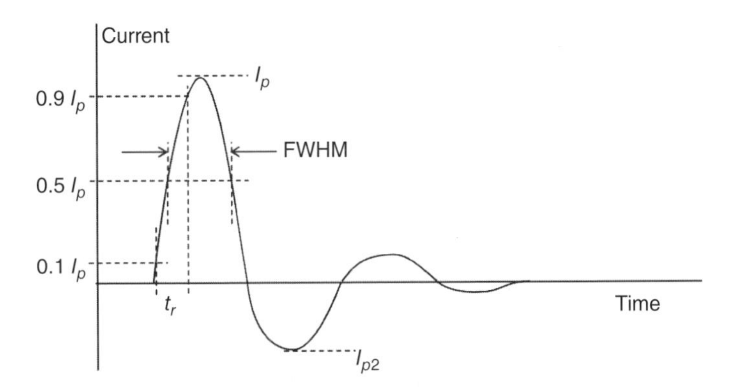

Figure 3.9 ANSI/ESDA/JEDEC JS-002 field-induced charged device model calibration waveform.

Table 3.9 ANSI/ESDA/JEDEC JS-002 CDM waveform parameters using small verification module measured with 1GHz bandwidth oscilloscope.

Test condition	Peak current I_p (A)	Rise time t_r (ps)	Full width half maximum duration FWHM (ps)	Undershoot I_{p2}
TC125	1.0–1.6	<350	325–725	$<0.7\,I_p$
TC250	2.1–3.1			
TC500	4.4–5.9			
TC750	6.6–8.9			
TC1000	8.8–11.9			

Table 3.10 ANSI/ESDA/JEDEC JS-002 CDM waveform parameters using large verification module and measured with 1 GHz bandwidth oscilloscope.

Test condition	Peak current I_p (A)	Rise time t_r (ps)	Full width half maximum duration FWHM (ps)	Undershoot I_{p2}
TC125	1.9–3.2	<450	500–1000	$<0.5\,I_p$
TC250	4.2–6.3			
TC500	9.1–12.3			
TC750	13.7–18.5			
TC1000	18.3			

TCxxx denotes the voltage stress level but does not correspond to the actual voltage applied to the field plate. This is because the plate voltage must be adjusted to give the required peak current values for each test condition during the calibration process.

The CDM test is quite sensitive to changes in experimental conditions including atmospheric humidity and contamination or oxidation of the verification module surfaces. At higher 1 kV and above, corona discharge may limit the voltage achievable on the device before discharge.

After test, devices are classified according to their failure level (Table 3.11).

A socketed device test (SDM) is easier to perform, but the result does not correlate well with FICDM results. In the SDM method, the device is placed in a test socket and subjected to discharges via a relay network. This technique has advantages for devices that have heat sink fins or uneven packages and so do not lend themselves to FICDM test.

3.2.8 Comparison of Test Methods

The discharge current and time characteristics of the ESD models are very different and yield different ESD withstand voltage results. The HBM, MM, and CDM ESD models can be view as special cases of a generalized ESD model in Figure 3.2. The waveform is generated by various combinations of the components L, C, and R (Table 3.12).

Table 3.11 ANSI/ESDA/JEDEC JS-002 CDM device classification.

Classification	Test condition failure range
C0a	Fails at less than 125 V
C0b	Survives 125 V but fails below 250 V
C1	Survives 250 V but fails below 500 V
C2a	Survives 500 V but fails below 750 V
C2b	Survives 750 V but fails below 1000 V
C3	Survives 1000 V

Table 3.12 ESD model parameters compared (Gieser and Ruge 1994) assuming a 10 Ω load in place of the device.

	500 V HBM	500 V MM	500 V CDM
C (pF)	100	200	10
L (nH)	10 000	750	2.5
R (Ω)	1500	10	10
Waveform type	Unidirectional	Variable, oscillating	Variable
Rise time (ns)	<10	<10	0.1
1/e decay time (ns)	150	<100	1
Peak current (A)	0.3	6.5	14
Peak power (W)	0.9	400	2000
Dissipated energy (μJ)	0.08	13	0.63
Stored energy (μJ)	12.5	25	1.25

At a time when the HBM, CDM, and MM models were under development, Gieser and Ruge (1994) compared the characteristics of the models (component values and characteristic waveforms may be slightly different from current model versions).

Geiser and Ruge noted the following:

- In the HBM discharge, most of the energy stored on the capacitor is dissipated in the series resistance.
- The series resistor and load form a voltage divider circuit. The applied voltage results in a voltage on the device that could stress oxide layers.
- If the stray component impedance is low, then the HBM discharge current is largely determined by the series resistance.
- The peak current achieved by HBM is achieved at much lower ESD voltage in MM and CDM. For the same ESD voltage, MM and CDM currents are about 10 times the HBM value.
- The rise and decay times for CDM are around two orders of magnitude faster than for HBM and MM.

- Assuming a resistive load, peak power dissipated in the load increases with the square of current. For CDM, peak power is in the kW range. MM is lower, in the 500 W region. HBM peak power is only in the watt region.
- The higher capacitance of MM stores about 20 times the energy of the CDM case for the same test voltage.
- In practice, the real HBM, MM, and CDM waveforms achieved in any tester are influenced by interactions with stray "parasitic" components, as well as the actual load impedance, that may be far from purely resistive.

According to Smedes (2009), at 2 kV the HBM peak current is 1.3 A, and the duration about 200 ns. At 200 V MM, the peak current is 3.8 A with a pulse width around 30 ns. A 500 V CDM ESD has peak current around 2–10 A and duration around 1 ns, depending on the device capacitance.

Failure modes for HBM and MM are generally found in the on-chip protection circuits, whereas CDM failures are usually gate oxide damage. The duration of the CDM pulse (<1 ns) is often less than required to trigger the protection circuits (Amerasekera and Duvvury 1995). In CDM ESD the direction of discharge current in internal circuit elements may be opposite to that of HBM and MM as the charge is stored internally and discharges to the outside world. The protection circuits may not be designed for this situation.

There is good correlation between HBM and MM damage stress levels, with the HBM withstand voltage about 12 times greater than MM (Kelly et al. 1995). More generally HBM withstand voltage is 10–20 times greater than MM. There is little correlation between CDM damage stress levels and the other models.

3.2.9 Failure Criteria Used in ESD Susceptibility Tests

The selection of failure criteria in ESD susceptibility tests can have a big influence on the failure threshold result (Gieser 2002). For ESD qualification of the product, a full functional production test of the AC and DC parameters is necessary to ensure it meets all the product specifications. However, this may not detect "walking wounded" latent damaged devices that show weakness and could give early failures.

3.2.10 Transmission Line Pulse Techniques

Transmission line pulse (TLP) techniques primarily provide a means for the on-chip ESD protection developer to characterize the performance of their ESD protection networks (Gieser 2002; Smedes 2009; ESD Assoc. 2006). TLP allows the device to be stressed with a repeatable short duration rectangular waveform pulse with well-defined duration and peak current. The TLP waveform is typically 100 ns in duration and has a rise time of 10 ns, but in some test arrangements these can be varied. A TLP waveform is rectangular rather than the decaying shape of HBM. The stored energy is approximately the same as HBM giving a correlation of 1.5–1.8 amps for each 1 kV stress of HBM test (Duvvury C. personal communication). The current and voltage readings in each test are used to build a complete pulsed V-I characteristic. The most reliable failure criterion has been found to be a change in leakage current, and a leakage current measurement can be done after each test to evaluate device damage.

Correlation between TLP, HBM, and MM results have been studied and some empirically derived correlations established, but these may not be universally applicable.

Application of TLP to CDM and system-level ESD tests is less clear. A very fast transmission line pulse (VFTLP) method has been proposed to address this for CDM, in which the pulse width is reduced to 1–5 ns with rise time around 200 ps. This has provided improved results. Nevertheless, VFTLP remains a stress applied between two pins, whereas CDM is a single-pin connection stress, and the device internal current flow is therefore different.

3.2.11 The Relation Between ESD Withstand Voltage and ESD Damage

There are essentially two main types of ESD susceptibility in devices – energy and voltage susceptibility (Baumgartner ESD TR50.0-03-03). Nevertheless, there are various ESD parameters that affect the likelihood of damage.

- The ESD energy dissipated in an ESDS device
- The peak current in a discharge passing through an ESDS device
- The power developed in an ESDS device during a discharge
- The charge transferred in one or more discharges into an ESDS device
- Excess voltage developed across part of the device

These parameters may have different effects on different components, and real-world ESD can have widely varying characteristics. Because of this, devices that have similar ESD withstand voltages may have different susceptibility to damage from real-world ESD. Changing the ESD parameters can have a large impact on ESD risk even though the ESD source voltage is not changed. As an example, a 250 V CDM device may risk damage if it contacts a metal part when the voltage difference between them is 250 V. The damage is often due to the fast high-current transient that flows in the charged device ESD event causing an overvoltage of either device internal part. If, instead of contacting metal, the device contacts a high-resistance material, the ESD current is greatly reduced and the discharge slowed by the material resistance. In this circumstance, the device may not be damaged even at voltages far higher than 250 V.

In practice, it is doubtful whether it is appropriate to measure ESD withstand thresholds in terms of ESD source voltage as this is only indirectly related to the parameter causing damage to the device. The damage sustained is often due to energy dissipation in the device or discharge current, and protection devices divert and must withstand these currents. The current at which a device fails is often approximately equivalent for HBM and MM (Amerasekera et al; 2002). Even in CDM the failures are due to high current levels, which cause internal voltage stresses (Brodbeck and Kagerer 1998).

In a practical situation, the source voltage is only one of several factors affecting the ESD current and energy in the device. Because of this, it is difficult to directly relate measured ESD withstand voltages to ESD risk in a factory situation. Nevertheless, at the time of writing, it seems unlikely that specification of ESD susceptibility in any terms other than source voltage will become generally accepted in the foreseeable future.

3.2.12 Trends in Component ESD Tests

ESD withstand test methods will need to evolve in response to technological challenges such as increased component pin counts, reduced pin spacing, and the costs of testing complex devices (ESD Assoc. 2016). The MM test method is already no longer recommended for device qualification. The detailed HBM and CDM test methods are developing to increase the efficiency of these tests and reduce test time and stress of the test equipment.

There is continuing pressure to test component pins that could be subject to system-level ESD using the HMM waveform. This device test method has so far shown poor reproducibility and is under further development.

TLP testing is under development for evaluation of HBM, CDM, and system-level ESD susceptibility (ESD Assoc. 2006, 2016).

The standardized ESD models represent only a proportion of real-world ESD sources and threats. Other recognized threats include charged board events (CBEs) and cable discharge events (CDEs). CDE can be regarded as a form of system-level ESD, but they often impinge directly on devices connected directly to the connector terminals (see Section 3.5.3.)

3.3 ESD Susceptibility of Components

3.3.1 Introduction

ESD damage can happen in many ways according to the specific susceptibility of the many different types of devices and the types and characteristics of ESD sources that they encounter. The susceptibility of a device to ESD damage, and the damage effects, can be very variable for different ESD sources and levels. This section attempts to provide an informative overview to help understanding of this topic. For illustration and understanding, it gives some specific examples representing a small fraction of the wealth of research that has been done to understand ESD failures.

3.3.2 Latent Failures

McAteer (1990) defined a latent failure as one that occurs in use conditions because of earlier exposure to ESD that did not result in an immediately detectable discrepancy. This topic is one surrounded by controversy and disagreement. McAteer devoted a chapter to review of studies on latent damage and concluded the following:

- Latent ESD failures exist.
- The presence of slight parametric shifts or softened V-I characteristics in a device does not assure further degradation with time.
- System-level latent failures can result from parts that might be regarded as out of specification after ESD stress.

ESD-induced oxide breakdown can distort metal-oxide silicon field effect transistor (MOSFET) V-I characteristics, causing a reduction of transconductance or even loss of transistor action. Loss of transconductance has been found to be associated with an increase in supply current that can threaten battery life.

Reiner (1995) suggested that a high-resistance silicon melt ball could form around the site of gate oxide damage caused by CDM. This caused unstable leakage currents of the order 1 μA at 5 V. Subsequent heating could convert some of this material into higher conductivity crystalline silicon and change to permanent damage.

Hellstrom (1986) analyzed ESD failures in bipolar and metal-oxide silicon (MOS) devices. They observed latent damage through "birth," growth, and completion to component failure.

Baumgartner commented that most experts agree that there are latent defects caused by ESD in some technologies, but they disagree as to whether the device goes on to become a latent failure. Latent damage is likely to occur in a narrow window of ESD strength between no damage and complete failure. This window may be narrow, and the probability of ESD strength lying within the window may be correspondingly small (Beal et al. 1983). Failure analysis data suggests that latent failures are statistically insignificant. Infant failures of components may be more likely than latent failures.

Tunnecliffe et al. (1992) subjected 32 enhancement mode and 32 depletion mode field effect transistors (FETs) to HBM ESD of ±100 and ±200 V. Latent damage was identified, but they concluded that the risk to reliability was small as the latent failure window is narrow. It is perhaps more likely that the device would either be catastrophically damaged or remain undamaged.

In contrast, Gammill and Soden (1986) found that latent failures occurred in complementary metal-oxide silicon (CMOS) integrated circuits (ICs) both as a field failure and in a life test experiment. McKeighan et al. (1986) found that reversible charge-induced surface inversion caused CMOS switches to fail to turn on when addressed. A factor influencing this behavior included use of a highly insulating ceramic package. The charge-induced failures occurred during board handling or assembly or were field induced.

Some workers have found that degraded devices could fail parametrically or functionally when subjected to ESD (Taylor and Woodhouse 1986; Enoch et al. 1983) and recover after further events, becoming fully functional. Shorting of a p-n junction did not recover on further ESD event, but a dielectric short due to ESD could recover. Krakauer and Mistry (1989) found that an ESD event lower than the threshold for catastrophic failure could cause charge carriers to be injected into oxide to become trapped there. This could cause voltage threshold shift and changes to MOSFET drain current.

Cook and Daniel (1993) found that ESD applied to ASIC power pins could cause latent damage that compromised device latch-up immunity.

Anand and Crowe (1999) studied latent failures in Schottky diodes that caused loss of microwave transceiver sensitivity. The microwave diodes they studied had a junction area of 5–8 μm diameter. Using a reverse leakage test, they identified damaged devices that had higher leakage current than good devices. Faulty devices were not identified by the standard test procedure because the device specification did not include reverse leakage current.

Chen et al. (2009) found that one effect of reverse applied HBM and MM ESD on GaN light-emitting diode (LED) devices caused an increase in leakage paths that could have a cumulative and latent damage effect.

Smedes and Li (2003) reported that ESD can cause latent damage as well as permanent damage to interconnect structures, reducing electrothermomigration lifetimes by a factor of 100. At low level, ESD metal leads can melt and in cooling change their grain boundary structure. This can then lead to reduction in electromigration lifetime due to an increase in the metal resistance.

Laasch et al. (2009) found that ESD protection diodes stressed with multiple pulses suffered cumulative latent damage in the form of a metal alloy front that progressed with each pulse and eventually could short the p-n junction.

Sylvania (2009) gave examples of partial failure of an LED in a series-connected LED "coupon" due to damage to one of the LEDs. In the examples, one of the LEDs in the string is degraded causing reduced light output while remaining sufficiently conductive to allow the remaining LEDs to function. The operating life of the unit may be dramatically reduced.

Dhakad et al. (2012) described detection of a functional CDM ESD failure in an advanced CMOS IC where no obvious physical damage was present. The failing device was identified as a single transistor that had suffered gate oxide charge trapping degradation.

Overall, these studies suggest that while latent damage has been demonstrated, it appears to be usually a low-probability risk. Nevertheless, it can be a significant concern, especially in high-reliability applications or where a high cost or severe consequences make a risk of failure unacceptable. They also suggest that life test studies of advanced technology high-speed semiconductor devices stressed with HBM or CDM at levels just below their threshold values might give some insight into their reliability in the field (Duvvury C. Private communication).

3.3.3 Built-in On-chip ESD Protection and ESD Protection Targets

Until the 1970s most components were not susceptible to ESD damage. As component technology developed, the internal component sizes of ICs reduced. This led to them becoming more susceptible to ESD damage. In the 1970s this became a known problem after the introduction of large-scale integration (LSI) technology. The internal device technology continued to increase in ESD sensitivity with reducing internal component size, and new, more sensitive device technologies added to the problem.

To counter this trend, the semiconductor component manufacturers started to design in ESD robustness, at least to the circuits that connect to the device pins. ESD protection networks began to be added to divert ESD current flow away from the more sensitive internal components.

In the 1970s and early 1980s, the automotive industry began to implement ESD pass levels (Industry Council on ESD Target Levels 2011, 2010b). Various companies adopted different MM or HBM pass voltage level criteria. In response to these demands from customers, by the mid-1980s semiconductor companies began to set internal HBM withstand voltage standard requirements, with 2 kV HBM the most common. This target for MM ESD withstand of 200 V remained constant for more than 20 years until 2007. Over the same period, a 500 V target for CDM withstand had developed. In 2007 the Industry Council on ESD target levels published its White Paper I calling for a reduction of ESD targets to 1 kV HBM. The correlation between HBM and MM withstand predicts that this results in MM withstand between 30 and 200 V (Smedes 2011). This was followed by a proposal to reduce the CDM withstand targets to 250 V in White Paper 2. This trend in reducing target ESD withstand voltages seems set to continue into the foreseeable future.

Some component pins are particularly difficult to protect with on-chip protection networks due to their specialist function. Examples include radio frequency (RF) and high data rate pins or some types of analog input/output pins. ESD protection networks increase capacitance and reduce the quality factor of RF circuits. Added capacitance can degrade the performance of RF circuits. Simple low-capacitance protection networks must be used in these capacitance-sensitive circuits. Higher-frequency operation generally reduces the allowed "capacitance budget" for ESD protection. The capacitance of HBM ESD protection networks typically increases with ESD withstand voltage, so reducing the capacitance budget can force a reduction in ESD withstand voltage of the protection network.

Not all pins on a device have the same ESD withstand voltage. For example, the Burr Brown DAC8043 data sheet gives detailed information about the ESD performance of the device (Burr Brown 1993). This device has mixed digital and analog functions. The digital pins are stated as withstanding ±2500 V HBM. The analog pins are stated as withstanding only 1000 V HBM. The data sheet goes on to say that two pins, V_{REF} and R_{FB}, "show some sensitivity." What this means is not clear, but presumably they did not survive 1000 V HBM.

On-chip ESD protection typically uses a system of clamp circuits to divert ESD current away from the sensitive internal circuitry (Figure 3.10). The clamp components can be various types of components such as p-n or p-i-n diodes, bipolar or MOS transistors, or silicon controlled rectifiers (SCRs) (Amerasekera et al. 2002).

Operation depends on the design clamp style used. If clamp 2 is a reverse diode, then only clamp 1 and the power supply ESD clamp turn on for positive stress with respect to ground. If clamp 2 is a breakdown NMOS device with characteristics shown in Figure 3.11, then it is clamp 2 that offers protection (Duvvury C. Private communication). For a negative transient, clamps 2 and 4 and the power supply clamp turn on.

The clamp devices typically have V-I curves that have some, or all, of the elements shown in Figure 3.11. During an ESD event, the voltage across the device increases until the device turns on at V_{t1} and the current through the device increases. If the current reaches a level I_{t2}, a second breakdown and thermal failure can occur. So, ESD protection clamp circuits are also ESD susceptible as there is limit to the current and energy they can handle. The ESD withstand voltage in an HBM test or failure during a real ESD event is determined by the maximum current flowing during the test rather than the source ESD voltage.

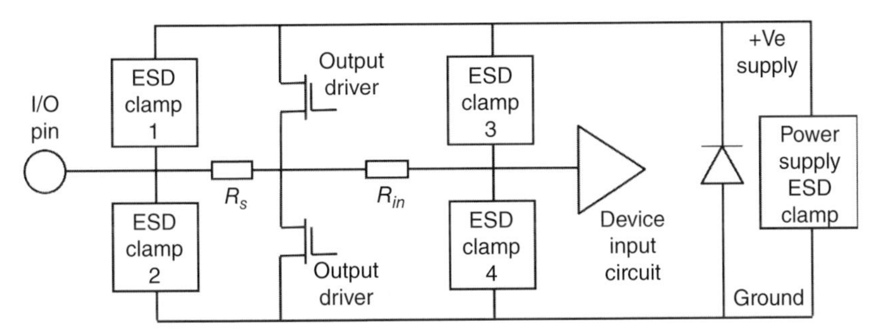

Figure 3.10 Typical clamp arrangement used in on chip ESD protection of an input/output (I/O) device pin.

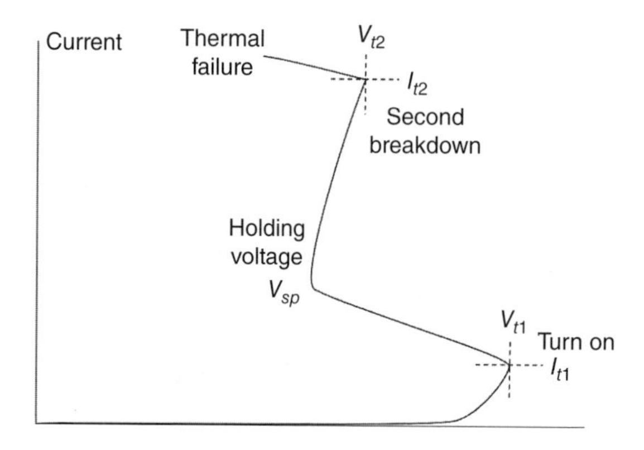

Figure 3.11 Elements of the V-I curve of a typical on chip clamp circuit.

3.3.4 ESD Sensitivity of Typical Components

The only ways to be sure of the ESD withstand voltage of a component are to measure it or to get a manufacturer's data on it. Unfortunately, many manufacturers do not give ESD withstand data on their component data sheets.

Nevertheless, some attempts have been made over the years to compile generic data to help guide ESD protection program design. Table 3.13 shows generic data based on that

Table 3.13 Range of typical HBM withstand voltages of components.

Device type	Typical HBM ESD withstand voltage range
MR heads, RF FETs, SAW devices	<10–100 V
MEMS	40 V–?
MOSFETs, laser diodes, PIN diodes	100–300 V
LEDs	50 to >15 000 V
MMICs	250 to >2 000 V
Pre-1990 VLSI	400–1 000 V
Modern VLSI	1000–3 000 V
Bipolar	600–8 000 V
Linear MOS	800–4 000 V
HCMOS	1 500–3 000 V
CMOS B Series	2 000–5 000 V
Power bipolar	7 000–25 000 V
Film resistor	1 000–5 000 V
Capacitors (low capacitance and breakdown voltage)	Depends on capacitance and breakdown voltage

Source: Based on IEC 61340-5-2/TS: 1999, component data sheets, and other sources.

given in IEC 61340-5-2:1998 with additions from other sources and data sheets. While this gives a general idea of the range of HBM ESD withstand voltage that might be expected for various technologies, it cannot be relied on for specific components.

In the 1980s and 1990s the Reliability Analysis Center published compilations of ESD susceptibility of a range of components including discrete and passive devices (Reliability Analysis Centre 1989a,b, 1995).

3.3.5 Discrete Devices

Discrete devices (individual transistors, diodes, or other simple devices) often do not have any on-chip protection networks. While many of these devices may have relatively high ESD withstand voltage, some can have low ESD withstand voltage. Voltage-sensitive devices such as small signal MOSFETs can have low HBM ESD withstand voltage. RF components are often particularly sensitive as their functionality requires small internal dimensions, and this often leads to ESD susceptibility.

3.3.6 The Effect of Scaling

As IC technology advances, the internal component dimensions are reduced. MOS technology components reduce in channel length, gate oxide thickness, and interconnect dimensions at every technology node (Duvvury and Gauthier 2011). Thin and small dimensions also increase the resistance of interconnects and other conductors, giving higher voltage drop for the same current flow. This results in higher ESD sensitivity of the components due to reduced energy, power, or current levels and oxide breakdown voltage for failure. This makes it increasingly difficult to meet HBM and CDM target ESD withstand levels. Clamping ESD protection circuits must maintain the voltage levels under transient ESD conditions below the breakdown stress of the internal circuitry. This may be less than 5 V in recent technologies.

3.3.7 Package Effects

Package differences have little effect on HBM withstand but can have considerable effect on CDM withstand voltage (Duvvury and Gauthier 2011). This is because each package introduces different package size and capacitance, bond wire inductance, and other factors. The package capacitance plays an important role in determining the charged device ESD stress.

Some modern packages such as ball grid arrays (BGAs) are much less vulnerable to human body contact and ESD than older packages such as dual in line (DIL). Some types with embedded and close spaced pins are difficult if not impossible to stress with human body ESD in practice and are in any case normally handled by automated processes. Charged device ESD is the main concern for these devices.

3.4 Some Common Types of ESD Damage

3.4.1 Failure Mechanisms

ESD-related failure mechanisms depend on the device type and technology. They typically include (MIL-HDBK-263 sec 50, Analog Devices UG-311 2014, Linear Technology n.d.)

- Thermal secondary breakdown
- Metallization and interconnect or resistor melting
- Breakdown of thin dielectric layers
- Gaseous arc discharge
- Surface breakdown
- Bulk breakdown

Thermal secondary breakdown, metallization and interconnect melting, and bulk breakdown are dependent on energy dissipated in the discharge through the device.

Breakdown of thin dielectric layers, gaseous arc discharge, and surface breakdown are due to excess voltage stress of some part of the device. In a high-resistance capacitive structure such as a MOS capacitor or MOSFET gate, conduction of sufficient charge into the structure can raise the voltage to the breakdown level.

According to Analog Devices (2014), most ESD failures occur due to conductor or resistor melting, dielectric damage, or junction damage or contact spiking.

Conductor or resistor melting occurs in thin metal or polysilicon interconnects and thin-film, thick-film, or polysilicon resistors. The high current flowing due to ESD causes local heating in the conductor or resistor material, leading to fusing and open circuit. This type of damage is commonly due to HBM ESD because a charged person represents a high energy ESD source. For thick- and thin-film resistors, partial melting and a change in resistance value are possible. This can lead to parametric failure of the IC. These damage mechanisms are largely found by HBM ESD testing rather than any vast evidence gathered from field failures. In contrast, for CDM ESD the field failures correlate well with observed damage for sensitive devices (Duvvury C. Private communication).

Dielectric breakdown occurs when a thin insulating dielectric layer such as silicon dioxide or nitride is stressed by an applied voltage that exceeds its time dependent dielectric breakdown strength. This results in punch through of the dielectric. results. This type of failure is typical of CDM damage because the fast rise time results in high voltages occurring within the chip. The high current that flows on breakdown can lead to formation of a melt filament of silicon.

Junction damage and contact spiking can occur when a p-n junction is subjected to avalanche breakdown and secondary breakdown. Avalanche breakdown occurs when the reverse breakdown voltage of a reverse biased p-n junction is exceeded. Secondary breakdown can then occur at a point where the junction material is sufficiently hot. The high current that flows is channeled through this site and causes high adiabatic local heating, increasing current flow. Ultimately, the silicon can melt if the temperature exceeds 1415 °C. If heating is sufficient, adjacent contact metal can melt and migrate. The junction is shorted by resistive material.

Melted junction material has, on solidification, a modified dopant profile, crystal structure, and electrical characteristics. This typically gives a soft reverse breakdown characteristic. This can result in a small or large leakage current increase and change in device parameters. Melting of contact material typically leads to failure of the corresponding IC pin.

3.4.2 Breakdown of Thin Dielectric Layers

MOS technology is widely used to fabricate discrete MOSFETs, ICs, MOS capacitors, and devices with metallization crossovers. Typical structures have a thin layer of insulating oxide separating two conductive layers. The resistance across the layer is normally very high and the leakage current between the conductors very low.

The breakdown voltage reduces with the thickness of the oxide layer. The voltage difference required to break down thin oxide layers used in modern components can be quite small. Dielectric breakdown can occur when the voltage across the dielectric exceeds a time-dependent dielectric breakdown value (Analog Devices 2014). The result can be a gate-source or gate-drain short in a MOSFET or a resistive leakage path between a metallization track and underlying semiconductor regions. According to Analog Devices, this is the main result of charged device ESD failures as the fast rise time is most likely to result in excessive internal voltages. Breakdown tends to occur at high field points such as corners, edges, or steps in the dielectric.

If breakdown occurs, a puncture defect is made in the insulating layer. If sufficient voltage is applied across the dielectric, the breakdown field strength is exceeded, and breakdown occurs. This type of damage is voltage dependent, requiring sufficient voltage to be applied across the oxide layer. After breakdown, high current flows through the breakdown point resulting in adiabatic heating of the breakdown region. If the energy in the discharge is sufficient, a small amount of semiconductor can be melted.

The results of the breakdown may, however, depend on the energy available in the discharge. Melted material can fill the puncture, short-circuiting the two conductors. This will normally cause a detectable component failure due to a short circuit or high leakage current. A lower energy discharge may, however, simply leave a clear puncture that may leave the component functioning but weakened. The same point may fail in future at a lower voltage, or an increased leakage current may result.

Dielectric breakdown damage has also been reported in microelectromechanical system (MEMS) devices (Sangameswaran et al. 2009).

3.4.3 MOSFETs

According to Infineon (2013), there are three basic ESD failure modes in MOSFETs. These are junction damage, gate oxide damage, and metallization burnout. The most common failure in HBM is junction damage caused by an ESD transient of sufficient ESD energy and duration. A damaged junction often shows high reverse bias leakage or a short circuit.

Gate oxide damage is, however, the main category of ESD damage. This occurs when the gate is subjected to ESD causing the gate oxide to break down. This happens when the voltage across the thin gate oxide exceeds the breakdown field strength. For thin oxide layers, the breakdown voltage can be small, although for short duration ESD, a much higher ESD voltage may be required to achieve breakdown.

Another common failure mode is that ESD may lead to trapped charge in the gate oxide. This can cause a shift in the gate threshold voltage and functional failure. This can sometimes anneal itself after a few hours, and so trapped charge effects are not always permanent (Duvvury C. Private communication).

Metallization burnout can also occur, although this is often a secondary effect occurring after initial junction or gate oxide breakdown.

A MOS gate is a capacitive structure. If the gate is charged to over the breakdown voltage of the dielectric, breakdown occurs, and the gate is likely to be damaged. An HBM or MM test of this type of structure simply takes the charge in one capacitor (the ESD source) and dumps it into another (the MOS gate) (Figure 3.12). The final voltage on the gate can be calculated if the gate capacitance is known. Hence, the MM or HBM ESD withstand can be estimated if the MOSFET gate capacitance and breakdown voltage is known (International Rectifier n.d, Application note AN-986).

The initial charge Q_{ESD} in the ESD source capacitance C_{ESD} at a voltage V_{ESD} is

$$Q_{ESD} = C_{ESD}V_{ESD}$$

During the ESD event, this charge is shared between the source capacitance and gate capacitance C_g in parallel. The charge Q_{br} required to achieve the gate breakdown voltage V_{gbr} is

$$Q_{br} = (C_{ESD} + C_g)V_{gbr}$$

The threshold ESD voltage at which the gate reaches this breakdown voltage can be found when Q_{ESD} equals Q_{br}.

$$C_{ESD}V_{ESD} = (C_{ESD} + C_g)V_{gbr}$$

$$V_{ESD} = \frac{(C_{ESD} + C_g)}{C_{ESD}}V_{gbr}$$

So, the effective ESD withstand voltage is heavily dependent on the gate capacitance. If $C_g \ll C_{ESD}$, then V_{ESD} is around the gate breakdown voltage V_{gbr}. For small devices, this can be a few volts or tens of volts. If $C_g \gg C_{ESD}$, then V_{ESD} is around $V_{gbr}\ C_g/C_{ESD}$.

3.4.4 Susceptibility to Electrostatic Fields and Breakdown Between Closely Spaced Conductors

There are relatively few components that are susceptible directly to damage from electrostatic fields. It is very common that electrostatic fields induce voltage differences between a

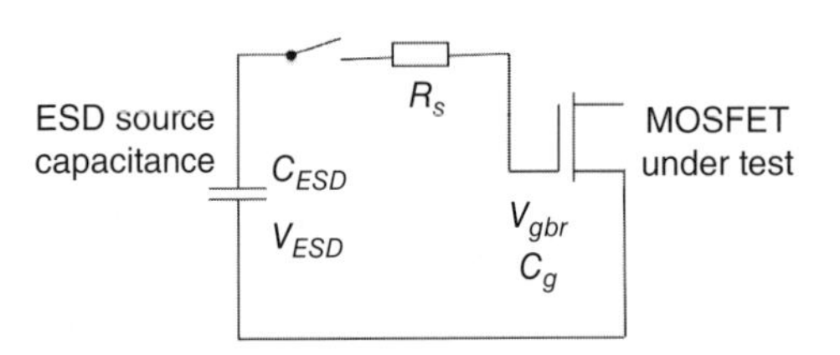

Figure 3.12 Charging of a MOSFET gate from an HBM ESD test source.

device and another object, and electrostatic field–induced charged device ESD occurs when they touch. Control of electrostatic fields is normally essential where charged device ESD is a concern.

Some types of components have closely spaced unpassivated conductors on a surface or with an air gap. Components containing isolated conductors at small distances apart will have voltage differences induced by changing electrostatic fields. These voltage differences could produce breakdown and discharges between them. Sparking can cause damage to the electrodes and transfer of material across the gap. Electromigration of materials under high electrostatic fields between conductors is also possible.

Photomask reticles used in semiconductor device manufacture have this type of structure. The mask consists of very fine metallic tracks with very fine (μm or nm) etched gaps on the surface of a quartz substrate. Electrostatic fields, for example due to charge on the reticle enclosure, cause voltage differences between the metallic tracks. These voltage differences between tracks can lead to breakdown, ESDs, and damage to the reticle at voltages below the Paschen minimum breakdown voltage (Rider and Kalkur 2008; Rider 2016). Damage to reticles leads to damaged components in the semiconductor wafer and reduction of component yield. Englisch et al. (1999) proposed a "canary" reticle test structure that could be used to simulate and evaluate the effects of electrostatic field exposure during reticle handling.

Other devices such as surface acoustic wave (SAW) filters that have closely spaced electrodes can also suffer damage (Mil HDBK 263B). Wallash and Honda (1997) reported that magnetoresistive (MR) recording heads could be damaged by electrostatic fields, due to breakdown of small gaps by voltage differences induced by the field.

Sangameswaran et al. (2009) found that MEMS capacitive switches could suffer both dielectric and air breakdown.

Wallash and Levitt (2003) found that ESD could still occur at very small submicron gaps less than the Paschen minimum by field emission. This could be a threat for photolithographic reticles, magnetic recording heads, MEMSs, and field emission displays.

3.4.5 Semiconductor Junctions

Semiconductor junctions are used to make bipolar transistors, diodes, junction FETs, PIN diodes, Schottky diodes, and thyristors. P-n junctions also occur as parasitic components in MOS technology and can be used in ESD protection networks.

An ESD current passing through a junction can cause intense local heating and even melting of the semiconductor material. Electromigration of the materials can also occur under the intense electrostatic fields created within the component (Analog devices 2014).

For ESD current passing through a reverse biased p-n junction, most of the power dissipation occurs in the junction. The ESD event is usually of very short duration compared to thermal time constants in device materials. The dissipation of energy in the device can usually be considered adiabatic with negligible diffusion of heat away from the heated parts. At breakdown, the current results in a hot spot at the junction. The resistance of the semiconductor material in this region reduces with increasing temperature, and this leads to more current flowing through the spot. Current flow increasingly focuses on the hot spot, and further heating occurs. Local temperatures in hot spots can approach or exceed the melting point of the semiconductor. This can result in changes in junction characteristics or short circuits. This is called *thermal secondary breakdown.*

This failure mechanism is dependent on the power dissipated in the junction. Junctions with high breakdown voltage and low leakage current may be more susceptible to ESD damage. Where hot spots do not develop to failure, migration of material under the electric field may cause a partially developed filament short circuit. This may increase leakage current.

When forward biased, the junction has low voltage drop, and the power developed by the ESD current is spread through the body of the device, and failure of the junction is less likely.

For most junction transistors, the emitter-base junction is more susceptible than the collector-base junction due to smaller junction dimensions. Junction FETs that have a high gate-source breakdown voltage and low leakage can be particularly susceptible. Schottky diodes and Schottky TTL components are more sensitive to ESD because they have thin junctions and metal present that may be carried through the junction.

Not all p-n junctions are susceptible to ESD damage. Transient suppressor diodes, zener diodes, power rectifier diodes, power bipolar transistors, and thyristors can be very robust to ESD.

3.4.6 Field Effect Structures and Nonconductive Device Lids

Some types of LSI and memory device (e.g. UVEPROMS) have quartz or ceramic packages that are highly insulating and can become highly charged (Mil HDBK 263B). Surface inversion or gate threshold changes can occur due to ions deposited by ESD. This can cause malfunction of the device that may be reversible in some cases by neutralization of the charge on the external device surfaces.

3.4.7 Piezoelectric Crystals

Piezoelectric crystals are used in oscillator crystals, delay lines, and SAW filters. The high voltages applied due to ESD can cause high mechanical forces that can damage or fracture the crystal (Mil HDBK 263B).

3.4.8 LEDs and Laser Diodes

LEDs and laser diodes can be extremely sensitive to ESD damage. LEDs can suffer catastrophic damage so that they do not light or conduct (Sylvania 2009; Nichia 2014; Osram n.d.). They can also suffer latent damage and degradation of reliability, remaining partly functional. In a LED string, one LED may fail allowing the remaining LEDs to function correctly. Nichia (2014) advised customers to test for ESD damage before assembly and give a method of detecting damage by measurement of forward voltage when a low current (<0.5 mA) is passed through the LED in a forward direction.

Indium gallium nitride (InGaN) blue and green LEDs have been found to be extremely sensitive to ESD damage (Avago Technologies 2007). Damaged devices can appear dim, dead, short, or with low forward or reverse voltage. Avago Technologies' InGaN LEDs are classified as Class 1x and Class 2 (HBM ESD withstand voltage 250–4000 V).

Talbot (1986) found that the ESD withstand voltage of nine various color LEDs mostly varied from 4–15 kV, although two "low current" types showed damage at 100–200 V. Devices that had been subjected to reverse breakdown often functioned normally in the forward direction. There was a drop in light output dependent on the ESD voltage. Failure mechanisms included junction burnout, nitride punch through, and metallization burnout.

Chen et al. (2009) found four different effects of reverse applied HBM and MM ESD on GaN LED devices, affecting leakage paths in the devices. Low-level ESD below 650 V reduced leakage but above 700 V leakage was increased by three to five orders of magnitude. Further application of ESD caused unstable behavior. The device still glowed in forward bias. When V-I measurements were made, the device was destroyed. They concluded that the increase in leakage paths could have a cumulative and latent damage effect.

3.4.9 Magnetoresistive Heads

MR heads are used in hard disk drives for reading data from the disk. They are among the most sensitive components currently in use, with ESD withstand voltages reported to be less than 5 V HBM. ESD damage to MR heads, and ESD control during manufacturing processes of these devices, has been the subject of intensive research over many years. There are many papers published on these subjects, which form a perennial topic in conferences such as the ESD Association Annual Symposium.

MR heads have susceptibility to both voltage and energy. ESD currents in the milliamp range can change or destroy the MR sensor. Thin insulating layers in the device may be broken down by low voltages (Wallash 1996).

3.4.10 MEMS

MEMS devices are used in modern electronics systems. Some unprotected MEMS have been found to been highly sensitive to ESD damage (Walraven et al. 2000, 2001). Sangameswaran et al. (2008, 2009, 2010a,b) found HBM withstand voltages as low as 40 V. Traditional electrical characterization was found to overestimate ESD robustness. Capacitive switch MEMs devices were found to suffer dielectric or air breakdown and mechanical failures in response to ESD. ESD failures were often detectable only through mechanical tests. Atmospheric gas, pressure, temperature, and humidity were all influential to ESD-induced breakdown.

3.4.11 Burnout of Device Conductors or Resistors

Most devices contain tracks and interconnects of metallization, polysilicon, or other conductive material to connect internal parts and components with polysilicon, thick- or thin-film resistors. High ESD currents can lead to intense local heating that can burn out a conductor, rather like a fuse (Analog Devices 2014). This may occur where a wire or interconnect dimension is reduced in cross section for some reason. Thick- or thin-film resistors can suffer partial melt resulting in a change in resistance and parametric failure.

3.4.12 Passive Components

ESD damage to passive components is less well documented but is certainly possible with some types of components. The effect of the ESD can be highly variable depending on the resistor type. Film resistors on an insulating substrate can be susceptible to ESD damage (MIL HDBK 263). Some types of resistor have been trimmed to the desired value using ESD (Vishay 2011). Szwarc (2008) commented that the sensitivity of resistors varies from hundreds of volts to a few tens of kilovolts.

Thick film resistance changes have been reported to be dependent on voltage rather than energy. In contrast, thin-film resistors can be susceptible to the energy in the discharge, showing only small change until an energy threshold is reached. According to MIL HDBK 263, carbon film, metal oxide, and metal film resistors at low tolerance and power ratings can be susceptible to ESD. With a 0.05 W 0.1% part, placing the resistors in a polythene bag and rubbing this with another polythene bags was sufficient to change the tolerance of the resistors.

Chase (1982) found that tantalum nitride thin-film resistors used in filters in hybrid circuits could be damaged by ESD stress voltages as low as 1 kV HBM and tantalum capacitors as low as 400 V HBM. Resistors were damaged nearly linearly in response to multiple discharges. For CDM, resistors and capacitors were damaged above 2000 V. Damage to the resistors depended on the resistor design.

As time goes on, components used in surface mount assembly PCBs have become increasingly small. Taminnen et al. (2014) have found that very small 01005 resistors and capacitors showed some sensitivity to ESD.

3.4.13 Printed Circuit Boards and Assemblies

PCBs and assemblies containing ESDS are susceptible to ESD damage due to the sensitive devices they contain. According to the MIL-HDBK-263B, assemblies and modules containing ESDS are often as sensitive as the most sensitive component they contain. One of the longstanding myths of ESD is that components, once assembled into PCBs, are no longer ESD susceptible (Danglemeyer 1999). In practice, the risk of ESD damage is hard to predict, but it, and the sensitivity of the component to ESD, may be reduced or increased (Boxleitner 1990). The ESD threat, characterized in terms of voltage, peak power, or energy dissipated in a component may vary unpredictably over two orders of magnitude depending on the ESD source, the discharge point, and the design of the PCB. The threat to ICs mounted on PCBs can be significantly greater than to a device before it is mounted on a PCB.

A modern PCB may contain many ESDS components that have varying levels of susceptibility. The PCB typically has many conductive tracks interconnecting the ESDS components, all of which have inductance and capacitance. Injecting ESD current into one point of the board can cause a rapid transient change in voltage of that point of hundreds of volts relative to other tracks on the PCB. Transient currents flow through the track and through any components connected to it. The resulting ESD stress to components on the PCB is difficult to predict. Boxleitner found no correlation between device ESD withstand voltages and the immunity of the device mounted on a PCB.

Boxleitner did find a correlation between the ESD source and the level of threat. ESD from the combined charged PCB and person holding the PCB created the greatest peak power and energy levels in the discharge. Discharge to connector or device pins gave the greatest threat, with the least threat for ESD to long PCB traces between devices. The threat was worse for PCBs with no ground plane and less for PCBs with ground and power planes. Floating PCBs represented a lesser threat than those connected to an external ground.

Often modules and assemblies are designed with protection network circuits on the connector pins emerging to the outside world. These by design will give a level of protection against ESD entering the assembly by this route. They will not, however, protect against ESD arising from direct contact with other parts of the assembly.

Shaw and Enoch (1985) found that type 74 373 octal latch ICs failed in the voltage range 250–2500 V due to charged PCB ESD transients, compared to 1000 to >4000 V for HBM and CDM testing. They found the charged PCB damage voltages were not related to CDM or HBM failure levels but were dependent on the capacitance of the PCB.

Olney et al. (2003) found that ICs that are relatively robust as components could be damaged by charged board ESD (see Figure 2.19). As PCB capacitance is higher than device capacitance, the energy stored for a given voltage is much greater. They commented that charged board ESD damage could be mistaken for EOS damage and that this should be considered before a conclusion of EOS is drawn in failure analysis. They gave guidelines on how to avoid charged board ESD failures.

Paasi et al. (2003) studied the behavior of charged PCBs to evaluate ESD sensitivity of devices on the board with respect to discharge current and energy. They concluded that energy-sensitive devices could be at risk at lower charge or voltage levels when on a PCB than before assembly. They based their evaluation on HBM and MM device data. They pointed out that the capacitance, voltage, and stored energy of a PCB vary with movement thorough the manufacturing line. For a given PCB charge, a low capacitance high voltage condition gives higher stored energy and ESD current than a high capacitance low voltage condition and can represent a higher risk of ESD damage.

Gärtner et al. (2014) concluded that a CBE was more likely in a PCB production line than a charged device ESD event. As peak discharge current is a critical parameter in CDM damage, they used it to evaluate CBEs. They showed that the peak current is not significantly higher than for a single device at the same voltage level. The total charge transferred is, however, significantly higher giving a comparatively long (10–50 ns) discharge. This seemed to represent a higher stress to devices than charged device ESD. Case studies have shown that devices can be damaged at a voltage level at which devices would have survived in the CDM test. Stresses encountered in the real world are typically lowered due to reduced PCB capacitance due to greater gap between the PCB and ground. Stresses are also reduced for field-induced stresses for a parallel external field compared to a perpendicular field.

3.4.14 Modules and System Components

Modules and system components can become damaged by ESD if this can occur to a part that is inadequately protected. Components and modules that are designed to operate within a larger system are often (and not always correctly) not considered ESD susceptible. They are often, however, designed with little or no ESD protection on connector pins because these are not intended to be exposed once assembled into the system. Before or during assembly into the system, they may be subject to ESD to the connector pins or on connection to cables. They may be assembled into the system in a facility that has little or no ESD control.

Many system components or modules are housed within polymer enclosures or potted with only connectors or flying leads exposed to the outside world. While these enclosures prevent direct contact and ESD to the internal circuitry, the enclosure may become itself charged during handling and transport, especially if packaged within plastic packaging. The internal circuits can attain high voltages by induction especially when packaging is removed. ESD can then occur when connection is made to the connectors or flying leads.

3.5 System-Level ESD

3.5.1 Introduction

System-level ESD issues are largely addressed as part of the topic of EMC in operating electronic systems. In this context, it is the impact of ESD on powered and operational equipment that is usually of concern. This subject is covered by Williams (2001) and Montrose (2000).

Typical operating environments for electronic equipment are uncontrolled areas from a static electricity view. Personnel in these areas can charge to over 10 kV through normal activities such as walking or rising from a chair (Wilson 1972; Brundrett 1976; Smallwood 2004; Talebzadeh et al. 2015). In some environments, other types of source are possible such as charged metal stretchers, beds, or other mobile equipment in a health care environment (Viheriäkoski et al. 2014) or charged cables.

Electrostatic shocks can be felt by personnel if their body voltage exceeds about 2000 V (2 kV) and they discharge to a conductive object such as a metal item or equipment or another person. The sensitivity of the body varies from person to person and over the parts of the body.

In many regions of the world and industries, electronic equipment must be shown to be immune to ESD of this type expected in their operating environment. This has led to the development of ESD tests such as the IEC 61000-4-2. Electronic equipment marketed within Europe must pass ESD immunity tests measured using the IEC 61000-4-2 standard ESD source. Equipment is tested at 2, 4, 6, 8, or 15 kV ESD stress levels according to the requirements of relevant equipment and market-related standards. It is assumed that the equipment will withstand in service ESD from personnel charged up to specified levels without serious malfunction. Allowed loss of function is defined as part of the tests (Williams 2001).

3.5.2 The Relationship Between System Level Immunity and Component ESD Withstand

There is, however, some overlap between the topics of system-level ESD immunity and component ESD susceptibility. It has often been assumed, mainly erroneously, that system-level immunity depends on component ESD withstand (Ind. Co. 2010a, 2012). This has led to system designers placing ESD withstand requirements on components used in the system.

System ESD failures can be classified as hard or soft. Hard failures are those in which physical damage has occurred that is not recoverable. Often, system-level failures are soft failures that represent upset or temporary malfunction of the system in a way that is recoverable.

Ind. Co. (2010a) found that there is rarely correlation between system-level ESD immunity and component HBM ESD withstand level. Component ESD withstand data is obtained with the component in an unpowered state and represents hard failures of the component. System failures are often soft failures and occur with the system in a powered state. The ESD waveforms used in obtaining component ESD withstand data have significant differences with those used in system ESD immunity evaluation, and the test environment is very different. In practice, system ESD immunity is dependent on the system design including PCB design and on-PCB ESD protection, as well as individual component response to ESD transients. Component ESD tests do not reflect the conditions that occur for devices during system-level ESD events.

There may be some components, e.g. those that are directly connected to external connector pins, that do require some ESD withstand capability. The Industry Council on ESD Target Levels discussed this area in its White Paper 3 and proposed a system-efficient electrostatic discharge design (SEED) approach to understanding system-level ESD needs regarding ESD robust components.

3.5.3 Charged Cable ESD (Cable Discharge Events)

Charged cable ESD, often known as CDEs, can be considered to be a system-level ESD event, although they often impinge directly on devices connected to connectors. These can occur when a charged cable is plugged into an electronic system connector. Stadler et al. (2006) found that these often give rectangular current pulses resembling TLPs with current levels of several amps.

Stadler et al. (2017) used SPICE simulation and measurements to investigate risks due to ESD from cables plugged into charged USB3 ports. The investigated risks included charging of the cable shield, charging of an internal conductor while the shield was floating, and a charged person touching the shield. Typical discharges from a 3 m cable were 20 ns in duration and resembled discharges from coaxial cables. A discharge from a charged data line resulted in a 2.5 A peak current. They found that not all USB cables have the shield connected to ground or the shell of the connector. The worst case occurred when the shield was already grounded but the data line became charged. With a charging voltage of 1000 V the peak discharge current exceeded 13 A. This risk, however, was thought to be significant only for some applications that were not USB compliant and had exposed data lines. They concluded that a stress of a few amps for about 20 ns duration was sufficient for testing USB protection.

3.5.4 System-Efficient ESD Design (SEED)

The Industry Council on ESD Target Levels has proposed a SEED approach to system design for ESD immunity. This approach recognizes that high-component ESD withstand voltages are not generally required for effective system ESD immunity design. Robust system ESD design can be achieved by providing the interactions between the ESD stress and the complete system design are understood and addressed. The SEED approach recognizes that

- System ESD immunity specification requirements are understood to be a separate issue to component ESD withstand.
- The system ESD failure mechanisms must be understood to allow effective design of system immunity.
- System design and component design share responsibility for system ESD protection.
- Design strategy should differentiate between system external and internal component pins and the stresses that result to them in ESD tests.
- Placing robust ESD protection on-chip for component pins directly connected to external connector pins may not ensure a robust system. A better design strategy may be to use external ESD protection clamps and understand their interaction with component internal ESD protection.
- TLP can be used to characterize on-board and on-chip ESD protection in co-design of the system.

The SEED approach was proposed as a better system design philosophy, optimizing the balance between system cost and performance and reducing design effort. The philosophy and approach to implementation of ESD robust system designs and state of the art was further described in Part 2 of White Paper 3 (Ind. Co. 2012). A more comprehensive dealing of SEED is presented by Duvvury and Gossner (2015).

References

Amerasekera, A. and Duvvury, C. (1995). *ESD in Silicon Integrated Circuits*, 1e. Wiley. ISBN: 0471954810.

Amerasekera, A., Duvvury, C., Anderson, W. et al. (2002). *ESD in Silicon Integrated Circuits*, 2e. Wiley. ISBN: 0 470 9871 8.

Analog Devices (2014). *Reliability Handbook*. UG-311 Rev. D. Available from: http://www.analog.com/media/en/technical-documentation/user-guides/UG-311.pdf [Accessed: 10th May 2017]

Anand, Y. and Crowe, D. (1999). Latent failures in Shottky barrier diodes. In: *Proc. of EOS/ESD Symp. EOS-21*, 160–167. Rome, NY: EOS/ESD Association Inc.

Ashton, R. (2008). Reliability of IEC 61000-4-2 ESD testing on components. E E Times Available from: http://www.eetimes.com/document.asp?doc_id=1273265 [Accessed: 10th May 2017]

Avago Technologies. (2017) Premium InGaN LEDs - Safety Handling Fundamentals ESD Electrostatic Discharge Application Note 1142. Available from: http://www.avagotech.com/docs/AV02-0160EN [Accessed: 10th May 2017]

Baumgartner, B. (n.d.) ESD TR50.0-03-03. *Voltage and Energy Susceptible Device Concepts, Including Latency Considerations*. Rome, NY, EOS/ESD Association Inc.

Beal, J. Bowers, J. Rosse, M. (1983) A study of ESD latent defects in semiconductors. In: *Proc. EOS/ESD Symp. EOS-5*. Rome, NY, EOS/ESD Association Inc.

Boxleitner, W. (1990). ESD stress on PCB mounted ICs caused by charged boards and personnel. In: *Proc. EOS/ESD Symp. EOS-12*, 54–60. Rome, NY: EOS/ESD Association Inc.

Brodbeck, T. and Kagerer, A. (1998) Paper 4A.7). Influence of the device package on the results of CDM tests – consequences for tester characterization and test procedure. In: *Proc. EOS/ESD Symp*, 320–327. Rome, NY: EOS/ESD Association Inc.

Brundrett, G.W. (1976). A review of the factors influencing electrostatic shocks in offices. *J. Electrostat.* 2: 295–315.

Burr Brown. (1993) DAC8043 CMOS 12-Bit serial input multiplying digital to analog converter. www.ti.com.cn/cn/lit/ds/symlink/dac8043.pdf [Accessed: 10th May 2017]

Chase, E.W. (1982). Electrostatic discharge (ESD) damage susceptibility of thin film resistors and capacitors. In: *Proc. EOS/ESD Symp. EOS-4*, 13–18. Rome, NY: EOS/ESD Association Inc.

Chen, N.C., Wang, Y.N., Wang, Y.S. et al. (2009). Damage of light-emitting diodes induced by high reverse-bias stress 97-B-016. *J. Cryst. Growth* 311: 994–997.

Cook, C. and Daniel, S. (1993). Characterisation of new failure mechanisms arising from power pin stressing. In: *Proc. EOS/ESD Symp. EOS-15*, 149.

Danglemeyer, T. (1999). *ESD Program Management*, 2e. Clewer: Springer. ISBN: 0-412-13671-6.

Department of Defense. 1994 Military Handbook. Electrostatic discharge control handbook for protection of electrical and electronic parts, assemblies and equipment (excluding electrically initiated explosive devices) MIL-HDBK-263B 31st July 1994

Dhakad, H., Gossner, H., Zekert, S., Stein, B., Russ, C. (2012). Paper 3A.1. Chasing a latent CDM ESD failure by unconventional FA methodology. Proc. EOS/ESD Symp.

Duvvury C, Gauthier R. (2011) IC Technology Scaling Effects on Component Level ESD. Ch. 6 in Industry Council on ESD Target Levels (2011) White paper 1: A case for lowering component level HBM/MM ESD specifications and requirements. Rev. 3.0. Available from: www.esdindustrycouncil.org. [Accessed: 10th May 2017]

Duvvury, C. and Gossner, H. (2015). *System Level ESD Co-Design*. Wiley – IEEE. ISBN: 978-1-118-86190-5.

Duvvury C., Ashton R., Righter A., Eppes D., Gossner H., Welsher T. and Tanaka M, (2012) Discontinuing Use of the Machine Model for Device ESD Qualification. In Compliance Magazine, July 2012. Available from: https://incompliancemag.com/article/discontinuing-use-of-the-machine-model-for-device-esd-qualification [Accessed 6th March 2019]

Englisch, A., van Hesselt, K., Tissier, M., and Wang, K.C. (1999). CANARY: a high-sensitive ESD test reticle design to evaluate potential risks in wafer fabs. In: *Proceedings of the SPIE, 19th Annual Symposium on Photomask Technology, BACUS*, vol. II, 886–892. Available from: http://dx.doi.org/10.1117/12.373381 [Accessed: 10th May 2017].

Enoch, R.D., Shaw, R.N., and Taylor, R.G. (1983). ESD sensitivity of NMOS LSI circuits and their failure characteristics. In: *Proc. EOS/ESD Symp. EOS-5*, 185–197. Rome, NY: EOS/ESD Association Inc.

EOS/ESD Association Inc. (2006) Trends in Semiconductor Technology and ESD Testing. White paper II. ISBN: 1-58537-116-5

EOS/ESD Association Inc. (2009). ESD S5.6-2009. *ESD Association Standard Practice for Electrostatic Discharge Sensitivity Testing – Human Metal Model (HMM) – Component Level.* Rome, NY, EOS/ESD Association Inc.

EOS/ESD Association Inc. (2016) *ESD Association Electrostatic Discharge (ESD) Technology roadmap – revised 2016.* Available from: https://www.esda.org/assets/Uploads/docs/2016ESDATechnologyRoadmap.pdf [Accessed: 10th May 2017]

EOS/ESD Association Inc., JEDEC. (2012). ANSI/ESD STM5.2-2012. *ESD Association Standard Test Method for Electrostatic Discharge (ESD) Sensitivity Testing – Machine Model (MM) – Component Level.* Rome, NY, EOS/ESD Association Inc.

EOS/ESD Association Inc., JEDEC. (2017). ANSI/ESDA/JEDEC JS-001-2017. *ESDA/JEDEC Joint Standard for Electrostatic Discharge Sensitivity Testing – Human Body Model (HBM) – Component Level.* Rome, NY, EOS/ESD Association Inc.

EOS/ESD Association Inc., JEDEC. (n.d.) ANSI/ESDA/JEDEC JS-002-2014. *ESDA/JEDEC joint standard for electrostatic discharge sensitivity testing – Charged Device Model (CDM) – Device Level.* ISBN: 1-58537-276-5 Rome, NY, EOS/ESD Association Inc.

Gammill, P.E. and Soden, J.M. (1986). Latent failures due to electrostatic discharge in CMOS integrated circuits. In: *Proc. EOS/ESD Symp. EOS 8*, 75–80. Rome, NY: EOS/ESD Association Inc.

Gärtner, R., Stadler, W., Niemesheim, J., Hilbricht, O. (2014) Do Devices on PCBs Really See a Higher CDM-like ESD Risk? In: *Proc. EOS/ESD Symp.* Rome, NY, EOS/ESD Association Inc.

Gieser, H. (2002). Test Methods. In: *ESD in Silicon Integrated Circuits*, 2e (eds. A. Amerasekera and C. Duvvury). Wiley. ISBN: 0 470 9871 8.

Gieser, H. and Ruge, I. (1994). Survey on electrostatic susceptibility of integrated circuits. In: *Proc. ESREF Symp*, 447–455. Rome, NY: EOS/ESD Association Inc.

Hellstrom, S., Welander, A., and Eklof, P. (1986). Studies and revelation of latent ESD failures. In: *Proc. EOS/ESD Symp. EOS-8*, 81–91. Rome, NY: EOS/ESD Association Inc.

Industry Council on ESD Target Levels (2010a) White paper 3: System Level ESD Part I: Common Misconceptions and Recommended Basic Approaches. Rev. 1.0 http://www.esdindustrycouncil.org/ic/en/documents/7-white-paper-3-system-level-esd-part-i-common-misconceptions-and-recommended-basic-approaches [Accessed: 10th May 2017]

Industry Council on ESD Target Levels (2010b) White paper 2: A case for lowering component level CDM ESD specifications and requirements. Rev. 2.0. http://www.esdindustrycouncil.org/ic/en/documents/6-white-paper-2-a-case-for-lowering-component-level-cdm-esd-specifications-and-requirements [Accessed: 10th May 2017]

Industry Council on ESD Target Levels (2011) White paper 1: A case for lowering component level HBM/MM ESD specifications and requirements. Rev. 3.0. Available from: http://www.esdindustrycouncil.org/ic/en/documents/37-white-paper-1-a-case-for-lowering-component-level-hbm-mm-esd-specifications-and-requirements-pdf [Accessed: 10th May 2017]

Industry Council on ESD Target Levels (2012) White paper 3: System Level ESD Part II: Implementation of Effective ESD Robust Designs. Rev. 1.0 http://www.esdindustrycouncil.org/ic/en/documents/36-white-paper-3-system-level-esd-part-ii-effective-esd-robust-designs [Accessed: 10th May 2017]

Infineon. (2013). Preventing ESD Induced Failures in Small Signal MOSFETs. Application Note AN-2013-04 V2.0 http://www.infineon.com/dgdl/Infineon+-+Application+Note+-+PowerMOSFETs+-+Small+Signal+-+Preventing+ESD+Induced+Failures+in+Small+Signal+MOSFETs.pdf?fileId=db3a30433dfcb54c013dfe36f38d0295 [Accessed: 10th May 2017]

International Electrotechnical Commission. (1999) IEC 61340-5-2/TS:1999. *Electrostatics – Part 5-2: Protection of electronic devices from electrostatic phenomena - User guide.* Geneva, IEC.

International Electrotechnical Commission (2008). IEC 61000-4-2. *Electromagnetic compatibility (EMC) - Part 4-2: Testing and measurement techniques - Electrostatic discharge immunity test. Ed. 2.* Geneva, IEC.

International Electrotechnical Commission (2012). IEC 60749-27. *Semiconductor devices – Mechanical and climatic test methods – Part 27: Electrostatic discharge (ESD) sensitivity testing – Machine body model (MM). ed. 2.1*, ISBN 978-2-8322-0407-8 Geneva, IEC.

International Electrotechnical Commission (2013). IEC 60749-26. *Semiconductor devices – Mechanical and climatic test methods – Part 26: Electrostatic discharge (ESD) sensitivity testing – Human body model (HBM) Ed. 3* ISBN 978-2-83220-746-8 Geneva, IEC.

International Electrotechnical Commission (2017). IEC 60749-28. *Semiconductor devices – Mechanical and climatic test methods – Part 28: Electrostatic discharge (ESD) sensitivity testing – Charged device model (CDM). Ed. 1.* ISBN 978-2-8322-4139-4 Geneva, IEC.

International Rectifier. (n.d.). ESD Testing of MOS Gated Power Transistors. AN-986. http://www.infineon.com/dgdl/an-986.pdf?fileId=5546d462533600a40153559f9f3a1243 [Accessed: 10th May 2017]

Kelly, M., Servais, G., Diep, T. et al. (1995). A comparison of electrostatic discharge models and failure signatures for CMOS integrated circuit devices. In: *Proc. EOS/ESD Symp*, 175–185. Rome, NY: EOS/ESD Association Inc.

Krakauer, D.B. and Mistry, K.R. (1989). On latency and the physical mechanisms underlying gate oxide damage during ESD events in n-channel MOSFETs. In: *Proc. EOS/ESD Symp. EOS-11*, 121–126. Rome, NY: EOS/ESD Association Inc.

Laasch, I., Ritter, H.M., and Werner, A. (2009). Latent damage due to multiple ESD discharges. In: *Proc. EOS/ESD Symp. EOS-31*, 4A.4-1–4A.4-6. Rome, NY: EOS/ESD Association Inc.

Linear Technology (n.d.). ESD Protection Program. http://cds.linear.com/docs/en/quality/esdprotection.pdf [Accessed: 10th May 2017]

McAteer, O. 1990 Electrostatic discharge control. MAC Services In. ISBN 0-07-044838-8

McKeighan, R.E., Dailey, W., Pang, T. et al. (1986). Reversible charge induced failure mode of CMOS matrix switch. In: *Proc. EOS/ESD Symp. EOS-8*, 69. Rome, NY: EOS/ESD Association Inc.

Montrose, M. (2000). *Printed Circuit Board Design Techniques for EMC Compliance*, 2e. Wiley. ISBN: 0 7803 5376 5.

Nichia (2014) Handling of LED products. Application Note SE-AP00001B-E. Available from: www.nichia.co.jp/specification/products/led/ApplicationNote_SE-AP00001B-E.pdf [Accessed: 21 Feb. 2019]

Norberg, A. (1992). Modelling current pulse shape and energy in surface discharges. *IEEE Trans. Ind. App.* 28 (3): 498–503.

Olney, A., Gifford, B., Guravage, J., and Righter, A. (2003). Real- world charged board model (CBM) failures. In: *Proc. EOS/ESD Symp. EOS-25*, 34–43. Rome, NY, EOS/ESD Association Inc.

ON Semiconductor (2010). Human Body Model (HBM) vs. IEC 61000–4–2. App Note TND410/D Rev. 0, SEPT – 2010. Available from: http://www.onsemi.com/pub_link/Collateral/TND410-D.PDF [Accessed: 10th May 2017]

Osram (n.d.) ESD protection for LED systems. Available from: https://www.osram.com/ds/news/avoiding-damage-caused-by-electrostatic-discharge/index.jsp [Accessed: 21st Feb. 2019]

Paasi, J., Salmela, H., Tamminen, P., and Smallwood, J. (2003). ESD sensitivity of devices on a charged printed wiring board. In: *Proc. EOS/ESD Symp. EOS-25*, 143–150. Rome, NY: EOS/ESD Association Inc.

Reiner, J.C. (1995). Latent gate oxide defects caused by CDM ESD. In: *Proc. EOS/ESD Symp. EOS-17*, 311–321. Rome, NY: EOS/ESD Association Inc.

Reliability Analysis Centre (1989a) Electrostatic Discharge susceptibility data of microcircuit devices Vol. I. VZAP-2 Reliability Analysis Center P.O. Box 4700 Rome, NY 13440-8200.

Reliability Analysis Centre (1989b) Electrostatic Discharge susceptibility data of discrete/passive devices Vol. II. Reliability Analysis Center P.O. Box 4700 Rome, NY 13440-8200.

Reliability Analysis Centre (1995) Electrostatic Discharge susceptibility data of discrete/passive devices. VZAP-95 Reliability Analysis Center 201 Mill St, Rome, NY 13440.

Rider, G.C. (2016). Electrostatic risk to reticles in the nanolithography era. *J. Micro/Nanolithogr. MEMS MOEMS* 15 (2): 023501.

Rider, G. C., Kalkur, T. S. (2008) Experimental quantification of reticle electrostatic damage below the threshold for ESD. Proc. SPIE 6922, Metrology, Inspection, and Process Control for Microlithography XXII, 69221Y

Sangameswaran, S., De Coster, J., Linten, D. et al. (2008). ESD reliability issues in michromechanical systems (MEMS): a case study on micromirrors. In: *Proc. EOS/ESD Symp*, 3B.1-1–3B.1-9. Rome, NY: EOS/ESD Association Inc.

Sangameswaran, S., De Coster, J., Scholz, M. et al. (2009). A study of breakdown mechanisms in electrostatic actuators using mechanical response under EOS-ESD stress. In: *Proc. EOS/ESD Symp*, 3B.5-1–3B.5-8. Rome, NY: EOS/ESD Association Inc.

Sangameswaran, S., De Coster, J., Linten, D. et al. (2010a). Investigating ESD sensitivity in electrostatic SiGe MEMS. *J. Micromech. Microeng.* 20 (5): 055005.

Sangameswaran, S., De Coster, J., Chermin, V. et al. (2010b). Behaviour of RF MEMS switches under ESD stress. In: *Proc. of the EOS/ESD Symp*, 443–449. Rome, NY: EOS/ESD Association Inc.

Shaw, N.R. and Enoch, R.D. (1985). An experimental investigation of ESD damage to integrated circuits on printed circuit boards. In: *Proc. EOS/ESD Symp. EOS-7*, 132–140. Rome, NY: EOS/ESD Association Inc.

Smallwood, J.M. (2004). Static electricity in the modern human environment. In: *Electromagnetic Environments and Health in Buildings* (ed. D. Clements-Croome). Taylor & Francis. ISBN: 0 415 31656 1.

Smedes, T. (2009). ESD testing of devices, ICs and systems. *Microelectron. Reliab.* 49: 941–945.

Smedes, T. (2011) Machine Model – Correlation between HBM and MM ESD. Ch. 3 in Industry Council on ESD Target Levels (2011) White paper 1: A case for lowering component level HBM/MM ESD specifications and requirements. Rev. 3.0. Available from: http://www.esdindustrycouncil.org/ic/en/documents/37-white-paper-1-a-case-for-lowering-component-level-hbm-mm-esd-specifications-and-requirements-pdf [Accessed: 10th May 2017]

Smedes, T. and Li, Y. (2003). Paper 2A.6). ESD phenomena in interconnect structures. In: *Proc. EOS/ESD Symp. EOS-25*, 108–115. Rome, NY: EOS/ESD Association Inc.

Stadler, W., Brodbeck, T., Gartner, R., and Gossner, H. (2006). Cable discharges into communication interfaces. In: *2006 Electrical Overstress/Electrostatic Discharge Symposium*, 144–151. IEEE.

Stadler W., Niemesheim J., Stadler A., Koch S., Gossner H. (2017) Paper 3A1. Risk Assessment of Cable Discharge Events. In: *Proc. EOS/ESD Symp. EOS-39*. Rome, NY, EOS/ESD Association Inc.

Sylvania. (2009) ESD protection for LED systems. Application note. LED093. Available from: http://assets.sylvania.com/assets/documents/ESD.a9e90e9d-c91e-4ea3-9ebe-0be6e5570cd7.pdf [Accessed: 30th October 2017]

Szwarc, J. (2008) ESD Sensitivity of Precision Chip Resistors Comparison between Foil and Thin Film Chips. Vishay Available from: http://www.vishaypg.com/docs/60106/esdsensi.pdf [Accessed: 10th May 2017]

Talbot, J.W. (1986). The effect of ESD on III-Vmaterials. In: *Proc. EOS/ESD Symp. EOS-8*, 238–245.

Talebzadeh, A., Patnaik, A., Moradian, M. et al. (2015, 2015). Dependence of ESD charge voltage on humidity in data Centers: part I-test methods. *ASHRAE Trans.* 121: 58.

Taminnen, P., Sydänheimo, L., Ukkonen, L. (2014) Paper 9A.2 ESD Sensitivity of 01005 Chip Resistors and Capacitors In: *Proc. EOS/ESD Symp.* Rome, NY, EOS/ESD Association Inc

Taylor, R.G. and Woodhouse, J. (1986). Junction degradation and dielectric shorting: two mechanisms for ESD recovery. In: *Proc. EOS/ESD Symp. EOS-8*, 92. Rome, NY, EOS/ESD Association Inc.

Tunnecliffe, M., Dwyer, V., and Campbell, D. (1992). Parametric drift in electrostatically damaged MOS transistors. In: *Proc. EOS/ESD Symp. EOS-14*, 112–120. Rome, NY, EOS/ESD Association Inc.

Viheriäkoski, T., Kokkonen, M., Tamminen, P., Kärjä, E., Hillberg, J., Smallwood, J. (2014) 4B.2 Electrostatic Threats in Hospital Environment. In: *Proc. EOS/ESD Symp. EOS 36*

Vishay (2011) Resistor Sensitivity to Electrostatic Discharge (ESD). Vishay Document 63129. Available from: http://www.vishaypg.com/docs/63129/esd_tn.pdf [Accessed: 10th May 2017]

Wallash, A.J. (1996). Field induced charged device model testing of magnetoresistive recording heads. In: *Proc. EOS/ESD Symp. EOS-18*. 4B.2, 8–13.

Wallash, A. and Honda, M. (1997). Field induced breakdown ESD damage of Magnetoresistive recording heads. In: *Proc. EOS/ESD Symp. EOS-19*, 382–385. Rome, NY: EOS/ESD Association Inc.

Wallash, A., Levitt, L. (2003) *Electrical breakdown and ESD phenomena for devices with nanometer-to-micron gaps*. Proc. SPIE 4980 Reliability, Testing, and Characterization of MEMS/MOEMS II, 87 doi: http://dx.doi.org/10.1117/12.478191

Walraven, J.A., Soden, J.M., Tanner, D.M. et al. (2000). Electrostatic discharge/electrical overstress susceptibility in MEMS: a new failure mode. In: *Proceedings of the SPIE 2000*, vol. 4180, 30–39.

Walraven, J. A., Soden, J. M., Cole, E. I., Tanner, D. M., Anderson, R. R. (2001). Paper 3A.6 *Human Body Model, Machine Model, and Charged Device Model ESD testing of surface micromachined microelectromechanical systems (MEMS)*. In: *Proc. EOS/ESD Symp. EOS-23*. Rome, NY, EOS/ESD Association Inc

Wang, A.Z.H. (2002). *On-Chip ESD Protection for Integrated Circuits*. Klewer Academic Press.
Williams, T. (2001). *EMC for Product Designers*, 3e. Newnes. ISBN: 0 7506 4930 5.
Wilson, N. (1972). The static behaviour of carpets. *Text. Inst. Ind.* 10 (8): 235.

Further Reading

Agarwal S. (2014) Understanding ESD And EOS Failures In Semiconductor Devices.
Amerasekeera, E.A. and Campbell, D.S. (1986). ESD pulse and continuous voltage breakdown in MOS capacitor structures. In: *Proc. of the EOS/ESD Symp. EOS-8*, 208–213. Rome, NY: EOS/ESD Association Inc.
Bridgewood, M.A. (1986). Breakdown mechanisms in MOS capacitors. In: *Proc. of the EOS/ESD Symp. EOS-8*, 200–207. Rome, NY: EOS/ESD Association Inc.
Colvin, J. (1993). The identification and analysis of latent ESD damage on CMOS input gates. In: *Proc. of the EOS/ESD Symp*, 109–116. Rome, NY: EOS/ESD Association Inc.
Electronic Design. (2017) Understanding ESD And EOS Failures In Semiconductor Devices Available from: http://electronicdesign.com/power/understanding-esd-and-eos-failures-semiconductor-devices [Accessed: 10th May 2017]
EOS/ESD Association Inc. (2000a). *Technical Report - Transient Induced Latch-up (TLU) ESD TR5.4-01-00*. Rome, NY, EOS/ESD Association Inc.
EOS/ESD Association Inc. (2000b). *Technical Report - Calculation of Uncertainty Associated with Measurement of Electrostatic Discharge (ESD) Current ESD TR14.0-01-00*. Rome, NY, EOS/ESD Association Inc.
EOS/ESD Association Inc. (2002) *ESD Phenomena and the Reliability for Microelectronics* ISBN: 1-58537-046-0 Available from: https://www.esda.org/assets/Uploads/documents/ESD-Phenomena-and-the-Reliability-for-Microelectronics.pdf [Accessed: 2nd November 2017]
EOS/ESD Association Inc. (2008a). *Technical Report – Determination of CMOS Latch-up Susceptibility – Transient Latch-up – Technical Report No. 2. ESD TR5.4-02-08*. Rome, NY, EOS/ESD Association Inc.
EOS/ESD Association Inc. (2008b). *Technical Report for the Protection of Electrostatic Discharge Susceptible Items - Transmission Line Pulse (TLP) ESD TR5.5-01-08*. Rome, NY, EOS/ESD Association Inc.
EOS/ESD Association Inc. (2008c). *Technical Report for the Protection of Electrostatic Discharge Susceptible Items - Transmission Line Pulse - Round Robin ESD TR5.5-02-08*. Rome, NY, EOS/ESD Association Inc.
EOS/ESD Association Inc. (2011). *Technical Report For Electrostatic Discharge Sensitivity Testing - Latch-Up Sensitivity Testing of CMOS/BiCMOS Integrated Circuits - Transient Latch-up Testing – Component Level - Supply Transient Stimulation. ESD TR5.4-03-11*. Rome, NY, EOS/ESD Association Inc.
EOS/ESD Association Inc. (2012). *ESDA/JEDEC Joint Technical Report User Guide of ANSI/ESDA/JEDEC JS-001 Human Body Model Testing of Integrated Circuits ESDA/JEDEC JTR001-01-12*. Rome, NY, EOS/ESD Association Inc.
EOS/ESD Association Inc. (2013a). *Technical Report for Electrostatic Discharge Sensitivity Testing - Transient Latch-up Testing ESD TR5.4-04-13*. Rome, NY, EOS/ESD Association Inc.

EOS/ESD Association Inc. (2013b). *Technical Report for the Protection of Electrostatic Discharge Susceptible Items – System Level Electrostatic Discharge (ESD) Simulator Verification ESD TR14.0-02-13.* Rome, NY, EOS/ESD Association Inc.

EOS/ESD Association Inc. (2014a). *Technical Report for Electrostatic Discharge (ESD) Sensitivity Testing – Very Fast – Transmission Line Pulse (TLP) – Round Robin Analysis ESD TR5.5-03-14.* Rome, NY, EOS/ESD Association Inc.

EOS/ESD Association Inc. (2014b). *Technical Report for Relevant ESD Foundry Parameters for Seamless ESD Design and Verification Flow ESD TR22.0.01-14.* Rome, NY, EOS/ESD Association Inc.

EOS/ESD Association Inc. (2014c). *Technical Report for ESD Electronic Design Automation Checks ESD TR18.0-01-14.* Rome, NY, EOS/ESD Association Inc.

EOS/ESD Association Inc. (2015a). *Standard Practice for Electrostatic Discharge Sensitivity Testing – Near Field Immunity Scanning - Component/Module/PCB Level ANSI/ESD SP14.5-2015.* Rome, NY, EOS/ESD Association Inc.

EOS/ESD Association Inc. (2015b). *Technical Report for ESD Process Assessment Methodologies in Electronic Production Lines – Best Practices used in Industry ESD TR17.0-01-15.* Rome, NY, EOS/ESD Association Inc.

EOS/ESD Association Inc. (2016a). *Standard Test Method for Electrostatic Discharge (ESD) Sensitivity Testing – Transmission Line Pulse (TLP) – Component Level ANSI/ESD STM5.5.1-2016.* Rome, NY, EOS/ESD Association Inc.

EOS/ESD Association Inc. (2016b). *Technical Report for Electrostatic Discharge Sensitivity Testing – Charged Board Event (CBE) ESD TR25.0-01-16.* Rome, NY, EOS/ESD Association Inc.

EOS/ESD Association Inc. (1999) *ESD Association Technical Report - Can Static Electricity be Measured? ESD TR50.0-01-99*

International Organization for Standardization. (2008) Road vehicles – Test Methods for electrical disturbances from electrostatic discharge. ISO 10605:2008/Amd.1:2014(en)

King, W.M. (1979). Dynamic waveform characteristics of personnel electrostatic discharge. In: *Proc. of the EOS/ESD Symp. EOS-1*, 78.

Lin, D.L., Strauss, M.S., and Welsher, T.L. (1987). On the validity of ESD threshold data obtained using commercial human-body model simulators. In: *Proceedings of the 25th International Reliability Physics Symposium*, 77. IEEE.

Lin, N., Liang, Y., Wang, P., and Pelc, T. (2014). *Evolution of ESD process capability in future electronics industry.* In: *15th Int. Conf. Elec. Packaging Tech*, 1556–1560. IEEE.

McAteer, O.J., Twist, R.E., and Walker, R.C. (1980). Identification of latent ESD failures. In: *Proc. of the EOS/ESD Symp. EOS-2*, 54–57. Rome, NY: EOS/ESD Association Inc.

McAteer, O.J., Twist, R.E., and Walker, R.C. (1982). Latent ESD failures. In: *Proc. of the EOS/ESD Symp. EOS-4*, 41–48. Rome, NY: EOS/ESD Association Inc.

Paasi, J., Smallwood, J., and Salmela, H. (2003) Paper 2B4). New methods for the assessment of ESD threats to electronic components. In: *Proc. of the EOS/ESD Symp*, 151–160. Rome, NY: EOS/ESD Association Inc.

Smallwood J, Paasi J. (2003) Assessment of ESD threats to electronic devices. VTT Research Report No BTUO45-031160

Smallwood J., Taminnen P., Viheriaekoski T. (2014) Paper 1B.1. Optimizing investment in ESD Control. In: *Proc. of EOS/ESD Symp. EOS-36*. Rome, NY, EOS/ESD Association Inc.

Strauss, M.S., Lin, D.L., and Welsher, T.L. (1987). Variations in failure modes and cumulative effects produced by commercial human-body model simulators. In: *Proc. of EOS/ESD Symp. EOS-9*, 59–63.

Viheriäkoski T, Peltoniemi T, Tamminen T, (2012) Paper 4A3. Low Level Human Body Model ESD. In: *Proc. of EOS/ESD Symp.* Rome, NY, EOS/ESD Association Inc.

Vinson, J.E. and Liou, J.J. (1998). Electrostatic discharge in semiconductor devices: an overview. *Proc. IEEE* 86 (2): 399–420.

Voldman, S. (2009). *ESD Failure Mechanisms and Models.* Wiley. ISBN: 978-0-470-1137-4.

Vollman, S., Hui, D., Warriner, L. et al. (1999). Electrostatic discharge (ESD) protection in silicon-on-insulator (SOI) CMOS thechnology with aluminium and copper interconnects in advanced microprocessor semiconductor chips. In: *Proc. of the EOS/ESD Symp. EOS-21*, 105–115. Rome, NY: EOS/ESD Association Inc.

4

The Seven Habits of a Highly Effective ESD Program

4.1 Why Habits?

Habit: "settled or regular tendency or practice, especially one that is hard to give up"
(Oxford Dictionary 2017)

For effective electrostatic discharge (ESD) control, we need to set up practices that reduce the ESD risk to our electrostatic discharge–sensitive (ESDS) devices to an acceptable level. These practices can be ways of working but also involve using certain ESD-protective equipment to reduce ESD risk. If we can establish and maintain these practices so well, they become a habit, and then our ESD control program is likely to remain effective.

Many ESD threats occur while handling ESDS devices, for example during assembly processes. To protect against this sort of threat, we can set up a permanent or temporary ESD protected area (EPA) in which the ESD threats are controlled so that we can handle the devices and assemblies conveniently and relatively free from ESD risk.

Other ESD risks occur when an ESDS device is stored or transported in an uncontrolled unprotected area (UPA) where static electricity can build up and ESD sources arise. In this situation, we use ESD-protective packaging to enclose and protect the ESDS device from damage.

Of course, we need to know how effectively we are adhering to our established ESD program practices. Also, equipment is likely to fail from time to time under the day-to-day wear and tear it experiences during use. We need to detect when equipment fails, falls out of specification, or requires maintenance. For these reasons, a habit of checks and tests is required.

Finally, we need to be sure that everyone who is concerned with ESD control or must use the provisions of the ESD control program understands what they must do and not do. They may need to know what equipment to use and even how to check it is functioning correctly. They need to know what procedures to follow. It can be of great benefit if they watch out for noncompliance and correct them as they go. So, training will be needed to make sure they have the knowledge they need to fulfill these roles.

This chapter explores these habits, the reasons for them, and how to decide what should be included in our habitual practices. Many of them are incorporated into the design and specifications of equipment and materials designed for use in ESD control. We can enact

The ESD Control Program Handbook, First Edition. Jeremy M Smallwood.

some aspects of the "habits" by specifying special equipment and materials that are used in EPAs where ESD risks are carefully controlled.

4.2 The Basis of ESD Protection

There are two key strategies that form the basis of successful ESD protection practice.

- Handle unprotected ESDS devices only in an area in which the ESD risks are reduced to an insignificant level.
- In uncontrolled (unprotected) areas, protect ESDS devices by enclosing them within ESD-protective packaging that protects the ESDS devices from ESD risks.

These two strategies should be applied to all aspects of handling, storage, and transport of ESDS.

4.3 What Is an ESDS Device?

ESDS devices can be of many types and forms, from minute individual semiconductor devices such as transistors, diodes, or integrated circuits, to printed circuit boards, modules, or system components. They usually contain semiconductor devices of some sort, although other types of device (e.g. some types of resistors and capacitors) can have some ESD sensitivity. ESDS devices and their failure modes are discussed in more detail in Chapter 3.

The key factor in identification of an ESDS device is understanding that the item would be at some risk of ESD damage if handled in an UPA without precautions. An item is an ESDS device if it satisfies two criteria.

- The item contains parts that could be damaged by ESD.
- If handled in a UPA, there is a risk that potentially damaging ESD could find a route to the ESDS parts.

It follows that if either of these criteria is not met, the item can be considered as not being ESDS.

In building an electronic system, the risk of ESD damage and susceptibility to ESD often change considerably with build state. Let us take as an example the construction of a simple product that consists of a printed circuit board (PCB) within a housing. Many of the components that go onto the PCB are likely to be ESDS. The populated PCB is also likely to be ESDS and should be handled as such. However, once built into and protected by its housing, the final product may well be quite immune to normal ESD sources due to the protective barrier provided by the housing. In many cases, electromagnetic compatibility (EMC) regulations may require testing and demonstration of ESD immunity of the working system for market acceptance. There may, however, be some residual ESD susceptibility dependent on its design, for example susceptibility to ESD to connector pins, e.g. from connection of charged cables.

In some cases, the build stage at which the product can be considered no longer susceptible to ESD damage is not clear. This will then have to be decided by some evaluation of the ESD risks and susceptibility.

An item such as a PCB may be subject to ESD occurring to almost any part of it that may be touched by a person, tool, or other conductive item. If the same PCB is encapsulated or potted to become a module or subassembly, the number of ways in which ESD can occur to it, and hence ESD risk, is much reduced as the encapsulation can form an effective barrier to ESD to most parts of the assembly. This may not mean that the module is immune to ESD damage or needs no ESD protection. The module may have flying leads or connectors to the PCB within. The module may be susceptible to ESD to these leads or connections, unless designed for immunity. Tribocharging of the module surface can lead to high induced voltages occurring on the PCB within, and these can discharge if the leads or connector pins touch something conductive.

A product that is at a built state at which it is no longer ESDS can become susceptible to ESD damage again if modified or disassembled in some way. For example, a desktop computer would not normally be considered ESDS once it is fully assembled to the state at which the user would normally receive it. If, however, the covers are removed, PCBs containing ESDS parts may become accessible to touch, and appropriate ESD control procedures should be in place while operating on these PCBs. Once the covers are replaced and no ESDS parts are accessible, the system can again be considered not susceptible to ESD damage.

4.4 Habit 1: Always Handle ESDS Components Within an EPA

4.4.1 What Is an EPA?

As far as ESD protection is concerned, there are two types of area: EPAs and UPAs (see IEC61340-5-3:2015). In some companies, the EPA may be known by another name or acronym such as *safe handling area* (SHA). In this book, the terminology used in current standards is used. It doesn't matter what they are called – it is what happens within these areas that is important in preventing ESD damage to sensitive components.

Most areas are of course UPAs. In these areas, static electricity is uncontrolled and often rife and omnipresent. We are not necessarily conscious of this, because we are ourselves rather insensitive to static electricity. As we move around, we routinely develop voltages on our bodies of hundreds of volts. We don't feel voltages; we feel the current and energy in a discharge. We touch things or people and discharges occur, and we are oblivious to them. Only if these body voltages reach a few thousand volts do we start to experience these discharges as shocks and, even then, only if we touch a substantial conductor like another person or a metal filing cabinet. If we touch a resistive material, the discharge current is reduced to the point where we don't feel it.

We are even less sensitive to the voltages that arise on insulating materials like plastic packaging and stationary materials. Voltages of several thousand volts (1000 V = 1 kV) can arise on these materials without us noticing. As these voltages exceed 20 kV or so, small brush discharges can arise that we might hear as an occasional clicking sound, if the conditions are quiet. We might start to feel the "tickling" sensation of hairs on our skin moving in response to electrostatic attraction from the high electrostatic fields near these surfaces. When taking off fleece or other clothing made of man-made fibers, we may hear the crackling of small discharges, and, in low light conditions, we may see them as tiny flashes.

While we are insensitive to ESD and electrostatic fields, as we have seen, many electronic components are not. We must protect them against electrostatic effects, either by handling them within an EPA or by enclosing them in ESD-protective packaging. This section is about EPAs and the habits we need to develop to make them effective.

We can make EPAs in many different forms. They can be temporary, or they can be fixed facilities. A field service kit can be used to provide a temporary EPA for field use. A fixed facility may be a single workstation or may enclose many workstations or a whole room or work area (Figure 4.1). A machine, or part of a machine, where unprotected ESDS devices are handled may need to be part of an EPA.

So, what do we need to have to make an effective EPA? There are two basic aspects.

- There must be a clear boundary.
- Within the boundary, all ESD risks must be controlled to give insignificant ESD damage levels.

The need for a clear boundary is because we must be clear whether we are inside the EPA or outside it. Outside the EPA boundary, the ESDS device should never be taken out

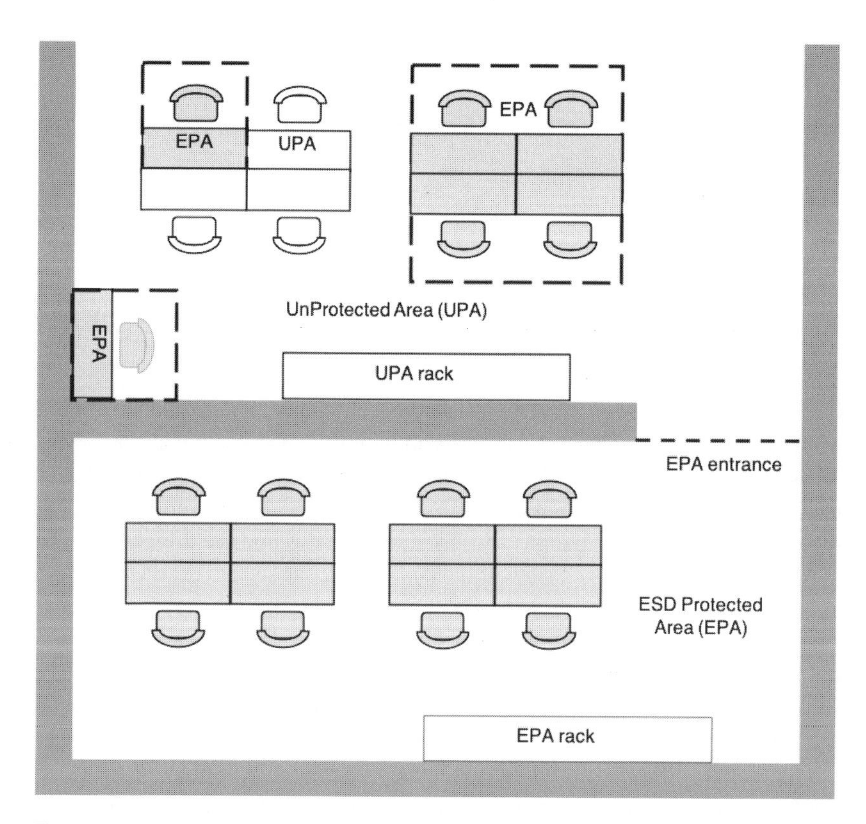

Figure 4.1 EPA and UPA.

of its ESD-protective packaging. To do so would be to expose it to risk of ESD. Inside the EPA, the ESDS device may be kept within its protective packaging, or it may be taken out to be handled or used in processes as required, because the ESD risks are controlled to a low level.

4.4.2 Defining the EPA Boundary

The first requirement for an EPA is for it to have a clear boundary so that everyone knows whether they are inside or outside the EPA. If we are not clear about the boundary, we cannot be clear about whether it is safe to take an ESDS device from its ESD-protective packaging. We will not know whether we should be taking the prescribed ESD control precautions and using ESD control equipment or not. So, lack of a clear boundary is likely to lead to noncompliance sooner or later, as well as possible ESD risk to ESDS.

For an EPA to be effective, all processes within the EPA must be evaluated and controlled if necessary. It is not sufficient to equip the area with common ESD control equipment (bench mats, wrist straps, and the like) if other major sources of ESD in processes are ignored.

It is usually beneficial to think carefully about the EPA boundary and include within the EPA only such processes as are necessary. The less is in there, the less needs to be equipped and evaluated for ESD control. Handling the unprotected ESDS device as little as possible can be an effective first ESD prevention measure. Minimizing the number of workstations and processes within the EPA can reduce the necessary expenditure on ESD control equipment, as well as reducing the burden of equipment checks and maintenance. It can be a good policy to exclude from the EPA any processes in which unprotected ESDS devices do not need to be handled.

Against this, it is sometimes more convenient to include certain processes that may give ESD risk within the EPA for convenience in moving between processes and operations. This can be acceptable provided the ESDS devices are not put at risk. A common example may be to have an office work desk area where papers may be kept and worked on. The solution to preventing ESD risk is to make sure unprotected ESDS are never brought into the office desk area. Conversely, the materials in the office area that might cause risk must never be taken to workstations where unprotected ESDS devices are handled. Managing this requires a high level of awareness and compliance in the personnel who work in these areas. This in turn requires some training to create the intended habits. It also requires some compliance verification to make sure the intended practice is successfully maintained.

4.4.3 Marking the EPA Boundary

There is no single method of marking an EPA boundary. What is most important is that the EPA boundary is recognized by, and immediately obvious to, the people entering and working in the area. This includes personnel who are not authorized to enter the EPA, perhaps because they have not had ESD training. Untrained personnel must clearly see that the area should not be entered.

It is often a good idea to restrict entry into the EPA to certain points rather than having an extended open boundary. This can minimize the amount of boundary marking required. A physical barrier (temporary or permanent) can go a long way to reduce unauthorized entry into an EPA.

EPA entry points should be marked with clear signage showing personnel approaching the entrance that they are about to enter an EPA (Figure 4.2). Signs should be obvious and eye-catching. Signs at around eye-level may be more easily noticed than signs positioned at low level or above an entrance.

It can also be a good idea to post signs visible to personnel leaving the EPA, warning that they are about to leave the EPA.

Figure 4.2 Examples of marking an EPA entrance. Source: C. Cawthorne.

Some companies establish electronically operated turnstiles or barriers at the EPA entrance that allow entry only to personnel who have passed personal grounding equipment (wrist strap or footwear) tests.

4.4.4 What Is an Insignificant Level of ESD Risk?

The level of ESD risk that is considered insignificant depends on the type of product and its market and on the possible consequences of a failure. At one end of the scale, a low-cost consumer product might merit a low level of care and expenditure on ESD protection with an acceptable number of ESD failures. For example, a low-cost insert for a musical greeting card or talking toy may be a throwaway item. A certain level of failure of these may be easily tolerable.

At the opposite extreme, a satellite must operate reliably once deployed, with no possibility of maintenance if a failure occurs. A failure would be catastrophic to the mission and be very costly. Automotive products are often made in high quantities, and a very low failure rate is demanded. A failure in service would be potentially catastrophic in this case also, with possible risk of injury or death to a driver and passengers. Aerospace and military applications are also areas where high reliability is demanded, and the consequences of a failure are dire.

The level of ESD risk that is considered insignificant is a matter for the user to decide, according to their view of their product and market needs.

4.4.5 What Are the Sources of ESD Risk?

There are two overall types of ESD risk that are controlled in an EPA.

- Direct ESD to or from the device
- Electrostatic fields that could lead to ESD to or from the device

Identification and evaluation of these is discussed further in Chapter 9. The sources of ESD include the following:

- Charged personnel
- Charged metal or conductive objects or materials
- Charged devices

The sources of electrostatic fields that are of concern are normally insulating materials such as plastics that easily become highly charged and retain their charge for long periods.

The risk of ESD damage is controlled by factors such as the following:

- The likelihood that ESD can occur to the ESDS
- The likelihood that the ESD current passes through the susceptible part of the ESDS device
- The likelihood that ESD energy, peak current, charge transferred in the discharge, or some other parameter exceeds a damage threshold of the ESDS device

It follows that the risk of ESD damage occurring can be reduced by reducing the risk of ESD occurring, as well as reducing the likely strength of ESD if it should occur.

There is a risk of ESD occurring whenever an ESDS device contacts another conductor at a different voltage. The strength of the ESD occurring can be minimized by decreasing the likely peak current, energy, and charge transferred in any ESD that may occur. The way we evaluate and control ESD risk is further discussed in Habits 2–5.

4.4.6 What ESD Protection Measures Are Needed in the EPA?

The ESD protection measures needed in the EPA will depend on a wide range of considerations including the processes and product, ESD susceptibility of the ESDS device, and other factors. They commonly include use of equipment such as the following:

- Packaging for ESD protection and control
- ESD control flooring or floor mats
- Personal grounding equipment (wrist straps or ESD control footwear and flooring)
- Bench mats or work surfaces, racks, and carts
- ESD control seating
- Garments
- Gloves and finger cots
- Tools

One common way of deciding what ESD measures should be implemented is to adopt the requirements of an ESD control standard (see Chapter 6). These normally list a range of control measures that address the most common ESD threats. A detailed evaluation of ESD risks is often not attempted. This approach has the advantage that it is easily achieved and needs the minimum of expertise. The ESD control program that results may be easily accepted by customers and found compliant with the standard by auditors.

This approach can, however, have disadvantages.

- ESD control measures may be included that address risks that do not in fact exist within the processes and facility.
- ESD risks may exist that are not addressed by the standard ESD control measures.
- The ESDS device is unusually sensitive, having ESD withstand voltage lower than the design withstand voltage level of the standard.

For these reasons, it may be necessary or preferable to undertake some level of evaluation of the ESD risks in a process and facility in preparation for determining the ESD control measures that are required. With fuller knowledge of the specific ESD risks, ESD control measures can be specified that address these more effectively, efficiently, and completely. Determination of ESD protection measures is further discussed in Chapter 10.

4.4.7 Who Will Decide What ESD Protection Measures Are Required?

An ESD control program that has no one leading and responsible for it is likely to fail through lack of attention and maintenance. Someone will need to develop and implement the ESD control program. Someone will need to document it, maintain it, test it, and train people to work within it.

These functions may of course involve several people taking different roles and responsibilities. Some companies have a committee of personnel within a site and, for a multisite organization, on different sites, working on a company ESD program.

Nevertheless, it is advisable to have a person responsible for coordinating, implementing, and maintaining the ESD control program at each site. In the main ESD control standards current at the time of writing, it is a requirement to have such a person, known as the *ESD Coordinator*. They do not have to do everything themselves, but they do have to make sure it gets done. So, they must have the necessary authority, backup, and resources to fulfill the role.

In some organizations, a committee rather than a person fulfills this role. Usually, the tasks required in implementing and maintaining the ESD control program are delegated wholly or in part to other people in the organization. For example, specially trained technicians may do routine testing of the equipment in the EPA, and a designated trainer may do some or all the training.

4.5 Habit 2: Where Possible, Avoid Use of Insulators Near ESDS

4.5.1 What Is an Insulator?

What is a insulator? In this book, I define an insulator as any material or object on which electrostatic charge cannot move away quickly enough to avoid significant electrostatic charge buildup or voltage differences occurring. This is a pragmatic rather than academic definition, and it reflects the way these terms tend to be used in industry in practice. It is deliberately a definition that can lead to differences between electrical characteristics that might be defined as insulating in different industrial contexts. This is what happens in practice. So, in electrostatic fire and explosion hazards avoidance in industrial processes, an insulator may be in general considered to be a material having resistance over 100 MΩ (see IEC60079-32-1:2013). Even within the 60079-32-1 document, the word *insulating* has a wide range of different definitions for different products or materials. An enclosure classified as insulating has volume resistance of 100 GΩ or greater, whereas a hose classified as insulating has resistance greater than 1 MΩ (see Section 1.7.5).

So, charge does not move around through or on an insulator easily and may remain for a significant time. This behavior gives the following effects:

- The voltage is unlikely to be the same at every point on the surface of the insulator. When voltage differences occur, electrical currents will not flow quickly enough in response to prevent the differences.
- If ESD from the insulator occurs, only a small amount of the charge and energy stored on the insulator may be delivered during the discharge.

The apparently perplexingly flexible definition of an insulator can be explained by revisiting our simple model of electrostatic charge buildup (Figure 4.3). In Chapter 3 we found that the resistance (or resistivity) had a strong effect on two important things.

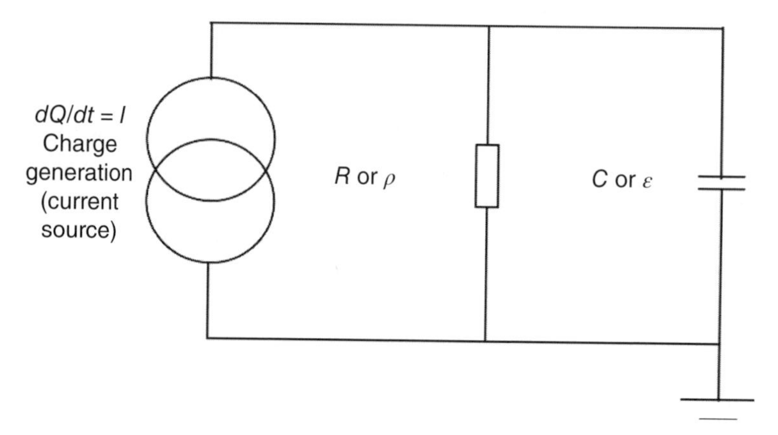

Figure 4.3 A simple electrical model of electrostatic charge buildup, revisited.

- The voltage developed in response to an electrostatic charging current, governed by the product of charging current and resistance (or material resistivity)
- The charge or voltage decay time governed by the product of resistance R (or material resistivity ρ) and capacitance C (or material permittivity $\varepsilon = \varepsilon_r \varepsilon_0$).

Typically, we wish to prevent a risk by either keeping the voltage produced below a certain value at which some hazard may occur or making sure that any voltages arising are dissipated quickly before they have chance to cause a hazard.

In an EPA where manual handling is the norm, things don't usually happen very quickly. If there are no strong continuous electrostatic charging mechanisms present, it will often be sufficient to make sure that charge and voltages produced during normal contact between materials is dissipated within a few seconds. So, materials will often be acceptable if the decay time $\tau = \rho \varepsilon_r \varepsilon_0$ is of the order of a few seconds. Given that many materials have relative permittivity ε_r around 2–3, and $\varepsilon_0 = 8.8 \times 10^{-12}$ Fm^{-1}, then setting an upper limit of material volume resistivity of 100 GΩm (10^{11} Ωm) gives theoretical decay times around 1.8–2.6 seconds. This approximately correlates with the upper limits usually chosen for ESD-protective packaging materials, although these are usually expressed as surface or volume resistance rather than resistivity (see Section 1.7).

4.5.2 Essential and Nonessential Insulators

Insulating materials can be thought of as being of two types. Essential insulators are those that are a necessary part of the process or product. Without these, we cannot make the product or do what we need to do with it.

All other insulators that are not a necessary part of the product or process are nonessential insulators. These should be kept sufficiently far away from ESDS devices that ESD risk is reduced to an acceptable level. Many organizations find it easiest to make sure that they do not enter the EPA. If they are allowed into the EPA, then the proximity of insulators to ESDS devices must be effectively managed. This requires careful definition of where and how they may be used, training of personnel, and some compliance verification.

Table 4.1 Some examples of essential insulators and nonessential insulators.

Item	Nonessential	Essential
PCB substrate and component plastic packages		Yes
Product components and parts made from plastics		Yes
Plastic packaging not designed for ESD control	Yes	
Personal items, coffee cups, lunch boxes	Yes	
Parts of test jigs and fixtures	Parts that do not need to be made from insulating materials	Parts that must be made from insulating materials
Papers	Papers that are not required to be present or used during the process	Papers that are required to be present or used during the process

Table 4.1 gives some examples of essential and nonessential insulators that can be found in many facilities. Some types of insulator are clearly essential (e.g. the PCB and components). Others are less easily categorized (e.g. parts of test jigs and production papers). In some cases, these may be essential, and in others nonessential.

4.5.3 Remove Nonessential Insulators from the Vicinity of ESDS

Electrostatic fields due to charged insulators introduce a risk of ESD. This is because they induce voltages on any isolated (ungrounded) conductors in the field. Any isolated conductors (including any ESDS devices) within the field will be at some unknown voltage that will in general be different from any other conductor nearby in the field (see Section 4.7 for discussion of conductors and grounding). If two conductors become sufficiently close or touch, ESD will occur between them. If one of the conductors is part of an ESDS device, this ESD risks causing ESD damage.

The electrostatic field from a charged object depends on the distance from that object. In Chapter 2 we saw that a uniform field is produced between two large parallel plate conductors. If an electrostatic field meter is inserted into an aperture in one of the plates, it will measure the field that is produced (Figure 4.4). The electrostatic field E is easily calculated from the voltage difference between the plates V and the distance between them d.

$$E = \frac{V}{d}$$

If the high-voltage plate is moved toward, or away from, the field meter plate, the electrostatic field is increased, or decreased, proportional to $1/d$.

This simple equation tells us that the closer an electrostatic field source is to an ESDS devices, the greater the electrostatic field. As electrostatic field is an indicator of ESD risk, the ESD risk is also greater. If we set a limit of say 5 kV m^{-1} for the field, this limit is achieved by different voltages at different distance (Table 4.2). The closer a voltage source is to the ESDS device, the more concerned we will be about it and the lower the tolerable voltage limit.

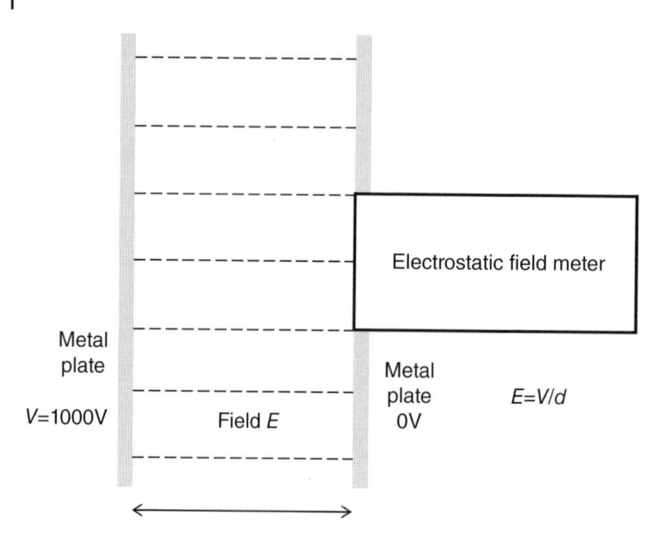

Figure 4.4 Electrostatic field between two parallel conducting plates.

Table 4.2 Voltages and distances between plane parallel electrodes giving the electrostatic field of $5\ \text{kV m}^{-1}$.

Voltage (V)	Distance giving $5\ \text{kV m}^{-1}$
10 000	2 m
5 000	1 m
2 500	50 cm
1 500	30 cm
500	10 cm
125	2.5 cm
50	1 cm

If the field meter plate is removed, leaving the field meter in position, the electrostatic field lines terminate at the field meter instead of the plate (Figure 4.5). We assume that the electrostatic field meter is constructed as an earthed conductor at 0 V. The field lines focus on the field meter and so a higher nonuniform electrostatic field exists at the field meter in this case. Because of this, we can no longer assume that as we change the distance between the field meter and the plate, the field will vary with $1/d$.

In practice, the assumption that the field decreases as $1/d$ often gives reasonable agreement with experiment (Stadler et al. 2018). Many field meters are calibrated to give a reading of surface voltage when at a set distance of 2 or 2.5 cm (1 in.) from a large flat metal sheet at a given voltage. In Figure 4.6 the voltage reading taken with the field meter at 2 cm distance from a flat metal plate at 1 kV. The field (voltage) readings at other distances are shown as a percentage of the field (voltage) reading at other distances. It is interesting to note that the field at 30 cm distance is only 7% of the value at 2 cm distance. Some standards require

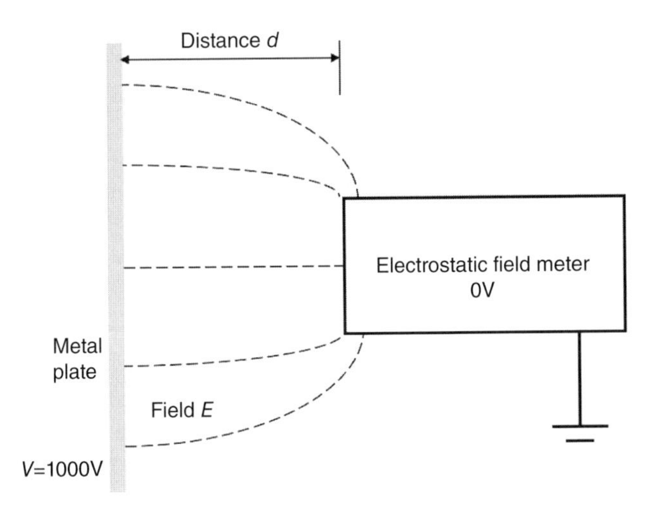

Figure 4.5 The electrostatic field between a field meter and a charged plate varies with distance.

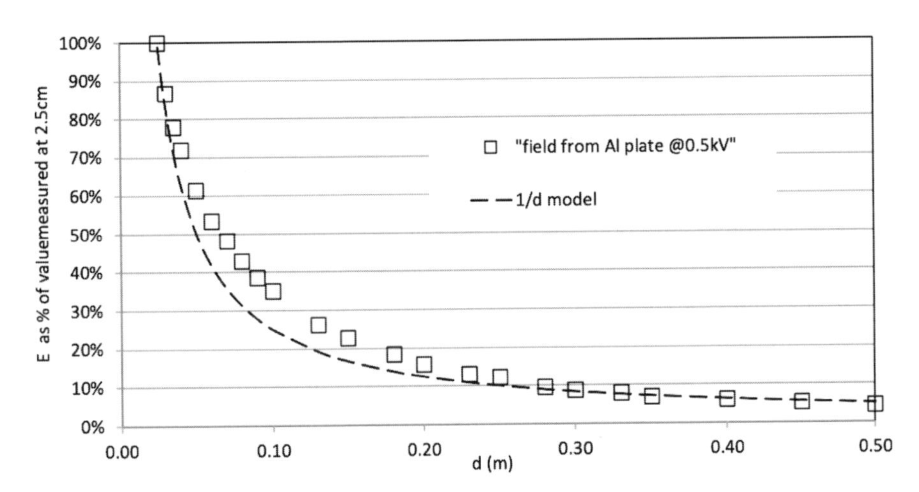

Figure 4.6 Electrostatic field meter reading variation with distance from a charged metal plate at 0.5 kV, expressed as a percentage of the value measured at 2.5 cm distance.

that any insulator that has a surface voltage >2 kV measured in this way should be kept at least 30 cm from any ESDS device. This practice ensures that the electrostatic field from the charged insulator experienced by the ESDS device is less than 7% of its close-up value.

Alternatively, standards often give a limit to the electrostatic field at the position of the ESDS, e.g. 5 kV m^{-1}. As most field meters are calibrated in terms of voltage, it is not immediately obvious how to measure this. However, any field meter calibrated in terms of voltage can be easily calibrated to find this field limit, using a metal plate raised to a set voltage to give the field as in Figure 4.5. For example, if a field meter is set up at 0.020 m from a metal plate, and 100 V is applied to the plate, the electrostatic field is 100/0.020 = 5000 V m^{-1}. The reading observed on the meter with this setup will depend on the design operational and calibration conditions of the meter. This does not matter, and it does not matter what the number is – it represents the field of 5 kV m^{-1}. Anything giving a reading above this value is producing a field above 5 kV m^{-1}, and anything less is a field <5 kV m^{-1}.

4.6 Habit 3: Reduce ESD Risks from Essential Insulators

4.6.1 What Is an Insulator?

For the purposes of this book, I define an insulator as any material or object on which electrostatic charge cannot move away quickly enough to avoid significant electrostatic charge buildup or voltage differences occurring (see Section 4.5).

In ESD work, a material of resistance above about 100 GΩ (10^{11} Ω) is usually considered an insulator. A material having resistance below this is usually considered a conductor (see Section 4.7) and can be used to control and avoid electrostatic charge buildup. However, different industries and disciplines have very different definitions or ideas about insulating material resistance.

4.6.2 Insulators Cannot Be Grounded

Inexperienced people in ESD often believe that the charge on an insulator can be controlled by grounding. This simply cannot work. The reason is obvious from my definition of an insulator. The charge cannot move from the insulator to any ground wire quickly enough to prevent static charge build up on the insulator. Sometimes, touching a ground wire to an insulator surface can reduce the surface charge level in the vicinity of the touch point by brush discharge (see Section 2.5.2). The charge level away from the touch point will often be unaffected by this as the charge cannot move across the surface easily or quickly.

4.6.3 What to Do About ESD Risk from Essential Insulators

ESD risk is in many cases a result of the electrostatic field from charged materials. The risk often arises because voltage differences arise between conductors in an electrostatic field if they are not equipotential bonded. One of these conductors is typically the ESDS device. If the ESDS device comes close enough to, or touches, another conductor at a different voltage, ESD will occur between them.

The classification of insulating items is arbitrary and a matter of opinion. The same item may be considered essential in one process and nonessential in another. For example, it may be easy to remove papers from one workstation process, but in another it may be difficult to proceed without them if they must be updated or signed on completion of process steps.

In most cases, the risk from insulators can be assessed by a simple process of evaluation such as the procedure given in Section 9.3.6. Most ESDS are not inherently sensitive to damage directly from electrostatic fields. The ESD risk is usually significant only if there are significant electrostatic fields and there is the possibility of contact between the ESDS device and another conductor within the field. If there is no contact with the ESDS within the field, it may not be necessary to control the field. If the insulator is sufficiently far from the ESDS, the electrostatic field arising at the position of the ESDS may be negligible. (Take care that the ESDS is not likely to be moved into a position where the field may be significant.) If the insulator is not likely to be handled or moved or become charged, the risk of a field arising may be negligible.

The ESD risk can in principle be controlled in several ways.

- Replacing the insulator with a grounded conductor
- Increasing the distance between the charged insulator and the position of the ESDS
- Containing the field from the insulator by shielding
- Preventing contact between the ESDS and other conductors within the field
- Reducing the charge and voltage level on the insulator e.g. by neutralization using an ionizer

The ESD control measure should be chosen as appropriate to the situation. Some possible examples are given in Table 4.3.

Electrostatic charging of the item should preferably be measured under worst-case conditions, which usually means under low-humidity (≪30% rh) ambient atmosphere. In practice, measurements may have to be done under ambient atmospheric conditions, as a humidity-controlled facility is usually not available. Nevertheless, an initial evaluation done under ambient condition at higher humidity may give useful first evaluation and should then be followed up and repeated when the weather conditions give lower humidity.

The question then arises – what level of charging can be considered negligible (Swenson 2012)? Unfortunately, this may not be easy to answer and depends on the withstand voltages of the ESDS being handled and other factors. For example, if it is charged device model (CDM) damage to the ESDS device that is of concern and if the voltages produced are lower than the CDM withstand of the ESDS device, they can be considered negligible. The voltage induced on a conductor can never exceed the voltage of the electrostatic field source. In practice, higher voltages may also be negligible, but evaluation of this may be more difficult.

Table 4.3 Some examples of essential insulators and possible ways of dealing with them.

Essential insulators	Possible ways of dealing with them
Product components and parts	Use an ionizer to reduce charge levels to acceptable value.
Parts of test jigs and fixtures that must be made from insulating materials	Treat with an antistat on regular basis and reduce charging with humidity control.
	Use an ionizer to reduce charge levels to acceptable value.
Papers that are required to be present and used during the process	Keep in static dissipative document holders.
	If required to be removed from holders (e.g. for annotation), do this on a separate work area a minimum of 300 mm from workstation where ESDS devices are handled.
	Use computer-based document displays designed to be ESD safe.
Computing equipment on the workstation	Position the equipment on a separate part of the workstation or well away from the likely position of ESDS.

Standards may also give requirements that can be used to evaluate fields from charged insulators. For example, the IEC 61340-5-1:2016 standard gives requirements that the electrostatic field at the position of the ESDS must be $<5\,\text{kV m}^{-1}$. Also, insulators charged to $>125\,\text{V}$ must be kept at least 2.5 cm from the ESDS, and if charged to $>2\,\text{kV}$ must be kept $>30\,\text{cm}$ from the ESDS. If these conditions are fulfilled, the electrostatic fields and voltages can be considered negligible for the purposes of this standard.

4.6.4 Using Ionizers to Reduce Charge Levels on Insulators

As previously explained, ionizers can be used to neutralize excess charge on surfaces. So, they can be used to neutralize charged insulators present in the EPA. For successful control of ESD risk, the limitations of ionizers must be understood.

Ionizers produce air ions at a given rate, and these drift to the charged surface at a rate determined by the electrostatic field (determined by the surface voltage) and the ion mobility (see Section 2.8). The surface charge on an insulator can be neutralized only at the rate at which these ions can arrive. The ion arrival rate and hence charge neutralization rate decreases as the surface voltage decreases (Figure 4.7). Moreover, as the ion mobility

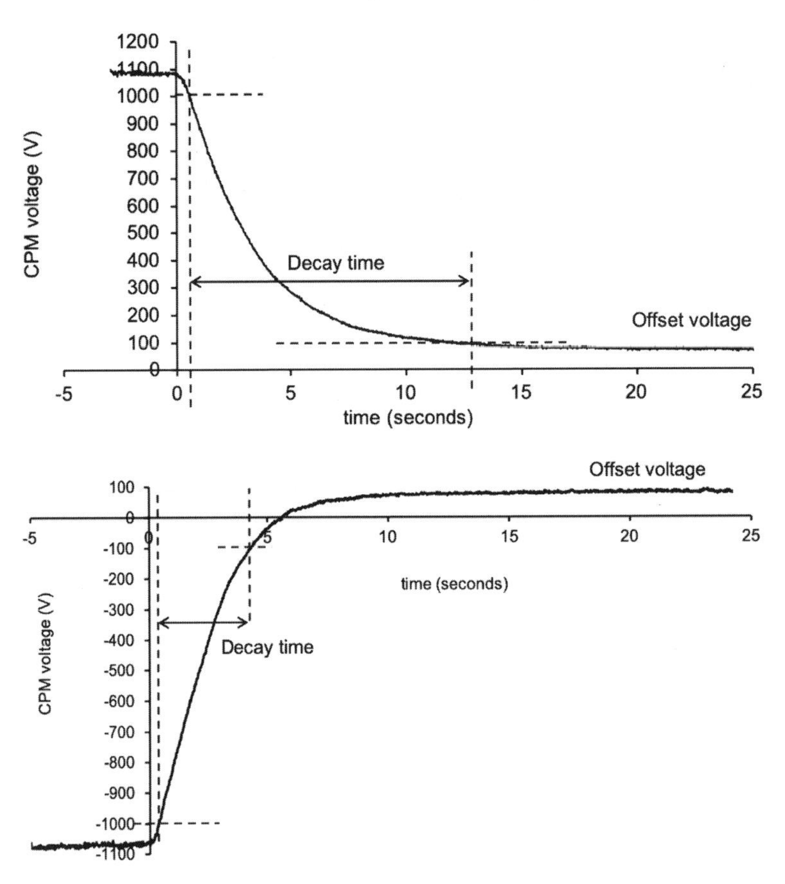

Figure 4.7 Ionizer charge decay curves showing decay time and offset voltage for positive (above) and negative (below) charge neutralization.

can be different for positive and negative ions, the charge neutralization rates can also be different, even for a well-balanced ion stream. For a poorly balanced ion stream the difference in ion concentration will also contribute to a difference in charge neutralization rate for each polarity. So, neutralization of one polarity can be significantly slower than the other polarity. As a result, as Figure 4.7 shows, it can take a significant time for a surface charge to be neutralized to a low level. In Figure 4.7, the negative polarity voltage is reduced to −100 V in <5 seconds. The positive polarity voltage requires about double this time to reduce to +100 V. In an assembly process, it may be necessary to wait several seconds until the charge has been reduced to a sufficiently low level to reduce ESD risk to an acceptable level.

The voltage decay time also varies with the position of the charged object relative to the ionizer. How it varies will depend on the type of ionizer used. This can be an important factor in choosing an ionizer for a process role. Typically, the charge decay time will increase as the distance from the ionizer increases. This is because the ion concentration in the air reduces as the ions spread out by mutual repulsion of like charges, and opposite polarity charges attract and recombine forming neutral particles.

Many ionizer types can also be quite directional in their effectiveness. For example, a fan ionizer blows ions in one general direction with the fan airstream. Its effectiveness can dramatically reduce outside the airstream.

Most electrical ionizers exhibit an offset voltage due to a small imbalance of the positive and negative ion density that they produce. The ionizer offset voltage does not usually cause any problems in neutralizing insulators. Small charge levels giving a few tens of Volts on insulators do not usually cause ESD risk except in handling extremely sensitive components. Standards often specify a maximum acceptable offset voltage or may leave it to the user to define an appropriate maximum for their application.

4.7 Habit 4: Ground Conductors, Especially People

4.7.1 What Is a Conductor?

What is a conductor? In this book, I define a conductor as any material or object that is not an insulator. I define an insulator as any material or object on which electrostatic charge cannot move away quickly enough to avoid significant electrostatic charge buildup or voltage differences occurring. So, a conductor is a material that allows charge to move away quickly enough to avoid static charge buildup.

These may seem vague nonspecific definitions, but they are deliberately so. Whether a material is considered a conductor or insulator often depends on the context or technology area. An electrical engineer might consider a material that has resistance of 1 GΩ or above to be an insulator. Even in electrostatics process hazards evaluation and prevention that might be considered the case. In ESD control work, a material of this resistance is considered a conductor and can be used to dissipate and control charge or ground conductors.

4.7.2 Conductive, Dissipative, or Insulative?

In ESD work, many people use the terms *conductive*, *dissipative*, and *insulative* as if they have generally accepted definitions. *Conductive* is often thought to apply to materials having resistance <1 MΩ (10^6 Ω). *Dissipative* is thought to apply to materials having resistance between 1 MΩ and 100 GΩ (10^{11} Ω). *Insulative* is thought to apply to materials having resistance over 100 GΩ (10^{11} Ω). Beware – these definitions are not universal. Within the 61340-5-1 and S20.20 systems of standards, these terms have standardized definitions only in some contexts. One is ESD-protective packaging materials (see Chapter 8), where an insulator is a material having surface or volume resistance greater than or equal to 100 GΩ. Even in this specific topic, the definition of a "conductive" packaging material has changed over recent years with updating standards. Because of this lack of clarity, it is unwise to specify materials in terms of conductive or dissipative. Instead, measurable parameters such as an acceptable range of surface resistance should be specified.

4.7.3 Properties of a Conductor

A conductor has some important properties for ESD control. These arise from the characteristic that charge can move around the conductor relatively easily.

- Under quasistatic conditions with no current flowing through the conductor, the voltage will be the same at every point on the surface of the conductor.
- If ESD from the conductor occurs, almost all the charge and energy stored on the conductor could be delivered from the conductor during the discharge

The first point arises from the fact that if a voltage difference occurred, a current would flow until no voltage difference is present. Equilibrium could be attained relatively quickly. In practice, the timescale in which it happens depends on the material characteristics, namely, resistivity and permittivity.

The second point arises from the fact that when a discharge is initiated from the material, the voltage at the point of discharge changes quickly as charge is conducted away. Voltage differences then occur across the material that cause currents to flow until the voltage across the material is again equalized. This can be the point at which charge stored on the material is exhausted. Because of this, a charged conductor is often a potent source of ESD.

4.7.4 Charge and Voltage Decay Time

In theory, a material has a voltage and charge decay time characteristic given by the product of the resistivity ρ and permittivity ε (see Section 2.3.3). In the case of a conductor of capacitance C and resistance to ground R, the charge or voltage decay time is given by the product RC. The shorter the decay time, the more quickly the voltage reaches equilibrium across the material. In the case of a grounded conductor or material, the time taken for the conductor or material to approach 0 V is governed by the charge decay time characteristic.

Many materials have permittivity around 10^{-11} Fm^{-1} and so resistance up to around 100 GΩ (10^{11} Ω) gives decay times of the order of a second or so. In practice and especially in manual processes, if any charge generated is dissipated within this time scale, it is unlikely to cause any significant ESD risk.

Small conductors in the workplace (e.g. hand tool bits) may have capacitance of the order of 10 pF. Resistance to ground as high as 100 GΩ will give charge or voltage decay times of the order of a second or so, which may be acceptable. Higher-capacitance items will need a lower resistance to ground to keep charge and voltage decay times acceptably short.

This situation may be different in automated processes, partly because continuous charge generation processes may be present. Machine movements may be much faster than in manual processes and so shorter charge decay times may be necessary to avoid significant charge buildup and ESD risk.

4.7.5 The Importance of Material Contact Resistance in Protecting ESDS

4.7.5.1 Reduction of Energy Delivered from Conductor in ESD

When an ESDS contacts a material and discharge occurs, the current flows through the ESDS device and also the material. A portion of the available energy in the discharge is absorbed in the material rather than in the ESDS device. The higher the resistance of the material, the higher the proportion of energy absorbed in the material and the less dissipated in the ESDS device. Where the energy dissipated in the ESDS device is important in the damage mechanism, reducing this can have a useful protective function. Having the ESDS device contact high resistance rather than low resistance material can give useful reduction in ESD stress.

4.7.5.2 Reduction of Peak Current in a Discharge

When an ESDS device makes contact with a material and discharge occurs, the peak current that flows is limited by the resistance and inductance in the discharge circuit. Where the resistance is high, this can be the main factor limiting the peak current in the discharge. Where the peak current in the discharge is important in the ESDS damage mechanism, increasing the resistance of materials with which it may make contact can significantly reduce the ESD stress.

This is a significant consideration in charged device ESD damage prevention. In CDM ESD susceptibility tests, it has been found that it is the peak current in the ESD that most often gives the device damage threshold (see Chapter 3). If the device contacts only high-resistance materials, the peak ESD current can be effectively limited to less than the damage threshold. Charged device ESD damage can be effectively prevented.

It is important to understand that the ESD peak current is limited by the resistance of the material at the point of discharge. Resistance at other parts of the circuit does not have the same effect. For example, if an ESDS is placed on a metal tray that is resting on a resistive workstation surface, a discharge between the ESDS device and the tray is not current limited by the resistance of the workstation surface. It is determined by the much lower impedance of the metal and spark. A risk of charged device ESD damage could arise. Similarly, adding resistance in a ground wire grounding a low-resistance or metal work surface would not give protection against high-current ESD (Wallash 2007).

4.7.5.3 Specification of a Minimum Material Resistance

Where these protective functions are important, it is common to specify a minimum resistance for ESD-protective materials that contact ESDS devices. Examples of this are in work surfaces and ESD-protective packaging materials. Where charged device or other similar ESD risks are a concern, a minimum material resistance on the order of 10 kΩ may be specified.

4.7.6 Safety Considerations

Where personnel are working with exposed continuous voltage sources (for example, powered system supplies) there may be a risk of shocks occurring if the person touches the power supply. In this scenario, safety can be a significant concern if high-voltage supplies are present. This can be a good reason to specify a minimum resistance acceptable in a ground path. Such safety issues may be subject to local or national safety regulations. If voltage sources are not present, it may be unnecessary to specify a minimum resistance in a ground path.

Typically, the ESD control standards do not specify a minimum resistance for safety purposes. This is because the standard is often concerned only with ESD control and regards safety as a matter for user specification. They may discuss the topic in user guides associated with the standard.

4.7.7 Elimination of ESD by Grounding and Equipotential Bonding

ESD occurs because two objects have a sufficiently high voltage difference between them to cause breakdown of the air gap between them (if any) and a current to flow. It follows that if we can keep voltage differences low, we get the following benefits:

- If there is insufficient or no voltage difference, ESD cannot occur.
- If ESD does occur, the energy, peak current, charge transferred, and other potentially damaging parameters are reduced in level.

If we connect two conductors electrically and they are at different voltages, a current will flow between them briefly until they are at the same voltage. At this point, current flow will stop, providing there is no externally applied voltage difference or current flow. At the initial time of contact, ESD has of course occurred, but thereafter no ESD can occur between them while they are at the same voltage.

Connecting two conductors to make sure that they are at the same voltage is called *equipotential bonding*. This is the main means of preventing conductors from becoming charged and an ESD source. In practice, we often equipotential bond all conductors to the earth. In this case, it is called *grounding* or *earthing* the conductors. This is particularly useful in many EPAs because they may already contain equipment that has been earthed for other reasons such as electrical safety.

4.7.8 Understanding the Grounding (Earth) System

4.7.8.1 Types of Ground

There are various ways in which EPA grounds may be implemented in practice. The main types are

- Equipotential bonding
- Electrical safety earth
- Functional earth

It is often thought that it is necessary to have a connection to earth for the elimination of voltage differences and ESD sources in an EPA to be successful. This is not so – we only need to have equipotential bonding of the conductors in the area. It would be perfectly possible to have successful control of voltage differences between conductors by equipotential bonding in an aircraft or other situation with no contact with earth. For this reason, in modern ESD control standards, equipotential bonding is treated equally with grounding. The term *grounding* is often used to include equipotential bonding as an alternative to other grounding methods, meaning connecting an item to the designated ESD ground.

In many EPAs, mains electrical safety earth is present, and many types of equipment are already connected to it for electrical safety. So, it often is most convenient to use mains electrical safety earth as the EPA earth for ESD control purposes. All items of noninsulative EPA materials and equipment are then electrically bonded to this.

In some facilities, mains electrical safety earth may not be available or for some reason it may be undesirable to bond to this earth. A separate "functional" ground such as an earth rod sunk into the ground may be used.

It is normally undesirable to have two different and separate grounds present in an EPA. If this occurs, they may be at different voltages and could become a serious source of ESD risk. All earths (grounds) in an EPA should be electrically bonded together to make sure that no significant voltage differences can occur between them.

4.7.8.2 The Grounding System

Reliable grounding of an item requires that a continuous electrical connection is established and maintained between the item and ground. This ground path may rely on several items of equipment or materials. Examples are

- A person grounded through ESD footwear and an ESD control floor
- A cart, chair, or rack grounded through an ESD control floor
- A hand tool, grounded through a person's gloved hand and body via a wrist strap

The requirements of each part of the system must be considered as part of the grounding system for all the items that must be grounded. Often the grounding requirements of one system may dictate the specification of a key part of the system. For example, the resistance requirement for a floor may be specified mainly by the need to achieve a chosen maximum resistance from a person's body to ground through footwear and flooring.

For reliable grounding to be achieved, the performance of each part of the system must be maintained under all the circumstances where grounding must be maintained. This means that each part of the system, or the system as a whole, must be tested from time to time. This is discussed further in Section 4.10.

4.7.9 Grounding Personnel Handling ESDS Devices

4.7.9.1 Basic Requirements for Grounding Personnel Handling ESDS Devices

People generate electrostatic charge continuously as they move around the environment. This is because their feet or clothing make and separate contact with other materials and surfaces as they move around. Many modern ordinary shoes have soles made of insulating materials, and outer clothes may also be made of insulating materials. Charges generated by contact with other materials can be conducted to the body or induce voltages on the body that can lead to ESD.

A basic requirement for grounding personnel is to control body voltage to make sure that it does not reach a level where damaging ESD could occur when ESDS device are touched. Current practice is to keep body voltage lower than the human body model (HBM) withstand of any ESDS devices that are handled. Current standards aim to safely handle 100 V HBM devices, and so the body voltage is controlled to less than 100 V. When devices of lower withstand voltage are handled, body voltage may need to be maintained below a lower level.

Considering the simple electrical model of Figure 4.3 for a person, the main way in which body voltage can be controlled is by providing a ground path with sufficiently low resistance from the person's body to ground through an appropriate grounding system. The two main grounding systems used are

- Wrist strap connected to the person's body and to an earth bonding point provided for the purpose
- ESD control footwear when standing on an appropriately specified ESD control floor

Other systems are occasionally used, with the principles remaining the same. The maximum resistance from body to ground is usually specified so that under all circumstances expected, body voltage generated is kept below a specified level.

4.7.9.2 Grounding Personnel via a Wrist Strap

For wrist strap grounding, it was established many years ago that the body voltage was maintained below 100 V if the resistance from body to ground was less than 35 MΩ. Many standards have therefore adopted this as an upper limit for an ESD control program handling devices down to 100 V HBM.

The wrist strap system in practice consists of a wrist band in contact with the skin, a grounding cord, and an earth bonding point connected to EPA ground. For the system to work correctly, all the parts must function reliably. Standards often specify resistance limits for the individual parts of the system that may be tested separately (e.g. from hand to wrists strap groundable point when wrist strap is worn, or resistance to ground of the earth bonding point) (see Section 6.5.12).

4.7.9.3 Grounding Personnel via Footwear and Flooring

For many years, the 35 MΩ limit used for wrist straps was also accepted as a limit for resistance to ground in grounding personnel via footwear and flooring. More recently, practice has moved away from this for various reasons.

For a standing person, much of the charge generation leading to body voltage is due to the contact between the footwear sole material and the floor material. Referring again to Figure 4.3, we can see that the charge generation rates combined with resistance to ground will have a strong effect on voltage buildup on the body. The triboelectric charge generation properties of the footwear-flooring combination have a major effect on body voltage, and different footwear and flooring materials with identical resistance can give very different body voltage performance. It is possible to select a footwear-flooring combination that gives body voltage within the required limits due to lower charge generation, despite their achieving a higher resistance from body to ground. So, modern ESD programs often allow this higher resistance level providing it has been demonstrated that the footwear and flooring used achieve the required body voltage performance.

It is important to realize that changing the footwear or flooring types for others of the same resistance would not necessarily give the same body voltage performance as the charge generation properties would be different. So, once a footwear-flooring combination has been evaluated and selected, it is necessary to use only this combination, or another that has likewise been shown to give the required performance. Unfortunately, the resistance of the footwear and the flooring measured separately have been shown to be a poor predictor of the footwear and flooring performance in combination (Smallwood et al. 2018).

If several floor types are to be used with the same footwear, then the performance of the footwear must be evaluated with each type of floor. Similarly, if several types of footwear are to be used, then each type of footwear must be evaluated in combination with the floors with which they will be used. Quite different performance can be found with different floor materials and at different atmospheric humidity (Figure 4.8). Most standard ESD control programs require the maximum body voltage produced during a walk test to be less than 100 V. This may be reduced when handling components less than 100 V HBM withstand voltage.

4.7.9.4 Grounding Seated Personnel

When seated, most people are likely to take their feet from the floor for some of the time. When they do so, contact between footwear and flooring is broken and so grounding by this method is unreliable. For this reason, most ESD control programs require seated personnel to be grounded via wrist straps. ESD control seating is not usually regarded as a reliable means of grounding personnel sitting on the seat (see Section 4.7.10.5).

4.7.10 Grounding ESD Control Equipment

4.7.10.1 General Considerations

The high resistance levels allowed for grounding conductors in ESD control may often surprise engineers new to ESD work, especially electrical engineers. Achieving a resistance to ground of around 1 GΩ, or in some cases even higher, may be sufficient to effectively ground items for static control purposes. In practice, grounding may be provided by a material (e.g. floor) rather than an installed permanent wire connection. As charge generation currents are small (microamps or less), grounding wires do not have to be thick to withstand the current. The reliability of the ground path is, however, an important consideration. This reliability is often governed by factors such as the following:

- Robustness of the grounding system components
- Contamination of contacting surfaces (e.g. floor, footwear soles, grounding wheels on chairs and carts)
- Human factors (e.g. deliberate or inadvertent unplugging of a grounding connector)

4.7.10.2 Work Surfaces

The purpose of providing a grounded static dissipative work surface is twofold. First, the work surface material itself should not become charged and give electrostatic fields that could lead to ESD risk. Second, the work surface provides a useful way of draining charge

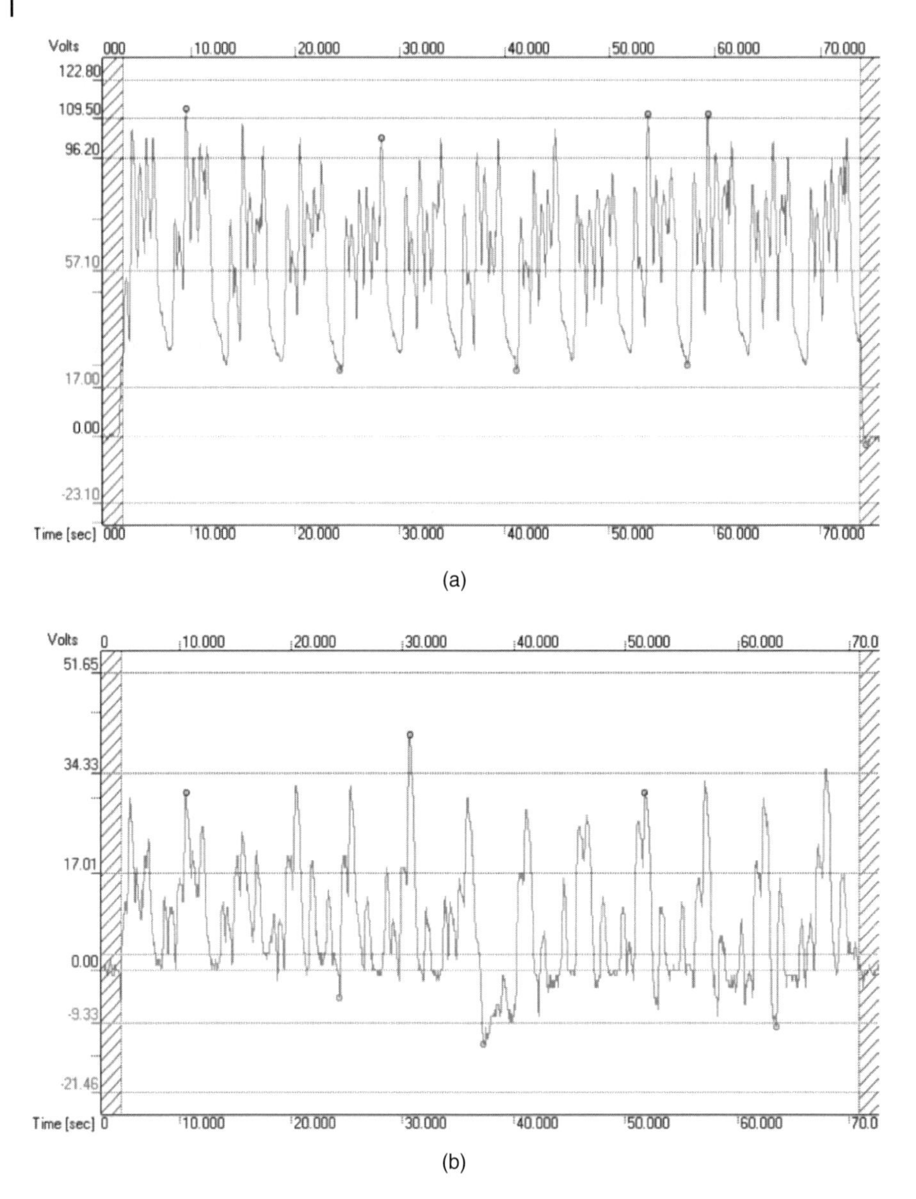

Figure 4.8 (a) Footwear 10 MΩ, "dissipative" floor 10 MΩ resistance to ground, 15% rh; (b) footwear 10 MΩ, "conductive" floor 900 kΩ resistance to ground, 15% rh; (c) footwear 10 MΩ, "standard vinyl tile" floor, 15% rh; (d) footwear 10 MΩ, "standard vinyl tile" floor, 50% rh. Body voltage generated using ESD control footwear with different types of flooring. Source: D. E. Swenson.

from any noninsulative material or object that is placed on it. This may include tools and components including ESDS devices.

Any isolated (nongrounded) conductor that is placed on the work surface can be expected to initially be at a different voltage. An ESD event will occur as the conductor approaches

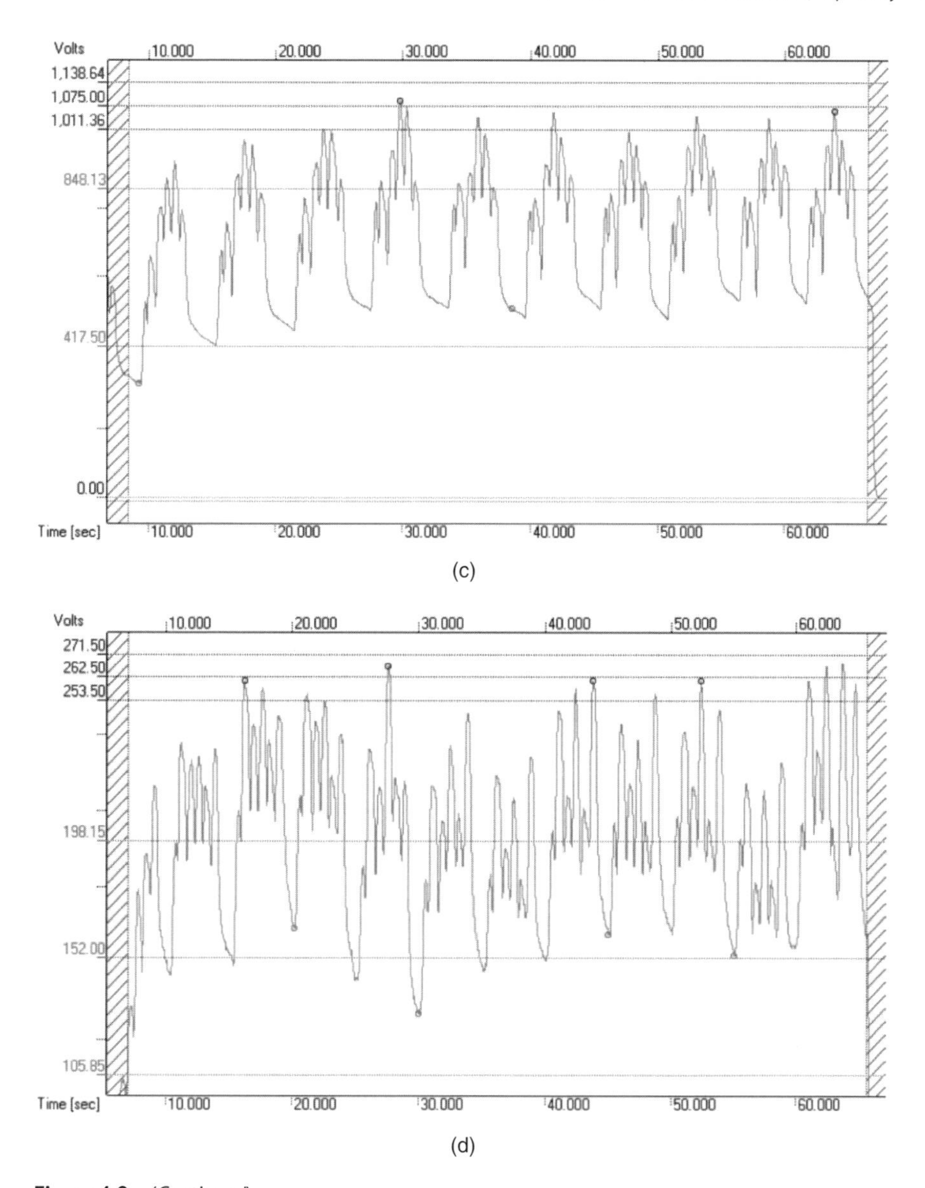

Figure 4.8 (*Continued*)

or touches the work surface. If the conductor is part of an ESDS, this gives a risk of charged device ESD damage. For this reason, where the ESDS devices handled are susceptible to charged device ESD damage, the work surface should be chosen to have a high surface resistance to limit the peak ESD current. Where the ESDS devices handled are not susceptible to charge device ESD, metal or low-resistivity surfaces may be used.

Work surfaces are usually either hardwired or connected via earth bonding plugs to the EPA ground system. A work surface material with point-to-point resistance and resistance from surface to ground less than 1 GΩ is normally considered adequate for ESD control in most EPAs. A minimum point-to-point resistance may be specified for safety or charged device ESD damage prevention. (See Section 4.7.5.3)

4.7.10.3 Floors

ESD control floors are often provided to give a convenient way of grounding personnel as well as carts, racks, chairs, and other free-standing equipment on the floor. The ESD control function of the floor is often misunderstood – it must operate as a system with all the items that it is intended to ground. So, in specifying the characteristics of a floor, it is necessary to consider the items that will be grounded by it. Typical grounding systems using the floor include the following:

- An operator's body grounded through footwear and flooring
- A cart (trolley) grounded from its surface through the chassis, wheels, and floor
- A rack grounded from the shelf surface through the frame, feet, and floor
- A seat grounded from the seat surfaces through the frame, feet, or wheels and floor

The resistance to ground of the system includes the resistance of all the parts of the ground path including the item being grounded, the floor, and the resistance of the contact between them. So, it might be expected that if the resistance of the individual parts of the ground path are measured, the total resistance to ground over the system would be the sum of the parts.

Unfortunately, this is not generally true, largely because the contact resistance between the floor and the item being grounded can be higher or lower than expected and can vary considerably. The contact resistance with the floor generally depends strongly on the area and pressure of the contact. The contact areas and pressures between the item and the floor can vary considerably. These pressures and areas are generally quite different from the pressure and area of a measurement electrode standing on the floor. So, the contact resistance between a shoe, wheel, or equipment foot and the floor are likely to be very different than that of a measurement electrode. Contamination or coatings on the contacting surfaces (e.g. dirt or polish) can also make a big difference to the effective contact resistance.

ESD control standards typically specify a maximum resistance from the floor surface to ground of 1 GΩ. Many do not specify a lower limit, as there is no minimum resistance requirement for ESD control purposes. In some applications, a minimum floor resistance to ground may be desirable, e.g. for safety in the presence of high voltages. The minimum body to ground resistance of the footwear and flooring system should be measured in qualification tests with the person standing on the ESD control floor wearing the footwear with which it will be used. This should be considered an essential part of the floor specification process.

Similarly, to a first approximation, the resistance to ground from the surface of an item grounded through feet or wheels through the floor can be assumed to be the sum of the resistance from surface to feet (wheels) and the floor surface to ground. In practice, this is often not correct for various reasons, including that the contact between feet (wheels) and floor surface can have significant resistance that is difficult to predict.

Nevertheless, if the resistance to ground of an item that is grounded via a floor is required to be below a certain value, the floor resistance to ground should also be below that value. It is often best to select a lower resistance from floor surface to ground to give some margin for added resistance from contact resistance. For example, if the resistance from body to ground of a person standing on a floor is required to be below 35 MΩ, the floor should be selected to give installed resistance to ground less than 35 MΩ. In practice, selecting a target value of, say, 10 MΩ gives some "headroom."

It is perfectly possible to have an effective EPA that does not have an ESD control floor, if one is not needed for grounding personnel or equipment. For example, a single workstation EPA in which a standing operator is grounded via a wrist strap may not need an ESD control floor.

Two such workstations nearby in the same room would effectively be two separate EPAs with uncontrolled space between them (Figure 4.9). An operator needing to transport an ESDS from one workstation to the other would need to consider using ESD-protective packaging to protect the ESDS. The operator would not be grounded when moving between the workstations, and this would lead to ESD risk. In contrast, if an ESD floor is provided, linking the two workstations, the operator can remain grounded when moving between the workstations. The two workstations can now be part of the same EPA, and moving an ESDS from one to the other does not take it through uncontrolled space. The ESD risk is greatly reduced, and the handling of the ESDS is made more convenient.

4.7.10.4 Carts, Racks, and Other Floor Standing Work Surfaces

Carts, racks, and other floor standing equipment may be grounded through floor on which it stands. Equipment that is not mobile may alternatively be connected to ground via snap-on ground cords or permanent hardwired connections.

Carts and racks on which unprotected ESDS devices may be placed are usually considered to be subject to the same point-to-point and resistance to ground requirement as work surfaces. Carts and racks that are to be grounded through the floor must usually be designed for the purpose, as they must have a continuous connection from the shelf surfaces to ground through the wheels or feet.

Equipment that is not designed for EPA use often includes insulating plastic joints or other parts that may isolate shelves from the frame and supports. Wheels and feet will often be made of insulating plastic material. In carts and racks designed for EPA use, the joints, wheels, and feet are made of conducting materials such as conductive or dissipative plastics or metal. It can be difficult to tell apart visually equipment or materials that are designed for EPA use from similar item that are not.

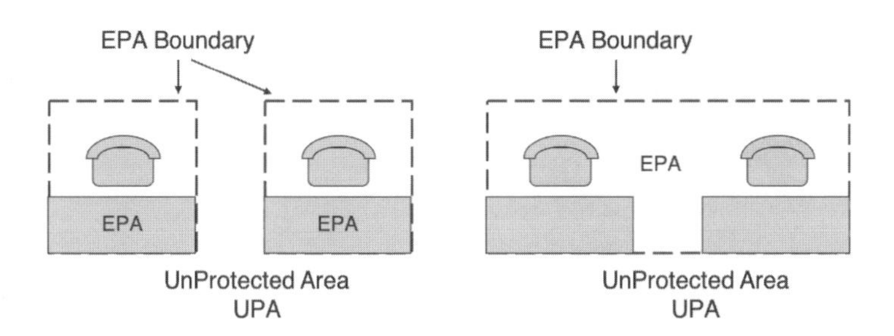

Figure 4.9 (left) Two EPA workstations separated by a UPA, (right) joined as single EPA using ESD control floor.

Wheels and feet are often prone to contamination from an accumulation of dust and dirt. This increases the resistance to ground over time and may ultimately cause it to exceed the specified resistance requirements. Regular cleaning may be required to bring the equipment back within specification.

Drag chains are sometimes used to ground carts, but these can be particularly prone to dirt collection and can be unreliable.

An example of a carts is shown in Figures 7.9.

4.7.10.5 Seats

It is a common misconception that ESD control seats are intended to ground personnel sitting in them. Although seats are sometimes used to do this, many are not designed for this purpose. They are generally designed to prevent the seat from becoming highly charged and an ESD risk (Figure 4.10). ESD control seats also reduce the charging of personnel sitting on the seat.

Like carts and racks, seats designed for EPA use must have a continuous conducting path from the various parts through the legs and feet or wheels to the floor.

Wheels and feet are often prone to contamination from an accumulation of dust and dirt. This may increase the resistance to ground over time and may ultimately cause it to exceed the specified resistance requirements. Regular cleaning may be required to bring the equipment back within specification.

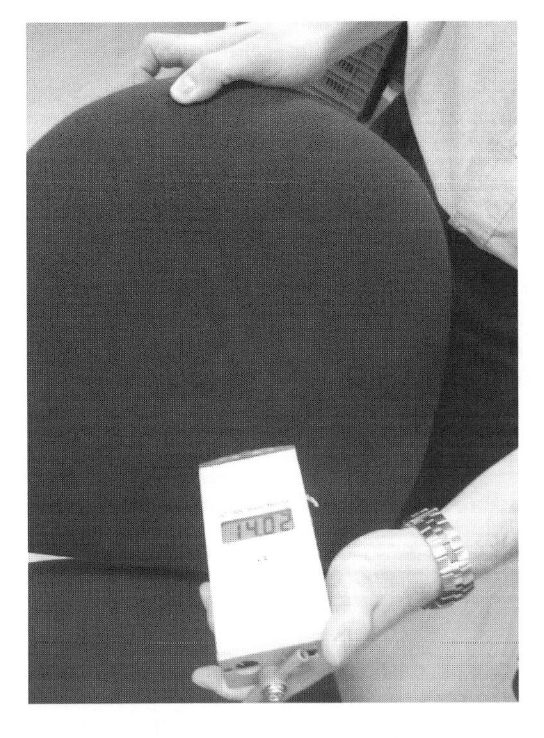

Figure 4.10 An ordinary seat can give high electrostatic fields, in this case showing 14 kV surface voltage.

4.7.10.6 Tools

Ordinary tools often have insulating handles and metal bits or other parts that are electrically isolated. These can reach a high voltage by tribocharging or induction and cause risk of ESD to any ESDS devices that they may contact.

Hand tools designed for EPA use typically have noninsulative handles that electrically connect the bit and any other metal parts to the user's hand. Any charge developed on the tool is dissipated safely to the user's hand and via their body and personal grounding to ground.

If a glove is worn, the glove must also be noninsulating to allow the tool to be grounded through the glove to the operator's hand (Figure 4.11).

Typically, the important characteristics for an ESD-protective tool include the resistivity of the material that may contact the ESDS devices and the resistance to ground via the normal ground path. A surprisingly high resistance to ground may be tolerable. For example, a tool that has capacitance 10 pF would have a charge decay time around one second if grounded with a resistance to ground of 10^{11} Ω. This would in most cases be sufficient to ensure insignificant ESD risk.

If low CDM ESD withstand voltage devices are to be handled, the resistivity of the material of any part of a tool that may contact an ESDS may need to be sufficiently high to limit ESD currents during any contact discharge (Wallash 2007).

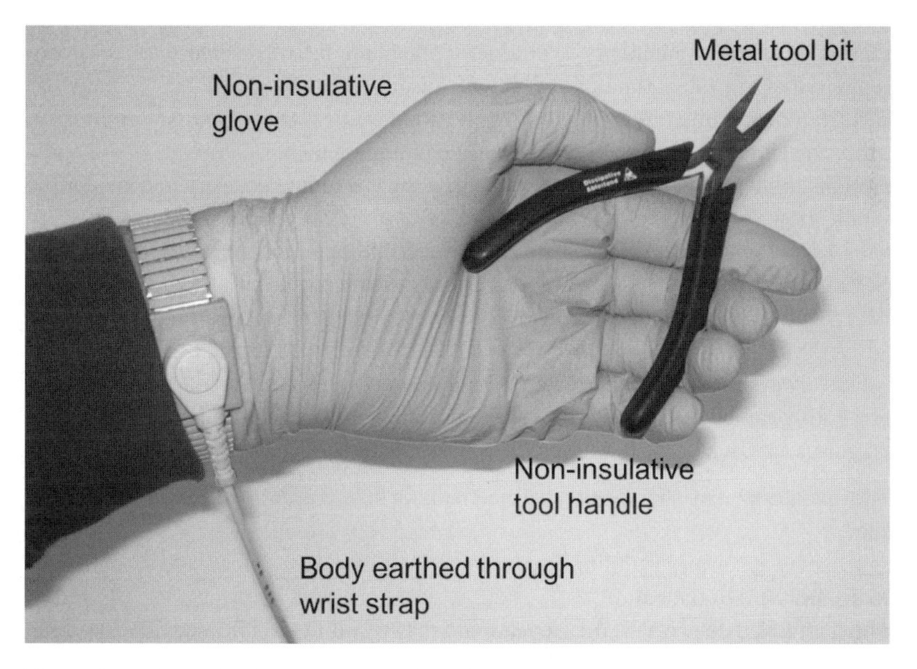

Figure 4.11 A hand tool designed for EPA use is grounded via the user's hand. If a glove is used, it must also be noninsulative to maintain the ground path.

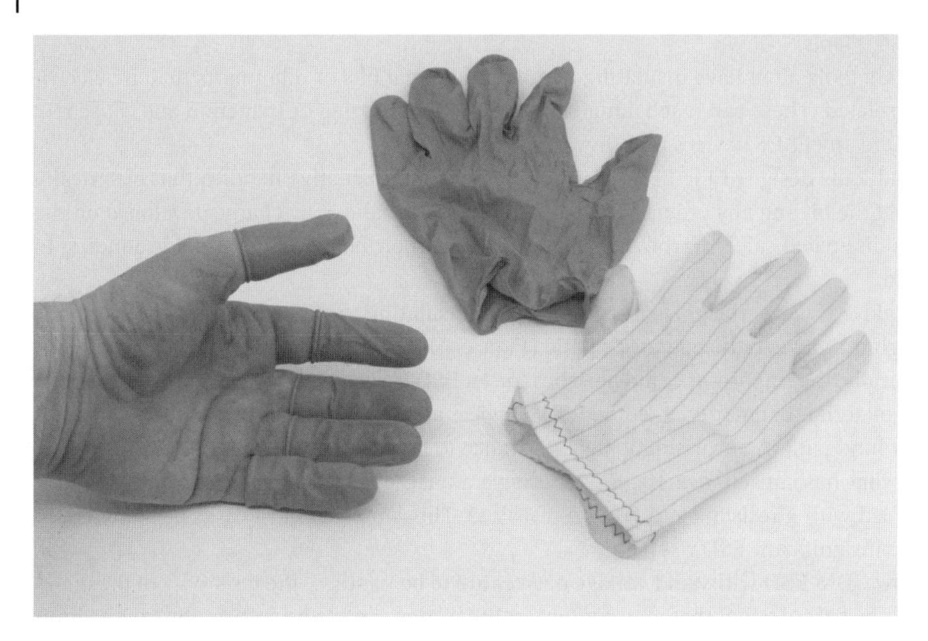

Figure 4.12 Gloves and finger cots.

4.7.10.7 Gloves and Finger Cots

In some operations or processes, the operator may need to wear gloves or finger cots to protect their hands or the components or product while being handled (Figure 4.12). Gloves are often required to protect the hands from chemicals, and oven gloves may be required to protect from heat while unloading product from an oven. If these are made from insulating materials, ESDS devices or tools held in the hand may risk becoming charged and a source of ESD to themselves or other ESDS devices. So, it is often necessary to consider the electrostatic properties of the gloves and finger cots used in these processes.

In some cases, gloves may be necessary to protect the operator for safety reasons, for example from heat or chemicals or high voltages, rather than to protect the product. In this case, the gloves are personal protective equipment (PPE). In Europe, use of PPE is subject to directives, and compliance with these is a legal requirement. Such safety and legal requirement always take precedence over ESD control, although often with careful thought, ways of working can be devised that meet legal requirements and minimize ESD risks.

When gloves are used while handling ESDS devices, the main ESD risks are

- The glove material may charge and give electrostatic fields very close to any ESDS device handled.
- The glove may act as an insulating barrier to prevent grounding of any tools, objects, or ESDS devices held in the hand.
- Touching an ESDS device with the gloved hand may result in the ESDS device becoming charged by triboelectrification.

Different materials will typically give different performance regarding these concerns. The risk of electrostatic fields can be eliminated, and grounding of handheld items can be ensured by use of static dissipative materials in the glove. This, however, will not eliminate the possibility that an ESDS device may become charged when handled due to tribocharging. The amount of charging of any item handled will depend on the choice of material used in the outer surface of the glove that comes into contact with the items handled.

4.7.10.8 ESD Control Garments

Many ordinary outer garment materials can easily become highly charged. These can be a source of strong electrostatic fields that contribute to ESD risks. Many ESD programs use ESD control garments to provide an outer clothing layer that does not give external electrostatic fields and acts to contain the electrostatic fields from clothing within (Figure 4.13).

It follows that these garments should cover all the clothing within. This is particularly important for the sleeves and front of the garment as these are normally the areas closest to the ESDS device. It also follows that coats should be fully fastened, as the underlying clothing is exposed by unfastened areas.

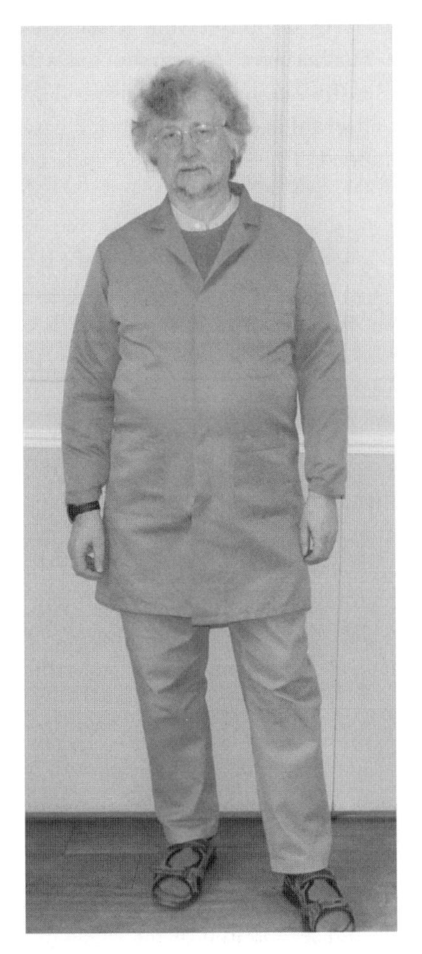

Figure 4.13 (left) ESD control garments should cover the clothing of the arms and body, (right) incorrectly worn.

Some types of garments are designed to maintain contact with the wearer's skin, usually at the wrists. This can be an effective way of grounding the garment material via the wearer's body (which must always be grounded).

An ESD garment should not contain any exposed ungrounded conductors of any significant size and capacitance that could possibly become charged and a source of ESD risk.

4.7.11 What If a Conductor Cannot Be Grounded?

Grounding a conductor is by far the best way of preventing it from becoming charged to a voltage that might cause ESD risk. If this is for some reason impractical and the conductor is likely to contact an ESDS, then it may be necessary to find some other way of reducing any voltage arising.

One possibility is to use an ionizer to neutralize charge on the conductor. This can work but suffers some disadvantages compared to grounding. Neutralization can be slow and may not adequately control charge produced by triboelectrification or voltage changes induced by fast-changing electrostatic fields nearby. Furthermore, the conductor will become charged to the ionizer offset voltage. For very sensitive components, this residual charge may be sufficient to cause ESD risk. The ionizer offset voltage will typically increase as time goes on, unless the ionizer is properly maintained. These issues are discussed further in Section 4.6.4.

4.8 Habit 5: Protect ESDS Using ESD Packaging

4.8.1 Don't Take Ordinary Packaging Materials into an EPA

Ordinary packaging materials such as paper, cardboard, polythene bags, and bubble wrap and polystyrene foam are designed for physical functions such as containment and physical protection of items. They are often made from highly insulating materials such as plastic or of materials that have unknown electrostatic charging and charge dissipation properties. These properties may vary widely between apparently similar materials (e.g. different papers or cardboards) and may be highly dependent on atmospheric humidity conditions.

These materials are often static generators or can isolate other items from ground, or their properties may at best be unknown and dependent on the weather and climate. In this case, so will be the ESD risk arising from their presence near an unprotected ESDS device. They are likely to have little or no ability to protect ESDS devices against ESD from external sources such as charged personnel or objects. If present in an EPA, electrostatic control is severely compromised. For this reason, ordinary packaging materials are best kept out of an EPA, or at least well away from any workstation or process in which ESDS devices are handled unprotected.

It follows that special ESD-protective packaging is needed for two reasons – first to protect ESDS from ESD risks when in an uncontrolled area and second to provide packaging materials that can be used within an EPA without compromising electrostatic control or introducing ESD risk.

4.8.2 The Basic Functions of ESD Packaging

ESD packaging can have several ESD control functions. First, it may be required to protect any ESDS device within from external electrostatic fields or direct ESD from sources in an external uncontrolled environment. Second, it must not itself provide ESD risk to the ESDS devices within. Third, it will be necessary for the packaging to go into an EPA, and it must not cause ESD risk to any unprotected ESDS devices in the EPA. Last, the internal surfaces may be required to prevent discharging of on-board batteries on a circuit board within the package. So, ESD packaging is typically designed with one or more of the following functions:

- Inner and outer surfaces minimizing electrostatic charging of the surface and any ESDS device in contact with it
- The surfaces dissipating electrostatic charge
- Surfaces in contact with ESDS devices having high enough resistance to minimize discharging of on-board batteries
- Shielding against electrostatic fields
- A barrier to direct ESD

These functions, and ESD packaging, are discussed further in Chapter 8. ESD-protective packaging materials can have one or more or all of these properties. The properties are in some cases relatively independent and in some cases highly dependent on atmospheric humidity.

4.8.3 Open ESD Packaging Only Within an EPA

At the beginning of this chapter we said that a key part of our ESD control strategy is that in UPAs, ESDS are protected by enclosing them within ESD-protective packaging. Unprotected ESDS should be handled only in an EPA. It follows that ESD-protective packaging should never be opened in an UPA.

Typical processes in which this principle is important are the goods in and stores areas and during transport. At goods in, shipments of components, assemblies, and materials typically arrive packaged within ordinary packaging materials for physical protection during transport. Within this ordinary packaging there may be an ESDS device within ESD-protective packaging, or there may be some non-ESD-sensitive item.

ESD-protective packaging is normally marked on the outside with symbols or warnings that it contains ESDS devices. At goods in, ordinary packaging may be stripped off only until these markings identify ESD-protective packaging. Typical markings in current use are discussed in Chapter 8.

ESD-protective packaging should never be opened in an UPA. If further opening of the ESD-protective packaging is required to access the ESDS, e.g. for inspection, the package must be taken into an EPA.

Ordinary packaging materials should not be taken into an EPA, as the act of removing ordinary packaging materials is likely to charge it highly. This would then cause ESD risk to any ESDS nearby.

In a stores area, non-ESDS items are often stored adjacent to ESDS items that are protected within ESD-protective packaging. There is no problem in doing this, providing the ESD-protective packaging is not opened in an UPA. If it is necessary to open the ESD-protective packaging to inspect, count, or remove components for a kit, this must be done in an EPA.

Kits of parts are often made up for product assembly processes. Where a kit is made up to go into an EPA, then every item that goes into the kit must be compatible with EPA requirements, even though the assembly operation may be done in an UPA. If this is not done, then the kit will introduce noncompliant items or materials and ESD risk into the EPA. So, the tote box containing the kit must itself be of ESD packaging material. ESDS devices will clearly be packaged in ESD-protective packaging. However, there may be many non-ESDS parts, assemblies, or materials also included in the kit. If these must be organized or protected in bags or other packaging, it must be ESD packaging suitable for EPA use. This packaging does not need to have an ESD-protective function to its contents, but it must not cause ESD risk to unprotected ESDS devices in the vicinity.

4.8.4 Don't Put Papers or Other Unsuitable Material in a Package with an ESDS Device

The materials of the inner surfaces of ESD packaging that contact ESDS devices have carefully designed and controlled properties to minimize ESD risks. In contrast, most grades of paper have unknown and poorly controlled electrostatic properties and become insulative at low humidity (see Chapter 8). It therefore is nonsensical to compromise the ESD risk control of ESD-protective packaging by including papers in the package in contact with the ESDS devices.

4.9 Habit 6: Train Personnel to Know How to Use ESD Control Equipment and Procedures

4.9.1 Why Train People?

The greatest ESD risk in manual handling and assembly processes is often from ungrounded personnel handling ESDS. Untrained personnel are likely to make actions that cause ESD risk, such as opening ESD-protective packaging and handling the contents outside an EPA. Untrained personnel are unlikely to be aware of the procedures and equipment to be used inside an EPA and how to use them correctly. If aware of ESD risks and control measures, they are likely to misunderstand and undervalue them. They may not know the differences between ESD packaging, tools, and equipment compared to ordinary packaging, tools, and equipment. Because of this, they are likely to bring noncompliant packaging, tools, and equipment into an EPA.

Conversely, a trained and aware person can avoid these mistakes. Furthermore, they are likely to recognize noncompliant packaging, tools, and equipment and can be trained to remove them from areas where they cause ESD risk. They can be trained to perform simple tests on essential equipment such as wrist straps and footwear and take corrective action when they are found faulty.

It is probably no exaggeration to say that untrained personnel can be the greatest ESD risk, and they can be converted into the first line of defense against ESD damage by effective

ESD training. So, ESD training can be an excellent investment. ESD training is discussed in greater depth in Chapter 12. An overview is given here.

For those involved in implementing ESD control, training and education can elevate the topic from "magic" to application of sound engineering principles. This helps them lay the foundation of an effective and efficient ESD control program through understanding ESD controls and systems.

4.9.2 Who Needs ESD Training?

The biggest ESD risk in manual handling and assembly operations is from charged ungrounded personnel touching the ESDS. It follows that it is essential that personnel working in an EPA reliably test and use their personal grounding equipment (wrist straps and footwear). Furthermore, personnel working in the EPA can be trained to prevent noncompliant materials and equipment from inadvertently being brought into the EPA and to recognize and correct common ESD control problems and noncompliances. Carefully trained personnel working in the EPA can, instead of being the greatest ESD risk, become an important part of the first line of defense against ESD damage.

Personnel who enter an EPA need to have ESD training relevant to their activities, before they enter the EPA. This can vary from minimal (e.g. for visitors) to extensive according to the person's role. As a minimum requirement, all personnel who handle ESDS must have training in use of the ESD control equipment and materials they will use. It may be beneficial for other personnel to have some ESD training, and some examples follow.

Managers may visit the EPA for a variety of reasons including showing visitors around. They may also have budget responsibility for expenditure related to ESD control. They will therefore need a working knowledge of ESD issues and the value of ESD control practice, relevant to their responsibilities. They may need sufficient knowledge of use of ESD control equipment (e.g. use of foot grounders, ESD garments) and EPA practices (such as to refrain from touching ESDS) for themselves and for escorting visitors.

Audit and test personnel are likely to need in-depth knowledge of ESD test and audit practice to test and audit the facility for compliance verification.

The ESD coordinator and other personnel responsible for the ESD program may need continuing development and update of their knowledge and skills. This may include how to specify, use, and evaluate ESD control materials and equipment, as well as comply with standards.

Purchasing personnel may not need to enter a EPA, but they may need to source ESD control equipment and materials as well as ESDS components for production requirements. They may need an awareness of ESD control practice to support these responsibilities.

Some subcontractors may need to enter the EPA to fulfill their contracts. These will need to have training in compliance with the company ESD control procedures. Other subcontractors may supply or process ESDS product or components. They will need to understand any ESD control requirements that are placed on them in handling the product or components they handle.

If cleaners are required to clean within the EPA, they will need instruction on the cleaning products, materials, and processes they are to use. They may also need instruction not to touch certain items or to avoid certain actions such as placing cleaning equipment in certain places or unplugging ground points to plug in cleaning equipment.

Visitors may need a simple set of instructions as to what they may or may not do, as well as instruction on using personal grounding equipment or garments.

Maintenance and facilities personnel may need specific instructions on how to undertake maintenance and other work in EPAs without causing ESD risk. There may be contention between ESD requirements and safety requirements (e.g. electrician's use of insulating footwear) that requires specific instructions to resolve. Safety requirements should always take precedence over ESD requirements.

4.9.3 What Training Do They Need?

The appropriate training content for personnel from different job roles is likely to vary according to their role and activities. This is discussed further in Chapter 12.

The training given should be clearly applicable to the situation in which the trainee works. General commercially available ESD training materials can be useful but can sometimes appear to be of little relevance if prepared for a situation very different from the trainee's experience and workplace.

4.9.4 Refresher Training

Refresher training is needed to reinforce understanding of ESD control practices and remind them of aspects they may have forgotten. It may also be necessary to give updates on changes in equipment, materials, or procedures used. These can occur due to changes in production techniques or processes, standards, or organization practices.

Refresher training can be an opportunity to redress aspects of ESD control that have been found to go wrong or be misunderstood in practice, e.g. incorrect use of personal grounding equipment or garments, or lack of discrimination between compliant and noncompliant packaging materials.

4.10 Habit 7: Check and Test to Make Sure Everything Is Working

4.10.1 Why Do We Need to Check and Test?

There are two main reasons to check and test equipment and materials. Before approval for use, they need to be specified and tested according to the specification to make sure that they will work as intended for the duration of their intended working life. After installation and commissioning, they need to be tested to ensure they continue to work as intended. Maintenance of equipment is necessary for prevention of equipment failures and as part of remedial actions.

Other checks may be needed of practice or procedure (e.g. correct wearing of wrist straps or foot straps).

Compliance verification is further discussed in greater depth in Chapter 9. An overview is given here. The test methods commonly used are discussed in Chapter 11.

4.10.2 What Needs to Be Tested?

Everything on which effective ESD control depends should be tested in some way. Each item or aspect to be tested needs to have defined a test methodology, pass criteria, a frequency of testing, and procedures for recording, disseminating, and keeping results.

Many of these tests will be of the nature of a measurement made of equipment, material, system, or installation characteristics. Examples of this are the resistance to ground from the surface of a floor or bench top and the resistance from a person's body to the end of a wrist band cord while the wrist band is worn.

4.10.3 ESD Control Product Qualification

Any ESD control equipment should be tested before selection for use to ensure that it will do the job that it is intended to do. Suitable tests methods must be found, and pass criteria decided.

In practice, for common ESD control equipment, suitable tests and pass criteria are given by the ESD control standards. ESD control product qualification can then be done using data sheet information or tests according to these standards.

ESD control equipment product qualification tests are often done only once during the process of selecting the product for use. Different tests are preferably done in a controlled (usually dry) atmosphere to check operation under expected "worst-case" conditions. More tests are often done to evaluate the equipment or material in some depth. The tests should aim to evaluate the suitability of the equipment or material over its intended life. The equipment or material is often tested in isolation, i.e. when not yet installed as part of the ESD control system.

In contrast, compliance verification tests should be simple and efficient in demonstrating the correct functioning of the ESD control equipment or material as part of the ESD control system.

4.10.4 ESD Control Product or System Compliance Verification

Once ESD control equipment is in use, regular testing is necessary to identify failed equipment or systems. Simple tests are usually used to test the operation of equipment in situ in the workplace. Often this forms a "system test." Examples include testing of the grounding of a chair through flooring using a "resistance to ground" test or testing of a wrist strap or ESD control footwear while worn by the operator.

These tests are normally done under normal operating atmospheric conditions in the workplace.

4.10.5 Test Methods and Pass Criteria

For commonly used ESD control equipment and materials, standard test methods and pass criteria are often provided in ESD control standards. These have the advantage of being widely recognized and accepted. Adoption of these methods and criteria will facilitate easy compliance with the relevant standard, if this is required.

For nonstandardized items, test methods and criteria must be defined during the ESD program planning process. Nonstandard test methods and pass criteria can also be used for items that are covered by standard methods and criteria. Modern ESD control standards such as IEC 61340-5-1 and ANSI/ESD S20.20 allow nonstandard test methods to be used, providing correlation of the results of the nonstandard test method with the standard test method are demonstrated (see Chapter 6). In most cases, it may be easier to adopt the standard test method.

Different test methods will usually give different results for various reasons. So, use of a different test method will usually require changing the pass criteria. A specified pass criterion should be applied only to results obtained using the test method with which it was intended to be used or a test method that has been demonstrated to reliably give the same results.

4.10.6 How Often Should ESD Control Items Be Tested?

The frequency of testing should be chosen to reflect the reliability and importance of the ESD control item or system concerned. Factors that may be considered in deciding this include

- The likely consequence and ESD risk to ESDS products associated with failure
- Likely permanence and ruggedness of the item
- Risk of performance change through contamination or other factors
- Failure history found by experience
- Risk of accidental human intervention (e.g. unplugging ground cord)
- Known item lifetime limitations

Some examples of how this might be applied follow:

- A wrist strap is essential for grounding a seated person who handles ESDS devices. The consequences of failure would be that the person would be ungrounded and a serious ESD risk to the ESDS handled. Because of this, wrist straps are normally tested each working day before use. Some organizations that handle high-cost items or require low ESD failure rates test even more frequently. Some organizations use an automated test system (e.g. turnstile access control) to exclude personnel from the EPA unless they pass wrist strap and ESD footwear tests.
- An ESD control floor that has been installed for some years in a clean area and has been regularly tested and found to give stable performance might have the frequency of testing reduced accordingly. A newly installed floor, or one in an area subject to dirt or other contamination, might need to be tested frequently to monitor performance changes. This would be especially important if it were used to ground personnel handling high-value ESDS devices or where a low failure rate is required.
- A permanently installed workstation surface mat that has hardwired ground connections and is in a clean process area might need relatively infrequent testing. A workstation mat that is grounded by plug-in connectors to a mains electrical point might need more frequent testing to ensure it has not been inadvertently disconnected. Similarly, workstation surface in a process that has a high risk of contamination would need frequent testing.

- An ESD control chair grounded through wheels to the floor might need frequent testing due to the risk of grounding failure due to contamination buildup on the wheels. This would be especially true in a dusty area. In a clean room facility, a longer time between tests could be adequate.

4.11 The Seven Habits and ESD Standards

It is not surprising that many of the aspects of ESD control described in this chapter form the basis of modern ESD control standards. These have been crystallized over many years in systems of requirements for ESD control equipment and procedures. The requirements of two ESD control standards, IEC 61340-5-1 and ANSI/ESD S20.20, prevalent at the time of writing, are discussed in Chapter 6.

The ESD control standards evolve over time, reflecting developments in ESD control practice. When preparing for compliance with a standard, obtain copies and refer to the standards for up-to-date compliance requirements.

4.12 Handling Very Sensitive Devices

At the time of writing, standard ESD control programs are designed to handle devices with ESD withstand voltage down to 100 V HBM and 200 V CDM (IEC 61340-5-1:2016, ANSI/ESD S20.20-2014). For devices down to about 500 V HBM and 250 V CDM (Industry Council on ESD Target Levels 2011), it is relatively easy to implement an effective ESD control program using standard ESD control measures alone. These basic ESD control measures can be summarized as

- Handle unprotected ESDS devices only in an EPA.
- In uncontrolled areas, protect ESDS devices using ESD-protective packaging.
- Keep voltages on personnel less than 100 V using personal grounding.
- Keep conductors at the same voltage using grounding.
- Reduce electrostatic fields in the EPA to a minimum by removing nonessential insulators and controlling the fields from essential insulators.
- Avoid contact between the ESDS and highly conductive materials.

As time goes on, more organizations are handling components and assemblies with ESD withstand below this level and some below 100 V HBM. These very sensitive components are sometimes known as "Class 0" devices. The 2016 ESD Roadmap (EOS/ESD Association 2016) predicts that the proportion of components in this range is likely to increase into the future (see Chapter 13). Handling these components requires a greater level of care and may require additional nonstandard ESD control measures. Multiple redundant control measures may also be used to ensure that control is maintained if a single measure fails.

When handling low ESD withstand voltage devices, specific ESD risks may need to be evaluated and addressed with specific control techniques. The technical requirements of standard ESD control programs may need to be tightened.

As an example, for handling 50 V HBM components, the body voltage developed on personnel should be limited to 50 V. In addition, it becomes more important that the body voltage developed during normal activities is verified (by ESD control product qualification and/or compliance verification). This means that the wrist band cords may need to have specification of a reduced upper limit of resistance. It becomes even more important to ensure that body voltage developed when grounded by ESD control footwear and flooring remains below 50 V under all working conditions. For reduced CDM withstand voltage, the electrostatic fields and voltages developed on insulators near the ESDS, and voltages developed on the ESDS itself, may need to be controlled to a lower voltage level.

For handling low CDM ESD withstand devices, it becomes important that voltages developed on the device by induction, conduction, or triboelectrification remain low. At the same time, the possibility of contact between the ESDS and highly conductive materials should be eliminated where possible. Materials contacting the ESDS should, where possible, be static dissipative (>10 kΩ surface resistance as defined in current ESD packaging standards; see Section 8.5).

4.13 Controlling Other ESD Sources

Standard ESD control programs largely address the most common ESD sources, namely, charged personnel, charged metal objects, and charge devices. Other ESD sources occur that are not necessarily well controlled by standard ESD control measures.

PCBs, assemblies, and modules can become charged and then discharge when they touch another conductor. The ESD susceptibility of these items is not usually tested. Some studies have found that damage previously thought to be due to electrical overstress can result from charged board ESD or charged cable ESD (Olney et al. 2003). Cables, PCBs, and assemblies can have rather high capacitance due to their size and so can store relatively high energy prior to discharge.

Charged cable ESD can arise when a cable is connected to a PCB, module, or assembly at a different voltage. This can occur if the cable is charged or the ESDS is charged or both. Cables can become charged by triboelectrification by contact with external surfaces (work surfaces or equipment) or packaging. High voltages can be induced on cable conductors by nearby electrostatic field sources. ESD can occur to connector pins when the cable is connected to an ESDS. A similar discharge occurs if the ESDS is charged when connected to the cable.

PCBs or assemblies are often potted in resin or enclosed within a plastic housing. Potted modules sometimes have flying lead terminations for power and I/O connections. When the housing becomes charged, high voltages can be induced on the PCB or assembly within. Similarly, electrostatic fields from nearby charged insulators can induce high voltages on the PCB or assembly. A discharge can occur when a cable is connected or a flying lead touches a metal item.

Evaluation and control of these ESD risks often requires analysis of the circumstances and evaluation of the possible ESD susceptibility of the ESDS. Simple specific ESD control measures can often be chosen to address specific ESD risks.

References

EOS/ESD Association Inc. (2014) ANSI/ESD S20.20-2014. *ESD Association Standard for the Development of an Electrostatic Discharge Control Program for – Protection of Electrical and Electronic Parts, Assemblies and Equipment (excluding Electrically Initiated Explosive Devices).* Rome, NY, EOS/ESD Association Inc.

EOS/ESD Association Inc. (2016) ESD Association Electrostatic Discharge (ESD) Technology roadmap – revised 2016. Available from: https://www.esda.org/assets/Uploads/docs/2016ESDATechnologyRoadmap.pdf [Accessed: 10th May 2017] Rome, NY, EOS/ESD Association Inc.

Industry Council on ESD Target Levels (2011) White paper 1: A case for lowering component level HBM/MM ESD specifications and requirements. Rev. 3.0. Available from: http://www.esdindustrycouncil.org/ic/en/documents/37-white-paper-1-a-case-for-lowering-component-level-hbm-mm-esd-specifications-and-requirements-pdf [Accessed: 10th May 2017] Industry Council on ESD Target Levels

International Electrotechnical Commission. (2013) PD/IEC TS 60079-32-1. *Explosive atmospheres Part 32-1. Electrostatic hazards, guidance.* ISBN 978-2-8322-1055-0, Geneva, IEC.

International Electrotechnical Commission. (2015) IEC 61340-5-3:2015. *Electrostatics - Part 5-3: Protection of electronic devices from electrostatic phenomena - Properties and requirements classification for packaging intended for electrostatic discharge sensitive devices.* Geneva, IEC.

International Electrotechnical Commission (2016) IEC 61340-5-1: 2016. *Electrostatics – Part 5-1: Protection of electronic devices from electrostatic phenomena - General requirements.* Geneva, IEC.

Olney, A., Gifford, B., Guravage, J., and Righter, A. (2003). EOS-25. Real-world charged board model (CBM) failures. In: *Proc. EOS/ESD Symp.*, 34–43. Rome, NY: EOS/ESD Association Inc.

Oxford Dictionary. Available from: https://en.oxforddictionaries.com/definition/habit [Accessed: 12th May 2017]

Smallwood, J., Swenson, D.E., and Viheriäkoski, T. (2018). Paper 1B.1. Relationship between footwear resistance and personal grounding through footwear and flooring. In: *Proc. EOS/ESD Symp. EOS-40.* Rome, NY: EOS/ESD Association Inc.

Stadler, W., Niemesheim, J., Seidl, S. et al. (2018). The risks of electric fields for ESD sensitive devices. Paper 1B.4. In: *Proc. EOS/ESD Symp. EOS-40.* Rome, NY: EOS/ESD Association Inc.

Swenson, D.E. (2012). Electrical fields: What to worry about? Paper 3B.6. In: *Proc. EOS/ESD Symp. EOS-34.* Rome, NY: EOS/ESD Association Inc.

Wallash, A. (2007). A study of "Soft Grounding" of tools for ESD/EOS/EMI control. 2B8-1. In: *Proc. EOS/ESD Symposium EOS-07*, 152–157. Rome, NY: EOS/ESD Association Inc.

Further Reading

EOS/ESD Association Inc. (2016) ESD TR20.20-2016. *ESD Association Technical Report - Handbook for the Development of an Electrostatic Discharge Control Program for the Protection of Electronic Parts, Assemblies and Equipment*. Rome, NY, EOS/ESD Association Inc.

International Electrotechnical Commission. (2018) IEC TR 61340-5-2. *Electrostatics – Part 5-2: Protection of Electronic Devices from Electrostatic Phenomena - User Guide*. ISBN 978-2-8322-5445-5 Geneva, IEC.

5

Automated Systems

5.1 What Makes Automated Handling and Assembly Different?

The American National Standards (ANSI)/Electrostatic Discharge (ESD) SP10.1 defines *automated handling equipment* (AHE) as "any form of self-sequencing machinery that manipulates or transports product in any form; e.g. wafers, packaged devices, paper, textiles, etc." One thing that makes automated assembly and handling different is that the ESD risks are mainly due to contact with machines and machine parts or part-built product assemblies. ESD risks due to manual handling are restricted to parts of the process where manual handling is performed. Manual handling areas should be set up according to the usual ESD control procedures.

Many people assume that they do not have to be concerned about ESD control within the automated equipment, as they assume it has been taken care of by the machine manufacturer. This can be incorrect for several reasons (Yan et al. 2009; Paasi 2004). First, although an ESD-aware manufacturer may have built-in basic ESD control, they may not have anticipated all ESD risks. Equipment manufacturers often include control measures, especially ionizers, with little apparent understanding of their operation and effectiveness. As an example, automated equipment moves quickly during operation. Ionizers used to neutralize charge may not have sufficient time to achieve this (Tan 1993).

In some cases, different materials such as anodized coatings are used for ESD control than in manual processes (see Section 5.6.3). These may have unusual or variable characteristics.

Second, due to the lack of manual handling, the usefulness of human body model (HBM) ESD withstand data is limited. Most ESD risks are instead due to contact between the ESD-sensitive (ESDS) devices and metal or nonmetal conductors. So, primarily charged device model (CDM) or sometimes machine model (MM) ESD withstand voltage data are more likely to be relevant. (At the time of writing, measurement of MM ESD withstand voltage of ESDS is falling out of practice.) The ESD withstand voltages of components are reducing year by year, so yesterday's ESD control may not be sufficient for today's devices and are unlikely to be sufficient for tomorrow's components (ESDA 2016a; Koh et al. 2013).

The ESD Control Program Handbook, First Edition. Jeremy M Smallwood.

So, if the machines used are not completely up to date, there is a chance that the ESD control may not be up to the standard required for handling the most sensitive of current components.

The ESDS devices of concern may be printed circuit boards (PCBs) as well as devices. Charged board ESD has been increasingly recognized as a significant source of damage. In recent years, it has been realized that damage mistaken for electrical overstress (EOS) has actually been caused by charged board ESD (Olney et al. 2003).

Third, adaption or modification of machines or processes can easily inadvertently introduce ESD risks. As in manual processes, failures of ESD control equipment can also result in ESD risks. Standards are not yet available for ESD control in automated equipment, although the ANSI/ESD SP10.1 gives some guidance on standard practices. In an automate process, automated ESD generators can arise that can provide automated damage of devices! If this occurs, the ESDS damage rate can be high.

The automated environment also brings some special challenges. Many components designed for automated handling are extremely small. The packaging for these has correspondingly small features, and this can make it difficult or impossible to test the packaging with standard test methods. Some components are so small that electrostatic charging can result in ejection of the component from the packaging in an uncontrolled manner by electrostatic forces (see Section 5.8).

Access into automated processes during operation may be highly restricted. So, it can be difficult to observe and diagnose potential ESD issues and make measurements during operation of the process. Special test and measurement methods may be required, either due to the form of the items measured or due to the restrictions of the measurement environment in AHE. Stepped process simulation can help enable measurements to be made.

5.2 Conductive, Static Dissipative, and Insulative Materials

It is difficult to write this chapter without making reference to materials categorized by their resistance as conductive, static dissipative, or insulative. It is important to realize that there is in general no fixed definition of these terms. In the case of the International Electrotechnical Commission (IEC) 61340-5-1 and ANSI/ESD S20.20, these terms are specifically defined in terms of surface and volume resistance ranges for ESD-protective packaging materials (see Chapter 8). These definitions are used also in this chapter.

In this discussion, a conductor may be any item made from conductive or static dissipative materials – in other words, any material that is not an insulator.

5.3 Safety and AHE

Working with AHE may bring specific safety issues relating to the equipment and processes. These may include risks due to moving machinery, electrical hazards, or high process or equipment temperatures. These must always be approached with due consideration of any hazards arising in the process, as well as any local regulations and practices that may apply.

AHE systems often work within interlocked protective enclosures. Making observations or measurements on operating systems can be difficult.

5.4 Understanding the ESD Sources and Risks

The main ESD sources in an automated production environment include

- Charged devices making contact with low-resistance conductors.
- Charged metal objects and machine parts making contact with ESDS.
- Charged PCB (or subassemblies or modules) making contact with low-resistance conductors.
- Charged personnel (in manual handling parts of the process) making contact with ESDS.
- For a few types of ESDS, electrostatic fields alone could give ESD risk. This is likely to be a concern particularly with voltage-sensitive devices in high impedance circuits and may be unusual.

In principle, the ESD risk could be understood by knowing the susceptibility of each type of ESDS to each ESD source. The risk could then be controlled by maintaining the parameters of ESD from each source below thresholds at which ESD damage are known to occur. To this end, HBM, MM, and CDM ESD susceptibility tests have been developed to allow reproducible measurement of the ESD susceptibility of components (see Chapter 3). The component ESD susceptibility is quoted as an ESD withstand voltage, which is the highest test voltage in the ESD source that did not result in damage to the component.

At first sight then, a strategy for risk evaluation and management might seem clear – if the real-world voltages in ESD sources can be kept below the withstand voltage levels, no ESD damage should be possible. Specification of ESD control measures would then be a matter of defining controls that can maintain the source voltages below the risk threshold defined by component ESD withstand voltages.

While this simple view and strategy is partially valid, it represents a significant oversimplification (see Section 9.3.1). First, component susceptibilities are not in fact normally damaged directly by the ESD source voltage, but to other parameters such as the peak discharge current or charge power or energy transferred to the device in the discharge. These parameters are related to the source voltage by circuit parameters such as inductance, resistance, and source capacitance. Real-world sources are unlikely to have the same capacitance, resistance, and inductance as the HBM, MM, and CDM models. This means that HBM, MM, and CDM discharges do not happen in the real world – they happen only in the component susceptibility test equipment. Real-world ESD is simply a variety of discharges from the human body, a metal part, a device, or other source. Each source has very different ESD waveform characteristics from the models and would yield a different withstand voltage (ESDA 2015; Gaertner and Stadler 2012). An ESDS might be more or less susceptible to damage from real sources than it is to HBM, MM, or CDM ESD. A simple action such as changing the position of an ESDS can change a relevant parameter such as its capacitance as well as the voltage and hence the ESD damage susceptibility. In specification of ESD controls and evaluation of ESD risks, it remains a major challenge to relate real-world risks to component ESD withstand data.

5.5 A Strategy for ESD Control

5.5.1 General Principles of ESD Control in AHE

The general principles of ESD control in AHE are not very different from in manual processes. Many ESD problems can be avoided by applying the recommendations of standards and textbooks, although in practice these may need some modification for use with AHE.

5.5.2 The Conditions Leading to ESD Damage

Review of the ESD sources shows that they have some simple commonality.

- ESD can usually occur only where contact occurs between an ESDS device and a conductor. The only exception is for ESDS devices where electrostatic fields could cause voltage differences that can cause damage within the ESDS device.
- ESD can occur only where there is a sufficient voltage difference between the ESDS device and the contacting conductor. The ESD risk increases with the voltage difference. The voltage difference that is "sufficient" is dependent on the circumstances and the component sensitivity.

The first point focuses our attention on parts of a process where an ESDS device makes contact with other conductors. These could be personnel, machine parts, tools, or other ESDS devices. Sufficiently far away from the vicinity of the ESDS device, where the risk of contact and the influence of electrostatic fields on it are insignificant, we do not usually need to take ESD control measures. It also makes clear that if we can reduce the number of occasions on which the ESDS makes contact with a conductor, we will reduce the number of opportunities for ESD to occur. This can in some circumstances reduce ESD risk and the burden of controlling it.

The second point highlights a need to control voltages on the ESDS device and on conductors that could contact the ESDS. The voltage difference between the ESDS device and any conductor that contacts it must be minimized, or at least kept below a risk threshold level. These voltage differences can be from two causes – triboelectrification (contact charging) and induction (voltages induced by nearby electrostatic fields) (see Chapter 2).

5.5.3 Strategies for ESD Control in Automated Equipment

From the previous analysis, we can see that we need concern ourselves only with the immediate region of the process in which an unprotected ESDS device can be present and the near vicinity of the ESDS device. So, the first task is to plot the path of ESD through the process. ESD control must be applied in a region around this critical path (Paasi 2004). Jacob et al. (2012) refer to this as the *zone of processing*, identifying this as the region in which ESDS devices are loaded, transported, or processed. SP10.1 recommends that in a critical path region within 15 cm of the ESDS device's critical path, all conductive machine parts should be grounded, and all insulators rendered static safe.

Within the critical region, ESD risks should be analyzed using visual observations, measurements, statistical failure data, or any other available relevant techniques and information. Critical parts of the process are where ESDS devices make contact with other items, especially when under the influence of electrostatic fields. Where contact is made, voltage differences between the ESDS and the contacting item must be controlled. Ideally ESDS devices should make contact only with static dissipative (see Section 5.2) materials. Chapter 9 looks in more detail at evaluation of ESD risks.

- Conductors that make contact with ESDS devices must be grounded, where possible.
- Conductors that cannot make contact with ESDS devices need not be grounded.
- If a conductor makes contact with ESDS devices and it cannot be grounded, the ESD risk must be evaluated. Other ESD controls must be devised if necessary.

Once the ESD risk points have been identified, they can be prioritized and suitable ESD control measures developed. The control measures will need to be documented and a compliance verification program developed. This may require test methods and equipment to be defined with suitable pass criteria and frequency of testing.

Paasi (2004) suggested it is useful to think in terms of "intimate" and "proximity" materials in AHE systems, as defined in ESD-protective packaging materials. From this view, intimate parts, materials, and surfaces are those that could come into direct contact with ESDS devices or PCBs during normal operation. Intimate parts of AHE include conveyor belts and rollers, supports, racks, and other parts that come into direct contact with the ESDS device. Tape and reel packaging is also of concern. Materials that come into contact with ESDS devices should be selected to minimize tribocharging of the ESDS device being handled.

Proximity parts, materials, and surfaces are in the critical region surrounding intimate parts, containing objects, materials, and surfaces that do not normally come into contact with the ESDS device. By analogy with packaging, intimate and proximity items should, if possible, not be made of insulating materials and should be grounded. This is to prevent electrostatic fields due to charged materials inducing voltages on ESDS devices within the critical path.

The materials, techniques, and equipment used to achieve these objectives are discussed in greater detail in Section 5.7.

5.5.4 Qualification of ESD Control Measures

Once ESD risk points have been identified and quantified through measurements, suitable ESD control measures should be defined. The effectiveness of these control measures should then be qualified and verified using measurements.

ESD control measures in AHE will often need to be developed specifically for the ESD risk that has been identified. So, the qualification process will be part of the process of specifying the control measure and checking that it works. Test methods, and pass criteria specific to the control measures, will need to be selected to demonstrate that the control measures work as intended. Sometimes these can be based on existing standard tests, but sometimes nonstandard tests will need to be defined.

5.5.5 Compliance Verification of ESD Control Measures

Most ESD control measures can fail for various reasons, so it is necessary to verify them on a regular basis. As with ESD control measures in manual processes, suitable test methods, pass criteria, and a frequency of testing should be defined and documented in a compliance verification program.

5.5.6 ESD Training Implications

While participation of personnel in automated processes is minimized, the need for training of different types should be considered.

Personnel who evaluate the AHE for ESD risks and develop ESD control measures will need an excellent level of expertise to successfully provide these functions. Standards and guidance documents are also regularly published or updated, and new technologies are developed. Understanding of ESD risks is continually improving, and the susceptibility of ESDS to these risks changes with technology development. ESD coordinators and personnel who are active in these areas will therefore need provision for developing and updating their skills and knowledge.

ESD control measures incorporated into AHE will often need special tests to be defined that may need specific setup and procedures for each type of equipment. Personnel who perform audit measurements will therefore need training in these test and measurement techniques.

5.5.7 Modification of Existing AHE

ESD risks can often be created inadvertently when equipment is modified (Paasi 2004). Analysis of ESD risks should be made as part of the process of design of any modifications so that ESD control measures can be specified and built into the modifications. Qualification and compliance verification tests may be needed.

5.6 Determination and Implementation of ESD Control Measures in AHE

5.6.1 Define the Critical Path of ESDS

The first step is to define the critical path (or paths) that ESDS takes through the AHE. These may include introduction of part-populated assemblies into the process as well as parts of the process where ESDS devices are loaded for mounting on an assembly. The size of the critical region around that path should then be decided. Electrostatic Discharge Association (ESDA) SP10.1 recommends a region of 15 cm around the critical path, but if it is feasible, a larger region up to 30 cm around the critical path could be considered for control of insulators. Further away from this, it is unlikely that significant risks are posed unless items can become very highly charged, as the electrostatic fields are much reduced over the distance (see Section 4.5.3).

5.6.2 Examine the Critical Path and Identify ESD Risks

The critical path should then be examined visually from the point(s) at which ESDS enter the process to the point(s) at which they leave the process. It is helpful to observe the ESDS traverse the critical paths. The objective is to identify ESD risks in terms of

- Points at which ESDS contact conductors including machine parts or personnel
- Places where potentially charged insulators are within the critical region

These points must be evaluated for ESD risk and, if necessary, addressed with ESD control measures.

One of the main ESD risks in AHE is charge device or charged board ESD. These can occur where a charged device or PCB makes contact with a low-resistance conductor or a charged ungrounded conductor makes contact with an ESDS device. There are four principal strategies for controlling these risks.

- Keep tribocharging of devices and PCBs below a risk threshold level by controlling contact between ESDS and other materials.
- Keep induced voltages on devices and PCB conductors below a risk threshold level by controlling exposure of ESDS to electrostatic fields.
- Reduce the potentially damaging effect of ESD by preventing contact between ESDS devices and low-resistance conductors.
- Ensure that ESDS devices where possible make contact only with grounded conductors. If the ESDS device makes contact with a conductor that cannot be grounded, the voltage difference between the ESDS device and the conductor should be controlled.

5.6.3 Determine Appropriate ESD Control Measures

5.6.3.1 General Control Measures

The following guidelines for typical ESD control measures are based on recommendations of ESD Association SP10.1 (ESDA 2016b) and should be applied within the critical region. The definitions of *conductive* and *static dissipative* used for packaging materials are used here (see Section 8.3.1, 8.8.1 and Table 8.3);

- All conductors should be grounded.
- Where operators may need to be grounded for participation in the process, designated wrist strap grounding points should be provided.
- All machine parts that may make contact with ESDS leads should have static dissipative surface and be grounded to reduce charged device ESD risk.
- All machine parts that are separated from the machine chassis by bearings should be grounded by some means that remains reliable during movement (SP10.1 recommends resistance to ground $<1\ M\Omega$).
- Surfaces that ESDS could be placed upon should meet the requirements for EPA workstation surfaces in the ESD program.
- Pneumatic and electrical lines should be constrained to avoid rubbing and tribocharging effects. Pneumatic lines within the critical region, if possible, should not be insulative and should be grounded. Alternatively, they could be shielded with a grounded metal braid shield.

- Wire bundles in the critical region should be shielded, e.g. with a grounded metal braid.
- Device pickup mechanisms such as vacuum cups, nozzles, and grippers should be made of conducting materials and should be grounded. The contact area and velocity should be minimized to reduce tribocharging of device packages.
- ESD grounding points should be directly connected to the equipment ground. SP10.1 recommends a resistance to ground of 1 Ω or less. Lower resistance might be necessary if high currents (e.g. for motors) or fault currents might be carried.
- Conductors that are relied upon to provide a ground path must be sufficiently robust to prevent accidental disconnection. The conductors should be braided wire where possible.
- For anodized surfaces, ensure the underlying substrate is grounded to the machine earth.

Methods of grounding moving parts suggested by SP10.1 include flexible conductors such as braided cables, metal bushes, graphite or beryllium copper commutators, and conductive greases within the bearings. One problem with moving parts is that resistance to ground measurements made when stationary may not represent the resistance to ground when moving. Oil films and other effects can result in intermittent loss of contact that leads to charge/discharge behavior.

5.6.4 Include ESD Control in New Equipment Specification

The ESD control development process should preferably start during equipment design and development (Paasi 2004). The equipment vendor should provide documentation of any necessary ESD control measurements and acceptance values.

ESD control should be included as part of any new automated equipment purchase specifications. It may be inadequate to assume that the equipment manufacturer has built in adequate ESD control (Yan et al. 2009; Tan 1993; Millar and Smallwood 2010), especially if low ESD withstand voltage devices are handled.

Whenever equipment is modified, the modified equipment should be carefully evaluated to make sure that ESD risks have not been inadvertently introduced. ESD control measures should be qualified using tests appropriate to their function and the intended conditions of operation.

5.6.5 Document the ESD Control Measures Used in the Machine

Once ESD control measures have been identified and specified, these should be documented. This is analogous to writing an ESD Control Program Plan for a manual handling or assembly process. If the ESD control measures are not adequately documented, it's likely that they will be omitted, fail, or be inadvertently rendered ineffective by some means in the future, reintroducing undetected ESD risk.

5.6.6 Implement Maintenance and Compliance Verification of ESD Control Measures

As in ESD control in manual handling processes, ESD control measures in AHE can fail or be compromised for many reasons. As examples, Millar and Smallwood (2010) found that all but one of eight YAC handler systems they examined had multiple failures of internal ground wires. They found that anodized soak boats suffered wear of the anodizing that allowed metal-to-metal contact between ESDS devices and soak boat substrate metal.

As with manual processes, all ESD control measures must be checked at regular intervals to detect any failures or damage that could cause ESD risk. This means that an inspection regime must be defined, with checks and tests performed at appropriate intervals designed to detect the failures. In some cases, it may be necessary to implement maintenance procedures to prevent failures.

Tests may be required using standard or custom measurements, with suitably chosen pass criteria. This compliance verification test program must be documented and established as part of the ESD control procedures.

5.7 Materials, Techniques, and Equipment Used for ESD Control in AHE

5.7.1 Grounding All Conductors That Make Contact with ESDS

As in manual handling of ESDS devices, all conductors that make contact with ESDS devices should be grounded if at all possible. Allowing contact between ESDS devices and an isolated (ungrounded) conductor should be a last resort adopted only where grounding is very difficult. The ground path must be continuous at all times during operation when ESDS devices might make contact with the conductor. A break in the ground path at any time can allow charge to build up on the conductor, which then becomes a potential source of ESD.

The conductor may be a conductive or static dissipative material (see Section 5.2), and the resistance to ground that can be achieved may depend on the materials used. Items that can be subject to charge generation due to movement (e.g. conveyor belts) should have sufficiently low resistance to ground (at all points) to prevent significant voltage buildup on them. The voltage buildup and resistance to ground should if possible be verified under operating conditions.

Grounding can, of course, be achieved either through a wired connection or through an electrically conducting material path. The ground path may include moving items such as bearings or sliding surfaces, providing the ground connection can be shown to be reliable under operating conditions.

In AHE, ground wires may often be used to link two parts that repeatedly move relative to each other. In this situation, ensuring the reliability of the connection can be a challenge. Ordinary wires may have short lifetime due to metal fatigue when repeatedly flexed. Braided wires may give a longer lifetime. Regular testing of the connection may be required to detect any failures occurring.

A maximum resistance to ground is normally specified for grounding of conductors. This gives a simple measurable parameter for verification of grounding. A maximum resistance to ground can be specified from two views. If the conductive item is subject to charging from a current source, e.g. due to machine motion, the maximum resistance to ground can be specified from the view of maintaining low voltage under condition of maximum charge generation current flow. For circumstances where current flow is low, maximum resistance to ground can be derived from charge and voltage decay time considerations.

In some cases, a minimum resistance to ground may also be specified for safe limitation of current under a fault condition, or to limit ESD current when charged ESDS devices contact the grounded material. For safety considerations, adding discrete resistance on the ground path may be acceptable. For reduction of charged device ESD current, the resistance must be inherent in the contacting material (see Section 4.7.5.2 and 5.7.5).

5.7.2 Isolated Conductors

Conductors in the critical region should always be grounded if possible. In AHE, it can sometimes be difficult to achieve grounding reliably due to moving parts preventing reliable connections. Ungrounded conductors can be evaluated and dealt with as described in Section 9.3.3.

5.7.3 Preventing Induced Voltages on ESDS Devices

Electrostatic fields in the vicinity of ESDS devices can induce voltages on the ESDS devices. If these voltages are present when the ESDS devices makes contact with another conductor, then an ESD will occur. The concern here is charged device or charged board ESD. It is therefore particularly important to control electrostatic fields in the critical region near positions where the ESDS devices make contact with other conductors. Low ESD withstand voltage devices will require voltage sources to be controlled below a lower voltage or field limit. The closer an electrostatic field source is to the ESDS devices at the contact zone, the lower the field or source voltage that can be tolerated (see Section 4.5.2). The sources of electrostatic fields that are often of most concern are charged insulators. Nevertheless, any significant voltages sources could be of concern, particularly if close to the contact zone. These could include fields from nearby wiring, pipes, or other machine components.

As with essential insulators in manual processes, the first step in treating an essential insulator within the ESDS critical path is to evaluate the ESD risk. If no contact is made between ESDS and other conductors in the vicinity of the insulator, there may be little ESD risk even if the insulator becomes significantly charged (Gaertner 2007; Yan et al. 2009). Nevertheless, it may be preferred to reduce ESD risk by reducing the electrostatic field experienced by the ESDS device.

If an insulator can become charged but is not near a part of the process where contact between the ESDS device and a conductor can occur, then no control is required unless the ESDS device is known to be susceptible to electrostatic fields. In practice, few ESDS devices may have this type of susceptibility. If the insulator cannot become charged, then no action is needed, except to verify that the insulator really cannot become charged under the full range of operating conditions. This may require making measurements under low-humidity conditions.

Where an insulator can become charged and could make contact with a conductor, an ESD control measure is likely to be essential. Control techniques could include, for example (Paasi 2004, SP.10), the following:

- Replacement or coating of the insulator with a conducting material, and grounding it
- Prevention of charging by some means (e.g. prevention of rubbing)
- Relocation outside the contact zone
- Shielding or suppression of the field from the insulator by positioning a grounded conductor between it and the ESDS device
- Use of an ionizer to neutralize charge
- Use of topical antistats to reduce charging

Replacing the insulator with a grounded conductor, moving it outside the critical contact zone, or preventing charging can be among the most effective solutions, if they are possible. Use of an ionizer is discussed in Section 5.7.9. Prevention of insulating surfaces from rubbing can be a useful low-cost way of preventing an insulator from becoming charged. This may require adjustment of the position of adjacent parts of equipment. It can be a particularly useful technique with cables, pneumatic lines, or other flexible insulating lines.

Perhaps the easiest but least reliable technique is to use a topical antistat to reduce electrostatic charging. These often effect a temporary solution but can quickly become ineffective. Tan (1993) used antistats to confirm that electrostatic charging of insulating spacers was the cause of some charged device ESD damage. He found that the lifetime of the antistat was much shorter in a varying (hot and cold) temperature environment than the accelerated life test had suggested, lasting only one day in practice. Often antistats rely on atmospheric humidity for their action, and so they can become ineffective under low-humidity conditions. In an enclosed space within AHE with heat sources present, local low-humidity regions can easily be established.

The effect of electrostatic field sources can also be reduced or eliminated by shielding the ESDS device from the field. This may be done by inserting a grounded conductor between the field source and the ESDS critical path. Tan (1993) gives several examples of using metal flanges, washers, copper tubes, or self-adhesive metal tapes for shielding purposes.

5.7.4 Reducing Tribocharging of ESDS Devices

It is often thought that an ESDS will not be charged by contact with a conducting material. This is not so – any contact between materials will separate electrostatic charge to some extent. Each of the contacting materials becomes charged by an equal amount and opposite polarity. If one of the surfaces is an insulating material (e.g. device plastic package), it will become charged. An insulating device package can be charged by contact with a conducting packaging material or a vacuum cup (Yan et al. 2009). Use of conducting materials for these contacting items merely allows the charge produced on them to dissipate, providing of course there is a route for the charge to be conducted away, i.e. the conducting material is grounded.

Tan (1993) gives an example of damage caused by tribocharging of a plastic device package body due to contact with a steel tool in a trim and form process. Even with the tools grounded, voltages up to 1100 V were measured on the devices.

It can be difficult to prevent charging by contact in this way. Choice of contacting materials can make a difference to the charge generation levels. Often the charge accumulated on the ESDS must be neutralized to bring it to an acceptable level before contact is made with another conductor.

Kim et al. (2012) found that triboelectrically generated charge on liquid crystal display (LCD) screens during production could not be adequately controlled by ionization (see Section 5.7.9). The glass panel substrate generated high charge levels and held charge for long periods. More than 40 material contact and separation actions occurred during the production process, including friction from photoresist coatings applied to the glass surface, deionized water spray rinses used to clean the surface, contact with rollers or belts, and pressure and separation of glass panels from vacuum chucks. Control measures they used to reduce ESD risks included increasing surface roughness and minimizing separation speed and vacuum pressure to control the charge generation of glass substrates. They found that using insulative instead of dissipative or conductive lift pin materials prevented ESD events. Specification of adequate separation distance and insulation thickness was also used to prevent air discharge between the glass and metal objects.

5.7.5 Using Resistive Contact Materials to Limit Charged Device ESD Current

Charged device ESD risk is greatest when the charged device makes contact with a low resistance conductor. This results in a high peak current discharge (so called hard discharge). One means of reducing this risk is to ensure that the device contacts a resistive material, rather than a highly conductive material. In this case, the peak discharge current can be limited by including sufficient contacting material resistance. Placing a resistor in the ground path is ineffective for this purpose (see Section 4.7.5.2).

There is some debate as to whether direct current (DC) resistance measurements on polymeric ESD control materials provide a realistic guide to ESD risk (Viheriäkoski et al. 2017). They found the "ESD resistance" due to frequency dependent impedance of the materials could give much lower ESD current and charged device ESD risk than would be expected from DC resistance measurements.

5.7.6 Anodization

Anodization is often used as a passivation and protection of machine surfaces. Although considered insulative by some, it can be used to provide a noninsulating surface coating on an aluminum alloy substrate (Bellmore 2001). The anodized layer may be 5–40 μm (0.0002–0.0015 in.) thick. Different anodizing processes can produce a wide range of surface resistance from <1 MΩ to over 100 GΩ.

If done after screw holds and threads are made, anodization can prevent good contact between metallic machine parts (Yan et al. 2009). The resistance through the joints may be increased to several ohms. Thick anodization can result in higher resistances. It can be difficult to remedy this after the machine has been constructed.

Anodization can be used to provide a hard-wearing static dissipative surface coating to reduce the risk of charged device ESD damage when device leads touch the surface (Yan et al. 2009; Smallwood and Millar 2010; Millar and Smallwood 2010).

Millar and Smallwood (2010) found that anodized soak boats experienced wear or damage to the anodized layer. Direct contact could then be occasionally made between the leads of devices in the boats and the underlying metal soak boat material. Smallwood and Millar (2010) measured anodized soak boat surface resistance using standard and nonstandard electrodes as well as resistance from the boat metal substrate through the surface and a supporting dissipative rail to ground.

The influence of applied test voltage was also measured. They found that the resistance of the anodized layer varied dramatically with the applied voltage and test electrodes used. At 1000 V the resistance through the anodized layer was <1 Ω, but at 100 V this increased to over 20 GΩ. Surface resistance from point to point was <1 GΩ using ESD S4.1 electrodes, but 30–96 GΩ using an S11.13 2-pin measurement. Other methods also gave varying results.

As variation in resistance could be expected to cause variations in charge decay time of the boat standing on the rail, this was also measured. The charge decay time varied with the remaining voltage on the boat, increasing as the voltage reduced. At 1000 V, the voltage disappeared immediately when the applied voltage source was removed. Decay times could be as long as 20 seconds at 100 V applied voltage.

Their results showed that large soft electrodes with high contact pressure such as S4.1 gave lower resistance values. The surface may have small regions of high conductivity, damage, or pin holes that show low resistance with these electrodes. Small area electrodes with hard surfaces are less likely to make good contact through low-resistance or damaged regions.

These results showed that when measuring the performance of an anodized layer, it is important to use test voltage and electrodes that simulate the working condition as far as is possible.

5.7.7 Bearings

Many moving machine parts are supported on bearings. The grounding of these machine parts is often through the bearing. If measured when stationary, the resistance through the bearing may be low. When the machine is running, the resistance through the bearing may be much higher or intermittent due to the oil film that separates the moving bearing surfaces. This problem can be avoided by fitting a bearing that is lubricated by a conducting (noninsulative) lubricant. Alternatively, a parallel ground path can be established through a commutator or other mechanism.

5.7.8 Conveyor Belts

Conveyor belts by their nature are often supported on wheels or rollers, which are themselves supported on bearings. If made of insulating materials, a belt will become charged.

Conveyor belts in the critical zone should be made of static dissipative materials and grounded. Grounding of the belt may not be easy as it may rely on conduction through the belt and bearings or other moving contacts. The resistance of the belt and belt resistance to ground may need to be set to a level comparable with work surfaces. They must at least be sufficiently low to prevent significant charging under operating conditions.

5.7.9 Using Ionizers to Reduce Charge Levels on ESDS Devices, Essential Insulators, and Isolated Conductors

Ionizers can be used to reduce the charge levels on essential insulators, ESDS devices, or isolated conductors in the critical path. To be effective, ionizers must be correctly positioned to reduce voltages before the point at which control is required and have sufficient time to neutralize the charge. This is typically before a point where the ESDS will contact another conductor.

The limitations of ionizers must be borne in mind when specifying this control measure.

- Ionizers take time to neutralize charge and reduce charge levels below the required threshold.
- Neutralization brings the final voltage to the offset voltage of the ionizer. This must be below the risk threshold for the device.
- Effects such as air flow can change the operational characteristics of charge neutralization in a process.

Unfortunately, equipment manufacturers may not have sufficient understanding to specify and fit ionizers that work effectively. Automated equipment moves quickly during operation. Ionizers are often used to neutralize charge but require time to achieve this (see Sections 2.8.2–2.8.6, 4.6.3, Tan 1993). Their effectiveness may be impaired by speed of operation compared to charge neutralization rate, and factors such as air flow can also affect their performance.

The ESD TR10.0-01-02 discusses use of ionizers in AHE. In a fast-moving process, charge neutralization time must be correspondingly short. The neutralization time will depend on the characteristics of the ionizer, the position of the item to be neutralized in the region around the ionizer, and other factors such as air flow.

Yan et al. (2009) reported a case in which an ionizer fitted by the manufacturer as an "ESD option" failed to prevent ESD damage. The ionizer was unable to neutralize charge on a device after release from a suction cup and before the device solder balls contacted a conductive transport boat. The problem was instead solved by anodizing the metal surface to give a static dissipative coating layer. The rate of charge neutralization drops as the electrostatic field and voltage on the item to be neutralized drops. It therefore takes longer to reduce voltages to a lower level.

One useful characteristic of charge neutralization by ionizers is that given sufficient time, all items neutralized tend to move toward the same voltage level – the offset voltage of the ionizer. If the task is to reduce the voltage difference between a charged isolated conductor and a charged ESDS that will make contact with it, the ionizer will move both items toward the same nonzero voltage. So, given sufficient time, the voltage difference between the items can be much less than the offset voltage of the ionizer.

The effectiveness of an ionizer in neutralizing charge on items should be verified in the process and circumstance in which it is used.

5.7.10 Vacuum Pickers

One area where it can be critical to reduce device charging is on vacuum pickers (Yan et al. 2009). The amount of charge on the device is critical at the time that it is released to make contact with another conductor. Where a device has an insulating package material, the package material that makes contact with the vacuum cup will become charged. If the vacuum cup material is also made of insulating material or is not grounded, charge will build up on the cup during successive operations. So, the cup should be made of a static dissipative material and grounded. Checking of the grounding of cups should be specified as part of compliance verification planning.

Nevertheless, some contact charging of the device package cannot be avoided (Yan et al. 2009). The residual charge on the device does not produce voltage on the device until separation between the cup and the device occurs. This may occur only when the device is a small distance from making contact with a conductor. Even highly efficient ionizers are unlikely to have much effect in the small time between release of the device and contact with the conductor. In this situation, specification of resistive contacting materials, if possible, may be necessary to avoid damaging ESD.

5.8 ESD Protective Packaging

ESD-protective packaging used in automated equipment often takes special forms such as tape and reel or JEDEC trays (see Chapter 8). These may be specifically designed to handle the devices used in the processes. Their shape and form may pose difficulties in measurement for verification of the packaging (see Section 5.9.2).

Components are supplied in tape and reel for placement on PCBs in a pick and place machine. ESDS devices should be supplied in ESD-protective packaging, but non-ESDS components are often supplied in tape and reel made from insulating materials. If ESDS and non-ESDS components are placed in the same operation, electrostatic fields from insulating tape and reel packaging can become a risk to the ESDS devices. One way of reducing this risk may be to separate ESDS and non-ESDS components in the placement machine.

Very small devices and their packaging can become sufficiently highly charged that they may be ejected from the packaging in an uncontrolled manner by electrostatic forces (Swenson 2018).

5.9 Measurements in AHE

5.9.1 Overview of Measurements in AHE

Measurements in AHE typically include

- Resistance to ground from parts of equipment, surfaces, conveyor belts, or other items
- Electrostatic field and voltage measurements on essential insulators or ungrounded conductors
- Voltage measurements on ESDS devices or small conductors
- Charge measurements on conductors or insulators
- Voltage decay time and offset voltage due to neutralization by an ionizer
- Detection of ESD using radiated electromagnetic interference (EMI) from ESD

In addition, measurements of ESD current are being explored by some as a means of understanding charged device ESD risk.

The principles and practice of measurement given in Chapter 11 should in general be applied but may need some modification for use in AHE. Standard electrodes and test voltages may not be appropriate for use in automated equipment or packaging used with it.

Measurements on machines are usually made in the ambient conditions in which they will be operating. If the conditions are variable, it is useful to make measurements under minimum humidity conditions. These will often be conditions under which maximum electrostatic charging will occur. Charging may often increase also with operating speeds. This may make it difficult to make measurements under likely worst-case operating conditions.

Test arrangements will often need to be improvised according to the machinery and conditions. It can be challenging to interpret the results.

Specific guidance on making measurements in AHE is given in ESD SP10.1.

Testing will normally be done in the operating environment. Testing under operating conditions is often preferable but may not be possible due to safety considerations or the difficulty in connecting instrumentation to moving machine parts.

5.9.2 Resistance Measurements

5.9.2.1 Overview of Resistance Measurements in AHE

Typical resistance measurements will include

- Point-to-point resistance of surfaces
- Resistance to ground from surfaces and machine parts

Different resistance measurement methods usually give different results (Smallwood and Millar 2010; Smallwood 2017; Smallwood 2018). Large standard electrodes may not be usable due to lack of large planar surfaces. Large soft electrodes with high contact pressure such as S4.1 gave lower resistance values. The surface may have small regions of high conductivity, damage, or pin holes that show low resistance with these electrodes. Small electrodes often give considerably higher results than large electrodes. Metal-faced electrodes often give higher results in contact with higher resistance hard materials than electrodes faced with conductive rubber.

Small-area electrodes with hard surfaces are less likely to make good contact through low resistance or damaged regions in anodized layers or highly variable ESD control materials. Where possible, it is perhaps best to choose measurement electrodes that simulate the actual working situation. This may require some ingenuity in devising a suitable nonstandard electrode.

Smallwood and Millar (2010) also found that test voltage could change the resistance results by several orders of magnitude. Low voltages typically give high resistance results (see Section 5.7.6). Low test voltages may be needed to make meaningful measurements where grounding is required to control voltages on conductors to low levels.

5.9.2.2 Resistance Meters

When measuring resistance to ground of metal machine parts, a suitable multimeter can be used. For measurements of ESD control materials, a 10 V/100 V meter should be used as per other resistance measurements on ESD control equipment (see Section 11.6).

5.9.2.3 Point-to-Point Resistance of Surfaces

Point-to-point resistance measurements may be used to evaluate the surface resistance of surfaces that the ESDS may contact in passing through the equipment. To reduce charged device ESD risk, it is often preferable to use a static dissipative contacting material rather than a low-resistance material such as metal (see Section 5.7.5).

5.9.2.4 Resistance to Ground from Surfaces and Machine Parts

Perhaps the most common measurement is of resistance to ground from a surface or machine part. The resistance should be measured from all designated ESD ground points to the machine chassis and earth. The resistance from all conductors within the machine critical path should also be determined.

Resistance from machine parts to machine earth is advised by SP10.1 to be <1 Ω. This can be measured using a multimeter that has capability to measure resistance down to 0.1 Ω.

For moving parts, a stationary test may not represent performance during operation in motion. This is because oil films in bearings and other factors may cause intermittent connection.

5.9.2.5 Resistance to Ground of a Potentially Isolated Conductor

The resistance to ground of a conductor can be measured using a 10 V/100 V high-resistance meter to determine whether it is isolated. Making a reliable connection to the conductor for measurement can be a challenge.

5.9.2.6 Measurements on ESD Protective Packaging

Standard concentric ring resistance test electrodes are suited for measurement of large planar surfaces (see Chapter 11). A standard two-pin electrode is described in IEC 61340-2-3 and ESD S11.13 that is suited for measurements on smaller features or moderately curved surfaces. Many of the ESD-protective packaging designs used with small ESDS components in automated processes, such as component tapes, have small features that cannot easily be measured using these electrodes. At the time of writing, this is an area of current development in standardization. Some nonstandard electrodes are available from some suppliers. Different electrodes will in general give different results in resistance tests (Smallwood 2018).

5.9.3 Electrostatic Field and Voltage Measurements

5.9.3.1 Voltage Measurement Instruments

Many low-cost electrostatic "voltmeters" are in fact electrostatic field meters calibrated in terms of voltage to make readings from large flat surfaces (see Section 11.6.12). They give correct surface voltage readings only with a large flat conducting surface held at the correct calibration distance from the meter. With a large flat insulating surface, these meters give a reading that is a net result of the varying surface voltage across the surface. Nevertheless, the results of these measurements on large surfaces are useful in that they are a valid indicator of ESD risk due to net electrostatic fields in the vicinity.

Field meter–based instruments are not usually suitable for making voltage measurements on small objects. Field meter–based instruments can sometimes be used with smaller and nonflat surfaces by calibrating the meter in use with a target of the same shape and form raised to a known voltage.

Voltages on ESDS devices or other moving items typically change with the position of the item due to changes in capacitance and charge levels. If the item is moving quickly or in a confined space, it can be difficult to mount a voltmeter in a way that can measure the voltage on it. In a fast-moving system, the response speed of the voltmeter needs to be fast enough to follow the voltage changes accurately. A data logger or storage oscilloscope operating in roll mode may be needed to record voltage changes as they happen. In practice, it is likely to be the voltage on the conductor immediately before contact with another conductor that is relevant for evaluating charged device ESD risk.

5.9.3.2 Voltages on Small Conductors or ESDS Devices

When evaluating ESD risk from charged devices or small ungrounded metal objects contacting ESDS devices, it may be necessary to make voltage measurements on small conductors or ESDS pins. This can be challenging, especially for very small items. Voltages on these items can be made using noncontact or contact electrostatic voltmeters with very high input impedance and low input capacitance (see Section 11.6.12 and 11.9.4).

5.9.3.3 Voltages on ESDS Devices

It is often not practical to measure charge on ESDS directly, so voltage or field measurements are used instead (Yan et al. 2009). Voltage measurements on ESDS devices can be done with contacting or noncontacting electrostatic voltmeters.

Tribocharging occurs whenever the ESDS is package contacted by another material. Charging of the conductors within the device can occur via contact with a charge source via the device pins. Any charge on the device package, or nearby electrostatic field sources, will induce voltages on the device conductors. The device voltage typically increases dramatically when it is separated from a surface against which it has become tribocharged.

Tribocharging of the device cannot be avoided, because contact between the device and other materials such as packaging or handler part is inevitable. ESDS packages are often made of polymer or ceramic insulating materials, and tribocharging occurs on insulating ESDS package parts even when the contacting materials are conductive or static dissipative.

The voltage on a device passing through a handling system is not constant, even if the charge on the device is unchanging. Voltages on noninsulative parts vary with charge level, proximity to electrostatic field sources, and conductor capacitance. The voltage V on a conductor is related to its charge Q but changes with the capacitance C of the conductor as

$$CV = Q$$

Capacitance varies with the orientation and proximity to other materials, especially conductors. So, the ESDS voltage is constantly changing as the device moves through a process. SP10.1 describes a method of calibrating voltage measurements with a known voltage applied to a conductor attached to a device.

Discharge occurs when the device pins touch another conductor at a different voltage. So, it is the voltage on the device immediately before contact is made that is of most interest. At this point, the voltage may have reduced considerably from a value when the ESDS device is far from the contact point, due to capacitance changes.

Voltage levels on insulative parts of an item are even more variable as the charge density on the surface of the insulator varies from point to point across the surface. There is no single value that can be ascribed to the surface of an insulator such as a device package material. Voltages measured are typically the net effect of charge density over a region on the surface and will change with proximity to conductors including any voltage measuring equipment used. This can be a useful measurement in ESD risk evaluation as in practice it is often also the net effect of charges inducing a voltage on a conductor that causes ESD risk.

5.9.4 Charge Measurements

The charge on an item is sometimes measured either directly, e.g. using a Faraday pail or transferred charge measurement, or indirectly using field or voltage measurements (Paasi 2004). In some cases, the charge measured on a device can be compared with an estimated charge threshold for failure calculated from CDM ESD withstand data.

5.9.5 Measurement of the Voltage Decay Time and Offset Voltage Due to Neutralization by an Ionizer

Standard measurement of the effectiveness of ionizers is done using a charged plate monitor (CPM) in which the measurement plate is 150 × 150 mm (see Section 11.6.13). The positive and negative polarity charge decay time and offset voltage are measured. This size of plate approximates the size of a semiconductor wafer or PCB. Many devices or machine parts that are to be neutralized in AHE are much smaller than this. It is questionable whether a CPM plate this size is representative of charge neutralization on small devices. Many non-standard CPM instruments are available that have smaller plates a few tens of millimeters in dimensions. It is possible that these give more representative indication of neutralization of similarly sized devices.

5.9.6 ESD Current Measurements

ESD current measurements are being explored by some as a means of evaluating charged device ESD risk (Bellmore 2004; Tamminen et al. 2017a,b). This is a difficult type of measurement that shows some promise but has not yet been shown to be achievable in practice.

5.9.7 Detection of ESD Using EMI Detectors

ESD occurring within operating equipment can be detected from its electromagnetic radiation (in other words, EMI) using a suitable detector (see Section 11.8.10) (Millar and Smallwood 2010). Some detectors are designed to differentiate between HBM and CDM ESD, but in real situations ESD usually has variable characteristics somewhat different from these models. ESD detectors will detect any ESD in the vicinity, including those that have nothing to do with ESD threat to semiconductor devices, such as operating contactors. So, it is necessary to critically evaluate any ESD detected and determine its relevance to ESD threat. Sometimes this can be done by direct visual observation. If an ESD is detected at a time that a device makes contact with a conductive item, ESD risk is indicated.

5.10 Handling Very Sensitive Components

As time goes on, devices having low HBM and CDM ESD withstand voltages are becoming more common (ESDA 2016a). Some have HBM withstand voltage less than 100 V HBM and 200 V CDM withstand voltage.

The ESD TR10.0-01-02 discusses measurement and control issues in handling very sensitive devices in AHE. This document defines *hard ground* as a having resistance to ground of 1 Ω or less, and *soft ground* as between 1 kΩ and 1 GΩ. Machine parts typically require hard grounding, but an ESDS device making contact with this low resistance may risk charged device ESD damage. AHE parts that may contact ESDS may require soft grounding. Charged device ESD damage is typically due to high ESD peak current flow, and it is the resistance (or impedance, see Viheriäkoski et al. 2017) of the material at the point of contact that typically limits ESD current in the discharge.

When handling low CDM ESD withstand voltage devices in AHE, the voltages in the critical region and on the device need to be kept to a low level. This means that if ionizers are used for voltage reduction, the ionizer may need to have greater capacity to reduce voltages in a shorter time, as well as having low-offset voltage (see Section 5.7.9).

CPMs used for characterizing performance of ionizers should be as representative as possible of the situation that they are simulating. In the AHE environment, this may mean use of a small low capacitance CPM plate. For simulating neutralization of low CDM ESD withstand voltage devices, it may be better to measure decay time over a lower voltage range (e.g. 200–20 V) than the standard CPM measurement (typically 1000–100 V).

References

Bellmore, D. (2001). Anodized aluminium alloys – insulators or not? In: *Proc EOS/ESD Symp. EOS-23*, 141–148. Rome, NY: EOS/ESD Association Inc.

Bellmore, D.G. (2004). Paper 4A.6. Characterizing automated handling equipment using discharge current measurements. In: *Proc EOS/ESD Symp. EOS-26*. Rome, NY: EOS/ESD Association Inc.

ESD Association. (2015). ESD TR17.0-01-15. *Technical Report for ESD Process Assessment Methodologies in Electronic Production Lines – Best Practices used in Industry*, Rome, NY, EOS/ESD Association Inc.

ESD Association. (2016a) ESD Association Electrostatic Discharge (ESD) Technology roadmap – revised 2016. Available from: https://www.esda.org/assets/Uploads/docs/2016ESDATechnologyRoadmap.pdf [Accessed: 10th May 2017] Rome, NY, EOS/ESD Association Inc.

ESD Association (2016b) ANSI/ESD SP10.1-2016. *Standard practice for protection of Electrostatic Discharge Susceptible Items – Automated handling Equipment (AHE)*, Rome, NY, EOS/ESD Association Inc.

Gaertner, R. (2007). Paper 3B.1. Do we expect ESD-failures in an EPA designed according to international standards? The need for a process related risk analysis. In: *Proc. EOS/ESD Symp. EOS-29*, 192–197. Rome, NY: EOS/ESD Association Inc.

Gaertner, R. and Stadler, W. (2012). Paper 3B.5. Is there a correlation between ESD qualification values and the voltages measured in the field? In: *Proc. EOS/ESD Symp. EOS-34*. Rome, NY: EOS/ESD Association Inc.

Jacob, P., Gärtner, R., Gieser, H. et al. (2012). Paper 3B.8. ESD risk evaluation of automated semiconductor process equipment – A new guideline of the German ESD Forum e.V. In: *Proc. EOS/ESD Symp. EOS-34*. Rome, NY: EOS/ESD Association Inc.

Kim, D.-S., Lim, C.-B., Oh, D.-S. et al. (2012). Paper 2B.1. Minimizing electrostatic charge generation and ESD Event in TFT-LCD production equipment. In: *Proc. EOS/ESD Symp. EOS-34*. Rome, NY: EOS/ESD Association Inc.

Koh, L.H., Goh, Y., and Lim, S.H. (2013). Reliability assessment of high temperature automated handling equipment retrofit for CDM mitigation. In: *Proc. EOS/ESD Symposium EOS-35*, 43–48. Rome, NY: EOS/ESD Association Inc.

*Proc. EOS/ESD Symp. EOS-32*Millar, S. and Smallwood, J.M. (2010). Paper 3B.2. CDM Damage due to Automated Handling Equipment. In: , 217–223. Rome, NY: EOS/ESD Association Inc.

Olney, A., Gifford, B., Guravage, J., and Righter, A. (2003). Real-world charged board model (CBM) failures. In: *Proc. EOS/ESD Symp. EOS-25*, 34–43.

Paasi, J. (2004). ESD control in automated handling. In: *6th International ESD Workshop in Dresden, Germany*, September 7–8, 2004. Rome, NY: EOS/ESD Association Inc.

Smallwood, J. (2017). A practical comparison of surface resistance test electrodes. *J. Electrostat.* 88: 127–133.

Smallwood, J. (2018). Paper 4B3. Comparison of surface and volume resistance measurements made with standard and non-standard electrodes. In: *Proc. EOS/ESD Symp. EOS-40*. Rome, NY: EOS/ESD Association Inc.

Smallwood, J.M. and Millar, S. (2010). Paper 3B4. Comparison of methods of evaluation of charge dissipation from AHE soak boats. In: *Proc. EOS/ESD Symp. EOS-32*, 233–238. Rome, NY: EOS/ESD Association Inc.

Swenson D.E. 2018. Private communication.

Tamminen, P., Smallwood, J., and Stadler, W. (2017a). Paper 1B.4. Charged device discharge measurement methods in electronics manufacturing. In: *Proc. EOS/ESD Symp. EOS-39*. Rome, NY: EOS/ESD Association Inc.

Tamminen, P., Smallwood, J., and Stadler, W. (2017b). Paper 4B.2. The main parameters affecting charged device discharge waveforms in a CDM qualification and manufacturing. In: *Proc. EOS/ESD Symp. EOS-39*. Rome, NY: EOS/ESD Association Inc.

Tan, W.H. (1993). Minimizing ESD hazards in IC test handlers and automated trim/form machines. In: *Proc. EOS/ESD Symp. EOS-15*, 57–64. Rome, NY: EOS/ESD Association Inc.

Viheriäkoski, T., Kärjä, E., Gärtner, R., and Tamminen, P. (2017). Paper 4B.3. Electrostatic discharge characteristics of conductive polymers. In: *Proc. EOS/ESD Symp. EOS-39*. Rome, NY: EOS/ESD Association Inc.

Yan, K.P., Gaertner, R., Wong, C.Y., and Ong, C.T. (2009). Automatic handling equipment-The role of equipment maker on ESD protection. In: *Proc. EOS/SED Symp. EOS-31*, 1B.2-1–1B.2-6. Rome, NY: EOS/ESD Association Inc.

Further Reading

Bellmore, D.G. and Bernier, J. (2005). Characterizing automated handling equipment using discharge current measurements II. In: *Proc. EOS/ESD Symp.*, 195. Rome, NY: EOS/ESD Association Inc.

Dangelmeyer, T. (1999). *ESD Program Management*, 2e. Clewer. ISBN: 0-412-13671-6.

EOS/ESD Association Inc. (2016). ESD TR20.20-2016, *ESD Association Technical Report - Handbook for the Development of an Electrostatic Discharge Control Program for the Protection of Electronic Parts, Assemblies and Equipment*. Rome, NY, EOS/ESD Association Inc.

ESD Association. (2002). ESD TR10.0-01-02. *Technical Report - Measurement and ESD Control Issues for Automated Equipment Handling of ESD Sensitive Devices Below 100 Volts*, Rome, NY, EOS/ESD Association Inc.

ESD Association. (2006). ANSI/ESD STM4.1-2006. *ESD Association Standard for the Protection of Electrostatic Discharge Susceptible Items – Worksurfaces - Resistance Measurements*, Rome, NY, EOS/ESD Association Inc.

ESD Association. (2014). ANSI/ESD S20.20-2014. *ESD Association Standard for the Development of an Electrostatic Discharge Control Program for – Protection of Electrical and Electronic Parts, Assemblies and Equipment (excluding Electrically Initiated Explosive Devices)*, Rome, NY, EOS/ESD Association Inc.

ESD Association. (2015) ANSI/ESD STM11.13-2015. *ESD Association Standard Test Method for the Protection of Electrostatic Discharge Susceptible Items – Two-Point Resistance Measurement*, Rome, NY, EOS/ESD Association Inc.

Halperin, S., Gibson, R., and Kinnear, J. (2008). Paper 2B-21. Process capability & transitional analysis. In: *Proc. EOS/ESD Symp. EOS-30*. Rome, NY: EOS/ESD Association Inc.

International Electrotechnical Commission. (2015) IEC 61340-5-3:2015. *Electrostatics. Protection of electronic devices from electrostatic phenomena. Properties and requirements classifications for packaging intended for electrostatic discharge sensitive devices*, International Electrotechnical Commission, Geneva.

International Electrotechnical Commission. (2016a). IEC 61340-2-3:2016. *Electrostatics. Part 2-3: Methods of test for determining the resistance and resistivity of solid materials used to avoid electrostatic charging*, International Electrotechnical Commission, Geneva.

International Electrotechnical Commission (2016b) IEC 61340-5-1: 2016. *Electrostatics – Part 5-1: Protection of electronic devices from electrostatic phenomena - General requirements*, International Electrotechnical Commission, Geneva.

International Electrotechnical Commission. (2018) IEC TR 61340-5-2:2018. *Electrostatics – Part 5-2: Protection of electronic devices from electrostatic phenomena - User guide*. International Electrotechnical Commission, Geneva.

Kietzer, G. (2012). Paper 2B.2. ESD risks in the electronics manufacturing. In: *Proc. EOS/ESD Symp. EOS-34*, 202. Rome, NY: EOS/ESD Association Inc.

Kim, D.-S., Lim, C.-B., Yoon, S.-H. et al. (2013). Paper 7B.1. Electrostatic control and its analysis of roller transferring processes in FPD manufacturing. In: *Proc. EOS/ESD Symp. EOS-35*. Rome, NY: EOS/ESD Association Inc.

Koh, L.H., Goh, Y., and Lim, S.H. (2013). Paper 1B.3. Reliability assessment of high temperature automated handling equipment retrofit for CDM mitigation. In: *Proc. EOS/ESD Symp. EOS-35*. Rome, NY: EOS/ESD Association Inc.

Koh, L.H., Goh, Y.H., and Wong, W.F. (2017). Paper 1B.2. ESD risk assessment considerations for automated handling equipment. In: *Proc. EOS/ESD Symp. EOS-39*. Rome, NY: EOS/ESD Association Inc.

Paasi, J., Tamminen, P., Kalliohaka, T. et al. (2002). ESD control tools for surface mount technology and final assembly lines. In: *Proc. EOS/ESD Symp. EOS-24*, 250–256. Rome, NY: EOS/ESD Association Inc.

Paasi, J., Tamminen, P., Salmela, H. et al. (2005). ESD control in automated placement process. In: *Proc. EOS/ESD Symposium EOS-27*, 203. Rome, NY: EOS/ESD Association Inc.

Steinman, A. (2010). Paper 3B3. Measurements to establish process ESD compatibility. In: *Proc. EOS/ESD Symp. EOS-32*. Rome, NY: EOS/ESD Association Inc.

Steinman, A. (2012). Paper 2B.4. Process ESD capability measurements. In: *Proc. EOS/ESD Symp. EOS-34*. Rome, NY: EOS/ESD Association Inc.

Steinman, A. (2014). Paper 1B.3. Measuring handler CDM stress provides guidance for factory static controls. In: *Proc. EOS/ESD Symp. EOS-36*. Rome, NY: EOS/ESD Association Inc.

Tamminen, P. and Viheriäkoski, T. (2007). Paper 3B.3. Characterization of ESD risks in an assembly process by using component-level CDM withstand voltage. In: *Proc. EOS/ESD Symp. EOS-29*, 202–211. Rome, NY: EOS/ESD Association Inc.

Tamminen, P. and Viheriäkoski, T. (2011). Product specific ESD risk analysis. In: *Proc. EOS/ESD Symp. EOS-33*, 97. Rome, NY: EOS/ESD Association Inc.

Welker, R.W., Nagarajan, R., and Newberg, C. (2006). *Contamination and ESD Control in High-Technology Manufacturing*. Wiley-Interscience/IEEE Press. ISBN-10: 0 471 41452 2, ISBN-13: 978 0 471 41452 0.

Yan, K.P., Gaertner, R., and Wong, C.Y. (2010). Paper 3B.1. ESD protection program at electronics industry – areas for improvement. In: *Proc. EOS/ESD Symp. EOS-32*. Rome, NY: EOS/ESD Association Inc.

Yan, K.P., Gaertner, R., and Wong, C.Y. (2013). Semiconductor back end manufacturing process – ESD capability analysis. In: *Proc. EOS/ESD Symposium EOS-35*, 30. Rome, NY: EOS/ESD Association Inc.

6

ESD Control Standards

6.1 Introduction

The facilities equipment and materials needed for use in electrostatic discharge (ESD) prevention are specified in various standards worldwide. The two main standards discussed in this book are IEC 61340-5-1 and ESD Association ANSI/ESD S20.20. IEC 61340-5-1 has also been adopted as a national standard in many countries, in Europe becoming the European Norm EN61340-5-1. Individual countries may have their own versions, and, in the United Kingdom, this is BS EN 61340-5-1.

For clarity, the 61340-5 series terminology is used, although the ESD Association (ESDA) series documents are in many cases nearly identical and are cross-referenced. In many cases, International Electrotechnical Commission (IEC) documents were based on ESDA standards that were published earlier. Where appropriate, some differences with the ESDA standards are noted.

6.2 The Development of ESD Control Standards

Before the days when ESD control became necessary in electronics manufacture, ESD control practices were already in use in gunpowder and explosives handling and some industrial processes where flammable materials were handled. Until the 1970s there was little need for ESD control in electronics manufacture and component handling (Ind. Co. 2011). ESD damage started to show as a problem in the late 1970s with the introduction of large-scale integration (LSI). The first ESD control programs were set up by industry companies, with little standardization or sharing of information on the subject.

One early standard was the US Military MIL-STD-1686 (Department of Defense (1980a), released with the handbook MIL-HDBK-263 (Department of Defense (1980b). Companies that supplied electronics to the military were required to comply with this standard, but others followed in-house ESD control procedures.

Some countries produced national standards such as the BS 5783:1984 in the United Kingdom. A European standard CECC00015 was introduced in 1991, providing a common standard for countries in the CENELEC Electronics Components Committee (CECC) and superseding BS5783 as the BS EN 100015-1:1992. At the time, CENELEC members included national electrotechnical committees of Austria, Belgium, Denmark, Finland,

The ESD Control Program Handbook, First Edition. Jeremy M Smallwood.

France, Germany, Greece, Iceland, Ireland, Italy, Luxembourg, Netherlands, Norway, Portugal, Spain, Sweden, Switzerland, and the United Kingdom. The EN 100015 had four parts (British Standards Institute 1992, 1994a,b,c). BS EN100015-1 gave the basic requirements for compliance, setting up ESD-protected areas (EPAs) and ESD control packaging. BS EN100015-2 gave additional requirements for low-humidity conditions less than 20% rh. Part 3 of the standard gave requirements for clean room conditions, and Part 4 for high-voltage areas. All these parts of EN 100015 were in turn superseded by the first version of EN61340-5-1.

Materials and equipment used in ESD control initially did not have standard test methods, procedures, or equipment to measure their properties. The variety of test methods and equipment used led to differences in results.

The ESD Association was formed in the early 1980s and commenced standards development for ESD control materials and equipment such as flooring, wrist straps, and work surfaces. Suppliers and users could use these to verify the properties and functioning of ESD control materials and equipment on a commonly recognized basis.

In 1995, the ESDA was given the task of replacing MIL-STD-1686 with an industry standard, leading to the development of ANSI/ESD S20.20-1999. Meanwhile in the 1990s the IEC had set up IEC Technical Committee 101 Electrostatics, to produce internationally applicable standards for test methods for evaluation of the generation, retention, and dissipation of ESD, ascertaining the effect of ESDs and methods of simulation of electrostatic phenomena. The scope of TC101 was later extended to include design and implementation of handling areas or procedures, equipment, and materials used to reduce or eliminate electrostatic hazards or undesirable effects and exclusions. Much of TC101's work has focused on electronics industry ESD control, although more generally applicable standards (e.g. for measurement of resistance of static control materials, properties of flooring, and footwear and flooring in combination) were developed for use in a wide range of applications. These were based on existing ESD Association, European, Asian, or industry standards. Electrostatic control equipment and material for other industries (e.g. properties of flexible intermediate bulk containers used in the chemical industry) are also developed by IEC TC101.

The first IEC 61340-5-1 published in 1998 was developed from the CECC100015 and other standards and had a less flexible approach compared to the ANSI/ESD S20.20 standard (EOS/ESD Association Inc 2014c). The first decade of the twenty-first century saw the globalization and increasing diversity of electronic industry processes and environments and increasing spread of the influence and use of the ANSI/ESD S20.20 worldwide. In response, the second edition 61340-5-1 adopted an approach based on ANSI/ESD S20.20 in the spirit of worldwide harmonization of ESD control standards for the electronics industry, albeit with some remaining differences. These have largely been eliminated with the publication of the 2016 edition of 61340-5-1 (International Electrotechnical Commission. 2016c). It is now widely considered that the 61340-5-1 and ANSI/ESD S20.20 standards are technically equivalent.

Some earlier standards gave guidance as well as requirements for compliance in the same documents. This was often confusing for the user, and many interpreted the guidance as aspects that must be implemented for compliance. Later standards removed guidance into separate documents IEC 61340-5-2 and ESD TR20.20 (International Electrotechnical Commission 2018; EOS/ESD Association Inc. 2016b). These user guides contained no requirements for compliance but gave an increasing amount of useful information and guidance to help develop and maintain an effective ESD control program.

These documents that do not contain requirements for compliance, but often contain guidance or information of a different type, are called "Technical Reports." In the IEC and ESDA systems they are designated by "TR" in the document number, for example ESD TR20.20 or IEC TR 61340-5-2. They can be extremely useful sources of information of many types. The references at the end of this chapter are a list of many of these.

6.3 Who Writes the Standards?

The ESD Association and IEC standards are written by volunteer experts working within Working Groups (WGs) at the ESDA and IEC Technical Committee 101 *Electrostatics*. The WG experts are usually volunteers who may or may not be supported by their employers or other organizations to work on ESD standards. In their professional life, they may be engaged in electronics manufacturing companies, ESD-protective equipment manufacturers, or suppliers, research, or consultancy work.

The ESDA has been developing test methods for ESD work since the early 1980s. Many of the currently used standard test methods were developed by ESDA and then adopted, with some revision or adaption, for the IEC system. They are then adopted as national standards in various IEC participating countries around the world. In most cases, the requirements of these standards and standard test methods specified in both systems are very similar.

The IEC was mandated by the World Trade Organization to draft electrotechnical standards to facilitate trade worldwide. IEC TC101 is responsible for electrostatic-related test methods and other documents. The IEC TC101 experts are nominated by their national standards committees (NCs) to work on projects of interest to their countries. As an example, in the United Kingdom, the BSI Committee GEL101 is the main NC following and participating in TC101's work. During development of a new document, drafts are returned to GEL101 on a regular basis for comment, and these comments are considered in the rewriting of a new draft. The process continues through various stages until a draft International Standard is achieved by consensus of all the participating NCs. In 2017, the participating members of IEC TC101 included 20 Participating members (who attend meetings and return comments on draft documents) and 13 Observer members (Table 6.1).

While most IEC TC101 member countries adopt the 61340-5-1 standard as their national standard, not all do. Notably, the United States maintains the ANSI/ESD S20.20 standard as its American National Standards Institute (ANSI) standard.

When a new IEC standard is published, it is also submitted to the European standardization organization CENELEC for possible adoption as a European Standard and to the standardization bodies of the participating countries. This is then voted on by European NCs participating in CENELEC. If adopted, it becomes a European norm (EN). In the United Kingdom, European Norms are normally then adopted by member countries. In the case of the United Kingdom, the UK standardization organization BSI, if adoption is agreed, publishes the document as a BS EN documents. So, IEC 61340-5-1: 2016 became EN 61340-5-1: 2016 in Europe and then published as BS EN 61340-5-1:2016 in the United

Table 6.1 IEC TC101 Participating (P) and Observer (O) members in 2017.

Argentina	O-Member
Australia	O-Member
Austria	P-Member
Belgium	P-Member
Bulgaria	O-Member
China	P-Member
Czech Republic	P-Member
Denmark	O-Member
Egypt	O-Member
Finland	P-Member
France	P-Member
Germany	P-Member
Hungary	P-Member
Ireland	P-Member
Israel	O-Member
Italy	P-Member
Japan	P-Member
Korea, Republic of	P-Member
Netherlands	P-Member
Norway	O-Member
Poland	P-Member
Portugal	O-Member
Romania	O-Member
Russian Federation	P-Member
Serbia	O-Member
Spain	P-Member
Sweden	P-Member
Switzerland	P-Member
Thailand	O-Member
Turkey	O-Member
Ukraine	O-Member
United Kingdom	P-Member
United States of America	P-Member

Kingdom. All these steps and procedures take an amount of elapsed time. It is often several years between commencement of a project at IEC and publication of an IEC standard and then several further months before publication as BS EN in the UK or other national standard!

6.4 The IEC and ESDA Standards

6.4.1 Standards Numbering

All standards are given a number according to a system. A list of ESDA and IEC test method publications used in ESD work and current at the time of writing is given in Chapter 11. The good news is that the average ESD practitioner in the electronics industry will not need to have copies of all of these! Many of the modern ESD test methods are similar in their ESDA and IEC forms. Beware, however, that there can also be some significant differences.

6.4.2 The Language of Standards

Standards use some words in a specific way. Many of the aspects described are "requirements" – which in standards terms means "you have to do this to comply with the standard." However, some aspects only have the status of "recommendations" – which means departing from their instructions does not represent noncompliance with the standard. It should, however, be born in mind that there may be good reasons for acting according to the recommendations.

In the standards, the difference between "requirements" and "recommendations" is often shown by the language used within the text. The word *shall* represents a "requirement," but the word *should* represents a "recommendation." For example, 61340-2-3:2016a Section 8.2.1 states "The contact material surface shall have a volume resistance of less than 10^3 Ω…" The *shall* means that for compliance the electrode manufacturer must select the electrode material with this characteristic.

In places, some standards give "examples," e.g. in 61340-2-1:2002 Figure 1 gives "Example of an arrangement for measurement of dissipation of charge using corona charging." International Electrotechnical Commission. (2015a) Examples are for assistance of the reader. Departing from the example is not necessarily a noncompliance, although it is often easiest to follow the example when making a test setup.

In 61340-5-1:2016, Annex A is labeled as "Normative." The word *normative* is another way of indicating that the test methods in this annex are part of the "requirements" of the standard. In contrast, a section marked as "Informative" means that the section can be considered merely informational or "recommendations."

IEC standards have "Normative references" sections that includes a list of standards that are deemed essential to the application of the standard. Where a reference to a standard in this section is dated (e.g. IEC 61340-4-1: 2003), this means that only this version must be used – later versions of the standard may exist but do not apply and may not even make sense in the context.

In contrast, an undated reference (e.g. IEC 61340-4-1 with no attached date) means that the latest and current version of the standard should be referred to.

Some IEC documents also are designated "TS," e.g. PD IEC/TS 61340-4-2 Electrostatics – Part 4-2: Standard test methods for specific applications – Electrostatic properties of garments (International Electrotechnical Commission 2013b). A Technical Specification document is one for which the required support was not achieved for the publication of an International Standard, the subject is still under technical development, or there is the future but not immediate possibility of agreement as an International Standard. It therefore resembles, but does not have the strength and status of an International Standard. Technical specifications normally reviewed within three years.

A third type of document is the Technical Report (TR), e.g. IEC TR 61340-5-2 Electrostatics – Part 5-2: Protection of electronic devices from electrostatic phenomena – User guide. These include data of a different kind from that normally published as an International Standard. TR documents must be entirely informative in nature and must not contain matter implying that it is normative.

Some of the ESD Association standards are also ANSI. For example, the 20.20-2014 standard is ANSI/ESD S20.20-2014.

Many of the ESDA documents referred to here are Standards identified by the letter *S* in the numbering system. These give "a precise statement of a set of requirements to be satisfied by a material, product, system, or process that also specifies the procedures for determining whether each of the requirements is satisfied" (DE Swenson 2017 private communication). The letters STM indicate a Standard Test Method (e.g. ANSI/ESD STM11.13-2015. ESD Association Standard Test Method for the Protection of Electrostatic Discharge Susceptible Items – Two-Point Resistance Measurement). These documents give "a definitive procedure for the identification, measurement, and evaluation of one or more qualities, characteristics, or properties of a material, product, system, or process that yields a reproducible test result."

The letters SP identify a "Standard Practice document (e.g. ANSI/ESD SP15.1 Standard Practice for the Protection of Electrostatic Discharge Susceptible Items – In-Use Resistance Measurement of Gloves and Finger Cots). Standard Practice documents give "a procedure for performing one or more operations or functions that may or may not yield a test result." Standard Practices may not give reproducible results.

Like the IEC system, the ESDA also has other types of documents that are not Standards. Some of these are TRs and give "a collection of technical data or test results published as an informational reference on a specific material, product, system or process." These documents do not have technical requirements; they are informational in nature. Examples are ESD TR20.20 and ESD TR53-01-06 (EOS/ESD Association Inc 2006b). The ESDA also has ADV, or "Advisory," documents (e.g. ESD ADV1.0-2017. ESD Association Advisory for Electrostatic Discharge Terminology – Glossary). These are usually older documents that would be published as SP or TR today.

There has been some effort to harmonize the ESDA and IEC series standards in various ways. So, some of the documents in the series are to some extent similar or near equivalent (Tables 6.2 and 6.3).

6.4.3 Definitions Used in Standards

The standards give specific definitions for some terms used within the standards. These may differ from document to document and from version to version. For example, the definitions of *conductive*, *dissipative*, and *insulative* often differ between standards in different industries and for different products such as footwear or packaging materials. There may also occur changes with different versions of standards. For example, the resistance range definitions of conductive and dissipative packaging materials have changed with

Table 6.2 Near-equivalent ESD control program standards in the IEC 61340-5-1 and ANSI/ESD S20.20 series of documents.

IEC 61340-5-1 series	ANSI/ESD S20.20 series	Content
IEC 61340-5-1 Protection of electronic devices from electrostatic phenomena – General requirements	ANSI/ESD S20.20 Development of an Electrostatic Discharge Control Program for – Protection of Electrical and Electronic Parts, Assemblies, and Equipment	General requirements for an ESD control program
IEC 61340-5-2 Protection of electronic devices from electrostatic phenomena – User guide.	ESD TR 20.20 *Protection of Electrical and Electronic Parts, Assemblies, and Equipment – Handbook.*	User guides for implementing an ESD control program according to 61340-5-1 and ANSI/ESD S20.20
IEC 61340-5-3 *Properties and requirements classifications for packaging intended for ESD sensitive devices.*	ESD Association. Packaging Materials for ESD Sensitive Items. ANSI/ESD S541	Properties and classifications for packaging for ESD-sensitive devices

Table 6.3 Near-equivalent test method documents in the IEC 61340-5-1 and ANSI/ESD S20.20 series.

IEC 61340-5-1 series	ANSI/ESD S20.20 series	Content
IEC 61340-5-4:2019	ESD TR53-01-06 Compliance Verification of ESD Protective Equipment and Materials.	Compliance verification measurement techniques
IEC61340-2-3 Methods of test for determining the resistance and resistivity of solid planar materials used to avoid electrostatic charging	ANSI/ESD STM11.11:2015b Surface resistance measurement of static dissipative planar materials. ANSI/ESD STM11.12:2015c Volume resistance measurement of static dissipative planar materials. ANSI/ESD STM11.13:2015d Two-point resistance measurement.	General surface and volume resistance test methods
IEC61340-4-3:2001 Footwear	ANSI/ESD STM9.1:2014b Footwear – Resistive Characterization.	Test methods for the resistance of footwear – not worn
IEC 61340-4-1 Electrical resistance of floor coverings and installed floors.	ANSI/ESD STM7.1:2013b Resistive Characterization of Materials – Floor Materials.	Measurement of resistance of floor materials before and after installation
IEC 61340-4-5:2004 Methods for characterizing the electrostatic protection of footwear and flooring in combination with a person	ANSI/ESD STM97.1:2015e Floor Materials and Footwear – Resistance Measurement in Combination with a Person. ANSI/ESD STM97.2:2016a Floor Materials and Footwear – Voltage Measurement in Combination with a Person.	Measurement of resistance and body voltage of person-footwear-floor system

Table 6.3 (Continued)

IEC 61340-5-1 series	ANSI/ESD S20.20 series	Content
IEC61340-2-3 Methods of test for determining the resistance and resistivity of solid planar materials used to avoid electrostatic charging	ANSI/ESD STM4.1:2006a Worksurfaces – Resistance Measurements.	Measurement of resistance to ground and to groundable point of work surfaces
IEC 61340-4-9:2016b Garments	ANSI/ESD STM 2.1 Garments.	Resistance measurement of ESD control garments
IEC61340-2-3 Methods of test for determining the resistance and resistivity of solid planar materials used to avoid electrostatic charging	ANSI/ESD STM12.1:2013a Seating – Resistive Measurement.	Measurement of resistance to groundable point of ESD control seating
IEC 61340-4-6:2015b Wrist straps.	ANSI/ESD S1.1:2013c Wrist Straps.	Requirements and measurement of wrist straps
IEC61340-2-3 Methods of test for determining the resistance and resistivity of solid planar materials used to avoid electrostatic charging	ESD SP9.2:2003 Footwear – Foot Grounders Resistive Characterization	Measurement of resistance of foot grounders
IEC 61340-4-7:2017 Ionization	ANSI/ESD STM3.1:2015a Ionization.	Measurement of performance of ionizers
IEC 61340-4-8:2014 Discharge shielding. Bags.	ANSI/ESD STM11.31:2012b Bags.	Measurement of performance of ESD shielding bags

different versions of 61340-5-1 and 61340-5-3 (International Electrotechnical Commission 2015c). When working according to a standard, the definitions used in that standard must be applied. Incorrect use of these terms can result in inadvertent use of noncompliant equipment or materials.

6.5 Requirements of IEC 61340-5-1 and ANSI/ESD S20.20 Standards

6.5.1 Background

The earlier editions of IEC 61340-5-1 and ANSI/ESD S20.20 were superseded by new versions during 2007. The requirements of these new standards were broadly similar – previously 61340-5-1 had been significantly different in its approach from ANSI/ESD S20.20.

IEC 61340-5-1 was adopted by CENELEC and replaced the first edition of EN61340-5-1: 2001 in Europe.

A new version of ANSI/ESD S20.20 was published in 2014. This was followed by a new version 61340-5-1 in 2016. These versions were current at the time of writing, and the following sections summarize the requirements of these documents. The documents continue to be updated in the light of technology and industry changes and the development of ESD protection and control knowledge and understanding as time goes on. The reader should of course check current versions of the standards for up to date information on them.

6.5.2 Documentation and Planning

The IEC 61340-5-1:2016 and ANSI/ESD S20.20-2014 standards require the organization to write four ESD plans.

- ESD Control Program Plan
- ESD Training Plan
- (ESD Control) Product Qualification Plan
- Compliance Verification Plan

The standard does not dictate in detail what should be contained in these plans – it is the responsibility of the ESD coordinator to specify this. The purpose and content of the plans is further explained in the following sections. Equipment used in the ESD program is normally expected to comply with the standard (where requirements are given) unless a tailoring statement is given explaining the rationale and technical justification for the departure is acceptable.

Although the requirements of 61340-5-1 are nearly identical, users seeking to comply with these standards are advised to obtain copies and refer to them. There may be small differences between the detail of the requirements and wording. Up-to-date copies should always be obtained as the standards are updated every few years as knowledge and technological needs progress and as ideas of best practice change from time to time. The following sections give an outline of the main requirements of the standards at the time of writing (2019).

6.5.3 Technical Basis of the ESD Control Program

The ESD control program designed according to these standards is intended to provide effective handling of ESD-sensitive (ESDS) devices with ESD withstand voltages down to 100 V HBM and 200 V charged device model (CDM). The scope of 61340-5-1 states that it "applies to activities that: manufacture, process, assemble, install, package, label, service, test, inspect, transport or otherwise handle electrical or electronic parts, assemblies, and equipment."

An ESD control program can be devised according to the standards for handling ESDS devices with lower withstand voltages. In this case, the program may need additional control elements or adjusted requirement limits for the specified equipment.

It is important to understand that these standards are not intended to be used in non-electronic applications and processes. Handling of, and processes involving, electrically initiated explosive devices, flammable liquids, gases, and powders are specifically excluded by the scope. Some of these nonelectronic applications may be covered by specific national regulations or other standards such as IEC 60079-32-1 International Electrotechnical Commission. (2013a).

6.5.4 Personal Safety

The standards recognize that use of some ESD control equipment could conceivably expose personnel to unsafe conditions under some circumstances. In other cases, safety or other regulations may apply to materials, equipment, or processes in which ESDS devices are also handled. In these cases, ESD control practices must be compliant with the relevant regulations, laws, or codes of practice. ESD control practices never supersede correct personal safety requirements.

6.5.5 ESD Coordinator

It is a requirement of the standards that a person must be appointed who has responsibility for implementing the requirements of the standard including establishing, documenting, maintaining, and verifying the compliance of the ESD program. Some organizations know the ESD Coordinator by other titles (e.g. ESD program manager).

6.5.6 Tailoring the ESD Program

Parts of the standard may not apply in all cases. Where a part of the standard has been assessed to be not applicable, requirements may be added, modified, or deleted as necessary. These tailoring decisions must be documented including a rationale and technical justification for the decision.

Tailoring statements are not required if the requirements of the ESD program are within the requirements of the standard. For example, if the maximum resistance to ground requirement of the ESD program for a floor is 10 MΩ and the requirement of the standard is 1 GΩ, the requirement of the standard is not exceeded, and no tailoring statement is required. On the other hand, if the maximum resistance to ground specified for chairs in the ESD program were 10 GΩ, and the standard required a maximum of 1 GΩ, a tailoring statement would be required for compliance.

6.5.7 The ESD Control Program Plan

IEC 61340-5-1:2016 states that "The ESD control program shall include all the administrative and technical requirements of this standard" and "The organization shall establish, document, implement, maintain, and verify the compliance of the program in accordance with the requirements of this standard."

The ESD Control Program Plan must cover

- Training
- Product qualification

- Compliance verification
- Grounding and bonding systems
- Personnel grounding
- EPA requirements
- Packaging systems
- Marking

The ESD Control Program Plan is the main document for implementing and verifying the ESD program and must be applied to all relevant areas. It should conform to internal quality requirements.

IEC 61340-5-1 requires that the ESD control program documents the lowest ESD withstand voltage that can be handled by the program. As it can be difficult to find the ESD withstand data on all components, it may be easiest to state the standard ESD withstand voltages of 100 V HBM and 200 V CDM in the ESD Control Program Plan.

6.5.8 Training Plan

The Training Plan must define

- The personnel required to have ESD training
- The type and frequency of training
- A requirement for maintaining training records
- Where the records are stored
- Methods used to ensure trainee comprehension and training adequacy

Initial and recurrent training must be provided to all personnel who handle or come into contact with ESDS devices. Initial training must be provided before personnel handle ESDS.

6.5.9 Product Qualification Plan

All ESD control items that are selected for use must be qualified. Test methods and requirements are given in the standards. Evidence for qualification may include

- Manufacturer data sheets
- Laboratory tests, including internal and third-party tests
- Ongoing compliance verification records for items that were installed before adoption of the standard

6.5.10 Compliance Verification Plan

The Compliance Verification Plan is intended to check that the requirements of the ESD program are met. This plan must document the following:

- The items to be measured and verified
- Pass criteria (measured parameter limits)
- Test methods used, including those not covered by the standard
- Frequency of measurements

Test methods may be used that differ from those given in the standard, but evidence must be provided in a tailoring statement that the results correlate with the reference standards.

Records must be kept providing evidence of compliance with the technical requirements of the ESD Program.

6.5.11 Test Methods

Most test methods called up by IEC61340-5-1 and ANSI/ESD S20.20 are similar or nearly identical. Compliance verification test methods are given in ESD TR53, and it is likely that a similar document will be produced in the IEC system in due course

Two types of tests are required – product qualification tests and compliance verification tests. Product qualification tests are to be used only when selecting equipment for use in the ESD Program. They would not be intended for regular checking and testing of equipment in use. Compliance verification tests are given for this purpose.

Compliance of equipment is evaluated based on measurements. The key measurements used are

- Resistance to ground (R_g) or to groundable point (R_{gp}).
- Point-to-point resistance ($R_{p\text{-}p}$).
- Packaging surface (R_s) or volume resistance (R_v) measurements. Packaging measurements using a miniature two-pin probe is treated as surface resistance.
- Packaging ESD shielding (bags).
- Measurements of electric field and potentials.
- Body voltage measurements.
- Ionizer decay time and offset voltage.

A summary of the requirements for equipment used in the EPA is given in Tables 6.4–6.8.

ANSI/ESD S20.20 also gives some requirements for soldering iron-tip resistance to ground, tip voltage, and tip leakage current (Table 6.9).

6.5.12 ESD Control Program Plan Technical Requirements

6.5.12.1 Safety

Technical limits given in these standards do not address lower resistance limits that may be required for safety. Due consideration should be given to whether such limits may be needed.

Table 6.4 Grounding requirements.

	IEC 61340-5-1:2016 requirements		ANSI/ESD S20.20-2014 requirements	
Grounding method	**Test method**	**Required limit**	**Test method**	**Required limit**
Protective earth	National electrical standard	National electrical code	ANSI/ESD S6.1:2014a	<1 Ω
Functional ground	National electrical standard	National electrical code	ANSI/ESD S6.1	<1 Ω
Equipotential bonding	See Tables 6.6 and 6.5	See Tables 6.6 and 6.5	ANSI/ESD S6.1	<1 GΩ

Table 6.5 Personal grounding requirements.

	IEC 61340-5-1:2016 requirements and test method		ANSI/ESD S20.20-2014 requirements and test method	
Measurement	**Product qualification**	**Compliance verification**	**Product qualification**	**Compliance verification**
Wrist band (not worn)	Interior $\leq 10^5\ \Omega$ Exterior $>10^7\ \Omega$ IEC 61340-4-6	Not specified	Interior $\leq 10^5\ \Omega$ Exterior $>10^7\ \Omega$ ANSI/ESD S1.1	Wrist strap system test is used
Wrist strap cord	$< 5 \times 10^6\ \Omega$ or user defined IEC 61340-4-6	Wrist strap system test is used	$0.8 \times 10^5 \leq R \leq 1.2 \times 10^6\ \Omega$ ANSI/ESD S1.1	
Wrist strap connection point (not monitored)	Not specified	$< 5 \times 10^6\ \Omega$	$R_g < 2\ \Omega$ ANSI/ESD S6.1	$R_g < 2\ \Omega$ ESD TR53 grounding and bonding systems
Wrist strap system test	Not specified	$< 3.5 \times 10^7\ \Omega$ IEC 61340-4-6	$< 3.5 \times 10^7\ \Omega$ ANSI/ESD S1.1 Sec. 6.11	$< 3.5 \times 10^7\ \Omega$ ESD TR53 wrist strap
Continuous monitors	Not specified	Not specified	User defined	Manufacturer defined limit ESD TR53 continuous monitors
Footwear	$< 10^8\ \Omega$ IEC 61340-4-3	Person-footwear system test is used	$< 10^9\ \Omega$ ANSI/ESD STM 9.1 ANSI/ESD STM 9.2	$R_g < 10^9\ \Omega$ ESD TR53 Footwear section
Person, footwear and flooring system test	Body voltage <100 V AND $R_g < 10^9\ \Omega$ IEC 61340-4-5	$R_g < 10^9\ \Omega$ IEC 61340-4-5 Periodic voltage test	Body voltage peak <100 V and floor $R_g < 10^9\ \Omega$ ESD STM97.1 and 97.2	Footwear $R_g < 10^9\ \Omega$ ESD TR53 Footwear section Flooring $R_g < 10^9\ \Omega$ ESD TR53 Flooring section
Person footwear system test	Not specified	$R_g < 10^9\ \Omega$ IEC 61340-5-1 Annex A		

Table 6.6 Requirements for benches, floors, and seats.

	IEC 61340-5-1:2016 requirements and test method		ANSI/ESD S20.20-2014 requirements and test method	
Measurement	**Product qualification**	**Compliance verification**	**Product qualification**	**Compliance verification**
Bench surface, racks, trolleys etc. surface	$R_{p\text{-}p} < 10^9\ \Omega$ IEC 61340-2-3 $R_{gp} < 10^9\ \Omega$ IEC 61340-2-3	$R_g < 10^9\ \Omega$ IEC 61340-2-3	$R_{p\text{-}p} < 10^9\ \Omega$ ANSI/ESD STM4.1 $R_{gp} < 10^9\ \Omega$ ANSI/ESD STM4.1 $R_g < 10^9\ \Omega$ ANSI/ESD STM4.1 < 200 V ANSI/ESD STM4.2:2012a	N/A $R_g < 10^9\ \Omega$ ESD TR53 worksurface ESD TR53 mobile equipment
Floor	$R_{gp} < 10^9\ \Omega$ IEC 61340-4-1	$R_g < 10^9\ \Omega$ IEC 61340-4-1	$R_g < 10^9\ \Omega$ ANSI/ESD STM7.1 $R_{gp} < 10^9\ \Omega$ ANSI/ESD STM7.1 $R_{p\text{-}p} < 10^9\ \Omega$ ANSI/ESD STM7.1	$R_g < 10^9\ \Omega$ ESD TR53 Flooring section
Chair	$R_{gp} < 10^9\ \Omega$ IEC 61340-2-3	$R_{gp} < 10^9\ \Omega$	$R_{gp} < 10^9\ \Omega$ ESD STM12.1	$R_g < 10^9\ \Omega$ ESD TR53 seating

Table 6.7 Requirements for ESD control garments.

	IEC 61340-5-1:2016 requirements and test method		ANSI/ESD S20.20-2014 requirements and test method	
Measurement	**Product qualification**	**Compliance verification**	**Product qualification**	**Compliance verification**
Static control garment	$R_{p\text{-}p} \leq 10^{11}\ \Omega$ or user defined IEC 61340-4-9 or user defined	$R_{p\text{-}p} \leq 10^{11}\ \Omega$ or user defined IEC 61340-4-9 or user defined	$R_{p\text{-}p} < 10^{11}\ \Omega$ ANSI/ESD STM 2.1	$R_{p\text{-}p} < 10^{11}\ \Omega$ ESD TR53 garments
Groundable static control garment	$R_{gp} < 10^9\ \Omega$ IEC 61340-4-9	$R_{gp} < 10^9\ \Omega$ IEC 61340-4-9	$R_{gp} < 10^9\ \Omega$ ANSI/ESD STM 2.1	$R_{gp} < 10^9\ \Omega$ ESD TR53 garments
Groundable static control garment system	Not specified	Not specified	$R_g < 3.5 \times 10^7\ \Omega$ ANSI/ESD STM 2.1	$R_g < 3.5 \times 10^7\ \Omega$ ESD TR53 garments

Table 6.8 Requirements for ionizers.

Measurement	IEC 61340-5-1:2016 requirements and test method		ESD S20.20-2014 requirements and test method	
	Product qualification	Compliance verification	Product qualification	Compliance verification
Ionizer neutralization (decay) time	decay <20 s (1000 V to 100 V) or user defined IEC 61340-4-7	decay <20 s (1000 V to 100 V) or user defined IEC 61340-4-7	User-defined ANSI/ESD STM 3.1	User-defined ESD TR53 ionizer
Ionizer offset voltage	±35 V IEC 61340-4-7	±35 V IEC 61340-4-7	±35 V ANSI/ESD STM 3.1	±35 V ESD TR53 ionizer

6.5.12.2 Grounding and Bonding Systems

All electrically conducting (noninsulative) items that might come into contact with ESDS must be electrically connected together or to ground. Three methods are given for this.

- Grounding using protective earth. This is the preferred option in which all electrically conducting items and personnel are connected to the electrical system protective earth.
- Grounding using functional ground. This covers the situation where electrically conducting items and personnel are connected to a functional ground such as a ground rod. It is recommended that this functional earth is connected to the protective earth.
- Equipotential bonding. Where a ground is not available, the electrically conducting items and personnel may simply be electrically bonded (connected together).

The terms *ground* and *grounding* are used whichever system is selected. Only one of these systems should be used and one grounding system present in the EPA. If different grounds were present, they could be at different voltages and cause ESD risk. A summary of the grounding requirements is given in Table 6.4.

6.5.12.3 Personnel Grounding

All personnel must be grounded or equipotential bonded when handling ESDS. Seated personnel must be grounded via a wrist strap system. Standing personnel may be grounded via a wrist strap or a footwear-flooring system. Where footwear-flooring is used (Table 6.5), personnel must wear footwear on both feet, and the maximum body voltage generation must be <100 V *and* the resistance from the person's body through footwear and flooring to ground must be <1 GΩ.

Earlier versions of IEC 61340-5-1 and ANSI/ESD S20.20 included an option for compliance based on a resistance from body to ground <35 MΩ without body voltage measurement. This was discontinued in these later versions, because it was found the resistance limit alone did not necessarily guarantee achievement of sufficiently low body voltage .

6.5.12.4 ESD-Protected Areas (EPAs)

ESDS must be handled only outside ESD-protected packaging when they are in an EPA. EPA boundaries must be clearly identified. Access to the EPA must be limited to personnel who have completed ESD training or are escorted by trained personnel.

6.5.12.5 Equipment Used in the EPA

Equipment commonly used in the EPA has measurable and verifiable technical requirements specified in the standards (Tables 6.6–6.8). These are measured using specified standard test methods (see Section 6.5.11 and Chapter 11).

6.5.12.6 Insulators

All nonessential insulators such as cups, packaging, and personal items must be removed from any workstation where unprotected ESDS devices are handled. (In ANSI/ESD S20.20 they must be removed from the EPA.)

Where process essential insulators are present, the ESD threat must be evaluated.

- The electrostatic field must not exceed 5 kV m^{-1} at the position where ESDS devices are handled.
- Where the surface potential of an insulator exceeds 2 kV, it must be kept at least 30 cm from any ESDS device.
- A new requirement that where the surface potential of an insulator exceeds 125 V, it must be kept at least 2.5 cm from ESDS devices.

Any ESD risk must then be mitigated by some means e.g. using ionizers.

6.5.12.7 Isolated Conductors

Requirements have been introduced covering isolated conductors. If a conductor that contacts an ESDS cannot be grounded, the voltage between the conductor and the ESDS must be reduced to <±35 V.

6.5.12.8 Hand Electrical Soldering and Desoldering Tools

ANSI/ESD S20.20 includes requirements for hand soldering and desoldering tools (Table 6.9). These do not appear in 61340-5-1.

Table 6.9 Requirements for soldering and desoldering hand tools (ANSI/ESD S20.20 only).

	IEC 61340-5-1:2016 requirements and test method		ANSI/ESD S20.20-2014 requirements and test method	
Measurement	Product qualification	Compliance verification	Product qualification	Compliance verification
Tip to ground resistance	Not specified	Not specified	<2.0 Ω ANSI/ESD S13.1	<10 Ω ESD TR53 soldering iron or ANSI/ESD S13.1 Sec. 6.1
Tip voltage	Not specified	Not specified	<20 mV ANSI/ESD S13.1	
Tip leakage current	Not specified	Not specified	< 10 mA ANSI/ESD S13.1	

6.5.13 ESD Packaging

ESD-protective packaging and package marking "shall be in accordance with customer contracts, purchase orders, drawing or other documentation." When these documents do not define ESD packaging, then packaging must be defined "for all material movement within protected areas, between protected areas, between job sites, field service operations and to the customer." Packaging materials classified as dissipative, conductive, insulating, electrostatic (electric) field shielding or ESD shielding, and shielding packaging are defined in related standards 61340-5-3 and ANSI/ESD S541 (Tables 6.10 and 6.11).

Table 6.10 ESD Packaging resistance classifications.

	IEC 61340-5-3:2015 requirements and test method		ANSI/ESD S541-2018 requirements and test method	
Classification	**Surface (Ω)**	**Volume (Ω)**	**Surface (Ω)**	**Volume (Ω)**
Packaging – Static dissipative	$10^4\ \Omega \le R_s < 10^{11}$ IEC 61340-2-3	$10^4\ \Omega \le R_v < 10^{11}$ IEC 61340-2-3	$10^4\ \Omega \le R_s < 10^{11}$ STM11.11 STM11.13	$10^4\ \Omega \le R_v < 10^{11}$ STM11.12
Packaging – conductive	$R_s \le 10^4\ \Omega$ IEC 61340-2-3	$R_v < 10^4\ \Omega$ IEC 61340-2-3	$R_s < 10^4\ \Omega$ STM11.11 STM11.13	$R_v < 10^4\ \Omega$ STM11.12
Packaging – insulator	$R_s \ge 10^{11}\ \Omega$ IEC 61340-2-3	$R_v \ge 10^{11}\ \Omega$ IEC 61340-2-3	$R_s \ge 10^{11}\ \Omega$ STM11.11 STM11.13	$R_s \ge 10^{11}\ \Omega$ STM11.12
Electrostatic (Electric) field shielding	$R_s < 10^3\ \Omega$ IEC 61340-2-3	$R_v < 10^3\ \Omega$ IEC 61340-2-3	"The user should determine whether or not a specific packaging configuration provides for a reduction of electric field strength at the position in the package where sensitive items are contained."	

Table 6.11 ESD packaging.

Classification	IEC 61340-5-3:2015 requirements and test method	ANSI/ESD S541-2018 requirements and test method
ESD shielding (bags)	< 50 nJ IEC 61340-4-8	< 20 nJ ANSI/ESD S11.31
Low charging (antistatic)	Not defined	User defined. Material having reduced amount of charge accumulation compared to standard packaging materials. ESD ADV11.2

In ANSI/ESD S20.20, where ESDS devices are placed on packaging and work performed on them, the packaging is considered to be a work surface and work surface requirements for resistance to ground apply.

6.5.14 Marking

Marking of ESDS devices, systems, or packaging must be according to customer contracts, purchase orders, or other documentation. Where these are not defined, then the need for marking should be considered in the ESD Control Program Plan. Where marking is needed, it must be documented in the ESD Program Plan.

References

British Standards Institute (1984) BS 5783:1984. *Code of practice for handling of electrostatic sensitive devices*. London, BSI.

British Standards Institute (1992) BS EN 100015-1:1992. *Basic specification. Protection of electrostatic sensitive devices. Harmonized system of quality assessment for electronic components. Basic specification: protection of electrostatic sensitive devices. General requirements*. London, BSI.

British Standards Institute (1994a) BS EN 100015-2:1994. *Basic specification. Protection of electrostatic sensitive devices. Requirements for low humidity conditions*. London, BSI.

British Standards Institute (1994b) BS EN 100015-3:1994.)*Basic specification. Protection of electrostatic sensitive devices. Requirements for clean room areas*.London, BSI.

British Standards Institute (1994c) BS EN 100015-4:1994. *Basic specification. Protection of electrostatic sensitive devices. Requirements for high voltage environments*. London, BSI.

British Standards Institute (2001) BS EN 61340-5-1:2001. *Electrostatics. Protection of electronic devices from electrostatic phenomena. General requirements*. London, BSI.

British Standards Institute (2016) BS EN 61340-5-1:2016. *Electrostatics. Protection of electronic devices from electrostatic phenomena. General requirements*. London, BSI.

Department of Defense. (1980a) MIL-STD-1686. *Standard Practice. Electrostatic discharge control program for protection of electrical and electronic parts, assemblies and equipment (excluding electrically initiated devices)*. Washington, D.C., DoD.

Department of Defense. (1980b) MIL-HDBK-263. *Military handbook. Electrostatic discharge control handbook for protection of electrical and electronic parts, assemblies and equipment (excluding electrically initiated devices*. Washington, D.C., DoD.

EOS/ESD Association Inc. (1999) ANSI/ESD S20.20-1999. *ESD Association Standard for the Development of an Electrostatic Discharge Control Program for – Protection of Electrical and Electronic Parts, Assemblies and Equipment (excluding Electrically Initiated Explosive Devices)*. Rome, NY, EOS/ESD Association Inc.

EOS/ESD Association Inc. (2003) ESD SP9.2-2003. *ESD Association Standard for the Protection of Electrostatic Discharge Susceptible Items – Footwear – Foot Grounders Resistive Characterization (not to include static control shoes)*. Rome, NY, EOS/ESD Association Inc.

EOS/ESD Association Inc. (2006a) ANSI/ESD STM4.1-2006. *ESD Association Standard for the Protection of Electrostatic Discharge Susceptible Items – Worksurfaces - Resistance Measurements*. Rome, NY, EOS/ESD Association Inc.

EOS/ESD Association Inc. (2006b) ESD TR53-01-06. *Technical Report for the protection of electrostatic discharge susceptible items - Compliance Verification of ESD Protective Equipment and Materials.* Rome, NY, EOS/ESD Association Inc.

EOS/ESD Association Inc. (2012a) ANSI/ESD STM4.2-2012. *ESD Association Standard for the Protection of Electrostatic Discharge Susceptible Items – ESD Protective Worksurfaces – Charge Dissipation Characteristics.* Rome, NY, EOS/ESD Association Inc.

EOS/ESD Association Inc. (2012b) ANSI/ESD STM11.31-2012. *ESD Association Standard Test Method for Evaluating the Performance of Electrostatic Discharge Shielding Materials – Bags.* Rome, NY, EOS/ESD Association Inc.

EOS/ESD Association Inc. (2013a) ANSI/ESD STM12.1-2013. *ESD Association Standard Test Method for the Protection of Electrostatic Discharge Susceptible Items - Seating - Resistance Measurement.* Rome, NY, EOS/ESD Association Inc.

EOS/ESD Association Inc. (2013b) ANSI/ESD STM7.1-2013. *ESD Association Standard for the Protection of Electrostatic Discharge Susceptible Items – Floor Materials - Resistive Characterization of Materials.* Rome, NY, EOS/ESD Association Inc.

EOS/ESD Association Inc. (2013c) ANSI/ESD S1.1-2013. *Standard for protection of Electrostatic Discharge Susceptible Items - Wrist Straps.* Rome, NY, EOS/ESD Association Inc.

EOS/ESD Association Inc. (2013d) ANSI/ESD STM2.1-2013. *ESD Association Standard for the Protection of Electrostatic Discharge Susceptible Items – Garments.* Rome, NY, EOS/ESD Association Inc.

EOS/ESD Association Inc. (2014a) ANSI/ESD S6.1-2014. *Standard for the Protection of Electrostatic Discharge Susceptible Items – Grounding.* Rome, NY, EOS/ESD Association Inc.

EOS/ESD Association Inc. (2014b) ANSI/ESD STM9.1-2014. *ESD Association Standard for the Protection of Electrostatic Discharge Susceptible Items – Footwear - Resistive Characterization.* Rome, NY, EOS/ESD Association Inc.

EOS/ESD Association Inc. (2014c) ANSI/ESD S20.20-2014. *ESD Association Standard for the Development of an Electrostatic Discharge Control Program for – Protection of Electrical and Electronic Parts, Assemblies and Equipment (excluding Electrically Initiated Explosive Devices).* Rome, NY, EOS/ESD Association Inc.

EOS/ESD Association Inc. (2015a) ANSI/ESD STM3.1-2015. *ESD Association Standard for the Protection of Electrostatic Discharge Susceptible Items – Ionization.* Rome, NY, EOS/ESD Association Inc.

EOS/ESD Association Inc. (2015b) ANSI/ESD STM11.11-2015. *ESD Association Standard for Protection of Electrostatic Discharge Susceptible Items – Surface Resistance Measurement of Static Dissipative Planar Materials.* Rome, NY, EOS/ESD Association Inc.

EOS/ESD Association Inc. (2015c) ANSI/ESD STM11.12-2015. *ESD Association Standard for Protection of Electrostatic Discharge Susceptible Items.*Rome, NY, EOS/ESD Association Inc.

EOS/ESD Association Inc. (2015d) ANSI/ESD STM11.13-2015. *ESD Association Standard Test Method for the Protection of Electrostatic Discharge Susceptible Items – Two-Point Resistance Measurement.* Rome, NY, EOS/ESD Association Inc.

EOS/ESD Association Inc. (2015e) ANSI/ESD STM97.1-2015. *ESD Association Standard Test Method for the Protection of Electrostatic Discharge Susceptible Items – Floor Materials and Footwear – Resistance Measurement in Combination with a Person.* Rome, NY, EOS/ESD Association Inc.

EOS/ESD Association Inc. (2016a) ANSI/ESD STM97.2-2016. *Floor Materials and Footwear – Voltage Measurement in Combination with a Person*. Rome, NY, EOS/ESD Association Inc.

EOS/ESD Association Inc. (2016b) ESD TR20.20-2016. *ESD Association Technical Report - Handbook for the Development of an Electrostatic Discharge Control Program for the Protection of Electronic Parts, Assemblies, and Equipment*. Rome, NY, EOS/ESD Association Inc.

EOS/ESD Association Inc. (2017) ESD ADV1.0-21017. *ESD Association Advisory for Electrostatic Discharge Terminology – Glossary*. Rome, NY, EOS/ESD Association Inc.

EOS/ESD Association Inc. (2018) ANSI/ESD S541-2018. *Packaging Materials for ESD Sensitive Items*. Rome, NY, EOS/ESD Association Inc.

Industry Council on ESD Target Levels (2011) *White paper 1: A case for lowering component level HBM/MM ESD specifications and requirements*. Rev. 3.0. Available from: http://www.esdindustrycouncil.org/ic/en/documents/37-white-paper-1-a-case-for-lowering-component-level-hbm-mm-esd-specifications-and-requirements-pdf (Accessed: 10th May 2017).

International Electrotechnical Commission. (1998) IEC 61340-5-1:1998. *Electrostatics - Part 5-1: Protection of electronic devices from electrostatic phenomena - General requirements*. Geneva, IEC.

International Electrotechnical Commission. (2001) IEC 61340-4-3:2001. *Electrostatics - Part 4-3: Standard test methods for specific applications – Footwear*. Geneva, IEC.

International Electrotechnical Commission. (2003) IEC 61340-4-1:2003+AMD1:2015 CSV. *Electrostatics - Part 4-1: Standard test methods for specific applications - Electrical resistance of floor coverings and installed floors*. Geneva, IEC.

International Electrotechnical Commission. (2004) IEC 61340-4-5:2004. *Electrostatics - Part 4-5: Standard test methods for specific applications - Methods for characterizing the electrostatic protection of footwear and flooring in combination with a person*. Geneva, IEC.

International Electrotechnical Commission. (2007) IEC TR 61340-5-2:2007. *Electrostatics – Part 5-2: Protection of electronic devices from electrostatic phenomena - User guide*. Geneva, IEC.

International Electrotechnical Commission. (2013a) PD/IEC TS 60079-32-1. *Explosive atmospheres Part 32-1. Electrostatic hazards, guidance*. Geneva, IEC.

International Electrotechnical Commission. (2013b) PD IEC/TS 61340-4-2:2013. *Electrostatics - Part 4-2: Standard test methods for specific applications - Electrostatic properties of garments*. Geneva, IEC.

International Electrotechnical Commission. (2014) IEC 61340-4-8:2014. *Electrostatics - Part 4-8: Standard test methods for specific applications - Electrostatic discharge shielding – Bags*. Geneva, IEC.

International Electrotechnical Commission. (2015a) IEC 61340-2-1:2015. *Electrostatics - Part 2-1: Measurement methods - Ability of materials and products to dissipate static electric charge*. Geneva, IEC.

International Electrotechnical Commission. (2015b) IEC 61340-4-6:2015. *Electrostatics - Part 4-6: Standard test methods for specific applications - Wrist straps*. Geneva, IEC.

International Electrotechnical Commission. (2015c) IEC 61340-5-3:2015. *Electrostatics. Protection of electronic devices from electrostatic phenomena. Properties and requirements classifications for packaging intended for electrostatic discharge sensitive devices*. Geneva, IEC.

International Electrotechnical Commission. (2016a) IEC 61340-2-3:2016. *Electrostatics. Part 2-3: Methods of test for determining the resistance and resistivity of solid materials used to avoid electrostatic charging*. Geneva, IEC.

International Electrotechnical Commission. (2016b) IEC 61340-4-9:2016. *Electrostatics - Part 4-9: Standard test methods for specific applications – Garments*. Geneva, IEC.

International Electrotechnical Commission. (2016c) IEC 61340-5-1:2016. *Electrostatics - Part 5-1: Protection of electronic devices from electrostatic phenomena - General requirements*. Geneva, IEC.

International Electrotechnical Commission. (2017) IEC 61340-4-7:2017. *Electrostatics - Part 4-7: Standard test methods for specific applications – Ionization*. Geneva, IEC.

International Electrotechnical Commission. (2018) IEC TR 61340-5-2. *Electrostatics – Part 5-2: Protection of electronic devices from electrostatic phenomena - User guide*. Geneva, IEC.

International Electrotechnical Commission. (2019) IEC TR 61340-5-4. *Electrostatics – Part 5-4: Protection of electronic devices from electrostatic phenomena - Compliance verification*. Geneva, IEC.

Further Reading

EOS/ESD Association Inc. (1995a) ESD ADV53.1-1995. *Advisory for Protection of Electrostatic Discharge Susceptible Items - ESD Protective Workstations*. Rome, NY, EOS/ESD Association Inc.

EOS/ESD Association Inc. (1995b) ESD ADV11.2-1995. *Advisory for the Protection of Electrostatic Discharge Susceptible Items - Triboelectric Charge Accumulation Testing*. Rome, NY, EOS/ESD Association Inc.

EOS/ESD Association Inc. (1999a) ESD TR13.0-01-99. *Technical Report - EOS Safe Soldering Iron Requirements*. Rome, NY, EOS/ESD Association Inc.

EOS/ESD Association Inc. (1999b) ESD TR15.0-01-99. *Standard Technical Report for the Protection of Electrostatic Discharge Susceptible Items-ESD Glove and Finger Cots*. Rome, NY, EOS/ESD Association Inc.

EOS/ESD Association Inc. (1999c) ESD TR50.0-01-99. *Technical Report - Can Static Electricity be Measured?* Rome, NY, EOS/ESD Association Inc.

EOS/ESD Association Inc. (1999d) ESD TR50.0-02-99. *Technical Report - High Resistance Ohmmeter Measurements*. Rome, NY, EOS/ESD Association Inc.

EOS/ESD Association Inc. (2000a) ESD TR2.0-01-00. *Technical Report - Consideration For Developing ESD Garment Specifications*. Rome, NY, EOS/ESD Association Inc.

EOS/ESD Association Inc. (2000b) ESD TR2.0-02-00. *Technical Report - Static Electricity Hazards of Triboelectrically Charged Garments*. Rome, NY, EOS/ESD Association Inc.

EOS/ESD Association Inc. (2001) ESD TR1.0-01-01. *Technical Report - Survey of Constant Monitors for Wrist Straps*. Rome, NY, EOS/ESD Association Inc.

EOS/ESD Association Inc. (2002a) ESD TR3.0-01-02. *Technical Report - Alternate Techniques for Measuring Ionizer Offset Voltage and Discharge Time*. Rome, NY, EOS/ESD Association Inc.

EOS/ESD Association Inc. (2002b) ESD TR4.0-01-02. *Technical Report - Survey of Worksurfaces and Grounding Mechanisms*. Rome, NY, EOS/ESD Association Inc.

EOS/ESD Association Inc. (2002c) ESD TR10.0-01-02. *Technical Report - Measurement and ESD Control Issues for Automated Equipment Handling of ESD Sensitive Devices Below 100 Volts*. Rome, NY, EOS/ESD Association Inc.

EOS/ESD Association Inc. (2003) ESD TR50.0-03-03. *Technical Report - Voltage and Energy Susceptible Device Concepts, Including Latency Considerations*. Rome, NY, EOS/ESD Association Inc.

EOS/ESD Association Inc. (2004) ESD TR55.0-01-04. *Technical Report - Electrostatic Guidelines and Considerations For Cleanrooms and Clean Manufacturing*. Rome, NY, EOS/ESD Association Inc.

EOS/ESD Association Inc. (2005) ESD TR3.0-02-05. *Technical Report - Selection and Acceptance of Air Ionizers*. Rome, NY, EOS/ESD Association Inc.

EOS/ESD Association Inc. (2011a) ANSI/ESD SP15.1-2011. *Standard Practice for the Protection of Electrostatic Discharge Susceptible Items – In-Use Resistance Measurement of Gloves and Finger Cots*. Rome, NY, EOS/ESD Association Inc.

EOS/ESD Association Inc. (2011b) ESD TR7.0-01-11. *Technical Report for the Protection of Electrostatic Discharge Susceptible Items – Static Protective Floor Materials*. Rome, NY, EOS/ESD Association Inc.

EOS/ESD Association Inc. (2012) ANSI/ESD S11.4-2012. *Standard for the Protection of Electrostatic Discharge Susceptible Items - Static Control Bags*. Rome, NY, EOS/ESD Association Inc.

EOS/ESD Association Inc. (2015a) ANSI/ESD S13.1-2015. *Provides electrical soldering/desoldering hand tool test methods for measuring current leakage, tip to ground reference point resistance, and tip voltage*. Rome, NY, EOS/ESD Association Inc.

EOS/ESD Association Inc. (2015b) ESD TR17.0-01-15. *Technical Report for ESD Process Assessment Methodologies in Electronic Production Lines – Best Practices used in Industry*. Rome, NY, EOS/ESD Association Inc.

EOS/ESD Association Inc. (2015c) ESD TR53-01-15. *Technical Report for the Protection of Electrostatic Discharge Susceptible Items – Compliance Verification of ESD Protective Equipment and Materials*. Rome, NY, EOS/ESD Association Inc.

EOS/ESD Association Inc. (2016a) ANSI/ESD SP3.3-2016. *Standard Practice for the Protection of Electrostatic Discharge Susceptible Items – Periodic Verification of Air Ionizers*. Rome, NY, EOS/ESD Association Inc.

EOS/ESD Association Inc. (2016b) ANSI/ESD SP3.4-2016. *Standard Practice for the Protection of Electrostatic Discharge Susceptible Items – Periodic Verification of Air Ionizer Performance Using a Small Test Fixture*. Rome, NY, EOS/ESD Association Inc.

EOS/ESD Association Inc. (2016c) ESD SP10.1-2016. *Standard practice for protection of Electrostatic Discharge Susceptible Items – Automated handling Equipment (AHE)*. Rome, NY, EOS/ESD Association Inc.

EOS/ESD Association Inc. (2017a) ESD ADV1.0-2017. *ESD Association Advisory for Electrostatic Discharge Terminology – Glossary*. Rome, NY, EOS/ESD Association Inc.

EOS/ESD Association Inc. (2017b) ANSI/ESD S8.1-2017. *Draft Standard for the Protection of Electrostatic Discharge Susceptible Items – Symbols – ESD Awareness*. Rome, NY, EOS/ESD Association Inc.

EOS/ESD Association Inc. (2019) *ESD Fundamentals Part 6: ESD Standards*. Available from: https://www.esda.org/index.php/about-esd/esd-fundamentals/part-6-esd-standards (Accessed 26th January 2019).

7

Selection, Use, Care, and Maintenance of Equipment and Materials for ESD Control

7.1 Introduction

Electrostatic discharge (ESD) control equipment can form a considerable part of the investment cost of an ESD control program. It is important to be sure that the equipment will do the task intended, for the intended life of the equipment. There may be many types of similar equipment on the market, and the type chosen should be appropriate for use in the facility and processes intended. Cleaning, care, and maintenance may be needed to ensure that the equipment continues to work correctly over its life.

Unless equipment is single use and disposable, it is important that any failures of equipment are detected, with failed equipment taken out of use for maintenance or replaced. So, most equipment will require compliance verification testing at regular intervals. Appropriate test methods and pass criteria must be defined. This is discussed further in Chapter 9, and some test methods are discussed in Chapter 11.

7.1.1 Selection and Qualification of Equipment

The first task in selecting ESD control equipment is to identify its intended functions. Aspects to be considered may include

- Physical or process task. What is the equipment there to do?
- ESD control functions and characteristics required.
- Safety considerations.
- Compliance with standards, if required.
- Tests or other methods of confirmation of functionality and ESD control characteristics.
- Intended life. Disposable or long-term use?
- Care and maintenance requirements.
- For long-term usage, compliance verification requirements and test methods. How will failure be detected?

The second step is to decide how selected items may be qualified to confirm that they will fulfill their intended function for their intended life. Product qualification can be done in various ways and should address all the aspects that have been identified important.

In the simplest case, it may be sufficient to check data sheets to ensure that the required functions and characteristics are addressed and that the product complies with relevant

The ESD Control Program Handbook, First Edition. Jeremy M Smallwood.

standards. The standards may be to do with ESD control or any other aspect such as safety. For example, footwear that is personal protective equipment (PPE) may, in Europe, need to be compliant with a standard such as ISO 20345.

In some cases, it may be advisable to obtain one or more specimens of items for evaluation or test. This may be done in-house or by a third party. As they are done only initially to qualify the product before use, the qualification tests may be more in-depth than compliance verification tests. They often are designed to check individual characteristics of parts of the equipment in detail. For example, in qualifying an ESD control chair, it may be desirable to check that the seat back and arms as well as the seat are made of static dissipative material and connected reliably together and to a groundable point, feet, or wheels. It may be necessary to establish that more than one foot or wheel is made of noninsulating material and capable of establishing a ground path to the floor.

In qualification tests, it is sometimes desirable to test the item isolated from its normal working situation in an uninstalled condition. For example, a work surface point-to-point surface resistance may be tested as well as the resistance from the surface to a groundable point. The latter test does not require connection to ground. Often, it will be desirable to qualify the item under "worst case" operating conditions. This usually means performing tests under dry atmosphere <30% rh.

Conversely, in some qualification tests, it is desirable to test the operation of the equipment as a system with other ESD control equipment. A good example of this is measurement of body voltage generated on a subject wearing ESD control footwear and the flooring with which it will be used. This test can be done only with representative samples of the footwear and flooring under qualification (see Section 7.5.4).

Where compliance with a standard is required, testing should be done according to the test methods required by the standard, and compliance with the standard pass criteria should be established. Standards will often specify qualification testing under dry atmosphere conditions. IEC 61340-5-1 and ANSI/ESD S20.20 usually require qualification testing at 12 ± 3% rh (International Electrotechnical Commission 1998, 2016b; EOS/ESD Association Inc 2014b).

7.1.2 Use

The intended use of equipment can be an important consideration, first in whether it is needed at all and second in selecting its required characteristics. ESD control equipment often acts as part of a system with other equipment. Careful thought should be given to how the equipment is used and if it will help to give a convenient working environment to the personnel in the area. Analysis of this can affect the combination of the ESD control equipment selected. ESD control practices that are selected to be convenient to the operator are more likely to be reliable in practice. This is discussed further in Chapter 10.

7.1.3 Cleaning, Care, and Maintenance of Equipment

It is important, when selecting ESD control equipment, to consider whether it may need specific cleaning, care, and maintenance procedures and materials. These may differ from similar equipment in unprotected areas. For example, ESD control floors and workstation surfaces may need different cleaning procedures and materials than those in unprotected areas, selected to preserve and enhance their ESD control properties.

Some equipment such as ionizers may need regular maintenance to keep their performance in the long term.

Equipment that relies on grounding through feet or wheels often collects dirt on these contact points, which can make the ground path fail or intermittent. Cleaning the wheels or feet can bring the equipment back into compliance.

7.1.4 Compliance Verification

Equipment that is not single use only will need to have its properties tested and verified on a regular basis to detect any failures or deviations from intended performance.

Compliance verification tests are normally simple efficient tests designed to test functionality and performance in the installed situation. Where equipment works as part of a system, it is often tested as a system. As these tests are repeated regularly and often many times during an audit, it is important that they are quick and effective.

For example, a simple resistance to ground test is used to simultaneously test the condition of a work surface or floor and its ground connection. For a chair standing on an ESD control floor, a simple resistance to ground measurement from the seat to the electrostatic discharge protected area (EPA) ground simultaneously checks the chair and the connection through its wheels and the floor to ground. Problems such as buildup of dirt on the wheels are detected as an excessively high resistance reading.

7.2 ESD Control Earth (Ground)

7.2.1 What Does the ESD Control Earth Do?

A primary objective in ESD control is to bring all electrically conducting equipment, materials, and personnel to the same voltage so that they cannot become charged and become an ESD source. This is done by connecting them to a common connection point called *ground*. This connection of all conductors to one common point is called *equipotential bonding* and establishes "zero volts" for the purposes of ESD control. Where conductors are connected, electrostatic voltage differences are quickly equalized between them. Elimination of voltage difference between eliminates the possibility of ESD occurring if contact is made between them.

7.2.2 Choosing an ESD Control Earth

The common connection point is commonly called *ground* and is usually, but not always, also connected electrically to the physical earth. One common and convenient way to do this is to connect to the mains electricity supply protective safety earth. A second method sometimes used is to connect to an earth rod buried in the ground for the purpose, known in the ESD control standards as *functional earth*. If it is not possible to connect to a convenient earth, for example in a vehicle or aircraft, then equipotential bonding alone will be sufficient for ESD control.

It is undesirable to have more than one earth present in a facility. If there are multiple earths and they are not connected together, they are likely to be at different voltages due to earth currents or other phenomena. Often, there can be a variety of equipment in the EPA that is already connected to mains protective "safety" earth. In this case, it is usually convenient to also use mains protective earth as the ESD control earth. More information on electrical power systems can be found in standards such as IEC 60364-1 and national electrical codes of practice and regulations (International Electrotechnical Commission 2005).

If there is no mains protective earth available in the EPA, it can be convenient to use a separate earth rod buried in the earth as the ESD control earth.

It is preferable to connect individual equipment (e.g. workstations) separately to ground in "star" connection form (Figure 7.1) rather than "chain" or series connection (Figure 7.2). This is because with the "chain" connection, failure of one connection can lead to several ESD control items being disconnected from earth.

An exception to this is where an ESD control item (e.g. ESD floor) is designed to provide a ground path for equipment (e.g. chairs or carts/trolleys) or personnel grounded through it as a grounding system.

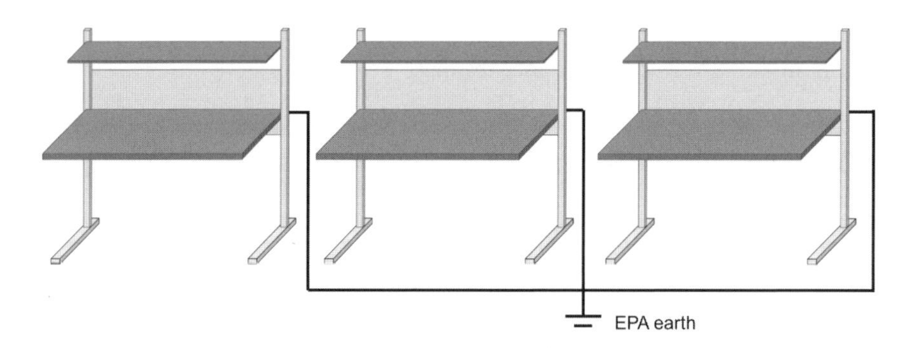

Figure 7.1 "Star" connection of ESD control benches.

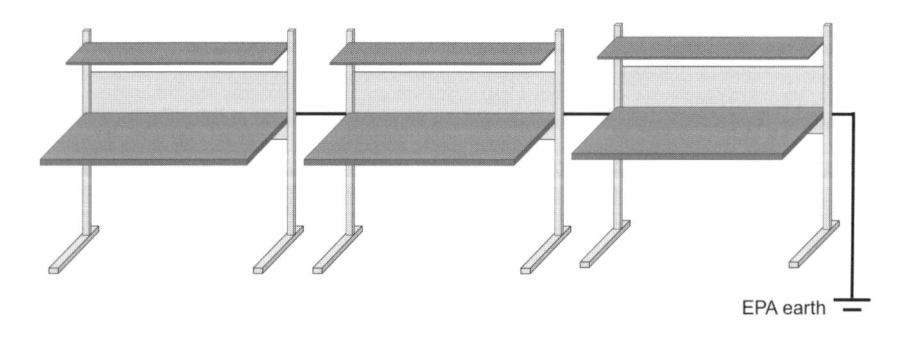

Figure 7.2 "Chain" or series connection of ESD control benches.

7.2.3 Qualification of ESD Control Earth

Where the electrical protective earth is used as ESD ground, the qualification requirements are likely to depend on national electrical code requirements. These should be tested by suitably qualified personnel. It is extremely important to verify that the connection is made to the earth and not live or neutral power connections.

Even where a temporary or semi-permanent connection to a mains power socket via an earthing plug is used, it is wise to check that the socket is correctly wired before use.

The electrical connection between ESD ground and the earth connection should be verified to be sufficiently robust and low resistance. ESD control standards 61 340-5-1 and S20.20 also give some requirements that can be applied if national regulations are not applicable.

7.2.4 Compliance Verification of ESD Control Earth

The electrical connection between ESD ground and the protective or functional earth connection should be verified to be sufficiently low resistance. National regulations may require this to be done by suitably qualified personnel.

7.2.5 Common Problems with Ground Connections

Mains sockets have occasionally been found to be incorrectly wired. Even where a temporary or semi-permanent connection to a mains power socket via an earthing plug is used, it is wise to check that the socket is correctly wired before use.

Plug-in mains ground connectors are sometimes unplugged to allow plug-in of other equipment and then inadvertently left disconnected. For example, a cleaner might remove a ground plug to plug in a vacuum cleaner, or an engineer might do so to connect some equipment.

Even hardwired ground connections can sometimes be disconnected by inadvertent and undetected damage. For example, moving a bench might cause undetected disconnection of a hardwired ground lead.

7.3 The ESD Control Floor

7.3.1 What Does an ESD Control Floor Do?

An ESD control floor, if required, can represent a large investment. It can also give significant benefit and convenience in ESD control.

ESD control floors are used to give a convenient way of grounding anything that stands on the floor, such as personnel, carts (trolleys), racks, chairs, and other free-standing equipment (see Section 4.7.10). ESD control floor coverings typically provide an electrical path between the floor surface and EPA ground via an underlying substrate material or grounding structure. Floor finishes or treatments may reduce the tendency for charge generation at the surface as well as providing an electrical path for charge to move to ground.

A floor also acts as a means of conjoining workstations or areas to make a single EPA with no uncontrolled area between them. This can improve the convenience of handling of electrostatic discharge–sensitive (ESDS) by eliminating the need for protective packaging for transport between workstations.

The floor operates as a system with the items that it is intended to ground. So, in specifying the characteristics of a floor, it is necessary to consider the items that will be grounded by it. Typical systems using the floor include

- An operator's body grounded through footwear and flooring
- A cart grounded from cart surface through the chassis, wheels, and floor
- A rack grounded from the shelf surface through the frame, feet, and floor
- A seat grounded from the seat surfaces through the frame, feet or wheels, and floor

Floors and floor covering materials of course have many non-ESD-related functions. They must withstand the physical wear and tear expected for the lifetime required of the floor. This may include considerations like use of forklift trucks and vehicles, or withstanding use of chemicals. Floor materials for use in clean areas may need to be selected for low contribution to particulate or other contaminants in the area. These factors must be included in the overall selection of the floor material.

7.3.2 Permanent ESD Control Floor Material

Rubber and vinyl materials are among the most widely used floor coverings, supplied either as sheet or as tiles. Some materials may be slippery when wet. Some may not withstand heavy traffic. Some materials may contain carbon or may outgas, making them unsuitable for clean room use.

Epoxy and polymeric coatings form a poured-on robust coating resistant to chemicals, abrasion, and vehicle traffic wear. They are seamless, can be easy to maintain, and are suitable for clean room use.

High-pressure laminates are usually used as raised floors or floor mats. As they are often sensitive to moisture, they may be unsuited for use where chemical or water spillage is likely or on concrete substrates where high moisture levels could arise. They may change considerably in resistance in response to changes in moisture levels.

Carpets may be attractive and absorb noise and can have lower maintenance requirement than resilient materials. They are available in sheet or tile form. In tile form, individual worn or contaminated tiles may be easily replaceable. Carpets may not, however, be suitable for use in clean rooms or where heavy soiling or wear, water, chemical spills or hot solder, or vehicle traffic may occur.

7.3.3 Semipermanent or Nonpermanent ESD Control Floor Materials

Semi- or nonpermanent ESD control floor materials include interlocking tiles, mats, floor finishes, paints, coatings, and topical antistats. They can provide a quick and easy floor covering solution, especially for small or temporary floor areas. They may have a relatively short life and require regular retreatment or replacement.

Floor mats can be an easy and portable solution, especially for small areas, for example around a single workstation. They can also provide a replaceable "sacrificial" surface around an area where high contamination or solder damage is expected. Comfort floor mats may be provided where operators are standing for long periods at workstations.

Some types of mat can tend to curl and ruck and form a trip hazard. They are often connected to ground via a wire attached to a bonding point, and this may act as a trip hazard or be prone to accidental disconnection.

Paints and coatings can be applied to existing floors such as concrete, giving easy application and coverage of large areas. They have intermediate lifetime and tend to wear, requiring reapplication. Some materials may not be suitable for clean areas due to their carbon loading or tendency to abrade and shed particles.

Floor finishes such as topical antistats are treatments that are easy to apply and can be applied to carpets. Their effect is usually temporary and requires regular refreshment. Their effect often depends on moisture from the air, and they may become ineffective in low relative humidity atmospheric conditions. Some finishes may tend to be slippery and cause slip hazard to personnel. Some may be susceptible to easily being washed off or worn away. Careless application and maintenance may result in unreliable performance. Some finishes may not be usable in clean room areas due to contamination issues.

7.3.4 Selection of Floor Materials

In selecting a floor material, the first considerations must be about how it will be used and the processes that will occur there. Considerations may include

- Is it to be a permanent area or a temporary area?
- What are the activities and processes anticipated?
- Will it be a new installation or a cover to an existing floor?
- Will it be a clean environment in which particle shedding or other contamination would be a concern?
- Are chemicals to be used in the area?
- Are there ergonomic considerations?
- What will be the maintenance and cleaning regime?
- Are there electrical considerations such as high voltage use?
- Are there possible safety considerations?
- Are there other safety or legislative requirements to consider?
- Will it need to support forklifts or vehicles or have other heavy usage considerations?

Where a floor may be used for handling solvents, explosives, or other flammable materials, the floor may need to be specified for safety in this use. Local safety regulations may take precedence.

7.3.5 Floor Material Qualification Test

Product qualification test of a floor should aim to test its performance as part of a system with the equipment with which it is intended to be used, under the lowest atmospheric humidity conditions likely in practice. Small specimens (e.g. 1 × 2 m) of coverings and coatings or mats may be tested under low-humidity laboratory conditions. The tests might include

- Resistance to a groundable point or simulated grounding structure, made at several points on the surface. This is to get an indication of the likely installed resistance to ground.
- Point-to-point resistance, made at several positions and orientations on the surface.
- Resistance from body to ground (or groundable point) of a person wearing selected ESD control footwear intended to be used in practice and standing on the flooring specimen.
- A walk test on the specimen by a person wearing the footwear intended to be used in practice to test the body voltage generation of the selected footwear-flooring combination(s).
- If carts, seats, or other equipment are to be grounded through the floor, it is wise to test that reliable grounding is achieved of the types of equipment intended to be used.

Suitable test methods are specified in standards such as IEC 61340-4-1, IEC 61340-4-5, ANSI/ESD STM 97.1, and ANSI/ESD STM 97.2. (International Electrotechnical Commission 2003/2015, international Electrotechnical Commission 2005, EOS/ESD Association Inc 2015, EOS/ESD Association Inc 2016)

In laboratory tests, the floor material is typically conditioned at the test humidity and temperature required by the test standard (e.g. 12% rh 23 ± 3 °C) for at least 24 hours before testing under the same conditions. In some standards, the required conditioning time may be as much as 48 or even 72 hours. The conditions may be determined by the test standard or by the user's requirements. One or more test specimens are mounted in a manner suitable for test as determined by the test method or standard. This may involve adding groundable points or a grounding method simulating the installed condition.

Resistance measurements are made using standard electrodes placed on the material surface. The measurement is made at 10 V for resistance less than 1 MΩ, and 100 V for resistance above 1 MΩ.

In a point-to-point measurement, two electrodes are placed on the surface 30 cm apart, or some other distance defined by the standard. The resistance between the electrodes is measured (see Section 11.8). This is repeated in several positions and orientations.

In the resistance to groundable point measurement, a single electrode is placed on the specimen surface. The resistance between this and a groundable point or simulated ground connection point is measured.

Where the floor is used for grounding personnel via ESD control footwear, qualification tests of the footwear and flooring in combination will be required (Section 7.5.4). In measuring the resistance from body to ground with a footwear-flooring combination, a test subject wears the footwear under test. They stand on the flooring specimen, holding a handheld electrode. The resistance between the handheld electrode and the floor specimen ground point is measured. The maximum resistance is usually required to be below a required value. Sometimes, the minimum resistance must also be above a required value, e.g. for safety reasons.

With the body voltage walk test, the test subject wearing the footwear under test walks on the flooring specimen under test. The floor specimen is grounded via the groundable point. The walk style and footstep pattern to be used by the subject is determined in some standards. The test subject holds a handheld electrode connected to an electrostatic voltmeter and means of recording the voltage. The highest peaks in body voltage during the test are recorded.

The final resistance to ground and performance of an installed floor material will typically depend to some extend on the underlying substrate, grounding method, and other installation conditions. For this reason, it can be advisable to test the proposed material as a small installed area before deciding to install a large area of material. Some materials (e.g. treatments and coatings) can by their nature be tested only in the installed condition. There may also be concerns over whether a treatment could affect other aspects such as the appearance of a treated material. In this case, it may be advisable to test a small area before treating the full proposed application area. Qualification test on an installed floor material will usually have to be done under ambient conditions. If possible, some tests should be done under worst-case (e.g. low-humidity) conditions. Testing typically includes

- Resistance to ground of the installed material, made at several points on the surface
- Resistance from body to ground (or groundable point) of a person wearing selected ESD control footwear intended to be used in practice and standing on the surface
- A walk test on the specimen by a subject wearing the footwear intended to be used in practice to test the body voltage generation of the selected footwear-flooring combination(s)

ESD control standards may require these tests to be done as part of their ESD control product qualification requirements. Where compliance with a standard is required, testing should be done according to the standard, and compliance with the standard pass criteria should be established.

7.3.6 Acceptance of a Floor Installation

After installation, a floor should be tested before use to make sure that it meets the required performance. These tests are normally done under ambient conditions. If possible, some tests should be done under worst-case (e.g. low-humidity) conditions.

The tests typically include resistance to ground test at several points on the surface. They may also include test of resistance from body to ground and a walk test by a subject wearing the footwear intended to be used in practice to confirm the body voltage generation of the selected footwear-flooring combination(s).

7.3.7 Use of Floor Materials

The purpose of an ESD control floor is to provide a convenient means of grounding personnel, racks, seats, carts, or other items standing on the floor. To get the best from an investment in an ESD control floor, it is important to make sure that full advantage of this is taken when specifying equipment for use on the floor within the EPA.

7.3.8 Care and Maintenance of Floors

Dirt and contamination on a floor can considerably change both its charge generation and its charge dissipation characteristics. Regular cleaning may be necessary to keep a floor working at its optimum.

Use of ordinary floor cleaning products can change and impair the characteristics of a floor. The cleaning processes and materials used should be specified according to the manufacturer's instructions to maintain good performance.

It is important to use a cleaning regime that will preserve and maintain the floor electrostatic characteristics. Use of incorrect cleaning materials or polish can leave a coating of wax or another contaminant material on it. This can seriously affect the surface charge generation characteristics as well as the surface contact resistance of the floor. Guidance should be obtained from the material manufacturer or supplier regarding suitable cleaning methods and materials.

Where water or liquid cleaners have been used, leave to dry well before retest of the floor, as the presence of residual moisture can dramatically affect the measured resistance value.

7.3.9 Compliance Verification Test

Floor compliance verification test is normally done using a simple resistance to ground measurement (see Section 11.8). A standard electrode is placed on the floor surface. A resistance meter is connected between this electrode and an ESD earth connection point. Standards often require a minimum number of measurements to be made on a floor or for a given floor area. It is good practice to also measure and note the atmospheric humidity in the area tested. This, together with the measurement results, will over time give an indication of variation of the floor resistance with humidity and show if the floor resistance to ground goes out of specification under low-humidity conditions.

The frequency of testing should reflect the likelihood that the floor performance is expected to change over time. For example, a permanent floor in a clean area, with a history of no change over a long time, may need to be tested infrequently. In contrast, a more temporary material subject to change, wear, or contamination may need testing on a more frequent basis. An area subject to extreme wear or contamination may need to be tested frequently. A mat subject to accidental disconnection may need frequent visual checks in between electrical tests.

IEC 61340-5-2:2018 suggested a typical time interval of three months between compliance verification tests.

7.3.10 Common Problems

Surface contamination such as dust, chemicals, or sprayed materials or use of incorrect polishes or cleaning materials can seriously affect the surface charge generation characteristics as well as the resistance to ground of the floor covering material. This can dramatically increase the resistance to ground of equipment and personnel standing on the floor, and the body voltage of walking personnel.

If a resistance to ground test has been failed, a first step is to make a visual check. Any surface dirt or contamination should be cleaned off using a suitable cleaning procedure. The test electrode itself should be checked, as this can also become contaminated. For a temporary installation such as a mat, the earth bonding system should be checked.

Where water or liquid cleaners have been used, they should be left to dry before retest of the floor. Moisture can dramatically affect the measured resistance value.

Electrodes can be initially cleaned of dust using a dry paper towel or clean cloth. If this is insufficient, an alcohol wipe (if compatible with the electrode material) can give stronger cleaning action while being quick in drying. The electrode should be allowed to dry completely before further measurements are made.

If areas of excessive resistance to ground are found on a floor, the area should be first cleaned to check whether it may be due to contamination. If this does not restore the floor performance, an evaluation should be made of the size of the area affected, whether the performance has changed, whether low humidity may be a factor, and whether there may be other areas in the process of deterioration. Depending on the results of this evaluation, remedial action may need to be considered.

Small areas of high resistance can sometimes be due to a previously undetected problem in a floor material or installation. If the area affected is very small and the effect on ESD control is minimal, then the position of the problem area may be noted, and the noncompliance might be acceptable as insignificant. A key question here is whether the noncompliant area prevents reliable grounding of an ESD control item, leading to ESD risk to ESDS devices.

7.4 Earth Bonding

7.4.1 The Role of Earth Bonding Points

Earth bonding (grounding) points provide a way of connecting ESD control equipment to EPA earth via earth bonding cords. They are available with, or without, built-in series resistors (Figure 7.3). They may have a variety of connector types for connecting ESD control equipment. The common connector types include 4 mm "banana" sockets and 10 mm circular studs.

Earth bonding points may be hardwired into a workstation or other equipment or may be provided as a plug-in unit to connect to electrical mains sockets. Some types can be screwed onto the underside of a bench and include a flying lead tail that can be hardwired to a convenient earthing point.

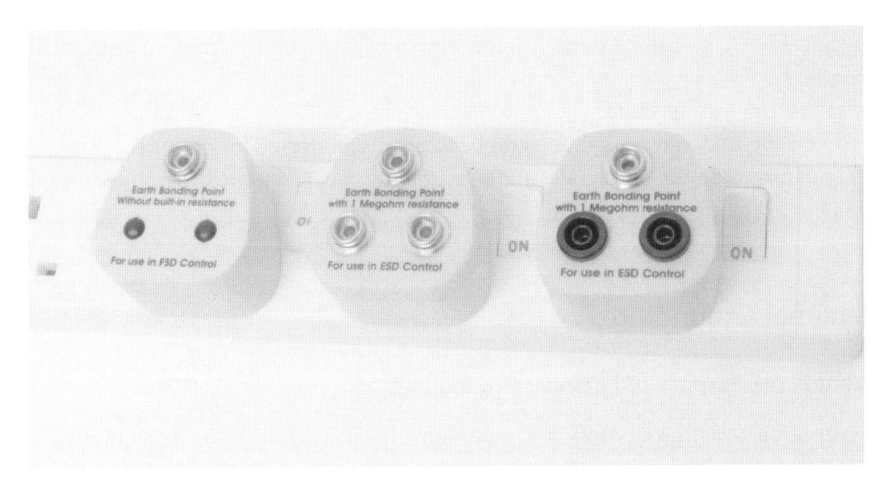

Figure 7.3 A typical earth bonding (grounding) plugs that can be used to connect to mains protective earth.

7.4.2 Selection of Earth Bonding Points

The main questions to be addressed in selecting earth bonding points are

- What will the bonding point connect to? Will it be hardwired in or make a temporary connection?
- What format connection sockets are required by the equipment to be connected to the bonding point?
- What is the maximum resistance to ground required through the bonding point?
- Is there a minimum resistance to ground required through the bonding point?

Where earth bonding connectors used for connecting measurement instruments to ground, a connector without a series resistor included is preferable. Any internal resistance is added to the measured resistance.

7.4.3 Qualification of Earth Bonding Points

Qualification should check that the connector form is as required and whether the correct value series resistor is included (if required).

7.4.4 Use of Earth Bonding Points

For earth bonding points that plug into mains electrical sockets, it can be wise to do a visual check that the plug is in place and connected before use. These items are often unplugged by personnel wanting to use the mains socket for other equipment.

Before using a mains socket to provide ESD earth connection, check that the socket is correctly wired. This can be done using a simple proprietary mains socket tester.

7.4.5 Compliance Verification of Earth Bonding Points

It is important to verify periodically that earth bonding points remain functional and within specification. This can be done using a simple resistance to ground measurement (see Section 11.8).

The frequency of testing chosen may depend on how reliable the bonding point is expected to be. Plug-in connectors that could be easily disconnected, or points that should be subject to damage, should be visually and functionally checked more often than rugged hardwired points that are more durable.

7.5 Personal Grounding

7.5.1 What Is the Purpose of Personal Grounding?

The purpose of personal grounding is to maintain the body voltage of personnel who handle ESDS at a low level so that they cannot be a significant source of ESD to ESDS devices (see Section 4.7.9). In a standard ESD program handling ESDS devices of 100 V human body model (HBM) withstand, the usual goal is to keep the body voltage below

100 V. To do this, a reliable and continuous electrical connection to ground must be established and maintained. Where devices of lower HBM ESD withstand are handled, it is wise to set a maximum body voltage as low, or lower than the device withstand voltage.

There are two main methods of grounding personnel – wrist straps and ESD control footwear with ESD control flooring. Other methods such as grounding via garments or seating are sometimes used. It is not necessary to use both wrist straps and footwear-flooring grounding, although providing both gives a useful redundancy that helps reliability of an important ESD control measure.

While personal grounding controls the voltage developed on the body, it does not necessarily eliminate voltages and fields from clothing and items worn or held by the person, unless these are designed to be grounded via the body. EPA equipment such as garments and tools are often designed to be grounded through the user's body via personal grounding.

More unusually, systems have been designed to ground personnel via special ESD control garments or seating although this is not currently common practice. It should not normally be assumed that use of an ESD chair or garment ensures that personnel are grounded adequately compliant with ESD control standards. Attempts to establish such as system should be tested rigorously.

Personal grounding is one of the most essential ESD control measures for manual handling of ESDS. Because of this, personal grounding is usually tested frequently, often every working day before work is commenced. An alternative is to use a constant monitoring system to monitor grounding.

7.5.2 Personal Grounding and Electrical Safety

Grounding of personnel handling ESDS devices is normally considered essential. In some processes where high voltages are present, such as in-circuit or burn-in test, personal grounding may conflict with safety considerations. Local electrical safety regulations should always be considered, and personal grounding strategies consistent with their requirements should be selected. Electrical safety is further discussed in Section 10.5.6.

The 61 340-5-1 and ESD S20.20 standards do not specify a minimum resistance to ground for personal grounding. Nevertheless, if there is a possibility of contact with high voltages, the user should consider specifying a minimum resistance to ground for the personal grounding system. A level of resistance is normally included in items such as wrist band ground cords, footwear, and flooring. The minimum resistance should be suitably specified according to the voltage sources present in terms of resistance level and voltage withstand. The level of resistance should limit the current possible due to possible contact with the voltage source to a safe level. Some earlier standards made recommendations for this – the IEC 61340-5-1:1998 recommended at least 750 kΩ per 250 Vac or 500 Vdc, to be increased pro rata for increasing voltage. If there are local safety regulations, it is important that these are observed and complied with. In some countries, a method such as ground fault circuit interrupters (GFCIs) or circuit breakers must be used to open the circuit if personnel contact high-voltage supplies.

Any protective resistors used must be able to withstand the voltages present. Failure modes should also be considered, with failure to open circuit normally being essential. Failure to a short-circuit condition may expose the user to unacceptable electrical shock risk.

If practical, the best option for both safety and ESD control is to prevent personnel from touching the ESDS system while high voltages are present. Handling the system with no voltages present is safe, providing there are no built-in energy sources such as high-voltage batteries or charged capacitors.

When working in high-voltage systems, personnel may be required to wear high-voltage protection gloves. These are an effective barrier to human body ESD at moderate levels as well as essential protection for electrical safety, although ESD risk due to triboelectric charging of the gloves may need evaluation.

7.5.3 Wrist Straps

7.5.3.1 Conventional Wrist Strap Systems

The purpose of the wrist strap system is to control body voltage by establishing a direct electrical connection between the wearer's skin and ground via an earth bonding point to keep body voltage below a required level. The wrist strap system consists of a wrist band, a ground cord, and an earth bonding point (Figure 7.4). The electrical and mechanical properties of the wrist band and ground cord parts are specified in standards such as IEC61340-4-6 and ANSI/ESD S1.1 (International Electrotechnical Commission 2015, EOS/ESD Association Inc. 2013). These standards specify tests and limits for electrical parameters such as wrist strap system resistance, wrist band inner, and outer surface resistance. They also specify tests and limits for mechanical parameters such as band size, breakaway force, ground cord extension, and bending life. The earth bonding point must also be specified, established, and checked.

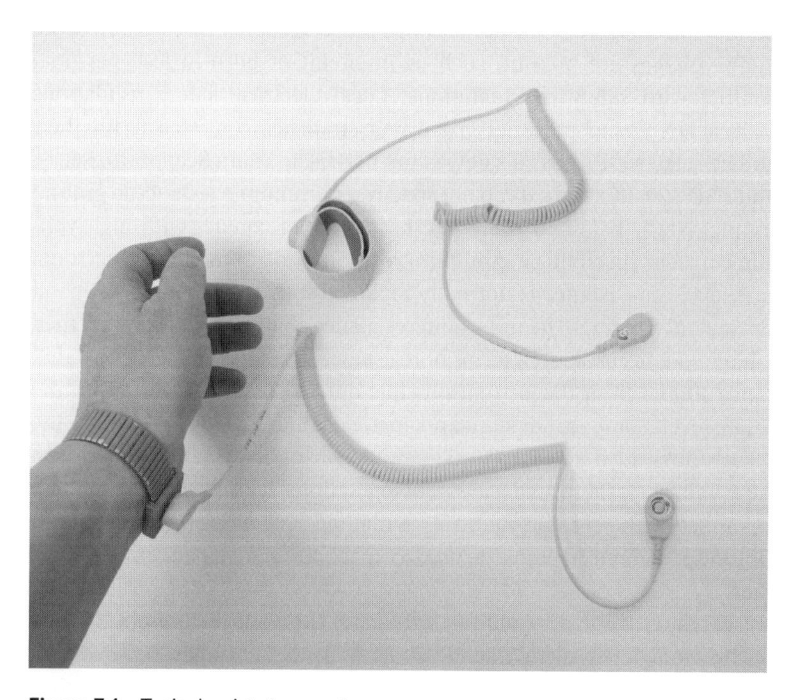

Figure 7.4 Typical wrist strap system.

The wrist band is a flexible band like a bracelet or watch strap that is worn in contact with the wearer's skin and must make good electrical contact with the skin. These bands are normally worn on the wrist but are occasionally worn in contact with other parts of the arms or legs. A ground cord makes electrical connection between the wrist band and an earth bonding point. The earth bonding point provides a point where the ground cord can be connected directly to EPA ground (earth).

Wrist bands come in various forms such as knitted or woven fabric, expanding metal bracelet, or resin bracelet. Stick-on patches are also available. Wrist bands usually include a hypoallergenic metal plate against the skin to promote good skin contact. The band inner surface is also usually conductive to give electrical contact around nearly all the inner surface of the band. The wrist band has a quick-release connection point for the ground cord. This is designed to give a reliable connection strong enough to prevent accidental disconnection but light enough for easy release. The disconnect force is normally in the range 13–36 N.

The ground cord is an insulated wire with a connector on each end, used to connect the wrist band groundable point to an earth bonding point. The ground cord includes various features for safety and other considerations. The cord usually contains a current limiting resistor at the wrist band end. Some types include a resistor at both ends. The cord wire must be designed to be flexible and withstand constant flexing and dragging over equipment and workstation edges. They are available in a range of length and colors and straight or coiled wire construction. The earth bonding point connector is commonly a snap connector or 4 mm plug, although any suitable connector may be used.

The wrist strap system earth bonding point can be any suitable and reliable direct electrical connection point to EPA earth. Dedicated earth bonding points are often provided on workstations or equipment. It is not good practice to clip a wrist strap ground cord to a workstation mat or other item that might introduce considerable addition resistance between the ground cord and earth. Dedicated earth bonding points may include some resistance providing the total wrist strap system resistance to ground does not exceed required upper limits. Current standards IEC 61340-5-1 and ESD S20.20 specify an upper limit for the wrist strap system resistance to ground of 35 MΩ although this may be changed according to the user's requirements.

7.5.3.2 Constant (Continuous) Monitor Wrist Strap Systems

A constant monitoring wrist strap system can be used to continuously test wrist strap grounding during use. They are often used where high value or very sensitive ESDS product are handled. The constant management system is designed to test the electrical connections between the person and wrist band and the ground cord and connection to the earth bonding point and set an alarm if any of these fail.

Constant monitoring systems are available in two types – single-wire systems and two-wire systems. Single-wire wrist strap monitors can be used with ordinary wrist strap systems, but two-wire wrist strap monitoring systems require a two-wire wrist strap system to be used. Constant monitors use an electrical current to measure the wrist strap system, and this can produce a low voltage on the wearer's body during use. This voltage may be unacceptable when handling the most sensitive (low ESD withstand voltage) devices.

Capacitance constant monitors use a single-wire wrist strap system. They detect the person-earth capacitance of the person wearing the wrist strap using an alternating current (AC) sensing signal applied to the system. Capacitance changes outside the set limits of the monitoring system cause an alarm. Some disadvantages are that the monitor must be set up for the individual using it and that false alarms may occur due to capacitance changes rather than failure of the wrist strap system.

Impedance constant monitors are single-wire systems that use similar principles to the capacitance monitor but detect circuit impedance changes rather than capacitance changes. This eliminates the need for adjustment and can reduce false alarms.

Resistance constant monitors use a two-wire wrist strap system, measuring the resistance of an out and return circuit through the wrist strap system and the wearer's body. An alarm sounds if the direct current (DC) resistance of this loop goes outside the accepted range. The system may use a constant or pulsed DC voltage up to 16 V to make the measurement. In most monitor designs of this type, the maximum voltage is evident only during an alarm condition. The voltage is generally considerably less under normal conditions when the person's resistance is well under the set limit of the monitor. Some users may experience skin irritation when a DC voltage is used.

Body voltage constant monitors use a two-wire system to measure the body voltage of the user. This system may not detect an open circuit that results in apparently zero body voltage. In some systems, an impedance or resistance monitor is incorporated to detect this situation.

Each type of constant monitoring system requires an appropriate tester to verify the operation of the system. The single- and dual-wire wrist strap systems used with the monitors can also be tested using resistance measurements.

7.5.3.3 Cordless Wrist Straps

I have never tested a cordless wrist strap that actually works in controlling body voltage to an acceptable level. Grounding needs a continuous connection between the body and ground, and a cordless wrist strap cannot provide this. Current standards require a maximum resistance is maintained between the body and ground, and cordless wrist straps cannot provide this connection.

Unconventional solutions like cordless wrist straps should always be carefully tested and evaluated before adoption for use.

7.5.3.4 Wrist Strap System Selection

In a standard ESD program designed to protect ESDS devices down to 100 V HBM ESD withstand, an upper limit of 35 MΩ is normally specified for the resistance from the wearer's body to the groundable point. This should be reduced pro rata for handling lower ESD withstand voltage devices. So, in an ESD program handling 50 V HBM devices, the maximum resistance from the wearer's body to groundable point should be reduced to 17 MΩ. In the manufacturing of magnetoresistive (MR) and giant magnetoresistance (GMR) heads for disk drives, the upper limit for personnel grounding is often set at 10 MΩ. When handling very sensitive devices, it would be wise to verify the maximum body voltages found in practice during wrist strap qualification.

The intended reliability and durability of the wrist strap system should be considered. The wrist band should be comfortable for the length of time the user is expected to use it. Discomfort might encourage users to remove the strap and forget or omit to wear it while handling ESDS.

Many wrist straps include a metal plate that contacts the skin under the ground cord bonding plug, improving contact reliability. The rest of the inside of the band is normally also conductive to give good all-round skin contact.

The length of ground cord, and whether it should be retractable, should be considered. The type of earth bonding point connector should normally be standardized throughout the facility so that there is no difficulty in using it in different areas as required.

If constant monitoring is to be used, it may be necessary to select a wrist strap system that is designed specifically for use with the monitoring system.

7.5.3.5 Wrist Strap System Use

Wrist bands must be a good comfortable fit and make good direct contact with the wearer's skin. They should never be worn over clothing such as shirt or ESD garment sleeves.

The wrist strap is effective only while it is connected to a working earth bonding point. Connection should be established before handling any ESDS and maintained constantly while handling ESDS.

When seated, it is normally essential to use wrist strap grounding. Grounding via footwear and flooring is unreliable in this situation as many seated personnel take their feet off the floor from time to time, breaking the ground connection.

7.5.3.6 Wrist Strap Qualification

A wrist band ground cord can be qualified by simple resistance measurement of the ground cord resistance. The types of connectors at each end of the cord should be confirmed to be suitable for the intended use.

The complete wrist strap system should be confirmed to give, when worn by typical users, system resistance acceptable within the ESD program requirements. Most organizations use a proprietary wrist strap checker for compliance verification to check that the wrist straps when worn pass this test.

A constant monitoring wrist strap system should be checked under typical operating conditions to make sure it detects common types of wrist strap system failure but does not give an unacceptable level of false alarms. The system should detect unworn or open circuit wrist strap systems. When working with very sensitive ESDS, the voltage typically induced on the wearer's body should also be tested to make sure it does not risk damaging the ESDS.

Where compliance with a standard is required, testing should be done according to the standard, and compliance with the standard pass criteria should be established.

7.5.3.7 Wrist Strap Cleaning and Maintenance

Wrist strap components (wrist band and cord) are normally replaced if they are found to fail a compliance verification test.

7.5.3.8 Wrist Strap System Compliance Verification Test

As a key ESD control measure, it is important that wrist strap grounding works reliably. All parts of the wrist strap system must be tested regularly to make sure they are working.

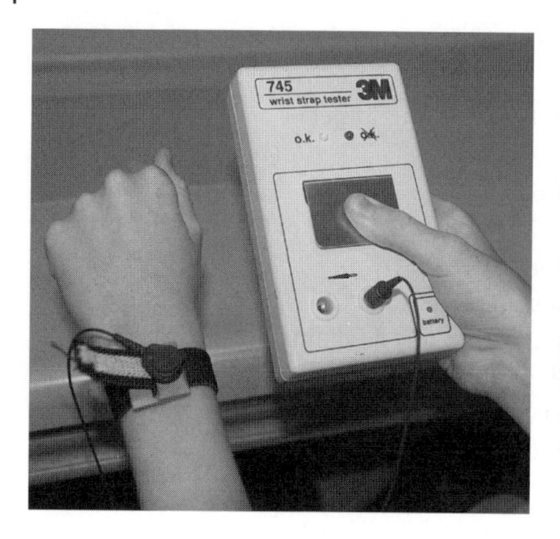

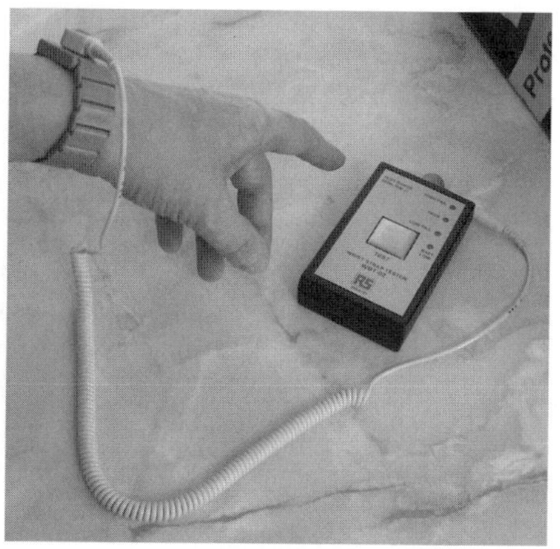

Figure 7.5 Examples of proprietary portable wrist strap testers. Source (left): D E Swenson.

The wrist band and cord are normally tested as a system as worn, using a proprietary tester (Figure 7.5). These are available as portable or wall-mounted testers, some of which can also test footwear. The tester does a simple resistance measurement to verify the resistance from the wearer's body through to the cord groundable point is within acceptable maximum and minimum limits. These limits are specified in the standards, although the user can select different pass criteria for their specific circumstances. The resistance of the skin to band contact is also tested in this measurement and can be highly variable. For a person with dry skin, this resistance can be high and result in failure of the test.

In practice, these tests are usually done by personnel for themselves using proprietary wrist strap checking instruments. These typically have built-in resistance checking, and display pass and fail as colored lights. Some sophisticated systems may use this to control access to the EPA via an entrance turnstile and automatically maintain compliance data. In other facilities, the user may be required to record their wrist strap checking pass result and relies on them to take remedial action if a failure is indicated.

With these instruments, the pass criteria are effectively provided by the instrument and may or may not be adjustable. It is important to ensure that the pass criteria are aligned with the ESD control program requirements. In current "off-the-shelf" testers, the upper limit is often set at 35 MΩ with a lower limit of about 750 kΩ. These limits, and the possibility of adjusting them if necessary, should be checked when qualifying a tester for use in an ESD program.

It is also possible and acceptable to measure the wrist strap resistance from body to the cord groundable point using a suitable resistance meter. The resistance meter must be selected to be safe for use in this way, with limited output current to prevent electrical shock risk. The measurement can be done with many types of instrument at 10–100 V test voltage. The voltage used may depend on compliance with a standard or safety regulations. Contact to the body should be made via a handheld electrode giving a reasonably large skin contact area. Holding a small-point probe may not give adequate contact area for reliable measurements.

It can be useful to keep failure data for the types of wrist straps in use, including for example identification of skin contact, ground cord or connector failure, as well as manufacturer and type data. This allows the reliability of different wrist strap systems and their components under operating conditions to be compared. The more reliable components can be identified and selected for future purchase.

Testing the EPA wrist strap earth bonding points must also be done periodically to ensure that any failure of these is discovered sufficiently quickly. This can be done using a simple resistance to ground test (see Section 11.8.3.5). A maximum resistance to ground for earth bonding points, and a suitable test method, should be specified in the ESD program's Compliance Verification Plan.

The frequency of wrist strap testing can be determined according to factors such as the level of use and established reliability of the wrist strap system. ESDS sensitivity and value of the product handled, and possible consequences and cost of product failure, are also relevant factors. If a wrist strap fails, any ESDS handled between failure and detection of the failure might be damaged.

As personal grounding is of major importance in ESD control, it is common for the wrist strap system to be tested at least each working day before the wearer commences handling ESDS devices. In some ESD programs, wrist straps are tested every time the wearer enters the EPA. When it is considered highly important to avoid risk of product damage and is wished to immediately detect any wrist strap system failure, constant monitoring can be specified.

7.5.3.9 Common Problems

Occasionally, personnel who have a dry skin do not make good low-resistance connection to the wrist band. Proprietary skin moisturizer lotions are available that can be used to counter this problem.

Some users attempt to ground the wrist strap by clipping to the edge of a bench mat or other point not designated as an earth bonding point. This is not good practice as it may introduce an unacceptable addition resistance of the mat to the wrist strap ground path.

Make sure the pass criteria of proprietary testers are aligned with the requirements of the ESD program.

7.5.4 Footwear and Flooring Grounding

7.5.4.1 The Importance of Footwear and Flooring in ESD Control

ESD control footwear and flooring systems are intended to keep the body voltage on personnel below 100 V during walking and other activities. A well-chosen footwear and flooring system can be used instead of wrist straps for reliable grounding of personnel who work while standing. For seated personnel, use of a wrist strap is required as there is a risk that the feet will be taken off the floor and contact with ground will be impaired or lost.

As has been previously explained, (see Section 4.7.9), the body voltage developed when walking is dependent on the charge generated by the walking action and the resistance through which it must pass in dissipation. Both charge generation and resistance to ground of a footwear-flooring system are dependent on both the footwear and the flooring in combination as they depend on the footwear-flooring material contact conditions. Both are necessary for successful body voltage control.

Ordinary street or work shoes often have soles made from insulating materials that block the flow of generated charge to ground. An ESD control floor gives the possibility of grounding personnel (and equipment) via the floor, but ordinary footwear blocks the flow of charge from the body to the floor. ESD control footwear provides the connection between the body and the floor via the footwear. The final resistance to ground is given by the combination of the contact resistance between the body and the footwear, the resistance of the footwear, the contact resistance between footwear and floor, and the resistance through the floor to ground.

7.5.4.2 Types of Footwear

There is a wide variety of ESD control footwear available to a modern ESD control program. This allows footwear to be selected according to a variety of needs and job roles within the ESD control program.

Heel and toe grounders are useful for fitting over street shoes or work shoes to give a ground connection (Figure 7.6). They typically have a conductive strap that goes over the shoe sole to contact the floor. This is attached to a conductive ribbon that can be tucked into the shoe, contacting the wearer through the socks or in contact with the skin. Low-cost one-use varieties are available that can be used by visitors.

Heel and toe grounders must be worn on both feet, as grounding depends on electrical contact of the body through the grounder to the floor. If only one grounder were worn, contact and grounding would be lost when that foot is raised from the floor. Toe grounders are arguably better than those that ground via the heel alone. The heels are lifted from the floor for more time during the walking action.

Booties and shoe covers are often used in clean areas for particle and contamination control as well as ESD control (Figure 7.7). These simultaneously contain particulates and provide grounding for the wearer.

ESD control shoes and boots are now available in a wide variety of styles (Figure 7.8). They are also available with built-in safety features such as steel toe caps. Many ESD control shoe types are hardly different in appearance from street shoes or work shoes. Many have a visible label or symbol indicating their ESD control function to help in their identification.

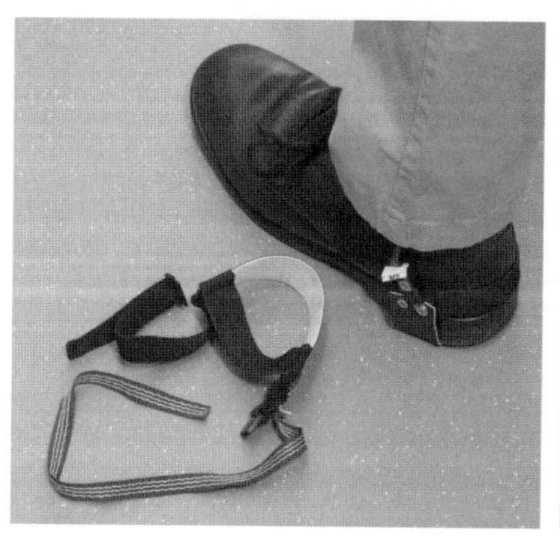

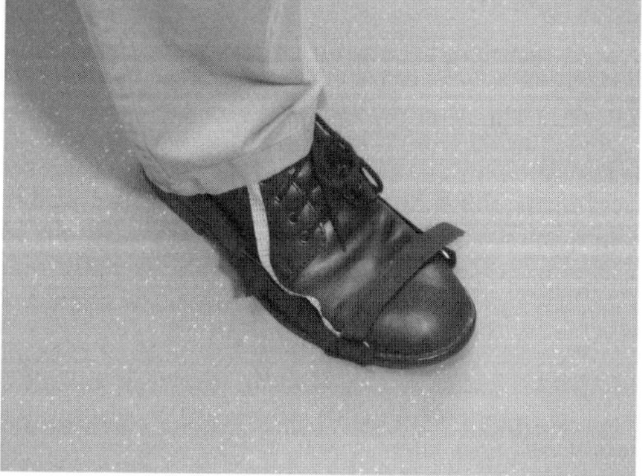

Figure 7.6 Heel and toe grounders.

Figure 7.7 Booties or shoe covers.

Figure 7.8 A wide variety of shoe and work boot styles are available.

7.5.4.3 Footwear and Safety

In addition to its ESD control function, footwear can have a personal protection safety function. In Europe, PPE including footwear must comply with the requirements of the PPE Directive. This means it must comply with the requirements of standards such as ISO 20345. These requirements cover a range of physical properties including toe protection, penetration resistance, leak proofness, water penetration and absorption, and cut resistance.

Electrical properties of footwear are classified under ISO 20345 as electrically "insulating," "antistatic," or "conductive" depending on their electrical resistance. Antistatic footwear has resistance between 100 kΩ and 1000 MΩ, and conductive footwear has resistance less than 100 kΩ measured using ISO 20344 (International Organisation for Standardisation 2011a,b). This test method is like, but slightly different from, test methods used for ESD control footwear. The conditioning and test atmosphere conditions may also be different. The result is that most types of antistatic and conductive footwear might be suitable for EPA use, but they are not tested using the test methods and conditions required for use in EPAs. PPE footwear that has not been qualified for EPA use should be tested and qualified using appropriate tests.

7.5.4.4 Selection of Footwear

There is a wide variety of ESD control footwear available to a modern ESD control program. This allows footwear to be selected according to a variety of needs and job roles within the ESD control program. When selecting footwear for an individual, their gender and personal preferences, job role requirements, and one-time or long-term usage can be considered. Care should be taken where safety requirements may also be necessary, especially in Europe where there may be a need to comply with regulations on PPE (European Union 2016). Guidance on selection, care, use and maintenance of PPE in Europe can be found in CEN/TR 16832 (European Committee for Standardisation [CEN] 2015).

The types of footwear offered should be qualified in conjunction with all the types of flooring with which they may be used to make sure they provide the ESD control performance required.

7.5.4.5 Qualification of Footwear

Footwear selected for ESD control use must always be qualified for use with every type of flooring with which it may be used. There are two basic types of qualification test.

The footwear must be tested to make sure that when worn by personnel, the resistance from body to ground required by the ESD control program will be achieved (See Section 11.8.3.4).

The footwear must be tested for body voltage generation when worn by a user walking on the floor types with which is it intended to be used. This is done using a walk test (see Section 11.8.9). In most ESD control programs, the limit of body voltage is chosen to be 100 V, reflecting the goal of handling devices down to 100 V HBM ESD withstand. If lower withstand voltage devices are handled, it may be necessary to use a lower body voltage limit. Examples of body voltage waveforms obtained with different floors and the same footwear are shown in Section 4.7.9.3 and Figure 4.8.

Initial qualification can be based on the resistance range of the footwear, tested unworn according to a suitable standard such as IEC 61340-4-3 or ANSI/ESD STM9.1 (International Electrotechnical Commission 2001; EOS/ESD Association Inc 2014a). Unfortunately, these tests do not predict the performance of the footwear when worn by a person standing on an ESD control floor. An important practical test is to determine the resistance from body to ground achieved when a test subject wearing the footwear is standing on the floor with which the footwear is to be used (IEC 61340-4-5:2004 or EOS/ESD Association Inc 2015) Selected footwear should also be qualified in conjunction with the floor surface material with which it will be used, to determine body voltage generated in a walk test (IEC 61340-4-5 or EOS/ESD Association Inc 2015). If the footwear is to be used with several types of floor material, it should be qualified for use with all of them.

Qualification tests should, where possible, be tested with the flooring under the worst-case conditions that can be envisaged during operation. This usually means testing under low-humidity atmospheric conditions in a laboratory. Where a minimum resistance for electrical safety is required, it may be necessary to establish performance under damp conditions.

Where qualification with an existing floor is required, it is often not possible to test the footwear-flooring combination in the laboratory. Qualification should be done under the worst-case conditions practical under the circumstances. Confirmation of performance under low-humidity atmospheric conditions when these arise is advisable. These conditions often occur in the winter in colder climates.

The footwear should also be checked with the intended compliance verification test equipment to make sure they reliably pass the intended compliance verification tests.

Where compliance with a standard is required, testing should be done according to the standard and compliance with the standard pass criteria should be established.

Sampling and test of incoming ESD control footwear is advisable.

7.5.4.6 Use of Footwear

Although a wrist strap probably provides the most reliable personal grounding, personnel who must stand or move around in their work can find wrist straps restricting or inconvenient. Using footwear-flooring grounding can provide an effective alternative for personnel who stand or walk around during their work.

Grounding of personnel through footwear and flooring is effective only while the person's footwear is in contact with an ESD control floor. If the floor-footwear contact is broken, grounding via footwear and flooring fails. This can happen, for example, if a person is seated and takes their feet off the floor or if contact is only through the heels and the person stands only on their toes. So, grounding through footwear-flooring is insufficient for seated personnel, and a wrist strap must be worn in this circumstance.

Heel or toe grounders must be worn on both feet. If a strap is worn only on one foot, ground contact is lost when this foot is taken off the floor.

The resistance to ground and body voltage limiting performance of a footwear-flooring combination can be changed by contamination on the footwear contact surface or the floor surface. Contaminants such as asphalt, paints, varnishes, or chemicals can provide an insulating barrier to current in the charge dissipation path. Water can have the opposite effect, soaking into a sole material, reducing the footwear-floor contact resistance, lowering the resistance to ground. This can be a problem if a level of resistance is required for electrical safety reasons.

For these reasons, some organizations do not allow ESD control footwear to be worn out of doors.

7.5.4.7 Compliance Verification of Footwear

Compliance verification of ESD control footwear is normally limited to measurement of the resistance to ground (sole of footwear) when the foot is in contact with a metal measurement plate. Proprietary testers are commonly available. These are often provided at the entrance to EPAs so that personnel can test their footwear before entry to the EPA.

A disadvantage of many testers is that they only give a pass/fail indication rather than a resistance value. Some automated data logging types also record the resistance value measured. If necessary, a resistance measurement may be made using a suitable resistance meter (see Section 11.8.3.3). The pass criteria of the footwear tester should be aligned with the requirements of the ESD Control Program Plan documents.

7.5.4.8 Common Problems

Heel or toe straps can, unnoticed by the wearer, become dislodged and lose contact with the floor during use. Some types of strap may have insufficiently long fixing straps to securely hold the strap in place for personnel with large feet or shoes. It may be difficult to use straps with some types of street footwear such as elevated heels or sandals.

Self-adhesive straps may not adhere well to the shoe if the wearer's shoe soles were dusty at the time of application.

ESD control shoes and boots often rely on moisture from the wearer's feet to establish contact from the skin through the socks to the shoe insole. If the wearer has dry skin or thick man-made fiber socks, this contact can be impaired or take some time to be established. The wearer may have to wear the shoes for several minutes before sufficient moisture is built up, especially under cool atmospheric conditions. Atmospheric humidity conditions also have an influence (Swenson et al. 1995).

Testers for wrist straps can sometimes be used to test footwear as well but may have wrong resistance range. Current standards require a maximum resistance of wrist straps as worn of 35 MΩ, whereas footwear as worn may have resistance up to 1000 MΩ.

In ISO 20345, safety footwear is classified as insulating, antistatic, or conductive. Some users may wear insulating footwear for electrical shock protection, whereas others may use antistatic or conductive safety footwear, for example for handling explosives or flammable materials. Others may have footwear specified for ESD control rather than personal protective use. Users may have poor awareness of these differences and may just think they are wearing "safety footwear" or "ESD footwear." This can result in personnel inadvertently wearing the wrong type of footwear when they enter EPAs or do particular jobs. This could be potentially unsafe if a person wore antistatic or conductive footwear while working with high voltages. It could result in an ESD risk if a person wore insulating safety footwear when within an EPA and handling ESDS devices. In the EPA, this can be guarded against by requiring that personnel verify their footwear before entry to the EPA.

7.5.5 Grounding via ESD Control Seating

A common misconception is that an ESD seat is designed to ground personnel sitting on the seat. In most ESD programs and current ESD standards, this is not so. Nevertheless, in some ESD control programs, seating has been successfully used to ground personnel. This is discussed further in Section 7.9.8.

7.5.6 Personal Grounding via an ESD Garment

Groundable static control garments are available that can be used to ground the wearer via the garment. These typically have a groundable point that can be connected to an earth bonding point. They are also made from a relatively low-resistance material and must make reliable contact with the body of the wearer. This is further discussed in Section 7.11.8.

7.5.6.1 Qualification of a Groundable Static Control Garment

Basic qualification of a garment fabric can be done on sample on the fabric. A point-to-point resistance measurement is usually used. The resistance of some fabrics can vary with direction depending on how conductive fibers lie (warp and weft) and so measurements should explore this possibility.

Qualification of a groundable garment should include tests to determine that all the panels of the garment are made of material that has point-to-point resistance less than the required upper limit and are electrically connected with the garment groundable point. The garment must make direct reliable contact with the wearer's body, and the wearer must be reliably grounded through the system.

When grounding personnel through a garment, tests should be made to establish that the body voltage is maintained below the level required by the ESD control program. Tests should include the lowest-humidity conditions likely to be found in practice. ESD control standards require that qualification is done at $12 \pm 3\%$ rh.

7.5.6.2 Compliance Verification of ESD Control Garments

ESD control garments used to ground personnel should be verified for grounding effectiveness before use. Tests should verify the resistance of the material and the ground path. If the garment is intended to ground the wearer's body, then the entire ground path from body to groundable point should be verified in the same way as for wrist straps. A proprietary tester can be used for this providing the pass criteria are aligned with the garment requirements in the ESD Control Program Plan.

7.6 Work Surfaces

7.6.1 What Does a Work Surface Do?

The purpose of providing a grounded static dissipative work surface is twofold (see Section 4.7.10). First, the work surface material itself should not become charged and give electrostatic fields that could lead to ESD risk. Second, the work surface provides a useful way of draining charge from any noninsulative material or object that is placed on it. This may include tools, assemblies, and components including ESDS.

The work surface provides an area in which the items being handled are brought to the same voltage by being connected to EPA ground so that they present little ESD risk. Any noninsulating item placed on the workstation surface can lose its charge to ground. The characteristics of the work surface are key in determining the nature of the inevitable discharge from any charged ESDS placed on the surface and ensuring that this discharge is not damaging to the ESDS device.

The simplest EPAs for field work may consist of little more than a portable workstation surface, personal grounding, and a connection to ground. In a simple case like this, the boundary of the work surface may define the boundary of the EPA.

7.6.2 Types of Work Surfaces

There are a wide variety of work surface materials available and in use. Some are made from a single material, whereas others have a multilayer structure.

Monolayer or homogenous materials have the same electrical properties through the bulk of the material. These materials typically give a point-to-point resistance that varies with the distance between the electrodes. The measured resistance to ground typically varies with the distance between the measurement point and the groundable point.

Multilayer materials may have two or three layers. The top layer normally has resistance between 10 kΩ and 1 GΩ. The underlying layer is normally low-resistance material. There may be an additional bottom layer. Because of the underlying low-resistance layer, the point-to-point resistance and resistance to groundable point measured on the mat surface are relatively constant.

High-pressure laminates are typically single or multilayer rigid materials applied to a substrate material. Some materials are affected by humidity. Their performance should be qualified under low-humidity conditions. If a minimum resistance is also specified, it should be tested under high-humidity conditions.

Mats and runners are usually flexible materials that can be used to cover surfaces that were not designed for ESD control use. They may be single- or multilayer materials. They may also be used as a sacrificial surface in areas of high contamination or to improve the characteristics of high-pressure laminate surfaces.

Portable and field service work surfaces are typically light, flexible materials that can conveniently be folded or rolled for inclusion in the tool kit of a field service engineer.

7.6.3 Selection of a Work Surface

In selecting a workstation surface, the first considerations must be about how the workstation is to be used and the processes that will occur there. Considerations may include

- Is it to be a permanent area or a temporary area?
- What are the activities and processes anticipated for the workstation, including the type and form of ESDS handled?
- What physical properties are needed?
- Is there a need to resist chemicals?
- Will it be a clean environment in which particle shedding or other contamination would be a concern?
- Are there electrical considerations such as use in the presence of high voltage, or handling of printed circuit boards (PCBs) with on-board batteries?
- Are there possible safety considerations?
- Are there ergonomic considerations?
- What will be the maintenance and cleaning regime?
- Are there other safety or legislative requirements (e.g. fire retarding qualities) to consider?

The activities and processes and types of ESDS handled, along with whether it is a temporary or permanent facility, will probably largely determine the type of work surface selected. For a field service workstation mat, compact storage and portability in a tool kit may be an overriding consideration.

Durability considerations include the hardness, abrasion, and tear resistance of the surface material. Aspects such as chemical (e.g. solvent), heat, and solder resistance may also be a concern.

One significant point will be to decide whether a soft and cushioning surface is needed for handling the ESDS device or a durable surface is needed for handling heavy or sharp items. Even on a permanent workstation there may be a reason to use a temporary work surface. Some ESD programs use a replaceable mat surface as a sacrificial cover where chemical contamination or physical damage is a problem. Others use a mat to modify the surface on some workstations, e.g. use of a cushioning mat on workstations that have a hard surface. For mats, curling with age or due to storage in rolled or folded form may be a concern.

Where items such as cutting mats or trays are placed on the workstation surface, they should comply with the requirements for worksurfaces.

If the workstation is in a clean room, this may restrict the work surface type to clean room–compatible materials.

In workstations that use high voltages or handle ESDS devices containing batteries or power sources, it may be necessary to use a work surface that has a specified minimum resistance. The minimum resistance may be specified for safety, power drain, or short-circuit elimination considerations. Where electrical safety is a concern, appropriate tests and specifications must be selected that comply with local regulations. DC resistance measurements as made for ESD control purposes may be inadequate for safety testing, especially where high-voltage AC is present.

In a test workstation, proximity or contact with a conductive work surface could adversely affect the operation of the ESDS under test.

A minimum resistance of the work surface may also be required to control a risk of charged device ESD to ESDS devices placed on the surface (see Section 4.7.5). A metal tray placed on a workstation surface can give charged device ESD risk if an ESDS device is placed upon it.

Appearance can be a consideration in selecting a work surface. Colors can be used to identify workstations for certain operations or for corporate identity identification.

7.6.4 Workstation Qualification Test

Workstation qualification test should address all the functionality, durability, chemical, electrical, and other characteristics identified in the selection process.

Work surface qualification is typically done using standard point-to-point and resistance to groundable point measurements (see Section 11.8). These should be done under controlled atmospheric humidity representing the worst case likely to be experienced in practice. In most cases, this means testing under low-humidity conditions to make sure the surface meets the upper resistance requirement under these conditions. If data sheet values are accepted as qualification, these should be obtained using the correct standards test methods at $12 \pm 3\%$ humidity.

If a minimum surface resistance is specified, it should be verified under high-humidity conditions.

Where compliance with a standard is required, testing should be done according to the standard and compliance with the standard pass criteria should be established.

7.6.5 Acceptance of Work Surfaces

Work surfaces should be tested after installation and before first use. A resistance to ground measurement can be used to verify that the work surface is grounded. A point-to-point resistance measurement can be used to check that the work surface material surface resistance characteristics are as expected from qualification data.

7.6.6 Cleaning and Maintenance of Work Surfaces

Work surfaces should be cleaned according to the manufacturer's instructions. Ordinary cleaning materials and procedures may not be suitable as they can leave a surface film of wax, silicone, or other materials that can seriously affect the performance of the surface.

If a surface has been cleaned with water or a liquid cleaner, it should be left to dry before testing as the residual moisture could affect the measurement results.

7.6.7 Compliance Verification Test of Work Surfaces

Current practice in compliance verification is to test the work surface resistance to EPA ground. Current standards specify an upper limit for this, although an ESD program can specify other pass limits to suit the expected operational performance of the selected items.

Visual checks can be a useful regular check of the ground connection, especially for temporary connections, e.g. bonding cords between a mat and a mains electricity ground connector.

The frequency of testing can depend on the expected reliability of the work surface material and its connection, and the possible consequences of failure on ESD security. With a permanently connected work surface in a clean area, with well-established history of measurements showing little change, a relatively long interval between testing may be acceptable. A bench mat placed on a surface for sacrificial use, known to become contaminated by process chemicals and connected by a temporary connection to an earth bonding point, may need to be tested very frequently.

7.6.8 Common Problems

Common failures can be due to failure or accidental disconnection of a ground connection or contamination of the bench surface, for example with insulative process materials.

Any surfaces on which ESDS may be placed (including trays or packaging) should comply with the requirements for worksurfaces.

Where other materials are placed on a worksurface, be careful that the ESD control properties are not compromised. A common mistake is to use a cutting mat made of insulating materials placed on an ESD bench, introducing a charged insulator risk. Another common example is a metal tray placed on the bench surface can introduce charged device ESD risk to ESDS devices placed on the tray.

7.7 Storage Racks and Shelves

7.7.1 Should It be an EPA Rack or Shelf?

Storage rack and shelf systems are used to store products and materials. Racks need no special ESD control qualities and do not need to be in an EPA if they are used to store ESDS device that are adequately protected in ESD-protective packaging for transport and storage outside EPAs.

Racks and shelves used to store unprotected ESDS inside an EPA must have ESD control qualities similar to workstation surfaces. They can also be used to store non-ESDS items and products, providing any packaging used has suitable ESD control properties and the proximity of such product does not introduce ESD risks. Non-ESDS items that are within secondary (non-ESD control) packaging or in themselves might cause ESD risk (e.g. insulating components such as enclosures) should be stored well away from exposed ESDS and EPA shelves.

Workstations often have shelves above them for storage of test and IT equipment, tools, materials, ESDS devices, and other items. If the shelf is used to store unprotected ESDS devices, it must be treated as an EPA workstation surface. This means it must have the electrical characteristics of a work surface and must be grounded. The usual ESD protective requirements of an EPA workstation, such as control of insulating materials, must be applied.

Workstation shelves that are not used to store unprotected ESDS can be treated as unprotected areas. They may be constructed from non-ESD control materials, providing they are sufficiently far from the position of any unprotected ESDS devices and no ESD risks are produced by this. It requires training and discipline on the part of the operators to ensure that unprotected ESDS are never placed on this shelf. The use of ESD control shelves in some workstations and non-ESD control shelves in others should be avoided if possible, as this can lead to confusion and placement of ESDS devices on non-ESD shelves. If necessary, signs can be used to identify shelves that differ from the norm. If unusual practices such as this are used, they should be carefully documented so that they do not form a non-compliance with the documented ESD Control Program. Additional training and compliance verification may be needed to ensure successful management of the risks.

Single- or multilevel shelves and racks are often provided for storage of ESDS and non-ESDS parts or product. The user will need to decide whether these should be specified as EPA shelves or not. The racks should be specified as EPA racks if unprotected ESDS devices are to be stored on them. If not EPA racks, then ESDS devices stored on them will need to be protected within ESD-protective packaging for transport and storage in an unprotected area.

EPA and non-EPA shelves should preferably not be mixed on the same rack as this may lead to confusion. This could result in storage of unprotected ESDS devices on unprotected area shelves or placing of insulative items on EPA shelves, causing ESD risks. EPA racks and shelves should be maintained a suitable distance from any unprotected area activities or storage. A distance of 0.5 m will normally reduce any electrostatic fields from charged insulative materials to a sufficiently low level to minimize ESD risks to all but extremely sensitive ESDS devices. A greater distance may give additional confidence.

Factors indicating that racks may be better specified as EPA might include

- It would be convenient to store ESDS devices without packaging them in ESD protective packaging.
- The items are to be taken to workstations in the same EPA after storage.
- The shelf can be sited for access without leaving an EPA. (If the floor between EPA workstations and the shelf is not an ESD control floor, transport to the shelf is effectively through an unprotected area.)
- Non-ESDS and insulative items can be stored on a separate shelf unit.

The rack may be better specified as an unprotected area if

- Most items to be stored are not ESDS.
- It is desirable to store items in secondary packaging or insulative items on the same shelves as ESDS.
- It is not inconvenient to package the ESDS within ESD-protective packaging for transport and storage in an unprotected area.
- The items will go out of the EPA after storage.
- Access to the shelf is via an unprotected area (e.g. across a non-ESD control floor).

Drawer systems and carousels used to store unprotected ESDS can often be considered to have the function of ESD control packaging and can be specified similarly.

7.7.2 Selection, Care, and Maintenance of Racks, and Shelves

When selecting EPA shelves and racks, similar considerations as for workstations may apply, including the following:

- What will the physical requirements be (e.g. strength, size)?
- What is the type and form of ESDS handled?
- Will it be a clean environment in which particle shedding or other contamination would be a concern?
- Are chemicals likely to be present?
- Are there electrical considerations such as high-voltage use or handling of PCBs with on board batteries?
- Are there possible safety considerations?
- Are there ergonomic and convenience considerations?
- What will be the maintenance and cleaning regime?
- Are there other safety or legislative requirements (e.g. fire retarding qualities) to consider?

The care and maintenance aspects are like those for workstations.

7.7.3 Qualification Test of EPA Shelves and Racks

Workstation qualification test should address all the functionality, durability, electrical, and other characteristics identified in the selection process. Qualification of the shelf surface is like workstation surfaces, using standard point-to-point and resistance to groundable point measurements (see Section 11.8). These should be done under controlled atmospheric humidity representing the worst case likely to be experienced in practice. In most cases, this means testing under low-humidity conditions to make sure the surface meets the upper resistance requirement under these conditions. If a minimum surface resistance is specified, it should be verified under high-humidity conditions.

Where compliance with a standard is required, testing should be done according to the standard, and compliance with the standard pass criteria should be established.

7.7.4 Acceptance of Shelves and Racks

Shelf and rack surfaces should be tested after installation and before first use. A resistance to ground measurement should be used to verify that the work surface is grounded. A point-to-point resistance measurement can be used to check that the work surface material surface resistance characteristics are as expected from qualification tests.

7.7.5 Cleaning and Maintenance of Shelves and Racks

Shelf and rack surfaces should be cleaned according to the manufacturer's instructions. Ordinary cleaning materials and procedures may not be suitable as they can leave a surface film of wax, silicone, or other materials that can seriously affect the performance of the surface.

If a surface has been cleaned with water or a liquid cleaner, it should be left to dry before testing as the residual moisture could affect the measurement results.

7.7.6 Compliance Verification Test of Shelves and Racks

Current practice in compliance verification is to test the shelf and rack surface resistance to EPA ground. Current standards specify an upper limit for this as for work surfaces, although an ESD program can specify other pass limits based on selected material characteristics.

Visual checks can be a useful regular check of the ground connection, especially for temporary connections, e.g. bonding cords between a mat and a mains electricity ground connector.

The frequency of testing can depend on the expected reliability of the shelf or rack surface material and its connection and the possible consequences of failure on ESD security. With a permanently wired-in shelf or rack surface in a clean area, with a well-established history of measurements showing little change, a relatively long interval between testing may be acceptable.

7.7.7 Common Problems

Common failures can be due to failure or accidental disconnection of a ground connection or contamination of the shelf or rack surface, for example with insulative process materials.

Some types of moveable shelf or rack systems can be difficult to distinguish visually from similar systems not designed for EPA use. An example of this may be metal racks that have plastic components joining the shelves to the frames. This can lead to unsuitable units being inadvertently substituted for EPA shelf units during a work area rearrangement.

7.8 Trolleys, Carts, and Mobile Equipment

7.8.1 Types of Trolleys, Carts and Mobile Equipment

Trolleys, carts, and mobile equipment are used to store and move product, materials, or equipment within an EPA or between EPAs. They may be simple mobile work surfaces or shelves or specialist items such as PCB racks on wheels (Figures 7.9).

Mobile equipment and carts must not cause ESD risk to any ESDS device that they come close to, may contact, or are used to store or transport. This usually means that the cart must be made of noninsulating materials, and at the times and positions where ESD product are loaded and unloaded, they must be grounded. At the times when the cart is ungrounded, personnel should not be able to touch any unprotected ESDS on the cart, even if the personnel are grounded. Personnel may safely handle ESDS on the cart only if they are equipotential bonded to the cart.

Where carts have shelves on which unprotected ESDS devices may be placed, the shelves should have a similar specification to EPA work surfaces. Where a mobile rack system or similar equipment is effectively mobile ESD-protective packaging, requirements like ESD-protective packaging can be applied.

Figure 7.9 Typical example of an ESD control cart (trolley).

Figure 7.10 An ESD control seat.

Mobile equipment and carts can be grounded through noninsulative wheels, a drag chain, or some other means to an ESD control floor. These can provide a convenient means of grounding the cart if it is standing on the ESD control floor and has the advantage that it needs no operator action. Qualification tests should demonstrate that grounding is effective when the equipment is standing on all the ESD control floors with which the cart will be used.

Where there is no ESD control floor (e.g. at a stand-alone workstation), carts and mobile equipment should be grounded via an earth bonding wire before and during loading and unloading.

7.8.2 Selection, Care, and Maintenance of Trolleys, Carts, and Mobile Equipment

In selecting a suitable mobile equipment or cart, considerations may include the following:

- Will the cart hold ESDS devices, or is it to be only used for tools, process equipment, or materials?
- Will the cart work in one EPA or many EPAs?

- Are chemicals likely to be present?
- What will the physical requirements be (e.g. strength, size)?
- Will the cart remain in the EPA, or will it go through unprotected areas or even outside between buildings? Will covers or additional protection be needed for this use?
- What is the type and form of ESDS devices handled?
- Are unprotected ESDS devices to be placed on the cart surface? If yes, does the ESDS device have charged device ESD susceptibility?
- If used for storing or transporting ESDS, will they be within ESD protective packaging, or will the cart provide the required level of ESD protection?
- When within EPAs, will an ESD control floor be available for grounding the cart? Will a ground wire be required for grounding the cart at any stage?
- Can the cart ESD control characteristics be measured using standard electrodes, or will special electrodes or techniques be required?
- Do the materials have permanent ESD control properties, or do they have coatings or other control materials that might wear or degrade?
- Will it be a clean environment in which particle shedding or other contamination would be a concern?
- Are there electrical considerations such as high-voltage use, or handling of PCBs with on board batteries?
- Are there possible safety considerations?
- Are there ergonomic and convenience considerations?
- What will be the maintenance and cleaning regime?
- Are there other safety or legislative requirements to consider?

Where wheels are used as groundable points, it is advisable to make sure that two or more noninsulative wheel groundable points are provided. This will provide a level of redundancy to ensure that grounding is maintained even if a patch of contamination on a floor or wheel prevents good contact. Some types of wheels, feet, or grounding chains do not reliably make good contact with all types of ESD control floor.

Where cart wheels act as the groundable points, regular cleaning will usually be necessary to keep them free of accumulated dirt contamination. Shelves may also need cleaning if contamination is known to impair their properties.

7.8.3 Qualification of Trolleys, Carts, and Mobile Equipment

Carts that have shelves on which unprotected ESDS devices may be placed should be qualified as for ESD control work surfaces. A point-to-point resistance measurement should be used to check the surface properties. If charged device ESD is a concern for unprotected ESDS devices placed on the surface, consider specifying a minimum point-to-point surface resistance.

A resistance to groundable point measurement should be made to confirm the ground path between all shelves and other parts and the wheels or groundable points. It is also wise to do a resistance to ground test with the cart standing on the floor with which it will be used to make sure the wheels make reliable contact through the floor to earth.

Resistance to ground tests should preferably be done with the cart standing on every floor with which it will be used to check for reliable grounding.

For mobile PCB racks or other items that are specified like ESD control packaging, the exposed surfaces should be tested to confirm that they are static dissipative or conductive. Resistance to groundable point tests should be made to ensure that the surfaces are bonded to the groundable points.

Any covers, doors, or sides to the cart or mobile equipment should be tested to make sure they do not have exposed insulating surfaces that could charge up and give significant electrostatic fields.

Where compliance with a standard is required, testing should be done according to the standard, and compliance with the standard pass criteria should be established.

7.8.4 Compliance Verification of Trolleys, Carts and Mobile Equipment

Compliance verification of carts and carts that have shelves or work surfaces is typically by testing the shelf surface resistance to EPA ground while standing on the EPA floor. Current standards specify the same upper limit for this as for work surfaces, although an ESD program can specify other pass limits.

Equipment such as mobile PCB racks may have surfaces specified as for ESD protective packaging. Some of these, especially items such as slots or wells holding PCBs or other specific ESDS devices, may require testing of their surface resistance and resistance to ground. This may be difficult to do using standard electrodes as the materials may be molded with small and complex nonplanar surfaces. A suitable electrode system for testing these items might have to be designed. Any special test methods or equipment used should be documented as part of the Compliance Verification Plan.

The frequency of testing can depend largely on the reliability of the ground connection. With mobile equipment in a clean area, with a well-established history of measurements showing little change, a relatively long interval between testing may be acceptable. Any parts that have coatings that may be abraded or surfaces that may be damaged by normal wear and tear should be regularly tested. Areas of conductive coating that have become isolated through wear and may contact ESDS devices could be a serious source of ESD risk.

7.8.5 Common Problems

A common type of failure is excessively high resistance to ground due to contamination of wheels with a buildup of accumulated dirt.

The wheels may make contact with the floor only over a small contact area. With some types of floor material having sparsely distributed conductive elements, contact between the wheels and conductive elements in the floor can be intermittent.

Contamination of shelf or rack surface, for example with insulative process materials, can impair its ESD control properties.

Some types of mobile equipment can be difficult to distinguish visually from similar systems not designed for EPA use. This can lead to unsuitable units being inadvertently substituted for ESD control shelf units during movement around a facility. Where this is a risk, the equipment should be clearly marked in some way to differentiate the ESD-compliant and noncompliant types.

7.9 Seats

7.9.1 What Is an ESD Control Seat for?

An ordinary chair not designed for ESD control usually has wheels or feet, structural parts, and cover material made of insulating materials. The covering material and structural part may become very highly charged in contact with the user's outer clothing and by movement of the chair over the floor. A person sitting on a highly charged seat might, if inadequately grounded, rise to high induced voltage and be a source of ESD risk. The design of an ESD control seat addresses these problems (see Section 4.7.10.5).

Wheeled seats can get charged by rolling action of the wheels on the floor. As well as being a strong source of electrostatic fields, ordinary chairs can generate internal ESD that can radiate electromagnetic interference (EMI) and interfere with electronic equipment (Smith 1993, 1999).

An ESD control seat has a noninsulating covering material that can dissipate the charge generated by contact with the user's clothes. The structure of the chair is designed so that an electrical connection is provided between the covering material and exposed chair parts through to ground, usually via the feet or wheels or a grounding means such as a drag chain.

A common misconception is that an ESD seat is designed to ground personnel sitting on the seat. In most ESD programs and current ESD standards, this is not so. One reason for this is that it can be difficult to ensure reliable connection between the user's body and the seat through layers of intervening clothing. It would be difficult in many cases to ensure that a sufficiently low-resistance ground connection and low body voltage is maintained. Sitting personnel must normally be grounded via a wrist strap (see Section 7.4). Nevertheless, personnel have sometimes been successfully grounded via seating. This is discussed further in Section 7.9.8.

7.9.2 Types of ESD Seating

A range of seating types are available including chairs and stools of many different heights and style (Figure 7.10). Many are designed to be grounded through an ESD control floor via conductive wheels or feet. Some may have additional grounding points such as bonding points for ground cords.

The seat and back cover surfaces may be made of noninsulating fabric or vinyl. Chair arms may be provided, made of noninsulating plastics. Metal footrests are often provided.

Fabrics used in upholstered chairs often have conductive fibers woven into the fabric.

Vinyl upholstery materials often have a thin conductive layer below the surface.

7.9.3 Selection of Seating

The seat style is primarily selected for its function and convenience to the user in the EPA workstation that it is intended to be used. Safety considerations (e.g. stability during working activity) should be considered.

The seat should be constructed using noninsulating materials for all exposed surfaces that might become charged by rubbing during use, especially those contacting the user's body. These surfaces should all be connected to the groundable point.

Where wheels or feet are used for grounding to an ESD control floor, it is advisable to make sure that two or more noninsulative wheels or feet are provided. This will provide a level of redundancy to ensure that grounding is maintained even if a patch of contamination on a floor, wheel, or foot prevents good contact.

ESD control seats can be almost indistinguishable in appearance from ordinary seats. If confusion with non-ESD control seating could be possible during use, consider identifying the seats with appropriate markings.

7.9.4 Qualification Test of Seating

Qualification testing of seating should include testing to ensure that the exposed surfaces that might become charged are not insulative and are electrically connected to the groundable points (wheels, feet, or earth bonding points). This should be done under dry humidity conditions. Where compliance with a standard is required, testing should be done according to the standard, and compliance with the standard pass criteria should be applied.

7.9.5 Cleaning and Maintenance of Seating

Build-up of dirt on the wheels or feet is a common cause of failure of compliance during verification tests. Where cart wheels or feet act as the groundable points, regular cleaning will usually be necessary to keep them free of accumulated dirt.

7.9.6 Compliance Verification Test of Seating

Compliance verification of seating is normally done using a simple resistance to ground measurement from an electrode placed on the seat surface (see Section 11.8.1). This tests the ground path from the seat surface through the grounding point (e.g. wheels and floor) to the EPA earth.

7.9.7 Common Problems

For seats grounded through feet or wheels on an ESD control floor, a common cause of grounding failure is the buildup of contamination and dirt on the wheels.

Wear and damage to the seat surfaces, especially those in contact with the user's body, can impair the ESD control properties of the materials. Repairs to ESD control seats should never be made using insulative materials.

7.9.8 Personal Grounding via ESD Control Seating

A common misconception is that an ESD control seat is designed to ground personnel sitting on the seat. In most ESD programs and current ESD standards, this is not so. One reason for this is that it is difficult to ensure reliable connection between the user's body and the seat through layers of intervening clothing. Ground connection through wheels or feet can also be unreliable as contamination can build up on the wheels/feet. It would be difficult in many cases to ensure that a sufficiently low-resistance ground connection and low body voltage is reliably maintained.

The resistance to ground through many types of seat is rather high for grounding personnel, with up to $10^9\ \Omega$ being allowed by current standards. In contrast, the limit required for wrist strap grounding is set at 35 MΩ. Current standards require that sitting personnel must normally be grounded via a wrist strap.

If seating is to be used for grounding personnel, these potential problems must be addressed. The first difficulty is to ensure that reliable connection can be established between the chair and the user. This may be easier in warm climates where fewer clothes are worn because more skin may be exposed on the body where it will contact the chair. In warmer temperatures, a greater level of body moisture helps make contact through clothing fabric. In colder climates where several layers of clothes may be worn and low-humidity conditions may arise, it can be difficult to establish reliable contact between the body and the chair.

The second difficulty is to ensure that the chair provides reliable grounding to the ESD control floor surface. A chair foot or wheel may contact a very small area of the floor and is prone to buildup of dirt between the foot/wheel and the floor. This is particularly so in areas where dust and contamination may be a concern.

As current standards do not use seats for grounding personnel, compliance would require a tailoring statement giving the technical basis and rationale, backed by data, establishing that the practice is acceptable. Qualification and compliance verification test methods and pass criteria for suitable seating would need to be established as these are not provided by standards in this case. For these reasons, a study would need to be done to establish the reliability of grounding personnel through seating and establishing suitable tests and pass criteria. The variation of resistance from the user's body to ground and body voltage should be studied over a range of temperature and humidity conditions. Cold and dry conditions may provide a "worst case." This and any other anticipated worst-case conditions should be carefully studied. The results of the study should be documented in a report that can be summarized and referenced in the ESD program documents and held on file for later reference. An explanation of the practice and reference to the test report should be included in the ESD program documents.

If seats are used for grounding personnel, then the body voltage must be maintained less than the maximum required in the ESD program. This should be established using body voltage measurements under simulated worst-case operating conditions including low-humidity air conditions. One problem is that unless outer clothing is specified (e.g. by use of ESD coats), the charge generation properties of outer clothing against the seat cover material can be highly variable with clothing material changes.

If several types of seat are used, all these should be tested and qualified, and each seat type to be used should be type tested to demonstrate that body voltage is adequately controlled and establish the range for acceptable resistance to ground. Seats purchased for use should then be restricted to the types tested and qualified.

Suitable test methods for compliance verification and pass criteria should be considered. A simple resistance to ground test of the seat would not establish grounding of the user. A resistance test from the body of the person sitting on the seat to ground would be needed to verify adequate connection (Section 11.8.3). A body voltage measurement based on Section 11.8.9, made with the subject simulating normal working activity, would also be advisable.

7.10 Ionizers

7.10.1 What Does an Ionizer Do?

Ionizers are commonly used for reducing charge levels, voltages, and electrostatic fields on essential insulators in use near ESDS devices (see Section 4.6.3). The primary means of controlling voltages on conductors is to ensure that they are reliably grounded, but where grounding cannot be used, it may be possible to use an ionizer to reduce voltages on conductors to an acceptable level. Reduction of electrostatic fields and voltages can be important for reducing ESD risk and for reducing electrostatic attraction (ESA) of contaminant particles to items in clean areas (Section 4.7.9).

An ionizer works by "spraying" positive and negative charges into the air as ions (see Section 2.8). Opposite polarity ions drift under electrostatic fields to the surfaces that are the sources of the fields. Arriving at the surface, they neutralize excess charges of the opposite polarity or reduce the voltage by accumulating there.

The ions are produced from the air present in the environment. Producing air ions in an ionizer does not produce any contamination and so ionization is suitable for clean room use, although the erosion of emitter points in electrical ionizers should be considered.

Use of ionization to neutralize charge and voltages suffers two main disadvantages. First, it takes time to neutralize the charge or voltages, limited by the rate of arrival of ions at the surface being neutralized. This can take many seconds. If charges are being generated at a greater rate than can be neutralized by the arrival of air ions, the charge neutralization is unsuccessful.

Second, depending on the type of ionizer used, the balance of positive and negative ions produced by most types of ionizer is often not exactly balanced. This results in charging of all ungrounded items or materials within the ionized region to a small offset voltage. The user must establish that this offset voltage is insignificant for their purpose. The ionizer imbalance typically changes with time and age.

Opposite polarity air ions attract each other due to their opposite charge. This means that they tend to drift toward each other and recombine to form neutral molecules or particles. The rate of arrival of ions, and hence charge or voltage reduction, can be highly dependent on distance between the ionizer and the surface, ionizer orientation, local air flows, and other factors. Effective neutralization occurs within a limited region around the output of the ionizer and typically takes longer for greater distance from the ionizer. If an item is within the ionized region for insufficient time, the surface voltages will not be completely neutralized and may remain to cause ESD risk. In specifying an ionization system, it is important to understand the size and position of the effective neutralization region, the neutralization time, and the offset voltage characteristics.

7.10.2 Ion Sources

There are various ionizer technologies in use. Most use high voltages applied to sharp-pointed needles within the ionizer to produce air ions by corona discharge. The rate of ion emission from electrodes typically varies somewhat between electrodes and for positive and negative polarity. Nuclear ionizers use nuclear radiation sources to produce a well-balanced air ion source. X-ray ionizers use "soft" X-rays for the same purpose.

Nuclear ionizers commonly use a polonium 210 nuclear source. This is packaged for nuclear safety to prevent escape of the nuclear material but allows emitted alpha particles to be released. These alpha particles (helium nuclei) collide with and split gas molecules in the air to produce equal numbers of positive and negative ions and a well-balanced ion source. Nuclear ionizers do not contain high voltages and so do not produce electrostatic fields. They are unusual in that they produce a completely balanced ion stream and so have zero offset voltage. One disadvantage is that the rate of ion generation is limited by the rate of nuclear disintegration. The ionizer may be effective only for a short range although airflow can be used to help distribution of the ions produced.

AC ionizers use AC high voltages applied to corona needle electrodes to produce positive and negative ions alternately from the same electrodes. The high voltage AC is usually at mains power frequency. As the ions of both polarities are produced at the same electrodes, they can quickly recombine. A fan or other air movement system is often used to transport the ions to the intended effective neutralization region.

DC systems have positive and negative DC high-voltage sources applied to separate corona discharge needles. These needles are placed at some distance apart. Loss of ions due to recombination is typically less than for AC systems. Lower air flow may then be used to transport the ions through the intended region of effective neutralization. If the ionization electrodes are far apart, the ion balance close to the electrodes may be poor, resulting in high offset voltages close to the ionizer.

Pulsed DC ionizers used the same electrodes with both polarities of high voltage alternately to generate both polarities of ions. The voltage is applied with a relatively low alternating pulse rate, usually below 10 Hz. This results in a lower recombination rate than AC systems, but the offset voltage may cycle negative and positive especially close to the ionizer emitter electrodes. One advantage is that at the high voltage pulse rate, delivery of ions can be adjusted to improve performance over the intended effective neutralization region.

Soft X-ray ionizers use X-ray sources to ionize the air, producing a well-balanced stream of positive and negative ions. The ions are produced along the beam of the X-ray, which can extend a meter or so from the ionizer. Personnel must be shielded from exposure to the X-ray radiation. Advantages of the system are that well-balanced source of ions is produced in a defined region without electric fields or air flow.

7.10.3 Types of Ionizer System

Ion sources can be built into various types of ionizer intended to be effective in regions of various sizes or in different types of installation or situation. They can also be used in many ways in processes or automated systems.

Room ionization systems are used when neutralization of charge sources is required over a large production area or room, rather than localized to a workstation or process. Room ionizer systems use multiple electrode systems as bars or grids, positioned to cover the required area. They can use AC, DC, or pulsed DC ion sources.

Laminar flow bench ionizers are used to create a workstation in which contamination is controlled, usually within a larger uncontrolled work area. Ionizers are used to control electrostatic charging that can cause ESD and contamination (particle attraction) within these workstations. AC, DC, pulsed DC, or nuclear ion sources may be used.

Work surface ionizers are used to control electrostatic charging in specific work surface regions. AC, DC, pulsed DC, or nuclear ion sources may be used. Common types include fans to take ions into the controlled region.

Where electrostatic control is needed in a small specific region within a process or machine, point-of-use ionizers may be used. Compressed air nozzles may incorporate ionizers to reduce electrostatic charge during blow-off particle removal. They may include AC, DC, pulsed DC, X-ray, or nuclear ion sources.

Some systems use feedback to adjust ion balance and improve performance. These rely on sensors to sense ion balance in the region around them. The sensor can assure balance only in a limited region around it. Where balance is required over a large area, it may be better to use several systems, each having their own sensor controlling smaller regions, rather than a single system controlling a large area. Use of a feedback-controlled system can reduce maintenance by automatically adjusting for changing emitter efficiency.

7.10.4 Selection of Ionizers

It may seem an obvious statement that an ionizer is normally required only if there is an identified task for it to do. Nevertheless, ionizers have often been used in a "sticking plaster" approach to try to solve or prevent real or imagined electrostatic control issues. They will rarely give value for money if used in this way.

Perhaps the most common task for an ionizer system is to reduce charge and voltage levels on process required insulators. Controlling voltages on ESDS devices or isolated conductors may also be required. These insulators or conductors may be part of the product being manufactured or handled, or they may be part of the process equipment. Understanding the electrostatic issue that is to be addressed is an important first step in selection of a solution.

A key first requirement for an ionizer system is that it must be able to deliver a sufficient quantity of ions quickly enough to neutralize charge on the target object to sufficiently low level in the timescale required in the process. In real situations, charge generated by the process or handling as well as preexisting charge must be neutralized. If the charge is not neutralized quickly enough, the ESDS devices may be put at unrecognized risk and the benefit of ionization is compromised, or the process may have to be slowed (by introducing waits or reducing process speed) to allow time for effective neutralization. The latter is often unacceptable. The necessary timescale for neutralization can vary from several tens of seconds or longer (in a manual process or one organized to give built-in waiting times) to a second or less in a fast automated process.

Selection of an ionizer needs consideration of a variety of environmental and process factors, as well as the ability of the ionizer to reduce the charge and voltages to the required levels within the required timescale. The size of the region to be controlled is an important consideration. Often, an ionizer system must be engineered to suit the application. It must always be evaluated and qualified within the actual operating area.

Ionizer performance is enhanced if airflow takes the ions into the working area to the neutralization target. In regions where the airflow is limited or blocked, the effectiveness of an ionizer can be considerably reduced. Some types of ionizers include fans to assist air flow. The work area should be arranged to remove any obstruction to air flow to the neutralization target.

Electrical ionizers use high voltages and sharp points. If insufficiently guarded, these can give electrical shock or accidental injury risks to personnel. Electrical ionizers generate a small amount of ozone that must not be allowed to become a threat to the health of personnel or to processes. Ionizers must meet local and national electrical, X-ray or nuclear safety, ozone generation, or other regulations. Electrical ionizers also generate EMI that may interfere with sensitive electronic equipment.

In summary, when selecting an ionizer, it may be necessary to consider the following:

- The application and task of the ionizer
- ESDS device withstand voltage
- Required ionizer offset voltage
- The size and position of the region to be controlled
- Whether the process can be modified to make the ionizer's task easier or more reliable
- High voltages and electrostatic fields generated, and the regions and levels of ion imbalance, especially near ionizer electrodes
- Environmental aspects such as airflow and blocks to ionized airflow
- Use of airflow to enhance neutralization
- Safety issues and regulations
- Maintenance requirements
- Installation, operation, and maintenance costs as well as equipment cost
- Need for associated facilities such as air supply
- Cleanliness and contamination, especially where air supplies and clean areas are involved
- Ozone and EMI effects (for electrical ionizers)
- For large systems, power distribution requirements

Nuclear ionizers should not be used in a chemical environment without consulting the manufacturer.

7.10.5 Qualification Test of Ionizers

Once a task has been identified for an ionizer and an ionizer system has been selected for the task, then it should be demonstrated that the system does indeed fulfill the role required. This may involve a variety of standard and nonstandard tests. These should include specific tests to show that the neutralization task works in practice in the working environment. In-situ tests may be the only way to finally determine that an ionizer can do the task intended. Once equipment is qualified, simpler acceptance tests may be required to check equipment purchased works as selected.

Basic ionizer performance can be evaluated using standard tests using a charged plate monitor (CPM) (see Section 11.8.8). Measurement of the charge decay time and offset voltage under repeatable fixed conditions form the basis of comparison of different ionizers. Tests should be made of neutralization of charge of both polarities. Where compliance with a standard is required, testing should be done according to the standard, and compliance with the standard pass criteria should be established.

Tests should also be performed with the ionizer working in the process and environment in which it will be used, testing performance at positions that represent the range of working positions of the ESDS devices or other items to be neutralized. Specific and nonstandard tests should be devised to demonstrate that the ionizer reliably fulfills the task intended. Tests might include, for example, investigation of the charge decay time and residual voltages on an insulating item, an ESDS device, or an isolated conductor within the process under conditions as close as possible to operating conditions. During in-situ qualification, it may be useful to determine a suitable test arrangement or CPM location for compliance verification testing. In some circumstances (e.g. within operating machines), it may be difficult or not viable to make measurements under working conditions.

It may be necessary to use nonstandard CPM instruments as well as other instruments such as electrostatic voltmeters or field meters to quantify operation under real conditions. As well as electrostatic tests, it may be necessary to test other aspects of ionizer performance such as ozone buildup, particulate contamination, or EMI.

Where the ionizer is protecting against low-probability ESD events that may happen in running production circumstances, the final proof of effectiveness may be demonstration that it solves the identified problem and does not introduce another.

7.10.6 Cleaning and Maintenance of Ionizers

Electrical ionizers contain high voltages and needle electrodes and should be cleaned or maintained only by competent personnel according to the manufacturer's instructions.

Electrical ionizer emitter points need regular cleaning and maintenance or replacement. The frequency required for this may depend on the emitter materials and ionizer design as well as the operating environment. Variation of rates of deterioration of the emitter performance affects both the decay time and the ionizer balance and offset voltage.

Radioactive ionizers experience gradual decay of their strength due to decay of the radioactive isotope. They require replacement once the strength has dropped to an unacceptable level. The radioactive source is normally returned to the manufacturer for replacement and disposal. Radioactive ionizers may also require leak test for safety.

X-ray ionizers may need periodic return to the manufacturer for replacement of the X-ray source. Functionality can be tested by measurement of decay time and offset voltage as for other ionizer types.

7.10.7 Compliance Verification Test of Ionizers

Standard test methods are available for using a CPM to measure the decay time and offset voltage of ionizers for compliance verification (see Section 11.8.8). A repeatable test arrangement or locations should be used, preferably determined during ionizer in-situ qualification, to ensure that verification tests give results representative of the performance required at the position of the item to be neutralized.

The test frequency should be chosen appropriately according to the circumstances and experience. In a situation where high-value and high-sensitivity components are handled, it may be desirable to do functional tests and quantitative tests relatively frequently. Visual checks can show correct orientation of movable fan ionizers. Less frequent checks may be

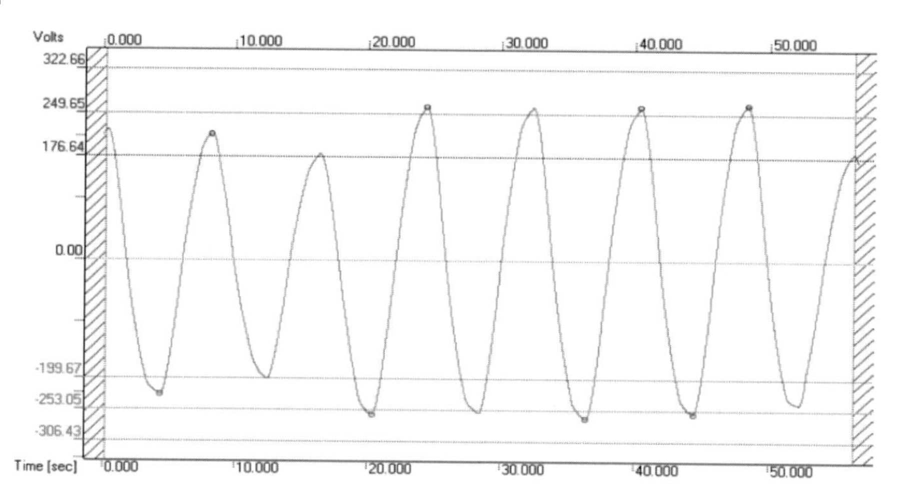

Figure 7.11 Typical ceiling mounted pulsed DC ionizer charging of an object at tabletop height. Source: D E Swenson.

used in a situation where the performance is less critical or where verification history has shown reliable and consistent performance.

7.10.8 Common Problems

Inadequate maintenance of electrical ionizers can result in excessive offset voltage developing due to changes in emitter efficiency. The result is that items in the vicinity become charged to the offset voltage and can cause undetected ESD risk. Isolated conductors that might contact the ESDS device can provide the most serious risk. I have seen ionizers that charge items to nearly 1000 V due to neglect. Pulsed dc ionizers can also induce time varying offset voltages (Figure 7.11).

Chemical contamination or excessive humidity can affect the high-voltage sources of electrical ionizers, causing internal leakage currents that can reduce the ion output and lengthen neutralization times and increase offset voltage.

7.11 ESD Control Garments

7.11.1 What Does an ESD Control Garment Do?

A well-designed and correctly worn ESD control garment can protect ESDS devices handled by the wearer from electrostatic fields and ESD from highly charged outer clothing. Where coats or coveralls are required for cleanliness or other reasons, use of a garment specified for ESD control prevents the garment from being a source of electrostatic fields or discharges (see Section 4.7.10.8) and prevents electrostatic fields and discharges originating from clothing under the garment from affecting the ESDS device (Paasi et al. 2005a,b).

Use of an ESD control garment does not remove the need to ground personnel via a wrist strap or footwear and flooring.

7.11.2 Types of ESD Control Garments

Current ESD standards IEC 61340-5-1 and ANSI/ESD S20.20 define three classes of garment.

- Static control garments that provide ESD control without necessarily grounding the garment. These have point-to-point resistance specified in the standards.
- Groundable static control garments that have a defined groundable point. They have a point-to-point resistance and resistance to groundable point specified in the standards.
- A groundable static control garment system that can be used to ground the wearer via the garment. These have a point-to-point resistance and resistance to groundable point specified in the standards. They include a means of reliably connecting the wearer to ground via the garment such as an integrated wrist strap or conductive cuff electrically connected to the groundable point.

IEC 61340-4-9 describes the garment types and characteristics (Table 7.1) in more detail, together with the test methods used to qualify and verify the garment characteristics (International Electrotechnical Commission 2016a).

Table 7.1 Types and characteristics of ESD control garments according to IEC 61340-4-9:2016.

Description and use of garment	Garment type	Tests used	Resistance range (Ω)
Garment has some electrostatic field suppression properties	Static control garment	Resistance point to point	$<10^{11}$
Garments with designated grounding point	Groundable static control garment system (garment in combination with a person)	Resistance point to point Resistance to groundable point	$<10^{9}$
Garments in continuous electrical path with the wearer but not used as the primary ground path	Groundable static control garment	Resistance point to point Resistance to groundable point	$<10^{9}$
Grounded with dual paths to ground via continuous monitoring equipment that requires two separate paths to ground	Groundable static control garment system (garment in combination with a person)	Resistance point to point Resistance to groundable point Integrated wrist strap in accordance with IEC 61340-4-6	$<10^{9}$ $<3.5 \times 10^{7}$
Grounded via a single wire continuous monitoring equipment	Groundable static control garment system (garment in combination with a person)	Resistance point to point Resistance to groundable point Integrated wrist strap in accordance with IEC 61340-4-6	$<10^{9}$ $<3.5 \times 10^{7}$
Garment used as primary ground path for personnel	Groundable static control garment system (garment in combination with a person)	Resistance point to point Resistance to groundable point Integrated wrist strap in accordance with IEC 61340-4-6	$<10^{9}$ $<3.5 \times 10^{7}$

Disposable garments, as the name implies, are intended to be used a small number of times. They are typically made from nonwoven materials. ESD control properties are often provided by finishes or treatments that have a limited life. Their properties may be dependent on atmospheric humidity and be impaired under low-humidity conditions. They may be useful in processes where contamination gives a short garment life and a need for frequent replacement.

Reusable topically treated garments may require retreatment after each cleaning. Treatments typically depend on atmospheric humidity and become inadequate under low-humidity conditions. Some materials have conductive fibers but also require topical treatment. For these materials, the properties may be less dependent on humidity.

Garments with permanent ESD control properties should maintain their ESD control properties for their intended life. They usually have a grid or stripes of conductive fibers woven into the fabric (Figures 7.12 and 7.13). If not grounded, exposed conductive fibers could represent an ESD source.

Static control garments can be made in various materials. Some are disposable, whereas others are reusable. They may be made from homogenous untreated fabrics or coated fabrics or may have conductive threads aligned in one direction only or in two directions as a grid.

Groundable static control garments include a means of grounding the conductive fibers in the fabric through one or more groundable points. The panels of these garments should be electrically connected. Fabrics that have a grid pattern may produce the most reliable connection. Connection between panels may be improved by including additional non-insulative (electrically conducting) material in the seams. The garment can be grounded via the wearer's body through conductive wrist cuffs or other areas of direct

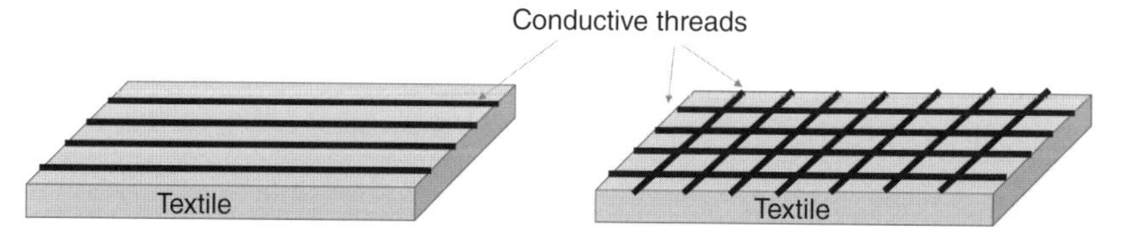

Figure 7.12 Stripe (left) and grid (right) conductive fiber patterns (Paasi et al. 2005a,b).

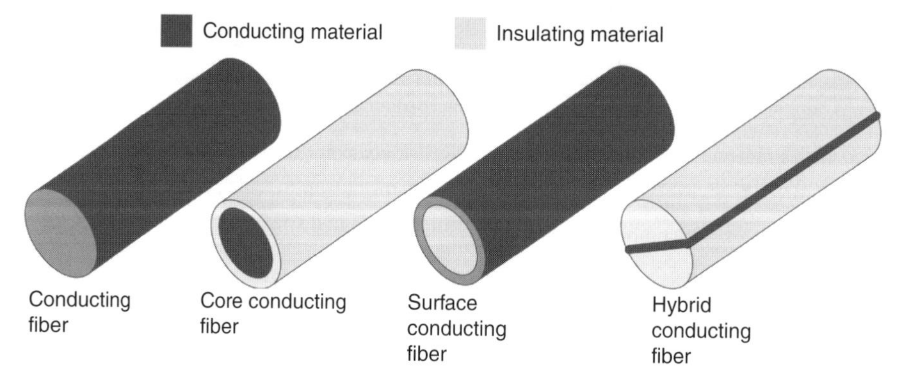

Figure 7.13 Examples of ESD control garment fiber types (Paasi et al. 2005a,b).

contact with the wearer's skin, providing the wearer is grounded through a wrist strap or footwear-flooring system. Alternatively, grounding may be achieved via a ground cord.

Fabrics that have a stripe pattern rather than grid may not have a good connection between the conductive fibers. These may need topical treatments or conductive material in seams to improve connection between the fibers.

Fabrics that depend on topical treatments can degrade unevenly and leave areas of fabric isolated from the ground connection.

Some experts have been dissatisfied with measurement of garment materials only on the basis of material resistance. Buried conducting fiber materials are usually rejected by resistance measurements as the measurement electrodes do not make contact with the conducting fiber. These simple tests do not characterize many of the properties of garment materials (Paasi et al. 2004, Baumgartner 2000). Potentially relevant factors that are ignored by the resistance test method include

- The triboelectric charging propensity of the fabric
- The effect of grounding of the fibers on the protective performance of the fabric
- ESD risk from unearthed conducting fibers in the material
- ESD risk from charged insulating areas of nonhomogenous material
- Effectiveness of protection against electrostatic fields from underlying charged fabrics
- The effect of the grounded wearer on the protective performance of the garment

In the early 2000s the ESTAT garments European Project researched the importance of these factors and produced recommendations for use and test of ESD-protective garments, considering surface and core conducting fiber and stainless steel thread materials. Their recommendations have not been widely adopted in ESD standardization, but their final report makes interesting reading (Paasi et al. 2005b). Their results broadly confirmed the value of simple resistance measurements in evaluating groundable garment materials.

They found that the ESD-protective fabric and garment characteristics were very dependent on the fabric and garment design, atmospheric humidity, and whether the garment and the wearer were grounded. The main functions of an ESD control garment are

- Reduction or elimination of electrostatic fields produced by clothing worn beneath the garment
- Prevention of ESD originating from the clothing worn beneath the garment
- The ESD control garment should not itself charge up and cause external electrostatic fields or be a potential source of damaging ESD

Charging of ESDS devices by accidental rubbing against the garment fabric could not be easily averted by choice of fabric but could be reduced by minimizing the risk of rubbing, e.g. by avoiding loosely fitting sleeves. Loose garments have lower protective value than close fitting garments. ESD could occur from garment materials with ungrounded conducting threads or sufficiently large insulating areas (over 20 × 20 mm). Most effective ESD control was achieved for materials having surface resistance in the range 100 kΩ–100 GΩ.

Paasi et al. (2005b) classified garments as Class A (using electrically continuous, low charging, static dissipative, or conductive materials and grounded) or Class B (low charging but need not have measurable electrical conductivity, and grounding not required). These approximately correspond with the "groundable static control garment" and "static control garment" classifications of IEC 61340-4-9 (Table 7.1).

Grounding the conducting parts of the groundable garment (e.g. to the grounded wearer) improves the performance of the garment both in terms of external electrostatic fields and as a potential source of ESD.

7.11.3 Selection of ESD Control Garments

ESD control garments may be required for protection of ESDS devices against the possible effect of ESD. Where flammable materials are handled, for example solvents, the garments may be classed as PPE and in some countries subject to regulation. Guidance on selection, care, use, and maintenance of PPE in Europe can be found in CEN/TR 16832.

The first question in the process of selecting an ESD control garment is, is it necessary to specify them? There is no simple answer to this. When developing an ESD program, the ESD coordinator will need to evaluate the need based on many considerations.

In other cases, the ESD coordinator must evaluate whether ESD risks associated with ordinary everyday work clothes are likely to be significant in the context of the ESDS devices handled and other considerations. Various factors can be considered (Table 7.2).

While finding that ESD control garments are not always necessary, Paasi et al. (2005b) recommended that they should be seriously considered if the ESD withstand voltage of ESDS was 500 V charged device model (CDM) or 1 kV HBM or less. They should also be specified if over garments are required for contamination control or where indicated by high cost and consequences of ESD failures and high reliability required of product. For cleanroom applications, the chargeability of the material is the main consideration.

Table 7.2 Factors that might be considered in decision to issue ESD control garments.

Factors that might indicate ESD control garments are desirable	Factors that might indicate ESD control garments are unnecessary
Low CDM ESD withstand voltage components. High cost and low withstand voltage ESDS.	Components handled have moderate or high ESD withstand voltages and low or moderate cost.
Consequence of ESD failure is highly undesirable.	Consequence of occasional ESD failures is acceptable to some extent.
Clean area where ESA is undesirable.	ESA is not a significant issue.
Likelihood that high charge generating clothing that is not close fitting may be worn e.g. cool dry climate.	Likelihood that single layer low-charging clothing will be worn, e.g. warm moist climate.
High reliability product or market where low ppm failure rate is required.	Consumer market where moderate failure rates are accepted.
Use of ESD control garments may help establish ESD control culture or help demonstrate best practice and care (e.g. to visitors).	

The main ESD control risk from ordinary garments is that they might charge highly and give an electrostatic field near ESDS devices. This electrostatic field could induce high voltages on devices, leading to charged device or charged board ESD. ESA can also be a concern in clean areas, where electrostatic fields can encourage attraction of contamination to the items handled. In most cases, the areas of the body most likely to come close to ESDS are the front of the torso and arms. It is therefore particularly important to shield ESDS from electrostatic fields from clothing in these areas. The clothing of the lower body is often less important as it is less likely to come near to unprotected ESDS devices.

Woven materials can also release fibers that can give unacceptable contamination in clean areas. Coveralls used in clean areas can be made entirely from ESD control materials designed for low fiber release to reduce ESA of particulate contaminants.

Once the decision to use ESD garments is made, a suitable type must be selected. In some cases (e.g. clean room), this will be a relatively simple matter of specifying an ESD control version of a garment that is required for a specific purpose (e.g. contamination control or personal protection against hazards). Ordinary personal protective clothing may be made from insulative materials that can be a strong source of electrostatic fields.

An ESD control garment should cover the underlying garments completely over the key areas of the body that need to be covered. As a guide, where the clothes are likely to come within 30 cm of ESDS devices, they should be covered by an ESD control garment. The material should be designed to reduce the electrostatic fields generated by underlying garments. It should not itself be a source of electrostatic fields or ESD. Materials used should not contain low resistance fibers or metal parts that could charge and become a source of ESD or could be an electrical safety concern.

According to Paasi et al. (2005b), a dense grid of conductive fibers (e.g. 5 × 5 mm) gives improved electrostatic field control. For materials with conductive stripes, the stripes should be less than 10 mm apart. For a grid fabric, the grid squares should be 20 × 20 mm or less. They found that core conductive fiber fabrics cannot give good grounding of the fibers under dry air conditions, but this does not necessarily preclude their value in ESD protective garments. They did not recommend use of a garment as the ground path of the wearer.

In some countries (e.g. in Europe), garments that have a PPE function are subject to regulations that require they comply with specific safety-related regulations. Individual organizations may also have safety rules that must be followed when selecting garments for use within that organization.

7.11.4 Qualification Test of ESD Control Garments

Where compliance with a standard is required, testing should be done according to the standard, and compliance with the standard pass criteria should be established. Current standard test methods rely on surface resistance of the material to be determined using point-to-point resistance measurements (see Section 11.8.2). For garments that have a groundable point, a resistance to groundable point measurement should also be made. Resistance should be measured between panels as well as across panels. Where the garment is intended to make contact with the wearer's body, this should also be tested.

Many garment materials are affected to some extend by atmospheric humidity. Cotton is inherently hygroscopic, but man-made fibers such as PET and nylon are usually less so. Topical treatments are often highly dependent on humidity for their performance. So, resistance measurements should be made under controlled low-humidity conditions. Testing over the full range of temperature and humidity conditions likely in practice may be advisable.

Garments that use fabrics that have core conductive fibers cannot be successfully measured using surface resistance measurements, as the test electrodes do not make good contact with the buried core conductive fibers. No standard test methods are currently defined for measurement of these materials for use in electronics environment (Swenson 2011). The user would need to define suitable test methods and pass criteria for qualification of these materials.

Paasi et al. recommended that as well as surface resistance tests for qualification of garments, EN1149-3 Method 1 could be used to evaluate tribocharging, and EN1149-3 Method 2 evaluate induction charging of all types of garment materials (British Standards Institute 2004). Holdstock et al. (2003) have developed a "capacitance loading" test method that can be used to evaluate charging of garment materials.

Garments made from fabrics that claim permanent ESD control properties should have tests that compare measurements on new fabrics with measurements on fabrics after repeated cleaning simulating long-term use. IEC 61340-5-2 recommends typically 50 cleaning cycles should be used.

7.11.5 Use of ESD Control Garments

The garment size should be selected to provide a reasonable fit to the wearer while wearing their usual clothing. For normal ESD control use, the garment should cover the material of the clothing beneath the garment at least on the arms and torso. Clothing should be worn fastened and should not be worn loose and open, exposing the clothing beneath.

Groundable static control garments should be connected to ground before the wearer handles ESDS. They should remain grounded during all handling of ESDS devices by the wearer.

7.11.6 Cleaning and Maintenance of ESD Control Garments

Garments that have become damaged should be replaced or repaired according to the manufacturer's recommendations. A repaired garment should be shown to pass compliance verification tests before use.

Laundry should be according to the manufacturer's instructions. Thorough rinsing should be included to ensure that cleaning chemicals are removed. An auditable tracking system should be used to monitor cleaning. Compliance verification tests should be used to verify the ESD control properties at least on a sample basis after cleaning.

7.11.7 Compliance Verification of ESD Control Garments

Garments should be tested for compliance verification on a regular basis and especially after cleaning. Garment materials that rely on treatments or additives will usually need to be tested more frequently that those that include a conductive fiber matrix. For garment materials that have measurable surface resistance, standard point-to-point resistance test methods can be used (see Section 11.8.2). For groundable garments, the resistance to the groundable point should also be tested.

Resistance test methods are not suitable for verification of garment materials that use buried conductor fibers. If these materials are selected for use, suitable verification test methods and pass criteria will need to be defined.

The frequency of compliance verification tests should reflect the expected permanence of the garment fabric technology.

7.11.8 Personal Grounding via an ESD Garment

Groundable static control garments are available that can be used to ground the wearer via the garment. These typically have a groundable point that can be connected to an earth bonding point. They are also made from a relatively low-resistance material and must make reliable contact with the body of the wearer.

Basic qualification of a garment fabric can be done on sample on the fabric. A point-to-point resistance measurement is usually used. The resistance of some fabrics can vary with direction depending on how conductive fibers lie (warp and weft) and so measurements should explore this possibility.

Qualification of a groundable garment should include tests to determine that all the panels of the garment are made of material that has point-to-point resistance less than the required upper limit and are electrically connected to the garment groundable point. The garment must make direct reliable contact with the wearer's body, and the wearer must be reliably grounded through the system.

ESD control garments used to ground personnel should be verified for grounding effectiveness before use. Tests should verify the resistance of the material and the ground path. If the garment is intended to ground the wearer's body, then the entire ground path from body to ground should be verified before use in a similar manner to wrist strap test.

7.12 Hand Tools

7.12.1 Why Have ESD Hand Tools?

Hand tools include any tools that are held in the hand such as screwdrivers, pliers, lead cutters, tweezers, vacuum pick-up tools, etc. Tools that are not designed for ESD control often have metal parts that are electrically isolated by, for example, an insulating handle (see Section 4.7.10.6). These isolated metal parts can become charged, and if they contact an ESDS, then a potentially damaging ESD could occur. Most ESD tools are designed to ensure that all parts of the tool are grounded, usually through contact with the user's hand.

7.12.2 Types of Hand Tool

The types of hand tool of concern are mainly those that are likely to come into direct contact with ESDS during use. These may include tools such as pliers, cutters, and desoldering tools. Screwdrivers and adjustment tools may also be of concern.

Hand tools are normally made ESD safe by ensuring that metal parts have a grounding path through to the user's hand. This usually means that insulative parts of the tool have been replaced by noninsulative parts.

In some cases, when handling ESDS devices that have low CDM ESD withstand, it may be necessary to ensure that metal parts (even though grounded) do not contact the ESDS devices. The materials that contact the ESDS devices may need to be resistive and a minimum resistance be specified to reduce the strength of any discharge that may occur on contact with the ESDS devices.

7.12.3 Qualification Test of Hand Tools

Where compliance with a standard is required, testing should be done according to the standard (if specified), and compliance with the standard pass criteria should be established. Unfortunately, the current versions of 61340-5-1 and ESD S20.20 do not specify requirements or test methods for hand tools. So, the user will have to develop test methods and qualification criteria.

Many hand tools (e.g. hand cutters or pliers) consist of a metal part that in a non-ESD control tool would be insulated from the user's hand by a handle made of insulating material. In an ordinary tool, this can represent an isolated metal part that can become charged and then contact and discharge to the ESDS device. To prevent this, the insulating handle is replaced with a noninsulative material. The resistance between the tool metal part and the user's hand should be below a level that will allow any charge on the metal to dissipate in a short time. In most cases, a dissipation time of up to a second or so would be acceptable. If the capacitance C of the metal part can be measured or estimated, an upper limit of the resistance from the metal part to ground R can be specified as $RC < 1$. As an example, if the tool metal part has a capacitance of 10 pF, R can be up to $10^{11}\ \Omega$ and be acceptable. In practice, it is difficult to reliably measure high resistances. It may be better to specify a pragmatic lower resistance limit determined to be easily achievable with products available on the market. Many ESD control tools in practice have resistance <1 GΩ. If a CPM is available, charge decay measurements are often easier and more reliable than resistance measurements on high resistance tools.

An early version of 61340-5-1 (IEC 61340-5-1:1998) specified resistance and charge decay tests and pass criteria that could be adopted or modified for use.

Some types of hand tools, for example handheld instruments, have significant areas of exposed housing. If this is made of insulating material, it could become charged and give an electrostatic field that could cause ESD risk if brought close to an ESDS device. In this case, a charging test of the material under dry air conditions may be advisable to evaluate if unacceptable high voltage is generated and held on the material.

7.12.4 Use of Hand Tools

Hand tools are usually grounded by contact with the user's hand. If the user must also wear gloves, use of insulating glove materials would prevent grounding of the tools. These could then become charged conductors that present an ESD risk.

So, gloves used with hand tools must be selected to present sufficiently low resistance through the glove to provide acceptable grounding of the tools.

7.12.5 Compliance Verification Test of Hand Tools

Where compliance with a standard is required, testing should be done according to the standard (if specified), and compliance with the standard pass criteria should be established. Unfortunately, the current versions of 61340-5-1 and ESD S20.20 do not specify requirements or test methods for hand tools.

A simple test can be used to verify the resistance from a tool tip via the user's hand to ground (see Section 11.9.5.2). Alternatively, a CPM can be used in a voltage decay test (Section 11.9.8.1). If the tool is usually held in a gloved hand, the system of tool-glove-hand to ground can be verified in either of these ways in one measurement.

7.12.6 Common Problems with ESD Control Hand Tools

Possibly the most common problem experienced with ESD control hand tools is that these are confused with similar non-ESD control versions. This can result in non-ESD control tools being present and used in the EPA. This can be avoided if tools having suitable ESD control identification markings are used.

Use of insulating glove materials prevents grounding of the tools. These can then become charged conductors that present an ESD risk.

7.13 Soldering or Desoldering Irons

7.13.1 ESD Control Issues with Soldering or Desoldering Irons

Many of the issues of concern for soldering irons are to do with potential Electrical Over-Stress (EOS) damage due to injection of potentially damaging leakage currents and voltages rather than ESDs (EOS/ESD Association Inc.1999). Energy-sensitive devices are likely to be more susceptible to damage from these currents and voltages than voltage-sensitive devices, providing the soldering iron bit is grounded.

For a soldering iron to be safe, the tip to ground open circuit tip voltage must not exceed 20 mV, the tip to ground short circuit current must not exceed 10 mA, and the tip resistance to ground must not exceed 2 Ω.

7.13.2 Qualification of Soldering Irons

ESD S20.20 requires that for soldering or desoldering irons three parameters are measured. These are the tip resistance to ground (or groundable point), tip voltage, and tip leakage current. The IEC 61340-5-1 standard lists no requirements for soldering and desoldering tools.

7.13.3 Compliance Verification of Soldering Irons

ESD S20.20 requires that for soldering or desoldering irons only the tip resistance to ground (or groundable point) is measured. This is usually easily done with a multimeter (Section 11.9.6).

7.14 Gloves and Finger Cots

7.14.1 Why Have Gloves and Finger Cots?

There are two main reasons why items such as gloves or finger cots are used in processes (see Section 4.7.10.7). First, the product being handled may need to be protected from the oils, salts, microbes, or other contaminants present on the operator's skin. Gloves may also be used to improve grip when handling product, tools or other items.

Second, in some processes the operator may need to protect their hands from heat, cold, chemicals, sharp edges, high voltages, or other threats present in the process. In this case, the gloves may be classified in some countries as PPE and may be subject to regional, national, or organization PPE regulations and requirements or other safety regulations. Local safety requirements must always be observed.

Three possible ESD-related threats are introduced when gloves or finger cots are used. First, a potentially insulative material is introduced between any item held in the hand and the wearer's body. This is a concern in any circumstance where it is desirable for a handheld item to be grounded via the operator's hand, for example in use of ESD hand tools.

Second, for some types of gloves, the materials of the gloves could charge and give electrostatic fields that could lead to ESD risks to ESDS product handled.

Third, contact between the material of a glove or finger cots is likely to give some level of charging of the ESDS or other items handled. This can lead to charged device, charged PCB, or other ESD risks.

7.14.2 Types of Gloves and Finger Cots

There are many types of gloves and finger cots made from a variety of materials. PPE gloves are required for safety reasons in some countries and organizations for some processes. The local regulations and requirements for these must always be observed.

Gloves may be disposable or reusable. The condition of reusable gloves should be monitored and periodically verified to make sure they remain fit for use. Some types of gloves and finger cots may use topical treatments or antistatic additives. In some processes, these could cause unacceptable contamination of the items handled.

Latex, vinyl, and nitrile gloves or finger cots are typically disposable. They may use topical treatments or antistatic additives.

Fabric gloves may contain surface conductive or core conductive fibers. In activities where it is important to provide a grounding path between items held in the hand and the operator's hand, surface conductive fibers should be used. Core conductive fibers are unlikely to maintain this conductive path unless topical treatments are also used.

7.14.3 Selection of Gloves or Finger Cots for ESD Control

PPE gloves are required for safety reasons in some countries and organizations. The local regulations and requirements for these are beyond the scope of this book but must always be observed. Guidance on selection, care, use, and maintenance of PPE in Europe can be found in CEN/TR 16832, and the gloves may be subject to requirements given in EN16350 (European Committee for Standardisation [CEN] 2014).

Selection of gloves and finger cots should first identify and address the main reasons for needing them, e.g. personal protection or avoidance of product contamination. These may well dictate the basic form of glove to be selected. Other considerations may then include

- Disposable or multiuse
- Physical and safety protection considerations
- Need for grounding of handheld items
- Possible charging of items handled
- Cost and sensitivity of ESDS handled
- Ease and convenience of use
- Suitability for use in clean areas
- Possible consequences of contamination due to antistats

Some types of glove, e.g. thin latex and vinyl gloves can have surprisingly good properties for ESD control use even though they were not intended for this purpose. If tests confirm this is the case, qualified types may be usable in the ESD Control Program.

7.14.4 Qualification Test of Gloves and Finger Cots

Where compliance with a standard is required, testing should be done according to the standard, and compliance with the standard pass criteria should be established. Where they are not specified by standards, tests and pass criteria will need to be established as part of the ESD control program. Most current standards at the time of writing do not specify requirements for ESD control gloves and finger cots.

As well as qualification based on physical or safety functions, qualification based on ESD control functions is required. Physical and PPE properties will often be listed on data sheets. There are two types of test often specified for ESD control gloves and finger cots. First, resistance tests can be used to establish whether a grounding path exists between the wearer's hand and a conductive item held in the hand. These can be done directly as a resistance measurement or indirectly as a charge decay measurement. Gloves that comply with EN16350 have resistance $<100\,M\Omega$ through the material. Second, where ESDS devices are directly handled, electrostatic charging of items handled can be a concern. Some test methods that can be considered are given in Sections 11.9.7, 11.9.8.2, 11.9.8.3 and 11.9.9.

Theoretical upper limits for resistance through the glove and charge decay can be calculated in the first instance as for tools (see Section 7.12.3). In practice, it is difficult to reliably measure high resistances, so it may be better to specify a pragmatic lower resistance limit easily achievable with product available on the market. If a CPM is available, charge decay measurements are often easier and more reliable than resistance measurements of high-resistance glove materials.

7.14.5 Cleaning and Maintenance of Gloves

Reusable textile gloves may be cleaned according to the manufacturer's instructions. Topical treatments may need to be renewed.

7.14.6 Compliance Verification Test of Gloves and Finger Cots

Where disposable gloves are used, a compliance verification test might not be needed provided a suitable qualification procedure is adopted. For reusable gloves, compliance verification tests should be used to check that basic ESD control properties established during qualification have not been lost. Suitable test methods and pass criteria should be defined. The frequency of testing should consider aspects such as the following:

- The possibility of contamination
- Established experience of reliability and life
- Cost and sensitivity of the ESDS handled
- The likely consequences of changes to glove electrostatic properties

Perhaps the easiest test to verify a glove or finger cot is a resistance to ground test, done with the item worn by a grounded person (see Section 11.9.7)

7.14.7 Common Problems with Gloves and Finger Cots

Gloves and finger cots may often be supplied in ordinary polythene packaging. If taken into an EPA, this can be a source of noncompliant secondary packaging materials that lead to ESD risks. To prevent this, secondary packaging should be removed before the gloves are taken into an EPA.

7.15 Marking of ESD Control Equipment

ESD control equipment may be marked as a way of distinguishing them from similar non-ESD control items. Only items qualified for ESD control should be marked in this way. A marking symbol recommended by 61340-5-2 and ANSI/ESD S8.1 for identification of ESD control equipment is shown in Figure 7.14 (International Electrotechnical Commission 2018; EOS/ESD Association Inc 2017).

Figure 7.14 Symbol recommended for marking ESD control equipment. Source: Reproduced by permission of the EOS/ESD Association Inc.

References

Baumgartner G. (2000) ESD TR2.0-01-00. *ESD Association Technical Report – Consideration For Developing ESD Garment Specifications*. Rome, NY, EOS/ESD Association Inc.

British Standards Institute. (2004) BS EN 1149-3:2004. *Protective clothing – Electrostatic properties – Part 3: Test methods for measurement of charge decay*. ISBN 0 580 43736 1

EOS/ESD Association Inc. (1999) ESD TR13.0-01-99. *ESD Association Technical Report EOS Safe Soldering Irons requirements*. Rome, NY, EOS/ESD Association Inc.

EOS/ESD Association Inc. (2013) ANSI/ESD S1.1-2013. *Standard for protection of Electrostatic Discharge Susceptible Items - Wrist Straps*. Rome, NY, EOS/ESD Association Inc.

EOS/ESD Association Inc. (2014a) ANSI/ESD STM9.1-2014. *ESD Association Standard for the Protection of Electrostatic Discharge Susceptible Items – Footwear - Resistive Characterization*. Rome, NY, EOS/ESD Association Inc.

EOS/ESD Association Inc. (2014b) ANSI/ESD S20.20-2014. *ESD Association Standard for the Development of an Electrostatic Discharge Control Program for – Protection of Electrical and Electronic Parts, Assemblies and Equipment (excluding Electrically Initiated Explosive Devices)*. Rome, NY, EOS/ESD Association Inc.

EOS/ESD Association Inc. (2015) ANSI/ESD STM97.1-2015. *ESD Association Standard Test Method for the Protection of Electrostatic Discharge Susceptible Items – Floor Materials and Footwear – Resistance Measurement in Combination with a Person*. Rome, NY, EOS/ESD Association Inc.

EOS/ESD Association Inc. (2016) ANSI/ESD STM97.2-2016. *Floor Materials and Footwear – Voltage Measurement in Combination with a Person*. Rome, NY, EOS/ESD Association Inc.

EOS/ESD Association Inc. (2017) ANSI/ESD S8.1-2017 *Draft Standard for the Protection of Electrostatic Discharge Susceptible Items – Symbols – ESD Awareness*. Rome, NY, EOS/ESD Association Inc.

European Committee for Standardisation (CEN) (2014) EN16350-2014 Protective gloves - Electrostatic properties. Brussels, CEN.

European Committee for Standardisation (CEN) (2015) PD CEN/TR 16832:2015. *Selection, use, care and maintenance of personal protective equipment for preventing electrostatic risks in hazardous areas (explosion risks)* Brussels, CEN.

European Union (2016) Regulation (EU) 2016/425 of the European Parliament and of the Council of 9 March 2016 on personal protective equipment and repealing Council Directive 89/686/EEC. Available from: https://eur-lex.europa.eu/legal-content/EN/TXT/?qid=1484921753526&uri=CELEX:32016R0425 [Accessed 17th Aug. 2019]

Holdstock, P., Dyer, M.J.D., and Chubb, J.N. (2003). *Test procedures for predicting surface voltages on inhabited garments*. In: *Proc. of the EOS/ESD Symp. EOS-25*, 300–305. Rome, NY: EOS/ESD Association Inc.

International Electrotechnical Commission. (1998) IEC 61340-5-1:1998. *Electrostatics - Part 5-1: Protection of electronic devices from electrostatic phenomena - General requirements*. Geneva, IEC.

International Electrotechnical Commission. (2001) IEC 61340-4-3:2001. *Electrostatics - Part 4-3: Standard test methods for specific applications – Footwear*. Geneva, IEC.

International Electrotechnical Commission. (2003/2015) IEC 61340-4-1:2003+AMD1:2015 CSV. *Electrostatics - Part 4-1: Standard test methods for specific applications - Electrical resistance of floor coverings and installed floors*. Geneva, IEC.

International Electrotechnical Commission. (2004) IEC 61340-4-5:2004. *Electrostatics - Part 4-5: Standard test methods for specific applications - Methods for characterizing the electrostatic protection of footwear and flooring in combination with a person*. Geneva, IEC.

International Electrotechnical Commission. (2005) IEC 60364-1:2005. *Low-voltage electrical installations - Part 1: Fundamental principles, assessment of general characteristics, definitions*. Geneva, IEC.

International Electrotechnical Commission. (2015) IEC 61340-4-6:2015. *Electrostatics - Part 4-6: Standard test methods for specific applications - Wrist straps*. Geneva, IEC.

International Electrotechnical Commission. (2016a) IEC 61340-4-9:2016. *Electrostatics - Part 4-9: Standard test methods for specific applications – Garments*. Geneva, IEC.

International Electrotechnical Commission. (2016b) IEC 61340-5-1:2016. *Electrostatics - Part 5-1: Protection of electronic devices from electrostatic phenomena - General requirements*. Geneva, IEC

International Electrotechnical Commission. (2018) IEC TR 61340-5-2. *Electrostatics – Part 5-2: Protection of electronic devices from electrostatic phenomena - User guide*. Geneva, IEC.

International Organisation for Standardisation (ISO) (2011a) ISO 20344:2011. *Personal protective equipment – Test methods for footwear*. Geneva, ISO.

International Organisation for Standardisation (ISO) (2011b) ISO 20345:2011. *Personal protective equipment -- Safety footwear*. Geneva, ISO.

Paasi, J., Nurmi, S., Kalliohaka, T. et al. (2004). Electrostatic testing of ESD-protective clothing for electronics industry. In: *Proc. Electrostatics 2003 Conference*, Edinburgh, 23–27 March 2003. Inst. Phys. Conf. Ser. No. 178, 239–246.

Paasi, J., Nurmi, S., Kalliohaka, T. et al. (2005a). Electrostatic testing of ESD-protective clothing for electronics industry. *J. Electrostat.* 63 (6–10): 603–608.

Paasi J, Fast L, Lemaire P, Vogel C, Coletti G, Peltoniemi T, Reina G, Smallwood J, Börjesson A. (2005b) Recommendations for the use and test of ESD protective garments in electronics industry. Estat Garments Project VTT Research Report No. BTUO45-051338. Available from: http://estat.vtt.fi/publications/vtt_btuo45-051338.pdf [Accessed 11th Oct 2017]

Smith, D.C. (1993). A new type of furniture ESD and its implications. In: *Proc. of the EOS/ESD Symposium. EOS-15*, 3–7. Rome, NY: EOS/ESD Association Inc.

Smith, D.C. (1999). Unusual forms of ESD and their effects. In: *Proc. of the EOS/ESD Symp. EOS-21*, 329–333. Rome, NY: EOS/ESD Association Inc.

Swenson D E. (2011) Understanding core conductor fabrics. In: *Proc. Electrostatics* 2011. J. Phys. Conf. Se. 301012051 doi: https://doi.org/10.1088/1742-6596/301/1/012051

Swenson, D.E., Weidendorf, J.P., Parkin, D.R., and Gillard, E.C. (1995). Paper 3.6. Resistance to ground and tribocharging of personnel, as influenced by relative humidity. In: *Proc. of the EOS/ESD Symp. EOS-17*, 141–153. Rome, NY: EOS/ESD Association Inc.

Further Reading

EOS/ESD Association Inc. (2016) ESD TR20.20-2016. *ESD Association Technical Report - Handbook for the Development of an Electrostatic Discharge Control Program for the Protection of Electronic Parts, Assemblies and Equipment*. Rome, NY, EOS/ESD Association Inc.

Paasi J, Kalliohaka T, Luoma T, Soininen M, Salmela H, Nurmi S, Coletti G, Guastavino F, Fast L, Nilsson A, Lemaire P, Laperre J, Vogel C, Haase J, Peltoniemi T, Viheriäkoski T, Reina G, Smallwood J, and Börjesson A. (2004) Evaluation of existing test methods for ESD garments VTT Research Report BTUO45-041224 Available from: http://estat.vtt.fi/publications/vtt_btuo45-041224.pdf [Accessed 11th Oct 2017]

8

ESD Control Packaging

8.1 Why Is Packaging Important in ESD Control?

Ordinary packaging materials are made from a wide variety of materials (Figure 8.1). Many of these are made of insulating materials such as plastics that charge up easily giving electrostatic fields and electrostatic discharge (ESD) risks. Others such as papers and cardboard may have characteristics that vary widely with type and moisture content. Those that are not insulative at moderate humidity often become insulative at low humidity below about 30% rh.

Ordinary (non-ESD control) papers and cardboard have highly variable properties that depend on moisture content and atmospheric humidity as well as the grade of material. Figure 8.2 shows the variation of surface resistance of some ordinary papers found in the author's office, with atmospheric humidity. Many types of cardboard or papers have resistance that varies by several orders of magnitude and may become insulative under dry air conditions. Some types may be insulative even under moderate humidity conditions. The unpredictability of the characteristics of these materials makes them undesirable in an area where static electricity is to be controlled. For this reason, ordinary papers and cardboard are usually excluded from areas where electrostatic control is required. The moisture content of papers and cardboard can also be affected by processes. For example, papers passed through a photocopier may emerge in a dried and charged condition.

Cardboard can be formulated for ESD protective packaging use and is often used to make boxes. Papers formulated for use in ESD protected areas (EPAs) are also available.

ESD protective packaging is required for two ESD control purposes. First, there is a need for materials that have known and defined characteristics and can be relied on to provide little in the way of ESD risk when used in an EPA. Second, materials and packaging are required that can protect ESD-sensitive (ESDS) devices from ESD risks when outside the EPA during storage and transport.

ESDS devices, printed circuit boards (PCBs), and assemblies are protected within an EPA by reducing the ESD threats and sources to a level where ESD damage is insignificant. When the ESDS devices are to be taken outside the EPA, protective packaging is used to provide protection from these ESD threats and sources. The ESD protection function may be combined with other functions such as organizing the ESDS devices into batches or quantities, types or other groupings, and protection against physical damage or humidity.

The ESD Control Program Handbook, First Edition. Jeremy M Smallwood.

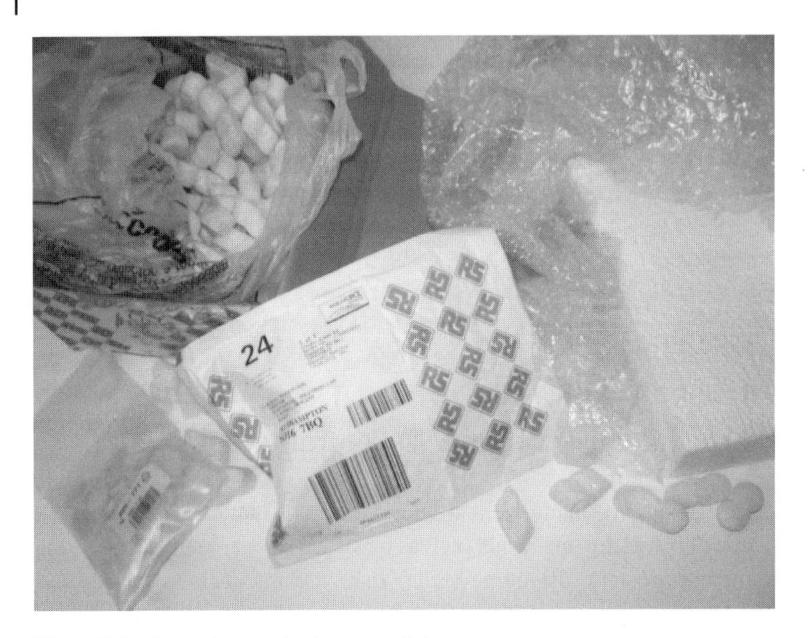

Figure 8.1 Secondary packaging materials.

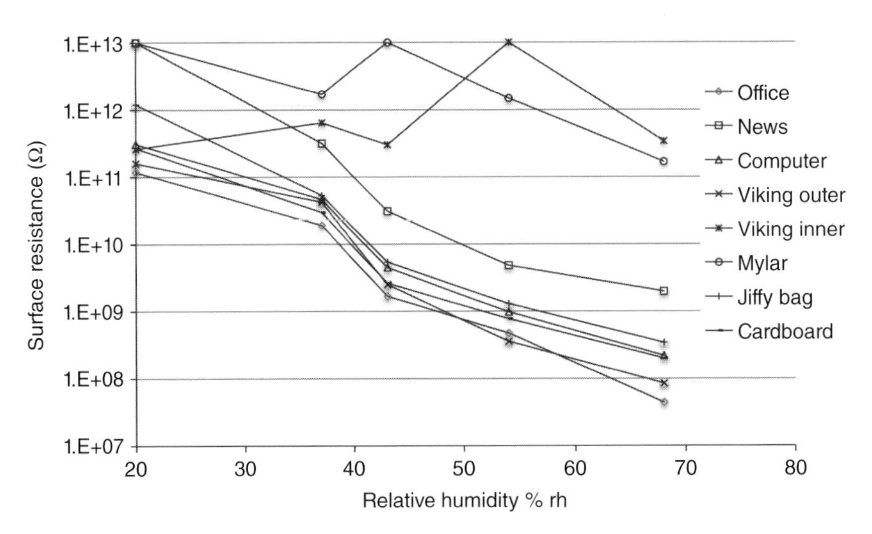

Figure 8.2 Variation of resistance of some papers with atmospheric humidity.

ESD protective packaging is now available in an extraordinary variety of types and forms (Figure 8.3) to suit different types of ESDS devices.

ESD packaging is so important that the International Electrotechnical Commission (IEC) and ESD Association (ESDA) standards systems have specific standards dealing with ESD packaging. In the IEC system, ESD packaging requirements are given by IEC 61340-5-3 (International Electrotechnical Commission 2015) and American National

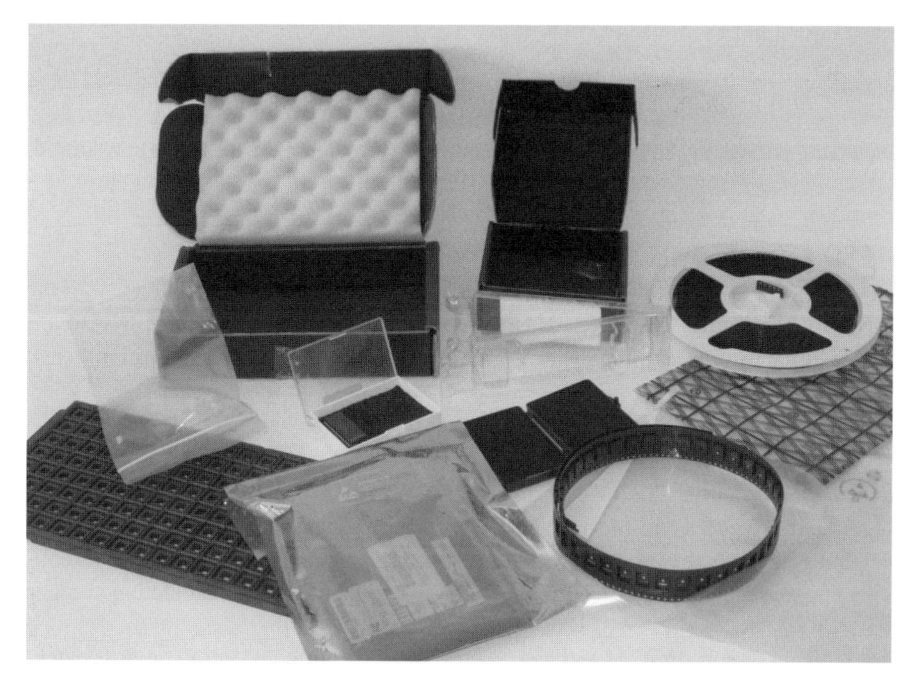

Figure 8.3 Some examples of the extraordinary variety of packaging types in use having ESD control functions.

Standards (ANSI)/ESD S541 in the ANSI/ESDA system. Nevertheless, the topic of ESD packaging remains one of the most poorly understood areas of ESD control.

8.2 Packaging Functions

ESD control packaging has one or more of the following major ESD protective functions:

- Protection against electrostatic fields.
- Protection against direct ESDs.
- Prevention of buildup of electrostatic charges due to tribocharging.
- The surfaces in normal contact with ESDS devices have sufficiently high resistance to avoid risk of discharge of on-board batteries or power sources, or charged device ESD damage.

A variety of ESD packaging solutions are available that have one or more of these properties. Packaging materials taken into an EPA workstation where exposed ESDS devices are handled should always be specified as ESD protective packaging. Ordinary (non-ESD protective) packaging often can easily become charged and cause ESD risks.

Packaging usually also has other functions that have nothing to do with ESD control but are nevertheless important, for example:

- Moisture barrier properties
- Transparency, so that the contents may be seen and identified or counted

- Physical protection
- Holding product ready for use in processes such as automated handling and assembly
- Bearing identification markings or other information

When selecting packaging, all the properties required at all stages of use must be considered.

8.3 ESD Control Packaging Terminology

In ESD control packaging, various terms have been defined by standards or are in common usage. Unfortunately, they are not always consistent even where standard definitions exist.

8.3.1 Terminology in General Usage

Intimate packaging is the ESD packaging material on the inside of a package that may contact any ESDS devices within the package. *Proximity* packaging is any ESD packaging material that surrounds the intimate packaging. *Secondary* packaging is any non-ESD control packaging that may be added around the ESD packaging for physical protection or other purposes.

Figure 8.4 shows an example of an integrated circuit (IC) protected within an ESD packaging box. The intimate packaging is the foam that holds the IC within the box. The box itself is proximity packaging. The two together give "ESD shielding" protection. Once sealed within the box, the device is fully protected against ESD risks and can be taken out of the EPA or put within secondary (non-ESD protective) packaging for transport or storage.

The electrostatic properties of the materials also have commonly used terms. These cover the resistive properties of the materials, their ability to attenuate electrostatic fields or act as a barrier to ESD, and their charge generating properties.

The electrical resistance of materials used in ESD packaging are often classified as conductive, static dissipative, or insulative, or variations on these terms. A *conductive* material has comparatively low electrical resistance. A *static dissipative* material has intermediate resistance. An *insulative* material has high electrical resistance and would not normally be

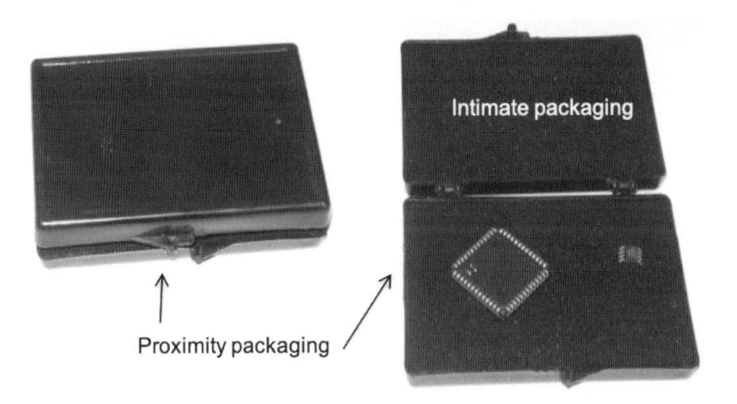

Figure 8.4 Intimate and proximity packaging.

used in exposed surfaces of ESD packaging as it may be prone to charge generation and retention. The exact resistance levels used for these classifications may vary with different ESD standards. The definitions used in IEC 61340-5-1 (International Electrotechnical Commission 2016b) and ESD S20.20 standards systems (EOS/ESD Association Inc. (2014, 2016)) at the time of writing are given in Section 8.8.

A material may be classified as "electrostatic (or electric) field shielding" if it attenuates electrostatic fields. A material that acts as a barrier to ESD may be classified as "ESD shielding." Different ESD packaging standards may vary in the detailed wording of these definitions and terminology.

The term *antistatic* has unfortunately been widely used to mean many different things. Many people refer to ESD packaging as "antistatic" packaging. In some ESD packaging standards, the term is not used at all because of the risk of confusion. In others, it may be used with specific meanings.

The property of low electrostatic charge generation is desirable in ESD control packaging materials. Materials that have lower charge generation properties than ordinary materials may be classified as "low-charging" materials. They may also be called antistatic materials by some.

These terms may have specific definitions in ESD packaging standards (Sec. 8.8 and Tables 8.). The resistance properties of the materials are defined in terms of the surface resistance R_s or volume resistance R_v measured using defined test methods. Electrostatic (electric) field shielding is also defined in these standards in terms of the material resistance. In contrast, ESD shielding is defined in two different ways. For bags, it is defined as the energy measured appearing within the package in response to a standard ESD waveform applied to the outside of the package in a standard test method. For packaging other than bags, *ESD shielding* is defined as a combination of criteria.

8.4 ESD Packaging Properties

ESD packaging must have defined properties, most of which can be tested by the user using standardized test methods. These test methods are given in Chapter 11. The ESD control properties of common concern are considered in the following sections.

8.4.1 Triboelectric Charging

The propensity of a material to generate and accumulate electrostatic charge is important as it is related to the likelihood that the material may cause electrostatic fields and voltage close to the ESDS. Another aspect is the propensity for a component to become charged by contact with the packaging material. These are separate but related issues.

Triboelectric charging happens whenever two materials come into contact. It happens whether they are the same material or different materials. The charging often becomes evident only when the charged materials separate. The amount of charge generated does, however, depend on the materials involved, their surface condition, and many other factors such as surface contaminants including moisture from atmospheric humidity. It follows that triboelectric charge generation is not a fixed property of the material but depends on the materials that contact it and the circumstances. Materials can be classified or ranked according to their relative charging polarity in the triboelectric series (see Section 2.2). Triboelectrification is, however, a highly variable effect and notorious for its poor reproducibility. Materials may be ranked differently by different experimenters in their triboelectric series (Cross 1987).

Some materials have been designed to be "low charging" or "antistatic" in that their propensity to generate electrostatic charge is less than other materials. The definitions used in IEC 61340-5-1 and ESD S20.20 standards systems at the time of writing are given in Section 8.8. It has been found difficult to standardize tests of this property due to the irreproducibility of results obtained. No agreement has been possible on pass criteria for a low-charging property, and it remains a comparative parameter.

Triboelectric charging is often confused with the ability of a material to hold an electrostatic charge and surface or volume resistance properties. There is no correlation between resistance and triboelectric charge generation properties. In tribocharging, two materials are involved in contact and charging. If one of these is an insulator, it will become charged even if the other contacting material is not an insulator. It is quite possible for triboelectric charging to be stronger when a material contacts a static dissipative or conductive material than it is when it contacts an insulating material.

8.4.2 Surface Resistance

The surface resistance of a material is related to its ability to dissipate electrostatic charge and prevent charge buildup. As the term suggests, it is a measure of the resistance across the surface of a material. This may be measured using a concentric ring electrode, miniature two-pin probe, or a point-to-point electrode system (see Section 11.8.4).

It is important to understand that the surface resistance of a material is unrelated to the charge generation properties. The fact that lower voltage buildup may be found with a lower resistance material is normally due to more rapid charge dissipation. The triboelectric charge generated by contact with the material may be the same or even greater than with a higher resistance material that shows high voltage buildup. This can be important where an insulating material such as a device package becomes charged in contact with a resistive packaging material. The charging of the insulative device package is due to the charge generation properties of the two materials in combination, not the packaging material resistive properties.

8.4.3 Volume Resistance

Volume resistance is a measure of the resistance through a material and hence the possibility that charge or current can flow through the material.

Many modern ESD packaging materials are made of nonhomogenous material, e.g. multilayered film or surface-coated cardboard. Surface resistance measurements only give the surface properties of the material. For a nonhomogenous material, each surface may have a material with different properties, and they may be separated by one or more other materials with other properties. The electrical resistance between inner and outer surfaces of a package may be of interest. It may be undesirable to have the inner and outer surfaces electrically isolated from each other. On the other hand, it may be desirable to have a high resistance or even insulating barrier layer that prevents significant conduction of direct ESD current between the inner and outer surfaces (see Sec. 8.4.5).

8.4.4 Electrostatic Field Shielding

Electrostatic field shielding refers to the ability of a material or system of materials to act as a barrier to electrostatic fields. It is important in situations where a sensitive component within a package must be shielded from the effects of electrostatic fields occurring outside a package. Electrostatic field shielding is not the same property as electromagnetic shielding in electromagnetic compatibility work.

The electrostatic field shielding property is related to the conductivity of the material, because it is related to the ease with which the charge can rearrange in response to the field, to equalize the potential across the material and hence nullify the field within the package. A package made from purely insulating material does not give any electrostatic field shielding effect because the charges cannot rearrange quickly in response to the field. A good conductor such as metal forms an excellent electrostatic field shield as the charges within it rearrange almost instantaneously. The field within a container made of an excellent conductor such as metal is zero because the voltage is quickly equalized around the conductor. This type of container is known as a *Faraday cage* (see Chapter 2).

For materials of intermediate resistance, the time taken for charges to redistribute around a container made of the material depends on the effective resistance and capacitive (charge storage) effects. The higher the resistance, the longer it takes for the charges to redistribute in response to a change in external field. Various consequences follow from this.

After a long enough time, the field inside a container made of resistive material is nullified just as it would be with a container made of a good conductor. If, however, the external electrostatic field is changed rapidly, a transient electrostatic field can arise within the container while the charges redistribute. The strength and duration of the field depend on the characteristic decay time of the charge redistribution. An example of this is shown in the case of the pink polythene bag response of Figure 8.5b.

If the external field is constantly varying, the effect within the container depends on the relative speed of variation compared to the characteristic decay times of the charge redistribution within the material of the container. If the external field varies slowly compared to the container characteristic decay times, good electrostatic field shielding may be seen. If the external field varies rapidly compared to the characteristic decay times, significant varying internal fields may arise. Figure 8.5 shows the internal field sensed within an ordinary polythene bag, a pink polythene bag, and a metallized ESD shielding bag during a step change in external electrostatic field of 5 $kV\,m^{-1}$.

In the case of the polythene bag, the material is a good insulator, and charge moves very slowly on its surface. The response shows the bag is effectively transparent to the field in the short term, although charge migration over a period of seconds starts to reduce the internal field. There is also an initial field due to charge on the bag surface before the external field is applied.

A package made of material such as pink polythene (see Section 8.7.1.1) has an effective resistance that depends highly on humidity can also have highly variable electrostatic

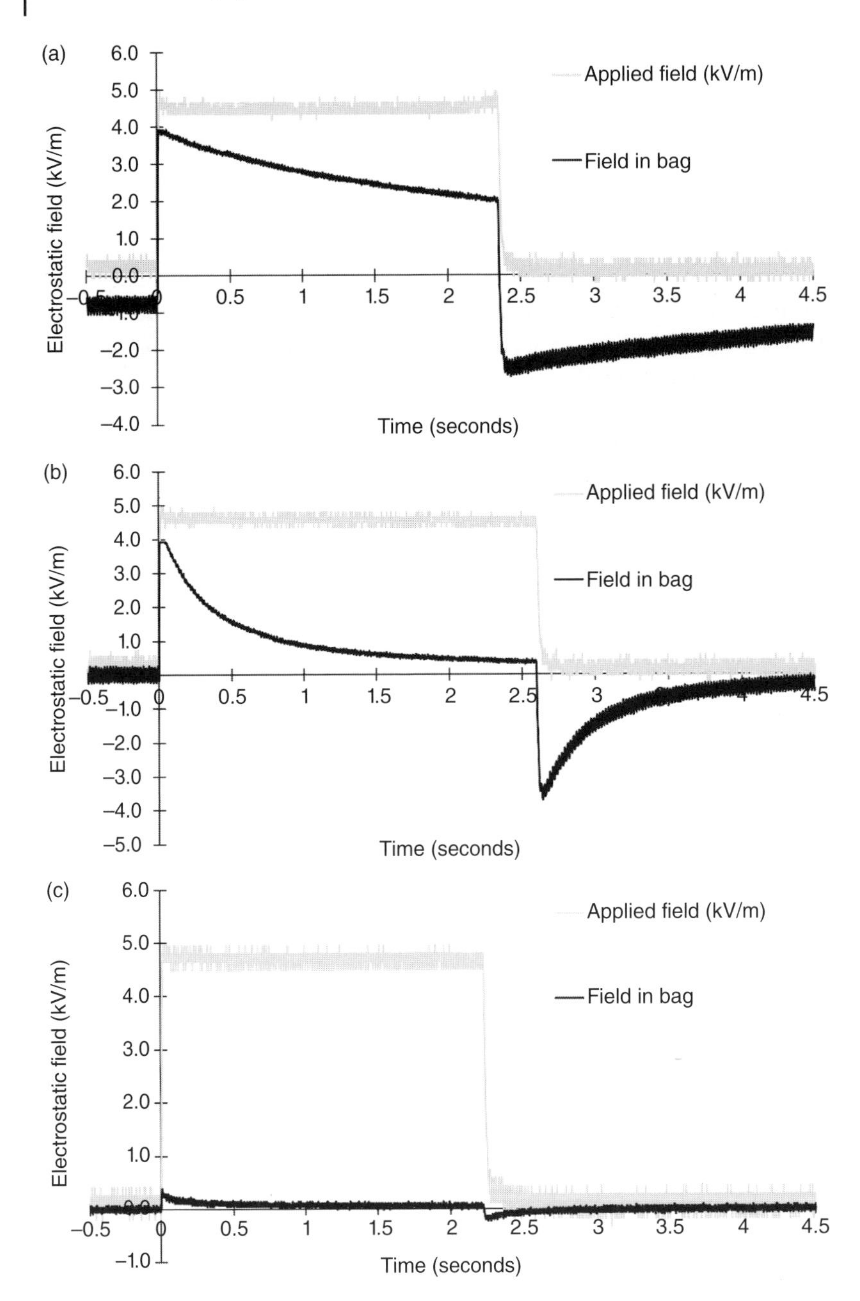

Figure 8.5 (a) Transient electrostatic field arising within an insulating package (polythene bag) due to external varying field. (b) Transient electrostatic field arising within an intermediate resistance package (pink polythene at 50% rh) due to external varying field. (c) Transient electrostatic field arising within a metalized ESD shielding bag at 50% rh due to external varying field.

field shielding effect. Under dry conditions, pink polythene can show little electrostatic field shielding. Under humid conditions, the same material can show considerable electrostatic field shielding although transient fields may arise within the package (Figure 8.5b). The moisture adsorbed by surface antistats allows slow charge migration to reduce the internal field. When the external field is removed, an internal transient field is produced as the charges again migrate to a new equilibrium.

In the case of an ESD shielding bag, there is a lower resistance metallized shielding layer that allows rapid movement of charge in response to the external field. So, the internal field transient is small (Figure 8.5c).

Until recently, ESD packaging standards, electrostatic field shielding materials have been classified according by their surface or volume resistance. The definitions and requirements given in IEC 61340-5-1 and ESD S20.20 standards systems at the time of writing are given in Section 8.8.

8.4.5 ESD Shielding

ESD shielding refers to the ability of a material (or system of materials) to act as a barrier to ESDs occurring to the material. It is important where an ESDS component must be protected against the effects of ESD occurring to the package. ESD shielding is not the same property as electromagnetic shielding in electromagnetic compatibility work.

An important element of ESD shielding packaging is that there must be a barrier preventing ESD that impinges against the outside of the package from penetrating and causing ESD risk to the ESDS devices on the inside of the package. The definitions and requirements given in IEC 61340-5-1 and ESD S20.20 standards systems at the time of writing are given in Section 8.8.

8.5 Use of ESD Protective Packaging

8.5.1 The Importance of ESD Packaging Properties

8.5.1.1 Charge Generation and Retention

The charge generation and retention aspect of ESD control packaging is important in preventing ESD risk within an EPA and to any ESDS within the package.

Within the EPA, electrostatic fields must be reduced or eliminated to prevent induced voltages on ESDS devices leading to ESD occurring when they contact other conductors. In most cases, it is important that the exposed surfaces of ESD control packaging material cannot generate and hold electrostatic charge so that it cannot be the source of electrostatic fields. The tendency to generate and hold charge depends on both the charge generation properties and the resistance (charge dissipation properties) of the material. In general, packaging brought into the vicinity of ESDS devices should not be insulating as it would charge up and cause ESD risk.

It is desirable to minimize the charging of the ESDS device by contact with ESD control packaging, as it could lead to charged device ESD risk when the ESDS device is removed from the packaging and contacts a conductive surface. Many ESDS devices have exposed insulating surfaces as well as conductive parts. These parts charge up by contact with the intimate packaging surface. Charging of the ESDS is dependent on the mutual charging properties of the ESDS materials and the intimate packaging material and is not related to the resistance of the intimate packaging material. Charge removal from the conducting parts of the ESDS device depends on the ESD packaging material resistance, but charge is not removed from the insulating parts of the ESDS device in this way. In some cases, it can be most important to minimize electrostatic charging of the ESDS device in this way, and that is not necessarily done by selecting a non-insulating material.

Where a charged device contacts a low-resistance material, the discharge that occurs can give charged device ESD risk. This can be minimized by avoiding contact between the ESDS device and low resistance materials. It is usually desirable for ESD control packaging to have surface resistance greater than a minimum value (at least 10 kΩ) for this reason.

For very small components, charge retention on the component or on the ESD control packaging can also result in uncontrolled ejection of the components from the package by electrostatic attraction or repulsion forces.

8.5.1.2 Electrostatic Field Shielding

Electrostatic fields could form a risk in situations where

- The ESDS is inherently susceptible to electrostatic fields.
- Induced voltages on conductors on the ESDS device could lead to excess voltage and break down across a part of the device.
- The field could give voltage differences and movement or contact between the ESDS device and other ESDS devices or conductors could lead to ESD occurring.

In practice, there are few components that are known to be susceptible to damage directly from electrostatic fields. An example is the photomask reticle, in which induced voltages between adjacent conductors can lead to ESD between them (See Section 9.5.1.2). Depending on the circumstances and susceptibility of the ESDS devices to damage from electrostatic fields, electrostatic field shielding may or may not be necessary for packaging protecting ESDS devices outside the EPA.

Effective electrostatic field shielding depends on whether the charge redistribution needed to balance field changes can happen as quickly as the field is changing. This depends on various factors including the resistance of the material. For response to rapidly changing fields, a low material resistance is required, although for slow-changing fields, a high resistance material can be effective.

8.5.1.3 Electrostatic Discharge Shielding

Most ESDS devices are susceptible in some way to damage from direct ESD to the ESDS device. This is usually the greatest risk to an unprotected ESDS devices. Most ESDS devices require protection against direct ESD when outside the EPA. ESD protective packaging for protecting these ESDS devices when outside the EPA must have ESD shielding properties.

8.5.2 Packaging Used Within the EPA

Within an EPA, ESD risk is normally eliminated by the design of the EPA and the ESD control equipment used within it. The primary requirement of packaging used within the

EPA is that it should not introduce ESD risk to ESDS handled there. For this reason, packaging used in the EPA must not have exposed insulating materials that could charge up and cause ESD risks. Exposed packaging surfaces must be low-charging, static dissipative of conductive. This applies equally to intimate and proximity packaging materials.

Clearly, secondary (non-ESD-protective) packaging usually contains exposed insulating materials or at least materials that have not been characterized for ESD purposes. These must be kept well away from ESDS devices due to the possibility they could cause ESD risk. It is therefore normal practice to keep these materials away from workstations where ESDS may be handled. Many ESD programs find it preferable to keep these materials out of the EPA altogether.

Occasionally ESD protective packaging could be required within the EPA to protect ESDS devices against specific threats that are not well controlled by standard ESD control measures. It is more usual that additional ESD protective properties will be required where ESD packaging will be used to protect ESDS devices taken out of the EPA.

Some types of ESDS may place additional requirements on ESD packaging materials that they may contact or be stored within. For example, for a PCB containing a battery, a minimum surface resistance may be specified for intimate packaging due to a concern that the battery charge may be drained by contact with the material.

8.5.3 Packaging Used to Protect ESDS Outside the EPA

The ESD packaging required to protect ESDS outside the EPA could vary according to the type and form of ESDS and specific way in which ESD could occur to it. Other requirements unrelated to ESD control may have a strong influence on the type of packaging selected. The minimum requirements for ESD packaging used in an EPA normally also apply, because the ESDS device must be placed into the package or taken out of the package within an EPA.

In general, an ESDS device such as a PCB or semiconductor component will need to be within ESD shielding packaging for transport or storage outside the EPA. This requires a barrier of high-resistance or insulating material to be built into the package in a way that does not cause ESD risk when within the EPA.

Sometimes the ESD risk is limited to specific threats that may be countered by simple protective packaging solutions. For example, a module may be housed within an enclosure that provides an effective barrier against ESD, but a residual ESD risk of contact with connector terminals or flying leads may be present. It may be sufficient to provide protective covers that prevent possible contact with connector terminals in this way.

8.5.4 Packaging Used for Non-ESD Susceptible Items

While a need for ESD packaging for the protection of ESDS devices is clear, it is less obvious why it should be used with nonsusceptible items. The key point is that any material brought into an EPA must not cause ESD risk to ESDS devices in the vicinity. Packaging made from insulating materials could cause ESD risk, so ordinary secondary packaging materials are normally excluded from the EPA. Any suitable type of ESD packaging could be used to package nonsusceptible components within the EPA.

8.5.5 Avoiding Charged Cables and Modules

ESDS devices that are contained within equipment and not exposed to the outside world can normally be assumed to be not susceptible to damage by ESD. A possible exception to this is where ESDS internal parts are connected to connectors and they may be subject to ESD to the connector, for example from connection of charged cables.

Cables are often supplied in polythene packages that can charge highly, especially under dry atmosphere conditions. When the cable is taken out of its packaging, it can acquire a high voltage of several kilovolts by triboelectrification. Highly charged packaging close to a cable can also induce high voltages on the conductive cores. This represents a highly energetic ESD source that can discharge on connection to equipment connectors. Cables can also be charged by handling (See Section 2.6.6).

A similar problem can occur when a module or equipment housed in an insulating enclosure is packaged in a polythene wrapper or other highly insulating packaging. High levels of charge can arise on the equipment housing or nearby packaging that can induce high voltages on internal PCBs. These may then discharge on connection to a cable or other equipment.

This problem can be reduced or avoided by packaging the cable or module in a low-charging material. Where low atmospheric humidity is not expected, pink polythene can be a suitable material. This material, however, loses its low charging properties below about 30% rh. Below this, humidity level another static dissipative material should be used. Pink polythene should always be used with caution – this material is further discussed in Section 8.6.2.

Where triboelectric or induced voltages may occur on the internal parts of a module, connecting or touching of potentially sensitive flying leads or connector pins to wiring or other conductive items should be avoided. It may be necessary to cover the connectors or leads to prevent accidental contact.

8.6 Materials and Processes Used in ESD Protective Packaging

8.6.1 Introduction

The basic materials used in ESD protective packaging are often polymers of some type. These may have some additive material or treatment or a surface coating to give them the desired ESD control properties.

8.6.2 Antistats, Pink Polythene, and Low-Charging Materials

Antistats are long surfactant type molecules that have a hydrophilic (water attracting) end and a hydrophobic (water repelling) end. The hydrophobic end is attracted to the polymer material, and the hydrophilic end attracts water from the air. Antistats used in pink polythene typically are ethoxylated fatty acid amines or amides – chemicals that give them their low-charging properties in conjunction with atmospheric moisture (Havens 1989). "Amine-free" materials typically use amides rather than amines.

Pink polythene is polythene that has been loaded or surface coated with a chemical antistat. The pink color is a colorant that was originally added to differentiate static control materials from ordinary secondary packaging. Nevertheless, a pink color does not guarantee that a material has low-charging characteristics. Users of these materials should be careful to ensure they know their characteristics before use.

In some types, a form of amine was used that caused oxidation of some metals and stress cracking in some plastics. For this reason, some manufacturers switched to amide-based antistats. An effect on device solderability has also been a concern.

The materials are made by compounding the antistat additive with the polymer base material, usually low-density polyethylene. Other polymers may be used to change properties such as stiffness and sealability. A material such as crushed silica or calcium carbonate is added to counter the tendency for the finished film to stick to itself. The finished material may have a greasy feel if the proportions of additive, polymer, and antiblock are out of balance. The additive may contaminate any surface that it contacts. The final material can be made into film, sheets, or foams or can be injection molded into trays or other forms.

For effective performance, first the antistat must diffuse to the material surface. Second, atmospheric moisture must be adsorbed to the surface. The amount of antistat at the surface depends on the material formulation but also can vary and reduce with age of the material, due to evaporation of the antistat. The amount of antistat present at the surface depends on the relative rates of evaporation and diffusion to the surface from the bulk material. The amount of atmospheric moisture available to be adsorbed to the antistat varies with atmospheric relative humidity. These factors mean that the material properties can change considerably with atmospheric humidity and age. At low humidity and extended age, the material may become ineffective. This means that many pink polythene materials have short shelf life.

Pink polythene materials have also been implicated in solderability concerns and cracking of polycarbonate materials. The latter is due to the additive dissolving into the polycarbonate material. Properties such as label adhesion and printability may be reduced.

8.6.3 Static Dissipative and Conductive Polymers

Static dissipative and conductive polymers are polymer materials that have been loaded with conductive particles such as carbon black, carbon fibers, metal fibers or flakes, metal-coated fibers, or metal powders (Drake 1996). More recently carbon nanotubes or fibers have also been used. Resistance levels from 100 GΩ to as low as 1 Ω can be obtained. The resistance of the material can be relatively free from dependence on atmospheric humidity. The base materials are commonly nylon (carpets or equipment parts), polyethylene (flexible packaging), polypropylene (tote boxes), acrylonitrile butadiene styrene (ABS, engineering plastic parts), polyvinylchloride (PVC) (flooring), and others such as rubber, polyimides, polyester, and phenolics.

The conductivity of the material is produced by contact between conductive particles in the insulating polymer matrix establishing continuous conductive paths through the material. The resistance of the material is not a linear relationship with the percentage loading of additive (Blythe and Bloor 2008). At low additive loading, the effect on resistance is small because there are few contacts between conductive particles and hence few conductive

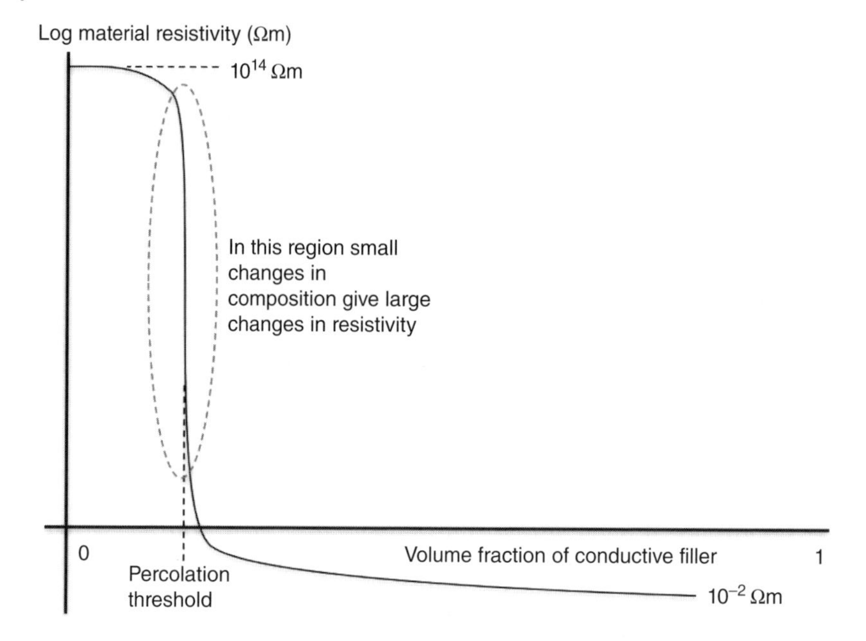

Figure 8.6 Typical variation of polymer resistance with conductive particle loading. Source: Blythe and Bloor (2008).

paths through the material. When the additive loading increases beyond a level, an abrupt reduction in resistance occurs (Figure 8.6). This level is known as the *critical loading* or *percolation threshold*. The level is influenced by the nature of the resin substrate, size distribution, and shape of additive particles and the quality of contact between the particles. Different fillers are added at different loading to produce a given resistance range. The percolation threshold occurs around 10–20% loading for spherical conductive particles but at lower concentrations for fibrous conductive materials.

The material resistance drops quickly with increasing conductive particle loading above a threshold loading level. At the same time, particle loading variation will inevitably be present due to variability of mixing. This means that there can be considerable variation of material resistance over short distances. Processes producing final packaging products can also affect the local additive loading and material conductivity and result in high resistance or insulating regions in the material.

The aspect ratio of the particles is the ratio of particle length to diameter. High-aspect ratio particles improve contact between particles and hence higher conductivity (lower resistance).

Adding the conductive particulate material influences the physical properties of the material, such as flexural modulus, tensile strength, hardness, viscosity, and heat distortion temperatures. Sloughing or particle shedding can be a problem and can prevent use in clean areas.

These materials are commonly used to make packaging products such as tote boxes, bags, tubes, bins, and trays by vacuum forming, injection molding, or film extrusion.

Carbon black is the most common additive used to produce conductive or static dissipative materials as they are cost effective. Carbon-loaded materials are normally black in color. Carbon black in low loading is used in materials not intended for ESD control, giving black color and other properties. At these low loadings, the materials remain highly insulating. So, it is important to understand that a black color does not necessarily imply a material has static dissipative or conductive properties or is suitable for ESD control use!

8.6.4 Intrinsically Conductive or Dissipative Polymers

Intrinsically conductive and dissipative polymers (ICPs and IDPs) are polymer materials that have intrinsic conducting properties such as polyaniline or poly pyrrole. These can be used as coatings or additives to other polymers. They can be used to make conductive fibers for use in fabrics.

Blending these materials with conventional polymers can result in stable materials in the static dissipative and conductive resistance ranges. They have been mixed commercially with ABS, polycarbonate, polyester, nylon, and polyurethane and can be translucent, nonsloughing, and available in many colors.

8.6.5 Metallized Film

Thin metallized films on a polymer substrate (e.g. polyester) are usually laminated in multilayer structures with other materials to give electrostatic field and ESD shielding and moisture barrier properties.

The metallization is typically aluminum. The metallization used in ESD shielding bags is typically thin to be transparent. Thicker metallization used for moisture barrier bags is typically opaque.

8.6.6 Anodized Aluminum

Aluminum is a highly conductive material, but anodization forms a resistive surface layer of alumina. Anodized aluminum can be used to make "boats" (holders) or machine parts for automated handling and tubes for components. The aluminum base can be grounded. The anodization layer is thin and can be damaged and may also have defects such as pinholes. The breakdown voltage of the layer can be as low as a few hundred volts and so a charged component at higher voltage could break down and discharge through the layer on coming into contact (Bellmore 2001; Smallwood and Millar 2010).

8.6.7 Vacuum Forming of Filled Polymers

Filled conductive and dissipative polymer sheets can heated and be vacuum formed into complex shapes over a solid mold. The process can result in changes to the conductivity of the material, especially where it is deeply drawn. Pockets can become isolated from the surrounding material.

8.6.8 Injection Molding

Filled conductive and dissipative polymer materials are often injection molded to form rigid shapes. The molding process often modifies the conductivity of the material, even producing nonconductive regions.

8.6.9 Embossing

Embossing can be used to change the surface profile of a material. This can in some cases be used to modify charge generation properties by reducing surface contact area.

8.6.10 Vapor Deposition

Vapor deposition is used to form a metallized coating on polymer surfaces, for example in ESD shielding bag or moisture barrier bag manufacture.

8.6.11 Surface Coating

Materials may be surface coated with other materials to give the desired surface properties. Examples include coating of a cardboard box with a conductive surface layer. With flexible film materials, coating can be done in moving web form.

8.6.12 Lamination

Several layers of materials with different properties may often be laminated together to form a package with the desired properties. Examples are ESD shielding bags and moisture barrier bags. Some laminated materials may be vacuum formed, but the forming process may alter the properties of the laminated layers.

8.7 Types and Forms of ESD Protective Packaging

Modern ESD protective packaging products take many forms. Packaging materials and products are used individually or as part of packaging systems. These systems are combinations of packaging materials that are used to give a required set of physical protections, ESD control, and other properties.

8.7.1 Bags

There are various types of ESD packaging bags on the market that have different combinations of physical and ESD control properties. They may be heat sealable or have a zip or self-adhesive seal. They are available in low-charging, static dissipative, or conductive or metallized ESD shielding materials. Low-charging bubble wrap bags are also available. Some materials are transparent, whereas others are opaque. Some examples of bags designed for ESD control or protection are shown in Figure 8.7. The bag properties specified in ANSI/ESD S11.4 (EOS/ESD Association Inc. 2012a) and MIL-PRF-81705-D standards are given in Table 8.1.

Figure 8.7 ESD packaging bags include pink polythene, black polythene, and metalized ESD shielding bags among other types.

Table 8.1 ESD control bags standard classification and properties specified.

Standard	Type	Electrostatic property						EMI shielding
		Low charging	Surface resistance or resistivity	Charge decay	Electrostatic field shielding	ESD shielding	Moisture barrier	
MIL-PRF-81705D and E[a)]	Type I		Yes	Yes	Yes	Yes	Yes	Yes
	Type II	Yes	Yes	Yes				
	Type III		Yes	Yes		Yes	Yes	Yes
ESD S11.4	Level 1	Yes	Yes		Yes	Yes	Yes	
	Level 2	Yes	Yes		Yes	Yes	Yes	
	Level 3	Yes	Yes		Yes	Yes		
	Level 4		Yes		Yes			
	Level 5	Yes	Yes					

a) Type II has been dropped from MIL-PRF-81705E

8.7.1.1 Pink Polythene Bags

"Pink" polythene is a low-cost low charging material that relies on moisture from the air to give it low-charging properties. It can give reduced charging of materials with which it is in contact but can leave a contaminating layer of antistat on the surface.

Originally pink polythene was given its pink color to distinguish for ESD control purposes it from ordinary polythene. Modern versions may be colorless or have various colors. It is a transparent material and so the bag contents can be easily seen.

The measured surface resistance can vary considerably with atmospheric humidity. Above about 30% rh, it is usually in the static dissipative range. Below about 30% rh, the measured surface resistance may rise above 10^{11} Ω (100 GΩ), and at worst the material behaves not very different from ordinary (non-ESD) polythene.

This type of bag does not have ESD shielding properties. At high atmospheric humidity, it may have a variable level of electrostatic field shielding, but this is not usually sufficient or reliable.

Pink polythene bags are mainly useful for replacing polythene secondary packaging for use in an EPA in applications where the humidity is not expected to drop below 30% rh. They may be used to enclose documents or components that are not susceptible to ESD damage and eliminate electrostatic charge generators from the EPA.

Pink polythene bags are an example of "TYPE II" from the U.S. military standard MIL-PRF-81705-D or Level 5 in ANSI/ESD S11.4 and "antistatic" in ANSI/ESD S541 (EOS/ESD Association Inc. 2018). The type was subsequently dropped from MIL-PRF-81705E.

8.7.1.2 Conductive (Black Polythene) Bags

Black polythene bags are made from carbon-loaded material and have relatively low surface and volume resistance around 10^3–10^4 Ω (1–10 kΩ). The material is black and opaque, and the contents cannot easily be seen without opening the package.

The material gives excellent electrostatic field shielding and charge dissipation. It does not necessarily give low-charge generation to materials in contact with it. As the material has low-volume resistance, a discharge to the outside of the bags can be conducted through the material to items within the bag. The material does not in itself have ESD shielding capability.

The charge may be transferred through the volume of the material to the device instead of around the material to ground.

Conductive bags are classified as Type 4 in ESD S11.4.

8.7.1.3 Metalized ESD Shielding Bags

ESD shielding bag materials have a multilayer structure that provides a barrier layer preventing direct ESD from being conducted through to the inside of the bag. The structure also contains a metallization layer that provides an electrostatic field shield and a path for ESD current to flow around the bag material. The material is generally sufficiently transparent to view the contents of the package to some extent, without opening the package.

There are two common types of this structure. In the "metal out" structure (Figure 8.8), a dissipative polyethylene inner intimate layer is bonded to a buried insulative polyester layer. This insulative layer forms a barrier to ESD current flow through the material. Outside this

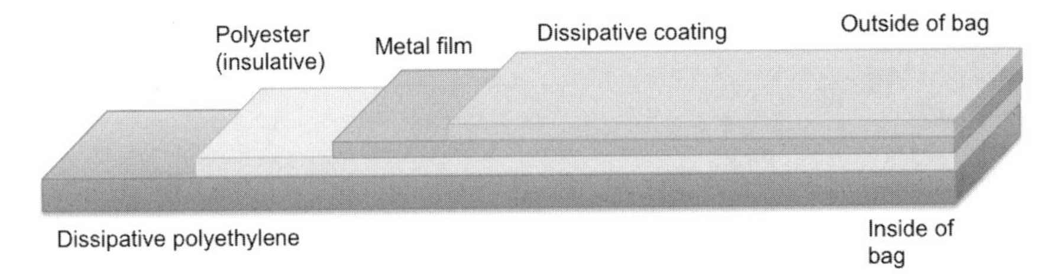

Figure 8.8 Typical "metal-out" ESD shielding bags material structure.

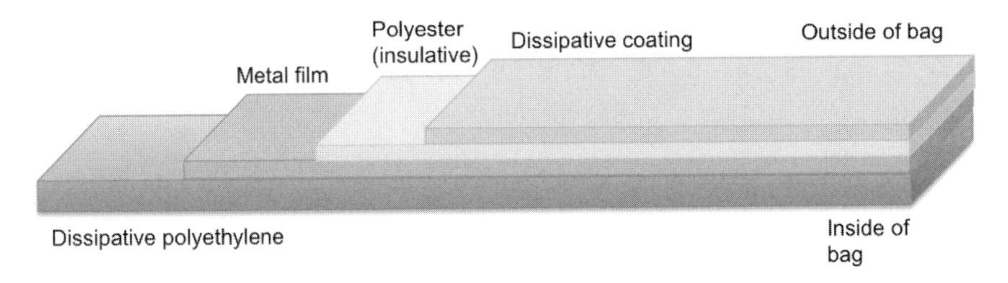

Figure 8.9 Typical "metal-in" ESD shielding bag material structure.

layer is a metallization layer. The metallization layer is protected by an outer dissipative coating.

In the "metal in" structure, the inner dissipative polyethylene layer has outside it the metallization layer, which is now buried (Figure 8.9). Outside this is a polyester barrier layer, and this is coated by a dissipative outer layer.

ESD shielding bags can be used to provide ESD shielding protection for ESDS components, boards, and assemblies in unprotected areas. They are classified as TYPE III under MIL-PRF-81705 and Level 3 in ESD S11.4 (see Table 8.1). The ESD shielding performance of bags is measured using an energy attenuation test (see Section 11.8.6) in which a standard human body model ESD stress is applied to the outside of the bag. The energy transmitted to a sensor in the inside of the bag is measured. Figure 8.10 shows results of this test for a selection of 16 pink, black, and ESD shielding bag types from the market.

8.7.1.4 Moisture Barrier Bags

Moisture barrier bags typically have static dissipative outer and inner surfaces with a low-charging treatment or coating. They typically provide electrostatic field and ESD shielding and include moisture vapor barrier protection (see Section 8.8.2). This protects moisture sensitive items and improves long-term storage.

Moisture barrier bags are typically similar in structure to metallized ESD shielding bags but are physically stronger than an ESD shielding bag (Figure 8.11). They may have one or more metallization layer that may be thicker than that of an ESD shielding bag. This metallization layer is often opaque rather than transparent. The moisture barrier bag is used when moisture barrier protection or maximum ESD shielding protection is needed. Moisture barrier bags are classified as Type I in MIL-PRF-81705 and Level 1 or Level 2 in ESD S11.4 (see Table 8.1).

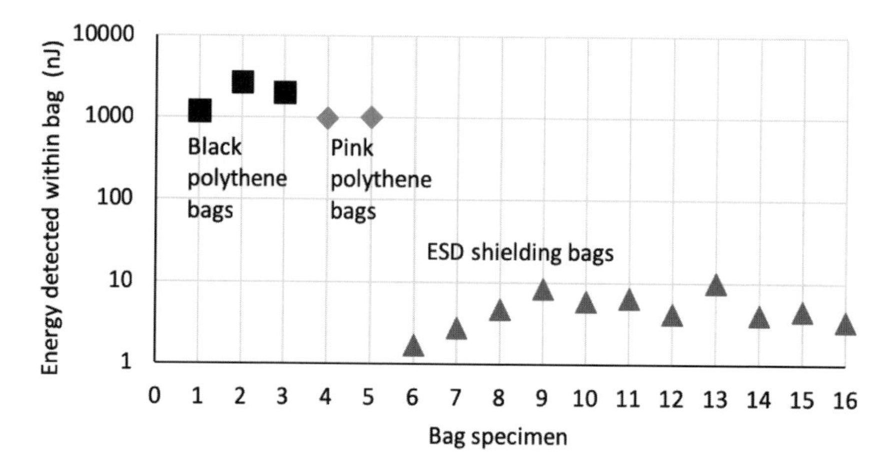

Figure 8.10 Comparison of ESD shielding capability of 2 pink polythene, 3 black polythene, and 11 ESD shielding bag specimens. Source: Smallwood and Robertson (1998).

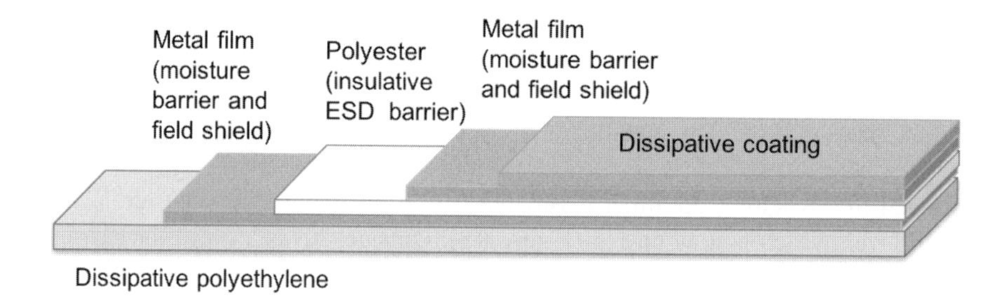

Figure 8.11 Typical moisture barrier material structure

8.7.2 Bubble Wrap

Bubble wrap material includes sealed gas bubbles for physically cushioning the package contents. It is available in low-charging material such as pink polythene blow molded from low-density polyethylene with antistatic additives. It is also available in colors other than pink.

8.7.3 Foam

Foams are available in insulative, low-charging, dissipative, and conductive materials. They may be used to cushion items within boxes or as a means of holding leaded package devices securely (Figure 8.12). A high resistance foam insert within a box or other package system can help give a barrier to ESD for ESD shielding.

8.7.4 Boxes, Trays, and PCB Racks

Boxes, trays, and PCB racks are commonly made from conductive or static dissipative carbon-loaded polymers (Figure 8.13). They may have covers or dividers and may be shaped

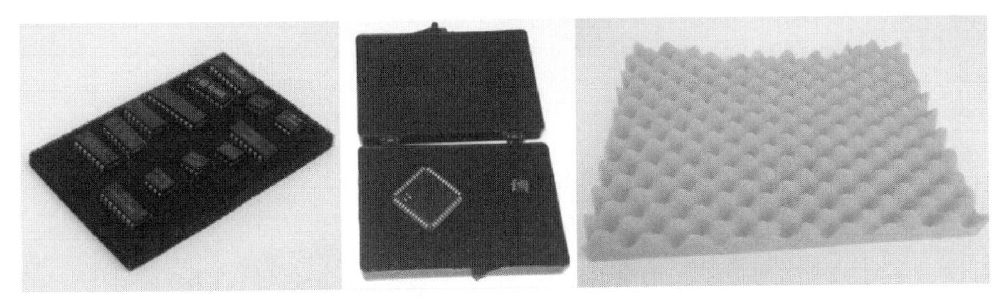

Figure 8.12 Foam is often used as intimate packaging or cushioning in packaging systems (left and mid) black carbon loaded static dissipative foam (right) pink low charging polythene foam.

Figure 8.13 ESD packaging boxes.

specifically to suit the product they are to contain. They may be designed for use with components or assemblies of finished products. They may be made from injection-molded materials such as polypropylene or vacuum formed from materials such as high-density polyethylene.

Shipping and storage boxes may also be made from cardboard, typically faced with conductive or static dissipative coatings and containing a foam insert.

Packaging systems based on boxes can include conductive layers or coatings to give electrostatic field shielding and air gaps, high-resistance materials, or even buried insulating layers as a barrier to ESD for ESD shielding.

Joint Electron Device Engineering Council (JEDEC) and waffle trays (Figure 8.14) are stackable trays containing multiple pockets in a matrix, designed to hold component package size and form made to dimensions defined by JEDEC standards. Trays may also be made to custom sizes.

Figure 8.14 JEDEC waffle tray.

8.7.5 Tape and Reel

Tape and reel packaging (Figure 8.15) is typically used to hold surface-mount semiconductor or passive components for automated assembly. The components are held in small pockets within the tape. These are then covered by a light cover tape held in place with

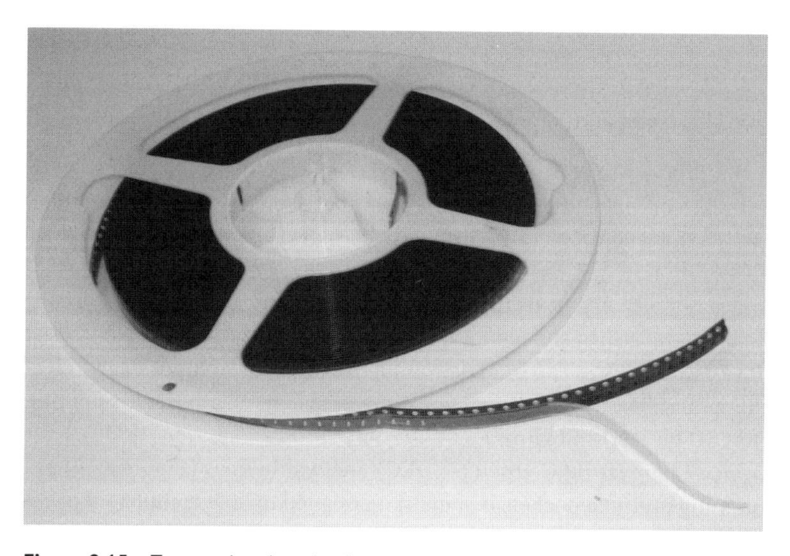

Figure 8.15 Tape and reel packaging.

adhesive. The tape is then wound onto the reel for transport, storage, and dispensing of components.

Tape and reel packaging is available in paper or polymer insulative materials for passive component handling, as well as in dissipative or conductive ESD control materials for use with ESDS. The reel itself can be made of insulative, dissipative, or conductive material. When used to contain ESDS devices, the tapes and reel should be made from static dissipative of conductive materials. The intimate surfaces of the tapes should be static dissipative or conductive and low charging.

The small feature sizes of the tapes make them impossible to test with current standard test methods that use large electrodes. The tape forming method can affect the resistivity of the material and can result in isolation of conductive or dissipative regions within the tape pockets.

8.7.6 Sticks (Tubes)

Stick magazines (Figure 8.16) are used for transport and storage of ESDS components or to feed devices to automatic placement machines for surface and through-hole board mounting. Several stick magazines loaded with components are often placed in boxes or bags to give defined component quantities. The devices are held within the stick using end pins or plugs, which should also be made of static dissipative material.

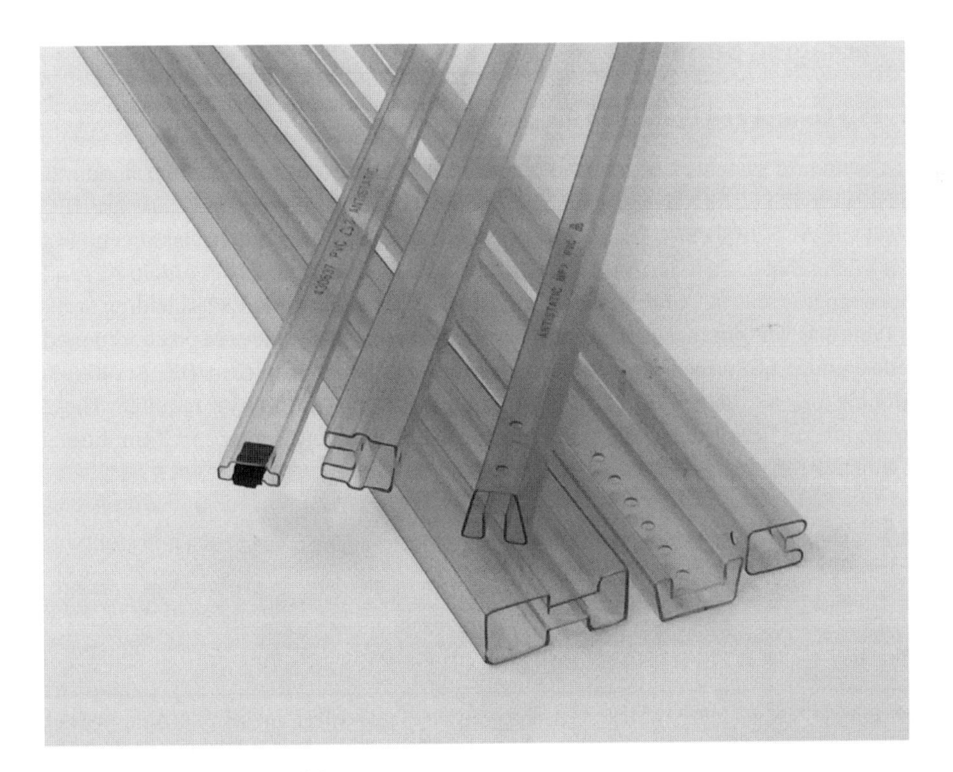

Figure 8.16 Component sticks.

Stick magazines or tubes are usually constructed of extruded rigid, clear, or translucent PVC. Metal sticks have also been used, and carbon-loaded static dissipative or conductive plastic tubes are available. The extrusion shape is tailored to component package types or shapes. The extruded magazines are surface treated, usually by dipping the tubes in a liquid solution. The pins and plugs are also treated.

8.7.7 Self-Adhesive Tapes and Labels

Conventional tapes and labels can generate high levels of electrostatic charge. Self-adhesive tapes and labels have been developed for applications where use in EPAs is required.

ESD tapes and labels typically have surface resistance in the static dissipative range up to 10^{11} Ω. A low-charging adhesive is used that also allows the charge generated by separating the label from its backing material to be dissipated quickly.

Tapes supplied on paper cores may be unsuited for use in clean rooms due to particulate contamination risk. Some are available on static dissipative plastic cores suitable for clean room use.

Beware that some self-adhesive "ESD tapes" are designed for sealing and marking packaging rather than for use in the EPA – these may be made of insulating materials unsuitable for use in an EPA.

8.8 Packaging Standards

8.8.1 ESD Control and Protection Packaging Standards

The ESD standard systems IEC 61340-5-1 and ANSI/ESD S20.20 have specific standards that deal with ESD protective packaging classification and properties. These are IEC 61340-5-3 and ANSI/ESD S541, respectively. These standards give specific classification of ESD packaging materials in terms of their electrostatic charging properties (Table 8.2) and resistance range (Table 8.3) and define electrostatic field shielding and ESD shielding properties (Table 8.4). For product qualification of packaging materials, they are preconditioned and tested under prescribed environmental conditions. For example, in IEC 61340-5-3, conditioning for ≥48 and test at 23 ± 2 °C and 12% ± 3% relative humidity is required. These conditions are set because dry air conditions typically represent "worst-case" conditions under which the material resistance is likely to be at its highest.

Table 8.2 Classification of packaging materials electrostatic charging properties in IEC 61340-5-3:2015 and ANSI/ESD S541-2019.

	Definition	
Classification	**IEC61340-5-3:2015**	**ESD S541-2019**
Low charging	Not defined	User-defined level that will ensure that ESDS items will not be charged excessively (produces unacceptable risk of discharge)

Table 8.3 Classification of packaging materials in IEC 61340-5-3:2015 and ANSI/ESD S541-2019 based on their surface and volume resistance range.

Classification or term	Definition	
	IEC61340-5-3:2015	ESD 541-2019
Dissipative	$R_s \geq 10^4\ \Omega$ and $< 10^{11}\ \Omega$ $R_v \geq 10^4\ \Omega$ and $< 10^{11}\ \Omega$	Provides an electrical path for charge to dissipate from the package $R_s \geq 10^4\ \Omega$ and $< 10^{11}\ \Omega$ $R_v \geq 10^4\ \Omega$ and $< 10^{11}\ \Omega$
Surface conductive	$R_s < 10^4\ \Omega$	Provides an electrical path for charge to dissipate from the package $R_s < 10^4\ \Omega$
Volume conductive	$R_v < 10^4\ \Omega$	Provides an electrical path for charge to dissipate from the package $R_v < 10^4\ \Omega$
Insulative	$R_s \geq 10^{11}\ \Omega$ $R_v \geq 10^{11}\ \Omega$	$R_s \geq 10^{11}\ \Omega$ $R_v \geq 10^{11}\ \Omega$

Table 8.4 Classification of packaging materials as electrostatic field or ESD shielding in IEC 61340-5-3:2015 and ANSI/ESD S541-2019.

Classification or term	Definition	
	IEC 61340-5-3:2015	ESD 541-2019
Electrostatic field shielding	Capable of attenuating an electrostatic field R_s or $R_v < 10^3\ \Omega$	See "electric" field shielding
Electric field shielding	See "electrostatic" field shielding	Attenuates electrical fields
Electrostatic discharge shielding (bag)	Capable of attenuating an electrostatic discharge. Calculated energy within a bag is <50 nJ when tested to 61340-4-8 (International Electrotechnical Commission 2014). (Edition 2: …or an equivalent test method modified to accommodate the product.)	Protects packaged items from the effects of static discharge that are external to the package and limits current flow through package Calculated energy within a bag is <20 nJ when tested to ANSI/ESD STM11.31 (EOS/ESD Association Inc. 2012b)
Electrostatic discharge shielding (other packaging)	Intimate packaging shall be conductive or dissipative. A barrier layer or a defined air gap attenuating ESD energy shall be included. No component of the packaging system shall cause ESD risk when taken within EPA.	User defined

These ESD packaging standards also give examples of markings that are recommended for use to identify packaging used in ESD control and protection. These are used to identify the packaging material or warn the user that ESDS devices may be contained within the package. This is discussed further in Section 8.10.

8.8.2 Moisture Barrier Packaging Standards

8.8.2.1 Handling, Packing, Transport, and Use of Moisture-Sensitive Devices

Automated assembly processes and the use of surface-mount devices (SMDs) led to damage such as cracks and delamination occurring due to from the solder reflow process. These problems arise because atmospheric moisture is absorbed into permeable component package materials. During solder reflow, SMDs are exposed to temperatures over 200 °C. Rapid moisture expansion and materials mismatch can result in cracking and/or delamination of material interfaces within the device.

IPC/JEDEC J-STD-033D describes the standardized levels of exposure for moisture/reflow-sensitive SMDs and the requirements for handling, packing, and shipping to avoid moisture/reflow-related failures (IPC, JEDEC 2018). The standard applies to all devices subjected to bulk solder reflow processes during PCB assembly, including plastic encapsulated packages, process sensitive devices, and other moisture-sensitive devices made with moisture-permeable materials (epoxies, silicones, etc.) that are exposed to the ambient air.

The standard calls for drying of moisture-sensitive devices followed by packing within moisture barrier packaging. IPC/JEDEC J-STD-033D calls for use of moisture barrier bags compliant with MIL-PRF-81705 Type I. The materials of these packaging materials are designed to limit moisture passage to $\leq 0.0310\,\mathrm{g\,m^{-2}}$ per day through the material. Moisture barrier packaging is marked with a symbol specified in MIL-PRF-18705D and ESD S11.4 (Figure 8.17). The materials are heat sealed after packing, with desiccant and a humidity indicator card.

8.8.2.2 MIL-PRF-81705

MIL-PRF-81705D 2004 gave specifications for heat-sealable electrostatic protective flexible barrier materials used for military packaging of ESDS for use by the US Department of

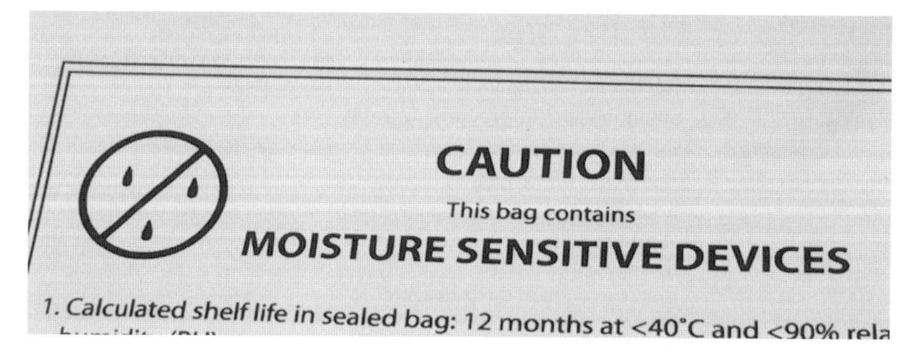

Figure 8.17 Example of moisture sensitive device packaging label specified in MIL-PRF-18705D and ESD S11.4.

Table 8.5 Applications of static control moisture barrier material classifications defined by MIL-PRF-81705D.

Type	Properties	Application
I	Water vapor proof, electrostatic protective, electrostatic, and electromagnetic shielding	Water vapor proof, electrostatic, and electromagnetic protection of microcircuits, and semiconductor devices, such as diodes, field effect transistors, and sensitive resistors
II	Transparent, waterproof, electrostatic protective, static dissipative	Use where transparency and static dissipation is required and contact with oil or grease is not contemplated
III	Transparent, waterproof, electrostatic protective, electrostatic shielding	Use where a transparent, waterproof, electrostatic-protective, and electrostatic field protective barrier is required

Table 8.6 Static control moisture barrier material requirements of MIL-PRF-81705D.

	Electrostatic property requirements				
Type	**Surface resistivity ρ_s (Ω sq^{-1})**	**Static decay**	**Electrostatic shielding**	**ESD shielding**	**Electromagnetic attenuation**
I	Inner: $10^5 \leq \rho_s < 10^{12}$ Outer: $\rho_s < 10^{12}$	<2 s	max 30 V peak	10 nJ	>25 dB
II	Inner: $10^5 \leq \rho_s < 10^{12}$ Outer: $\rho_s < 10^{12}$	<2 s			
III	Inner: $10^5 \leq \rho_s < 10^{12}$ Outer: $\rho_s < 10^{12}$	<2 s	Max 30 V peak	10 nJ	>10 db
Test method	ASTM D257	FED STD 101 method 4046A (superseded by MIL STD 3010 METHOD 4046)	EIA 541 2 probe method 1000 V	ANSI/ESD S11.31	Specified in standard

Defense. It classified materials as Type I, Type II, or Type III (Tables 8.5 and 8.6). Each type has two subclasses: Class 1 for unlimited use, and Class 2 for use on automated bag making machines only. The standard specifies physical characteristics such as seam strength, water vapor transmission rate, thickness, and transparency as well as electrostatic properties.

Type I – Water vapor proof, electrostatic protective, electrostatic, and electromagnetic shielding.
Type II – Transparent, waterproof, electrostatic protective, static dissipative.
Type III – Transparent, waterproof, electrostatic protective, electrostatic shielding.

The main properties and applications of the materials classified by 80715D are summarized in Table 8.5. The ESD-related requirements of these materials are summarized in Table 8.6. A new version, MIL-PRF-81705E:2009, dropped the Class II classification.

MIL-PRF-81705D specifies some different properties than those specified in ESDA and the IEC standard. First, the surface properties are specified in terms of surface resistivity measured according to ASTM D257 (ASTM International 2014), not surface resistance. Second, "electrostatic shielding" is specified in terms of voltage measured using the EIA 541 (Electronic Industries Assoc. 1988) two-probe method of ESD S11.31. Third, "static decay" is specified, measured according to FED STD 101 Method 4046A (superseded by MIL STD 3010 METHOD 4046 (Department of Defense 2002)). These differences are further discussed in Section 8.8.3.

8.8.2.3 ESD Association ANSI/ESD S11.4

The ESDA ESD S11.4 defines five levels of bags for static control purposes (Table 8.7), listed here:

Level 1 bags are moisture barrier bags for device packaging intended to protect items that will be subjected to reflow soldering. They provide moisture barrier properties preventing excess moisture absorption that can lead to device body cracking. They include ESD and electrostatic field shielding properties. The inner surfaces are low charging to avoid charge accumulation on or in the bag. The inner and outer surface resistance may differ for each surface and allows charge to dissipate from the inside or outside surface of the bag when the surface is grounded. They are typically a multilayer structure of metal foil and plastic layers containing or coated with an antistat.

Level 2 bags are moisture barrier bags for general packaging to protect items that will not be subjected to reflow soldering. They have ESD and electrostatic field shielding properties to protect electronic items from ESD. They have low-charging properties to avoid charge accumulation on or in the bag. Their inside and outside surface resistance may be different for each surface and allows charge to dissipate when the surface is grounded. They typically have a multilayer structure of metalized plastic and contain or are coated with an antistat.

Level 3 bags provide ESD and electrostatic field shielding properties intended to protect ESDS electronic items. Their inner and outer surface resistance may differ and allows charge to dissipate when the surface is grounded. They are typically a multilayer structure of transparent metalized plastic and plastic layers containing or coated with an antistat.

Level 4 bags are conductive and intended for protection of ESDS items. Their surface and volume resistance allow charge to dissipate from the bag when in contact with ground. They have electrostatic field shielding properties but do not provide ESD shielding. They are typically constructed from extruded plastic containing conductive materials and have the same surface resistance on inner and outer surfaces.

Table 8.7 Static control moisture barrier bags levels defined by ESD S11.4.

Level	Application	Structure	Property				
			Low charging	Electrostatic field shielding	ESD shielding	Surface resistance (Ω)	Moisture barrier
1	Device packaging: Items subject to reflow soldering	Multilayer metalized plastic	Yes	Yes	Yes <20 nJ	Yes. Inner surface may be different from outer surface Inner: $10^4 \le R_s < 10^{11}$ Outer: $R_s < 10^{11}$	Yes ≤0.002 g/100 in.^{2} /d
2	General packaging: Items not subject to reflow soldering	Multilayer plastic	Yes	Yes	Yes <20 nJ	Yes. Inner surface may be different from outer surface Inner: $10^4 \le R_s < 10^{11}$ Outer: $R_s < 10^{11}$	Yes ≤0.02 g 100 in.^{2} /d
3	"Static shielding" to protect ESDS	Multilayer plastic	Yes	Yes	Yes <20 nJ	Yes. Inner surface may be different from outer surface Inner: $10^4 \le R_s < 10^{11}$ Outer: $R_s < 10^{11}$	No
4	Conductive to protect ESDS	Conductive extruded plastic	Yes	Yes	No	Yes. $R_s < 10^{11}$	No
5	Static dissipative to protect ESDS	Extruded plastic with antistat	Yes	No	No	Yes. Inner and outer surface typically similar Inner: $10^4 \le R_s < 10^{11}$ Outer: $10^4 \le R_s < 10^{11}$	No
Test method					STM11.31	STM11.11 (EOS/ESD Association Inc. 2015a)	ASTM F1249 (ASTM International 2013)

Level 5 bags are static dissipative bags for protection of ESDS items. They have low-charging properties intended to avoid charge accumulation on or in the bag that could be damaging to ESDS items. Their inner and outer surface resistance is typically similar and is intended to allow charge to dissipate when the surface is grounded. They are typically made from extruded plastic containing or coated with an antistat.

8.8.3 ESD Control Packaging Measurements

ESD-related standard measurements of packaging materials and products fall broadly into four types, listed here:

- Resistance measurements to characterize the resistance across a surface or through a material
- Charge decay measurement
- Tribocharging measurement
- ESD shielding

Packaging materials may, of course, be also tested for many other non-ESD-related properties such as physical protection and moisture barrier properties.

Various standard measurement methods have been produced over the years, used by different ESD packaging standards. Test methods specified for use within the 61340-5-3 and ESD S541 standards are given in Section 11.5.

It is important to understand that in general, different test methods are likely to give different results due to variations in test electrode systems, procedures, and test voltages. It is important to make sure that when specifying and selecting materials for use within an ESD control program that the characteristics are specified and tested using appropriate test methods. Standards normally specify test methods that are to be used to make measurements to evaluate properties for compliance with the limits set in the standard.

The test methods required by 61340-5-3 and S541 are nearly identical in most cases. Table 8.8 gives the approximate equivalence of IEC and ESDA test standards. There may be some differences between the test methods given in the standards in some cases.

One apparent major difference is seen in the specification of material surface resistivity (measured according to ASTM D257) in MIL-PRF-81705 and surface resistance (measured according to IEC 61340-2-3 and ANSI/ESD S11.11) in the ESDA and IEC 61340 standards.

Table 8.8 Approximate equivalence of ESD packaging test standards in the IEC 61340-5-x and ESDA standards systems.

	Standards system test method standards	
Measurement	**IEC 61340-5-3**	**ANSI/ESD S541**
Surface resistance	61340-2-3	ESD S11.11
Volume resistance	61340-2-3	ESD S11.12
Point to point surface resistance	61340-2-3	ESD S11.13
ESD shielding	61340-4-8	ESD S11.31

Surface resistivity is defined as a material surface parameter that is corrected for the measurement electrode form (see Section 1.7.2). In contrast, surface resistance is the result of a measurement made on the surface, with no correction for electrode form. The electrodes specified in IEC 61340-2-3 (International Electrotechnical Commission 2016a) and ANSI/ESD S11.11 give a factor of 10 reduction in the resistance measured, compared to the surface resistivity of the material. So, a material with surface resistivity of 10^{12} Ω according to MIL-PRF-18705 would also be at the surface resistance limit of 10^{11} Ω as specified and measured according to IEC 61340-5-3 or ANSI/ESD S541.

8.9 How to Select an Appropriate Packaging System

8.9.1 Introduction

In selecting suitable packaging for an ESDS, many factors may need to be considered. These include the type of ESDS and likely ESD threats, the environment through which it will pass, physical and environmental protection requirements, and customer requirements. The overall cost of the packaging is likely to be a significant consideration.

8.9.2 Customer Requirements

The first consideration is, does the customer place mandatory requirements on the packaging that have been specified in contracts? If so, there may be only one option – to comply with these requirements.

Even where no contractual customer requirements exist, it may be necessary to consider likely customer and marketing needs. An example might be packaging a computer PCB for retail sale. The package may need to be transparent and show the product attractively in a retail display stand. It may be necessary to design the package according to the type of display stand on which it will appear. The package may need to display product features and description or marketing or technical information to help the buyer select it.

A package may need to be tamper proof or designed to help prevent counterfeiting. Alternatively, it may need to be designed to allow inspection of the contents in some way.

8.9.3 What Is the Form of the ESDS Device?

One of the first considerations is to examine the form of the ESDS device as this may eliminate some packaging possibilities or suggest others. ESDS devices may be components or devices, PCBs, modules or assemblies. The size, weight, and value of the ESDS could be important considerations. While small components may be packaged one or more in a bag, a large component or module may need a sizable or even custom-built box to contain it.

8.9.4 ESD Threats and ESD Susceptibility

ESD can occur when two conductors at different voltages touch (or come sufficiently close) so that a discharge can occur between them, and the ESD current that flows is sufficient to do damage. ESD can be prevented by ensuring that contact between the ESDS device and other conductors cannot happen, by ensuring that significant voltage differences do not exist between the ESDS device and conductors that it may make contact, or that when ESD occurs, its strength (peak current) is limited and discharge energy is absorbed by high resistance of the contacting material.

There are two types of ESD threats that should be considered. The main one is protection of the ESDS devices against direct ESD that may occur from an external ESD source or from a charged ESDS on contact with an external conductor.

Electrostatic fields are a second threat. While most ESDS devices do not suffer damage directly from electrostatic fields, they can set up the conditions under which ESD can occur by inducing high voltages on conductors within the ESDS device (see Section 2.4.4).

Where the ESD threats to the ESDS are not well understood, then it is best to protect the ESDS device against direct ESD to or from the ESDS device. Protection against electrostatic fields should also be considered. It may often be most cost effective to take this approach rather than analyze and evaluate the specific ESD threats to the ESDS device.

If the specific ESD threats to the ESDS device are well defined, it may be possible to choose ESD packaging that directly prevents those threats. An example is an ESDS electronic module built into a metal shielding housing, in which the only threat is the possibility of ESD to connector pins. In this case, it may be sufficient to protect the connectors against direct ESD, for example by use of static dissipative or conductive connector cover caps. In some cases, even insulative caps may be acceptable if these do not present ESD risk within EPAs.

There is a risk of ESD wherever a conductive part of an ESDS device can contact another conductor (person, equipment, or object) in the environment. The ESD threat, of course, depends on the susceptibility of the exposed ESDS part to the ESD source with which it makes contact. Possible sources of ESD include

- ESD from charged personnel
- ESD from charged packaging, objects, or equipment
- The possibility that the ESDS may become charged and discharge to an item that they touch

The number of ways in which ESD can enter an ESDS device can vary tremendously with the type and form of ESDS device. A PCB or multipin component can have many ESD entry points that could touch an external item or person resulting in ESD. Some types of modules may have few contact points limited to exposed flying leads or connector pins. In some cases, connector pins may be recessed in a connector housing to the extent that contact with them is highly unlikely. In other cases, a connector may extend in a way that it is extremely likely that it will be the first point of contact with another item.

The ESD susceptibility of an ESDS device, and ways in which ESD can enter the ESDS device, is likely to change with the build state. Ultimately, many ESDS products are protected against ESD by being built into a housing or potted, preventing ESD from occurring in all but a small number of well-defined ways. The final product may be considered insensitive to ESD. In some markets, e.g. Europe, the product may need to pass ESD susceptibility tests before it can be placed on the market. Some customers may have similar requirements. A change in build state can also introduce a new ESD risk. Potting or building a PCB into a plastic housing can prevent ESD into all parts except connector pins. At the same time, it introduces the risk that the housing can become highly charged and induce high voltages on the internal PCB, giving a risk of damaging "charged module" ESD on connection to a wiring loom or other external item.

8.9.5 The Intended Packaging Tasks

The packaging task may require consideration of various aspects such as the quantity of ESDS device housed, physical protection needs, package life, and recyclability.

In some cases, only one ESDS device is packaged, and in other cases multiple ESDS devices must be packaged. An example might be a tray designed to accept multiple PCBs for storage and transport.

Packaging may need to protect the contents against physical impact, crushing, and other physical threats. This requirement is likely to be greater for packaging that is intended to protect ESDS devices for transport outside the EPA and between sites, especially using commercial transportation networks.

The packaging may be used in transport or storage, or both. Storage may include tasks, such as display in a retail setting, that may place special requirements on the package. Packages may need to be designed for stacking in specific storage areas or using specific stacking and storage methods.

The packaging may be single use or multiple use. Packaging products meant for single use may have lower initial cost. Some types of packaging (e.g. bags) may be less physically robust than others (e.g. boxes and rigid or semirigid containers). The initial cost of these more robust packages may be relatively expensive in initial cost but may be the least expensive choice when all lifetime costs are considered. For reusable packaging, the cost of the material can be amortized over the anticipated number of times it can be reused. A means of testing for detecting failed or damaged packages may be required. Where multiple use is a possibility, a cost/benefit analysis should be made to evaluate the packaging material for the intended purpose over its lifetime. The cost of returning packaging for reuse may need to be considered with other lifetime costs. These costs may include collecting, sorting, cleaning, and preparing materials for the return shipment and reuse, testing costs, and transport costs.

Where packaging is to be reused, it may be necessary to test and verify the properties of each element of the package during its lifetime. Packaging surface and volume resistance properties can be evaluated by the user using simple test methods. Other types of packaging properties (e.g. ESD shielding) are difficult to evaluate without sophisticated equipment. For this reason, metallized ESD shielding bags may be considered one-use packaging. The cost of testing should be considered as part of the lifetime cost of reusing the packaging.

Some materials may have a limited shelf life that needs to be considered if items or the packaging material are to be stored for long periods.

Recycling and recovery of discarded or failed packaging materials should be considered.

8.9.6 Evaluate the Operational Environment for the Packaging

If the packaging is to be used only within an EPA, then the main ESD protective task may be to ensure that the packaging does not charge up and provide ESD risk to ESDS items.

In many cases, the packaging will need to protect the ESDS device for transport and storage outside the EPA. In this case, it must protect against the anticipated ESD threats as well as provide no ESD risks to ESDS devices while within the EPA.

Within an EPA, the package may be handled only by trained personnel, which may reduce the requirement, for example for physical protection. Outside the EPA the package may be handled by untrained personnel and must withstand the physical threats likely to be encountered when handled in the same way as any other package.

Within an EPA, it is often necessary to use ESD packaging for non-ESDS items. Secondary packaging must be kept well away from ESDS devices. Many ESD programs keep secondary packaging out of the EPA completely. Any suitable ESD packaging can be used to package non-ESDS within, and for entry into an EPA. For example, a conductive or static dissipative tote box can equally be used to contain non-ESDS items. Conductive or static dissipative bags can be used for non-ESDS items instead of ordinary polythene bags. Low-charging (pink polythene) can also be used for non-ESDS device inside or outside the EPA, providing the atmospheric humidity within the EPA is not expected to go below about 30% rh. Pink polythene should also be checked regularly to make sure it has not lost its properties due to loss of antistat.

If the package is required to enter a clean area, it may be necessary to select materials that will not give a risk of contamination. This may prevent use of some common ESD control materials such as carbon-loaded materials or pink polythene. Intrinsically dissipative or conductive polymer materials may be more acceptable in this case.

The physical and environmental threats are likely to be more severe where packages go out of the factory for transport between sites or to the customer.

Use of third-party courier or postage services may require a greater level of protection than use of company transport services. Packages may be subject to transport by air, ship, rail, or road during a journey.

Where the package crosses national borders, there may be a need to inspect the contents by personnel who are untrained in ESD control procedures.

The temperature and humidity likely to be encountered during transport and storage may be subject to considerable variation. The packaging may need to protect against wide temperature and humidity extremes during vehicle or air transport. Even within an EPA, the humidity may vary dramatically with season, weather, and the presence of heat sources such as ovens.

Packaging used within an EPA should always be selected to maintain its properties to the lowest humidity level that may be encountered in practice. Modern packaging standards usually require materials to be characterized at low humidity (e.g. 12% rh). Some materials (e.g. pink polythene) rely on antistats and atmospheric moisture for their properties. These materials be used with caution where the humidity is likely to drop below about 30% rh. Below this humidity, their properties may become little better than ordinary secondary packaging materials.

If the package is to take the items to a customer, then customer requirements may be a strong consideration. Some customers may only accept items packaged as specified in contracts or standards.

The way in which the package is used, stored, or handled at the customer's site may give specific requirements for the package and package marking. For example, a retail customer may require packaging that looks attractive for retail display of the product. The package may need to carry information on the contents to help the consumer select a product for purchase.

In some cases, it may be sufficient to protect the package contents from ingress of rain, for example during postage and transport. In other cases, the need for moisture protection may include maintaining low-humidity conditions within the package to avoid process problems at a later stage. High humidity can cause problems to electronic parts including corrosion and difficulties in soldering. It may be necessary to enclose parts susceptible to such problems within moisture barrier packaging.

The expected temperature range during transit and storage could be an important consideration. In a sealed package, the relative humidity level doubles with a 10 °C fall in temperature. If the temperature falls below the dew point (see Section 1.11) condensation may occur on surfaces within the package. A desiccant may be used to reduce the air humidity to a minimum, or the air can be evacuated during sealing to prevent this. The package may need to have a moisture barrier to prevent vapor ingress.

Removing the water in the internal space of the package with a desiccant can result in the "worst-case" low-humidity conditions where electrostatic charging may be a concern. Even without the presence of desiccant, low-humidity conditions can arise due to temperature changes. When ambient temperature increases, the internal relative humidity drops, halving with a 10 °C rise in temperature.

8.9.7 Selecting the ESD Packaging Type and ESD Protective Functions

8.9.7.1 Intimate Packaging

Intimate packaging is the material in contact with the ESDS device within the package. It must not charge up appreciably to avoid electrostatic risk to the ESDS device and other items in an EPA.

The material must not cause significant charging of the ESDS device that is in contact with it. At first sight, a packaging material that does not retain charge might be expected to also minimize charging of the ESDS device. However, charge accumulation on any material is the product of two characteristics – charge generation and the dissipation of charge from the material. Typically, the ESDS device has some insulating surfaces as well as conducting surfaces. Any charge generated by contact of these insulating surfaces with the packaging can be retained on the ESDS device. In contrast, the packaging material typically should be a noninsulative material and dissipate the charge produced on this material. So, the charge retained on the packaging is not correlated to the charge retained on the ESDS device. For low charging of the ESDS device, a material that will give low charge generation characteristics with the ESDS device should be selected.

It is impossible to avoid some charge generation by contact of the ESDS device with the intimate packaging material. For dissipation of charge generated by contact with the ESDS device, the intimate packaging material should be static dissipative or conductive (see Section 8.3.1 and Table 8.3). Any making or breaking of contact with the ESDS device will result in ESD at some level. If a high-resistance intimate packaging material is used, the ESD that occurs has its peak current limited by the high resistance of the material. The energy in the discharge is mainly absorbed in the intimate packaging material rather than the ESDS device.

It may also be necessary to set a lower limit to the surface resistance of the intimate packaging. For example, a PCB with on-board battery may require an intimate packaging of sufficiently high resistance to avoid discharge of the battery by contact with the packaging.

The physical design of the intimate packaging can also give benefits. For example, a purpose made rigid material can be shaped to hold the ESDS device firmly while introducing an air gap to minimize contact with the surfaces of the ESDS device (and hence charging) and at the same time act as a barrier to direct ESD. As another example, foam inserts within a box can be used to hold ESDS devices of a variety of size or form firmly with minimal movement and can virtually eliminate making and breaking of contact with ESDS leads.

8.9.7.2 Proximity Packaging

The proximity packaging design and materials will depend very much on the circumstances of usage. Table 8.9 shows the characteristics required of ESD packaging when used inside the EPA and outside the EPA. As ESD packaging must go into an EPA, it should not charge appreciably to avoid risk to ESDS items within the EPA. The surfaces that may become exposed within the EPA should be low charging and static dissipative or conductive.

Proximity packaging will often also be designed to give many of the physical and other protective functions required of the package for transport or storage of the ESDS, within or outside the EPA.

Where the package will go out of the EPA, the need for electrostatic field or ESD shielding should be considered (see Section 8.4.4 and 8.4.5). The most important of these is usually ESD shielding against direct ESD from ESD sources in the environment. In an unprotected area, these could include ESD from personnel, carts, and other substantial metal items.

ESD shielding is provided by a barrier preventing or attenuating ESD current flowing through and from the outside of the package to the ESDS device. The objective is to reduce to a low level the ESD current that can flow through to the ESDS device. This can be provided by an air gap, a layer of high resistance material, or even (as in ESD shielding bags) a buried layer of insulative material. It is important that any insulating material cannot be exposed to give ESD risk when in an EPA.

Table 8.9 Electrostatic characteristics of ESD packaging required inside and outside the EPA.

Packaging property	Inside EPA	Outside EPA
Low charging	Required	Required
Surface resistance (inner and outer)	Static dissipative or conductive	Static dissipative or conductive
Electrostatic field shielding	Not required	Optional, depending on ESDS failure modes
ESD shielding	Not required	Usually required, depending on ESDS failure modes

Electrostatic field shielding is given by inclusion of a layer of low-resistivity material in the package. This should enclose the whole intimate package forming a "Faraday cage" (see Section 2.4.3).

8.9.7.3 Packaging Systems

The functions of ESD packaging can be provided by one carefully designed packaging item (e.g. ESD shielding bag) or can be provided by a system of packaging materials that in combination provide the functions required. Figure 8.4 shows an example of a box packaging system. The foam intimate packaging that gives some degree of attenuation of ESD current that can flow to the ESDS gives physical cushioning to the ESDS and holds the device securely. The conductive box proximity packaging acts as a Faraday cage giving electrostatic field shielding and significantly reduces voltage differences within the box. It also encourages ESD current to flow around, rather than through the package. The combination gives both ESD shielding and electrostatic field shielding.

There may be many aspects of the packaging that require consideration for reasons unrelated to ESD control and protection. These may include physical protection against crushing, shock and vibration, strength, cleanliness, chemical contamination, plastic compatibility, marking and bar code reading, transparency, and testability.

The complete packaging system for transport and storage outside an EPA is likely to include secondary outer packaging to give some of the protection unrelated to ESD control.

8.9.8 Testing the Packaging System

In some cases, it may be necessary to test the packaging system to ensure that the contents will not become damaged in transit or storage. The tests may include physical tests such as vibration or drop tests as well as environmental tests such as rain, humidity, and temperature cycling.

Electrostatic and ESD testing and pass criteria should be selected to reflect the anticipated ESD threats and sensitivity of the ESDS. They may include aspects such as application of ESD to the package or electrostatic charging evaluation of the ESDS enclosed.

Packaging should be checked and tested before reuse. The first test that should be done is a visual check for damage during usage. ESD protective packaging should never be reused if there is visible damage that might affect its performance. Bags can easily be damaged by puncture by leads or sharp edges on components or PCBs within. Coatings can be torn off by removal of documents or labels attached to the outside of a package.

8.10 Marking of ESD Protective Packaging

There are various types of ESD packaging markings encountered in current practice. These inform the user that the package is likely to contain ESDS or give details of the type of packaging material. At the time of writing the IEC 61340-5-1 and ESD S20.20 standards require that marking is done in accordance with customer contract requirements if they exist. If no customer requirements are in place, then these standards require that the need for marking requirements must be determined and if necessary suitable markings defined.

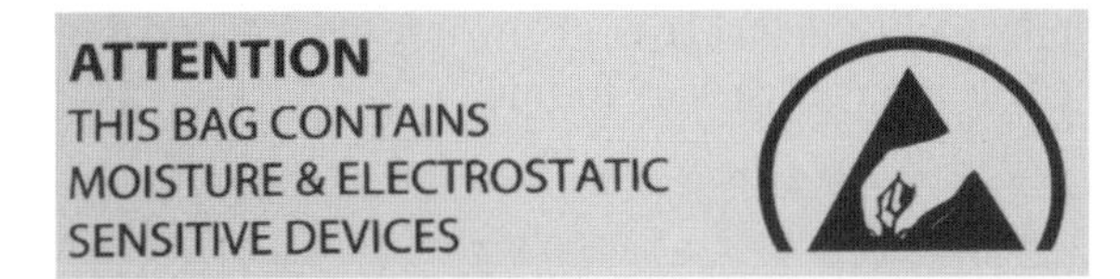

Figure 8.18 Example of packaging marking to identify ESD control packaging or materials according to IEC 61340-5-3, ANSI/ESD S541 and ANSI/ESD S11.4. Source: Reproduced by permission of the EOS/ESD Association Inc.

The ESD packaging standards IEC 61340-5-3 and ANSI/ESD S541 give examples of ESD packaging marking that can be used with ESD control and protective packaging. The outer surface of an ESD protective package should be marked with a symbol and/or wording identifying it as a package that is expected to contain ESDS devices (Figure 8.18). This is so that personnel handling the package can easily recognize that the package may contain ESDS devices and should not be opened further outside an EPA. Where a packaging system has several materials in layers, each material layer should be identified as an ESD control material. The "hand" symbol should also normally have a letter code beneath, indicating the main packaging function codes as follows: *S* for electrostatic discharge shielding, *F* for electrostatic field shielding, *C* for electrostatic conductive, and *D* for electrostatic dissipative.

Unfortunately, in practice packaging manufacturers are rather variable in how they mark ESD control packaging materials and products. Markings from older and superseded standards are often still seen. The letter codes showing the packaging function are often missing.

References

ASTM International. (2013) ASTM F1249 13 Standard Test Method for Water Vapor Transmission Rate Through Plastic Film and Sheeting Using a Modulated Infrared Sensor. West Conshohocken, PA, ASTM.

ASTM International. (2014) ASTM D257-14 Standard Test Methods for DC Resistance or Conductance of Insulating Materials. West Conshohocken, PA, ASTM.

Bellmore, D. (2001). Anodized aluminium alloys – insulators or not? In: *Proc EOS/ESD Symp. EOS-23*, 141–148. Rome, NY: EOS/ESD Association Inc.

Blythe, T. and Bloor, D. (2008). *Electrical Properties of Polymers*, 2e. Cambridge University Press. ISBN: 978-0-521-55838-9.

Cross, J.A. (1987). *Electrostatics Principles, Problems and Applications*. Adam Hilger. ISBN: 0-85274-589-3.

Department of Defense (2002) MIL-STD-3010). *Test Method Standard. Testing Procedures for Packaging Materials. Test Method 4046 Electrostatic Properties*, 33–41. Washington, D.C.: DoD.

Department of Defense (2004) MIL-PRF-81705D:2004). *Military Specification. Barrier Materials, Flexible, Electrostatic Protective, Heat Sealable*. Washington, D.C.: DoD.

Department of Defense (2009) MIL-PRF-81705E:2009). *Military Specification. Barrier Materials, Flexible, Electrostatic Protective, Heat Sealable.* Washington, D.C.: DoD.

Drake N. (1996) Polymeric materials for electrostatic applications. RAPRA Report ISBN 1-85957-076-3

Electronic Industries Assoc. (1988) ANSI/EIA-541-1988. *Packaging material standards for ESD sensitive items.* Washington D.C., USA, Electronic Industries Association.

EOS/ESD Association Inc. (2012a) ANSI/ESD S11.4-2012. *ESD Association Standard for the Protection of Electrostatic Discharge Susceptible Items - Static Control Bags.* Rome, NY, EOS/ESD Association Inc.

EOS/ESD Association Inc. (2012b). ANSI/ESD STM11.31-201. *ESD Association Standard Test Method for Evaluating the Performance of Electrostatic Discharge Shielding Materials – Bags.* 2 Rome, NY, EOS/ESD Association Inc.

EOS/ESD Association Inc. (2014). ANSI/ESD S20.20-2014. *ESD Association Standard for the Development of an Electrostatic Discharge Control Program for – Protection of Electrical and Electronic Parts, Assemblies and Equipment (excluding Electrically Initiated Explosive Devices).* Rome, NY, EOS/ESD Association Inc.

EOS/ESD Association Inc. (2015a) ANSI/ESD STM11.11-2015. *ESD Association Standard for Protection of Electrostatic Discharge Susceptible Items – Surface Resistance Measurement of Static Dissipative Planar Materials.* Rome, NY, EOS/ESD Association Inc.

EOS/ESD Association Inc. (2015b) ANSI/ESD STM11.12-2015. *ESD Association Standard for Protection of Electrostatic Discharge Susceptible Items*–Rome, NY, EOS/ESD Association Inc.

EOS/ESD Association Inc. (2015c) ANSI/ESD STM11.13-2015. *ESD Association Standard Test Method for the Protection of Electrostatic Discharge Susceptible Items – Two-Point Resistance Measurement.* Rome, NY, EOS/ESD Association Inc.

EOS/ESD Association Inc. (2016). ESD TR20.20-2016. *ESD Association Technical Report - Handbook for the Development of an Electrostatic Discharge Control Program for the Protection of Electronic Parts, Assemblies and Equipment.* Rome, NY, EOS/ESD Association Inc.

EOS/ESD Association Inc. (2018) ANSI/ESD S541-2018. *Packaging Materials for ESD Sensitive Items.* Rome, NY, EOS/ESD Association Inc.

Havens, M.R. (1989). Understanding pink poly. In: *Proc. EOS/ESD Symp. EOS-11*, 95–101. Rome, NY: EOS/ESD Association Inc.

International Electrotechnical Commission. (2014). IEC 61340-4-8:2014. *Electrostatics - Part 4-8: Standard test methods for specific applications - Electrostatic discharge shielding – Bags.* Geneva, IEC.

International Electrotechnical Commission. (2015) IEC 61340-5-3:2015. *Electrostatics - Part 5-3: Protection of electronic devices from electrostatic phenomena - Properties and requirements classification for packaging intended for electrostatic discharge sensitive devices.* Geneva, IEC.

International Electrotechnical Commission. (2016a). IEC 61340-2-3:2016. *Electrostatics. Part 2-3: Methods of test for determining the resistance and resistivity of solid materials used to avoid electrostatic charging* Geneva, IEC.

International Electrotechnical Commission (2016b) IEC 61340-5-1: 2016. *Electrostatics – Part 5-1: Protection of electronic devices from electrostatic phenomena - General requirements.* Geneva, IEC.

IPC, JEDEC. (2018) Handling, Packing, Shipping and Use of Moisture/Reflow Sensitive Surface Mount Devices. J-STD-033D. ISBN# 978-1-61193-348-2

Smallwood, J.M. and Millar, S. (2010) Paper 3B4). *Comparison of methods of evaluation of charge dissipation from AHE soak boats*. In: *Proc. EOS/ESD Symp*, 233–238. Rome, NY: EOS/ESD Association Inc.

Smallwood J M, Robertson C J. (1998) *Evaluation of Shielding Packaging for Prevention of Electrostatic Damage to Sensitive Electronic Components*. ERA Report 97-1079R, ERA Technology Ltd., Cleeve Rd, Leatherhead, Surrey, KT22 7SA

Further Reading

EOS/ESD Association Inc. (1995). ESD ADV11.2-1995. *Advisory for the Protection of Electrostatic Discharge Susceptible Items - Triboelectric Charge Accumulation Testing*. Rome, NY, EOS/ESD Association Inc.

EOS/ESD Association Inc. (2017) ANSI/ESD S8.1-2017. *Draft Standard for the Protection of Electrostatic Discharge Susceptible Items – Symbols – ESD Awareness* Rome, NY, EOS/ESD Association Inc.

EOS/ESD Association Inc. (2017). ESD ADV1.0-2017. *ESD Association Advisory for Electrostatic Discharge Terminology – Glossary*. Rome, NY, EOS/ESD Association Inc.

Fowler S. (2000) ESD protective packaging. Available from: http://www.esdjournal.com/techpapr/twenty1/intro.doc [Accessed 10th Nov. 2017]

Gale S F. (2006) Zero tolerance for ESD. Solid State Technology http://electroiq.com/blog/2006/09/zero-tolerance-for-esd [Accessed 29th Nov. 2017]

Huntsman J. R., Yenni D. M., Mueller G. E. (1980) Fundamental requirements for static protective containers. Nepcon West VI pp. 624–635

Huntsman, J.R. and Yenni, D.M. (1982). Test methods for static control products. In: *Proc. of EOS/ESD Symp. EOS-4*, 94–109. Rome, NY: EOS/ESD Association Inc.

Huntsman, J.R. (1984). Triboelectric charge: its ESD ability and a measurement method for its propensity on packaging materials. In: *Proc. EOS/ESD Symp. EOS-6*, 64–77. Rome, NY: EOS/ESD Association Inc.

International Electrotechnical Commission. (2018) IEC TR 61340-5-2:2018. *Electrostatics – Part 5-2: Protection of electronic devices from electrostatic phenomena - User guide*. Geneva, IEC.

Koyler, J.M. and Anderson, W.E. (1981). Selection of packaging materials for electrostatic discharge (ESDS) items. In: *Proc. EOS/ESD Symp. EOS-3*, 75–84. Rome, NY: EOS/ESD Association Inc.

Matisoff, B. (1997). *Handbook of Electronics Manufacturing Engineering*, 3e. Springer.

Swenson, D.E. and Lieske, N.P. (1987). Triboelectric charge-discharge damage susceptibility of large scale IC's. In: *Proc. EOS/ESD Symp. EOS-9*, 274–279. Rome, NY: EOS/ESD Association Inc.

Swenson, D.E. and Gibson, R. (1992). Triboelectric testing of packaging materials: practical considerations – what is important? What does it mean? In: *Proc. EOS/ESD Symp. EOS-14*, 209–217. Rome, NY: EOS/ESD Association Inc.

Texas Instruments. (2002) Electrostatic Discharge (ESD) Protective Semiconductor Packing Materials and Configurations. Application Report SZZA027A - April 2002

Vermillion R. (2014) The Silent Killer: Suspect/Counterfeit Items and Packaging. In Compliance. Available from: https://incompliancemag.com/article/the-silent-killer-suspectcounterfeit-items-and-packaging [Accessed 29th Nov. 2017]

Vermillion R. (2013) Pin Holes & Staples Lead to Diminished Performance in Metallized Static Shielding Bags. InCompliance. Available from: https://incompliancemag.com/article/pin-holes-a-staples-lead-to-diminished-performance-in-metallized-static-shielding-bags [Accessed 29th Nov. 2017]

Vermillion R. (2016) Have Suspect Counterfeit ESD Packaging & Materials Infiltrated the Aerospace & Defense Supply Chain? Interference Technology. Available from: https://interferencetechnology.com/suspect-counterfeit-esd-packaging-materials-infiltrated-aerospace-defense-supply-chain [Accessed 29th Nov. 2017]

Vermillion R. J., Fromm L.. (n.d.) A Study of ESD Corrugated. Available from: http://talkpkg.com/Papers-Presentations/Presentation/AHPStudy.pdf [Accessed 29th Nov. 2017]

9

How to Evaluate an ESD Control Program

9.1 Introduction

There are always several ways of looking at the "fitness for purpose" of an electrostatic discharge (ESD) program. As this phrase implies, it is necessary first to understand the various objectives that the ESD control program may be addressing. The primary purpose is presumably to control ESD risks and reduce ESD damage to an acceptable level.

A second purpose is often to help satisfy customers that the organization takes sufficient care in the manufacture or handling of their product commensurate with the market and reliability requirements. In this way, the ESD control program may form a positive contribution to the marketing of the product handled. Customers may audit the facility from time to time as part of their supplier quality assurance procedures. Some customers may require that their suppliers comply with an ESD control standard, and some of these may insist on compliance with their own preferred ESD control standard.

Another important aspect of evaluation is of the cost effectiveness of the ESD control program. It is usually desirable to attempt to get the maximum benefit from the least investment in resources (time and money). Evaluating the likely benefit from the ESD control program may be difficult but can be helpful in selecting the objectives of the program as well as deciding the level of resources to invest in it.

9.2 Evaluation of ESD Risks

9.2.1 Sources of ESD Risk

There are two overall types of ESD risk that are normally controlled in an electrostatic discharge protected area (EPA).

- Direct ESD to or from the device
- Electrostatic fields that could lead to ESD to or from the device or in some cases to damage to electrostatic discharge–sensitive (ESDS) devices

The ESD Control Program Handbook, First Edition. Jeremy M Smallwood.

Identification and evaluation of these is discussed further in Chapter 9. The usual sources of ESD risk include

- Charged personnel
- Charged metal or conductive objects or materials
- Charged devices

Other ESD risks may also occur, such as

- Charged board ESD
- Charged cable ESD
- Charged modules or assemblies

Understanding the ESD risks is an important step toward determining effective ESD control measures.

9.2.2 Evaluation of ESD Susceptibility of Components and Assemblies

The process of evaluating ESD risks must start with an evaluation of the ESD susceptibility of the ESDS component or assembly. In some cases, component human body model (HBM) and charged device model (CDM) ESD withstand voltage data can be obtained from device data sheets. Unfortunately, not all manufacturers' data sheets give information on this. Furthermore, the ESD susceptibility of printed circuit boards (PCBs), modules, and assemblies is not usually tested. So, the evaluation must often proceed based on some assumptions.

When handling ESDS devices of unknown ESD withstand voltage, it is often easiest to implement an ESD program with standard ESD control measures such as those given in IEC 61340-5-1 or ANSI/ESD S20.20. These are designed to control the most common ESD risks for devices of ESD withstand voltage as low as 100 V HBM and 200 V CDM.

If possible, any components that might have particularly low ESD withstand voltage or unusual ESD susceptibility should be identified. It is particularly important, if possible, to identify any components with ESD withstand voltage less than 500 V HBM or 250 V CDM, or those sometimes called "Class 0" devices. These should be handled with particular care. Components with ESD withstand voltage less than 100 V HBM or 200 V CDM may require special or adapted ESD control requirements.

The ESD susceptibility of the items handled usually changes through a production process. An ESDS component is typically assembled into a PCB assembly. This may then be assembled into a higher-level assembly or module. At some stage, this assembly is built into a final working product. The ESD risks and susceptibilities are likely to be different at each build stage. In many cases, it is convenient to implement standard ESD control measures for handling these items of unknown ESD susceptibility.

The final working system product may well be not considered to be ESD susceptible. So, at some build state the product becomes treated as not susceptible to ESD. The build state at which the product can be considered no longer susceptible to ESD is often debatable as it is rarely determined by design. Where the product has been subjected to, and passed, an ESD test as part of electromagnetic compatibility (EMC) immunity testing then this gives an assurance that it is also relatively robust to ESD damage. Nevertheless, there could be a remaining possibility of some level of susceptibility to ESD to untested parts. An example might be susceptibility to cable discharge to connector pins.

A final product that is not susceptible to ESD damage often relies on a housing or covers for protection of the ESDS internal parts. If the housing or covers are removed and the ESDS parts exposed, it may be necessary to again consider the item to be susceptible to ESD.

An assembly containing ESDS devices should be considered also to be ESDS devices unless there is good reason or evidence that it is adequately protected against ESD at that build level. The following possible ESD sources should be considered:

- Direct ESD from charged personnel to the ESDS parts of the product, especially via hand-held metal tools
- ESD occurring on contact with charged metal parts, chassis, or machinery
- The possibility that the internal parts could reach high voltage by induction or tribocharging leading to ESD on contact with another conductive item
- ESD occurring due to exposed terminations of a charged product contacting external conductive items
- Connection of charged cables or wiring looms to the assembly

Where only specific ESD threats occur in a process step, it may be sufficient to use specific ESD control measures rather than general ESD control measures. General ESD control measures may not adequately protect against specific ESD risks.

9.3 Evaluating Process Capability Based on HBM, MM, and CDM Data

9.3.1 Process Capability Evaluation

In recent times it has become a goal to be able to evaluate a process and determine its capability of handling devices in terms of HBM, machine model (MM), or CDM ESD withstand voltage levels. This has become more important as more devices are handled that have low ESD withstand voltages (ESD TR17.0-01 2015; Lin et al. 2014; Halperin et al. 2008). This is also linked to a trend toward reducing device on-chip ESD protection ESD withstand voltages toward levels that are not excessively above those easily handled by processes (Industry Council 2010, 2011; ESDA Association 2016b).

9.3.1.1 A Structured Approach to Process Evaluation

Evaluation of process risks in manual or automated processes is best done using a structured approach in which the ESDS path is followed from beginning to end through the processes (e.g. Gaertner 2007; Halperin et al. 2008; Jacob et al. 2012; ESD TR17.0-01 2015; ESD SP10.1 2016a). In every process step, the ESD risks are evaluated. This may include contact with personnel, contact with potentially charged metal objects, and the risk of the ESDS devices contacting a low-resistance conductor when in a charged state. Halperin et al. (2008) has called this *transitional analysis*.

9.3.1.2 Evaluate the Critical ESDS Path Through the Process

From the point of view of ESD risk, the positions that unprotected ESDS devices can take in the process define a critical path. The ESD risk occurs, and ESD threat must be evaluated, only in a region around this critical path (ESD SP10.1 2016a). ESD sources outside this region, a safe distance from the ESDS device, do not provide an ESD risk.

The responses to risks are usually specified in accordance with IEC 61340-5-1 and ESD S20.20 or the guidance of ESD SP10.1. For example, ESD SP10.1 recommends that in a region 15 cm around the ESDS critical path, all conductive machine parts should be grounded and insulative parts made static safe. S20.20 and 61340-5-1 require that the electrostatic field at the position of the ESDS device is $<5\,\text{kVm}^{-1}$, and items having surface voltage greater than 2 kV must be kept >30 cm from the ESDS device. The responses to risks may also need special measures not prescribed by these documents. These standards also require that

- Personnel handling unprotected ESDS devices must be grounded according to the standards.
- Metal objects contacting the ESDS devices must be grounded. If they cannot be grounded, the voltage difference between the ESDS devices and metal object must be reduced to below a hazard threshold level (± 35 V in 61340-5-1:2016 and 20.20-2014).
- To reduce charged device ESD risks, low-resistance conductors that contact the ESDS device may be replaced with intermediate- or high-resistance conductors.
- To address field-induced charged device ESD risk, nonessential insulators that might become charged are removed from the vicinity of the ESDS devices. Essential insulators are treated according to ESD risk.
- To address charged device ESD risks caused by contact charging or rubbing of ESDS devices, process steps that cause tribocharging of the ESDS devices are minimized.

Notice that all the ESD risks listed involve contact between the ESDS devices and a person or material. For most ESDS devices, where there is no contact, there is no ESD risk. The contact that causes ESD may, however, occur during a following step rather than the step at which charging occurs.

Electrostatic fields usually provide an ESD risk only where there is contact with the ESDS device in the presence of the field. There are, however, a small number of ESDS device types (e.g. reticles) that may be damaged by electrostatic field alone due to induced voltage differences causing breakdown of internal low-voltage insulating layers or air gaps. Smallwood (2019) has shown that damage could be caused to some voltage-sensitive devices such as metal-oxide-semiconductor field-effect transistors (MOSFETs) with high-impedance terminals, by charge injection from fast-changing electrostatic fields.

When evaluating processes, it's likely that the ESD specialist will need assistance from process specialists to ensure that ESDS movements are correctly evaluated, risk points identified, relevant measurements made, and suitable solutions found. Examples are given by Gaertner (2007), Halperin et al. (2008), Lin et al. (2014), and ESD TR17.1-01-15. The TR17.1-01 document draws on several earlier works.

TR17.1-01 following Jacob et al. identifies an evaluation procedure steps as follows:

- Assess potential risk
- Identify measurement points
- Make measurements
- Assess measurement results
- Define and implement corrective actions

The first step involves demonstration of the process under real conditions, with the goal of identifying potential risks and measurement points. Making measurement and evaluating the results presents its own set of challenges. Some of these are discussed by Lin et al. (2014).

9.3.1.3 Use of ESD Withstand Data in Process Evaluation

Unfortunately, it is not easy to correlate process capability with component HBM, MM, or CDM ESD withstand data. Some general guidance can be offered. In practice, the analysis and assumptions given in this section are approximate. Unfortunately, accurate methods of evaluating ESD risk in real production situations are currently unavailable. In some cases, the estimates given are likely to significantly overestimate the ESD risks to the device, but in other cases the risk may be overestimated or underestimated.

In principle, the ESD risk could be understood by knowing the susceptibility of each type of ESDS devices to each ESD source. The risk could then be controlled by maintaining the parameters of ESD from each source below thresholds at which ESD damage are known to occur. To this end, HBM, MM, and CDM ESD susceptibility tests have been developed to allow reproducible measurement of the ESD susceptibility of components (see Chapter 3). The component ESD susceptibility is quoted as an ESD withstand voltage, which is the highest test voltage in the ESD source that did not result in damage to the component.

At first sight then, a strategy for risk evaluation and management might seem clear – if the real-world voltages in ESD sources can be kept below the withstand voltage levels, no ESD damage should be possible. Specification of ESD control measures would then be a matter of defining controls that can maintain the source voltages below the risk threshold defined by component ESD withstand voltages (Steinman 2010, 2012).

- The body voltage on personnel should be kept below the HBM ESD withstand voltage.
- The voltage on metal objects that contact ESDS devices should be kept below the MM ESD withstand voltage.
- The voltage on devices should be kept below the CDM ESD withstand voltage.

While this simple view and strategy is useful and partially valid, it represents a significant oversimplification. Nevertheless, this may be the best starting point that we have currently, representing an estimated "worst-case" capability for many cases. Steinman (2010) suggested that the limits used should be 50 percent, rather than 100 percent of the respective ESD withstand voltages.

In the case of discharges between the ESDS devices and metal objects, the MM and CDM withstand voltages probably give a conservative limit to work to (Steinman 2010; Tamminen and Viheriäkoski 2007). As an example, if the ESDS device makes contact with a conductor of resistance >10 kΩ rather than a metal object, voltages on the ESDS devices considerably greater than the CDM ESD withstand voltage do not cause damage. In the case of human body ESD, the HBM withstand voltage probably gives a conservative limit in many cases but not necessarily all, depending on the particular type of sensitivity of the ESDS devices.

Component susceptibilities are not in fact normally to the ESD source voltage, but to other parameters such as the peak discharge current or charge power or energy transferred to the device in the discharge. These parameters are related to the source voltage by circuit parameters such as inductance, resistance, and source capacitance. Real-world sources are unlikely to have the same capacitance, resistance, and inductance as the HBM, MM, and CDM models. This means that HBM, MM, and CDM discharges do not happen in the real world – they happen only in the component susceptibility test equipment. Real-world ESDs are simply discharges from the human body, a metal part, a device, or another source. Each source has very different ESD waveform characteristics from the HBM, MM, and CDM models and would yield a different withstand voltage (TR17.0-01, Tamminen and Viheriäkoski 2007; Gaertner and Stadler 2012). So, ESDS devices might be more or less susceptible to damage from real sources than it is to HBM, MM, or CDM. A simple action such as changing the position of an ESDS devices can change a relevant parameter such as its capacitance as well as the voltage and hence the ESD damage susceptibility. Tamminen and Viheriäkoski (2007) found that the real charged device ESD risk in a process can be significantly less than suggested by the CDM withstand voltage. As components approach a flat grounded surface, the device voltage could in some cases reduce by 95 percent compared to the initial value away from the surface. This drop varied with component package, being least for tall components such as dual in-line (DIL) packages. Similar results were found with a component approaching a floating PCB. They were able in some cases to get a better estimate of ESD risk by taking account of the initial voltage and charge on the device, especially where the geometry of the device and environment were taken into account. They commented that in placing of a component, the resistive characteristics of solder paste may affect discharge characteristics and ESD risk.

Even in the well-controlled world of the device ESD susceptibility tester, there can be significant differences in parameters due to parasitic components in the tester. These can lead to variation in withstand voltages measured for the same device with different testers. Changing a device package type can also lead to changes in measured withstand voltage.

Some ESDS devices such as PCBs or assemblies are not normally tested for susceptibility using standard HBM, MM, or CDM ESD. Even if they were, the susceptibility would be likely to change for every possible contact or ESD entry point on the ESDS devices. A complex ESDS device such as a PCB thus could have a large number of possible contact points. Susceptibility to ESD could vary with each new component added to an assembly and with each ESDS position in an assembly process.

Some workers have suggested that it would make more sense to specify parameters such as peak current (Bellmore 2004; Smallwood and Paasi 2003), charge, power, or energy instead of the source voltage as withstand data parameters. This added complexity would at least allow use of relevant parameters in evaluation of ESD risk in real-world situations. Even this would not necessarily give the complete answer. It can be difficult to measure real ESD parameters in a process, especially in automated equipment. Inserting measurement equipment and sensors usually modifies the parameters that govern the ESD waveforms that are being measured. In fast-moving automated equipment, it can be difficult to add measurement equipment safely and without modification of the process.

A further complication is that it is not always easy to obtain ESD withstand data for all components. In practice, this may not matter too much as it is likely that the process evaluation will focus on the most sensitive components. As long as these are identified and the ESD control designed for handling them, then the less-sensitive components are probably adequately protected.

9.3.2 Human Body ESD and Manual Handling Processes

ESD risk from charged personnel is notionally related to the HBM withstand voltage of components. As a first approximation, it is assumed that the voltage developed on personnel handling a device should be limited to less than the HBM withstand voltage of the most sensitive device handled. So, in an ESD control program handling 100 V HBM devices, the body voltage of personnel handling ESDS devices should not exceed 100 V. If 50 V HBM devices are handled, the body voltage should be limited to less than 50 V.

If lower withstand voltage devices are handled, the allowed body voltage should be reduced accordingly. For example, for handling 50 V HBM devices, the maximum body voltage should be 50 V. The maximum wrist strap resistance to ground of 35 MΩ is determined to limit body voltage to <100 V. For 50 V maximum body voltage, this maximum resistance should be reduced pro rata to 17 MΩ.

When working with low HBM ESD withstand voltage devices, it is wise to directly evaluate body voltage produced during usual process and handling activities to ensure that body voltage does not exceed required maximum levels. This is because resistance to ground is only one factor related to body voltage. Charge generation is also highly important and is neglected if only resistance to ground is measured. Charge generation depends the materials and equipment used by the operator and can change dramatically if the type of material or equipment, or surface conditions (including contamination of surfaces) or atmospheric humidity, are changed. Evaluation of body voltage generation, for example with a change of footwear or flooring type or condition, can be especially important for safe manual handling of low HBM ESD withstand voltage devices.

9.3.3 ESD Risk Due to Isolated Conductors

The MM ESD test simulates a "two-pin" ESD risk that occurs if a charged external metal object can discharge causing current flow through the ESDS device and through a second pin to another conductor. (HBM and MM tests are "two-pin" ESD. Charged device ESD is a "one pin" because the device, once charged, discharges through a single pin.) ESD risk could occur, for example, if a charged external object discharges through the ESDS device to ground or in charged board ESD if the ESD current flows across the PCB through the ESDS device.

In principle, if the ESDS device is susceptible to the energy in the discharge, all the stored energy in the MM test capacitor is delivered to the ESDS device. This energy, E_{mm}, can be calculated from the MM test capacitor capacitance C (200 pF) and test voltage V by

$$E_{mm} = 0.5\, C\, V^2$$

The energy E_{mm} can then be assumed to be the maximum that the device will survive in discharge from another metal object. The maximum voltage V_s allowed on any other source of capacitance C_s can then be calculated from

$$E_{mm} = 0.5\, C_s V_s^2$$

As many real sources are much less than 200 pF, a considerably higher voltage than the MM ESD withstand voltage can often be allowed. In practice, the capacitance of the source is likely to be variable and can be difficult to measure.

This analysis depends on the ESDS device being susceptible to the energy in the discharge. If it is susceptible to the voltage or charged transferred in the discharge, then a different analysis would need to be applied (Paasi et al. 2003). For simplicity, it is tempting to specify that the maximum voltage between any external metal ESD source and the ESDS device should not exceed the MM ESD withstand voltage. This would, however, greatly overestimate the ESD risk, and it could be difficult to maintain these low voltages. In practice, the MM ESD susceptibility test is being discontinued by IC manufacturers.

As with charged device ESD, if the contacting material has significant resistance, this absorbs some of the discharge energy and to some extent protects an energy-susceptible ESDS device. If the ESDS device is voltage or charge susceptible, resistance in the contacting material may not have any protective effect.

9.3.3.1 What Is an Isolated Conductor?

A primary ESD control measure is that all conductors that may make contact with ESDS devices are grounded if at all possible. In unusual cases, it may be found that some conductors that make contact with ESDS devices cannot easily be grounded and have to be left electrically isolated. In this case, the ESD risk may need to be evaluated and controlled. The ESDS device itself is not usually grounded.

Although the standards give some requirements for dealing with isolated (ungrounded) conductors that touch the ESDS device (see Section 6.5.12.7), they do not actually give any definition of an isolated conductor, or criteria by which they may be identified. Future standards may address this issue, but in the meantime, if in practice the resistance from a conductor to ground is less than 1 GΩ (10^9 Ω), it can usually be considered to be grounded. If the resistance from the conductor to ground is greater than 100 GΩ (10^{11} Ω), it should be considered isolated. Between 1 and 100 GΩ, some evaluation may be required. In this case, if the voltage produced on the conductor during operation remains sufficiently low during normal operation, then it can probably be considered adequately grounded. The question is, what voltage is considered "sufficiently low"? The answer may depend on the sensitivity of the ESDS device or may be specified by ESD control standards.

9.3.3.2 How to Deal with Isolated Conductors

Figure 9.1 gives a simple flowchart for detecting and dealing with isolated conductors. Conductors that cannot make contact with ESDS devices can be disregarded as they do not provide an ESD risk.

The root of the ESD risk is that the voltage difference between the conductor and the ESDS device may be sufficient to give damaging ESD. This can be a type of charged device (one-pin) ESD, or more like ESD from a charged metal object (two-pin). If the conductor is a low-resistance material, it could represent a significant ESD risk. The voltage between the conductor and the ESDS device must be maintained at a sufficiently low level to control this risk.

The current standards IEC 61340-5-1 and ANSI/ESD S20.20 give some requirements for ungrounded conductors. IEC 61340-5-1 requires that the voltage difference between the

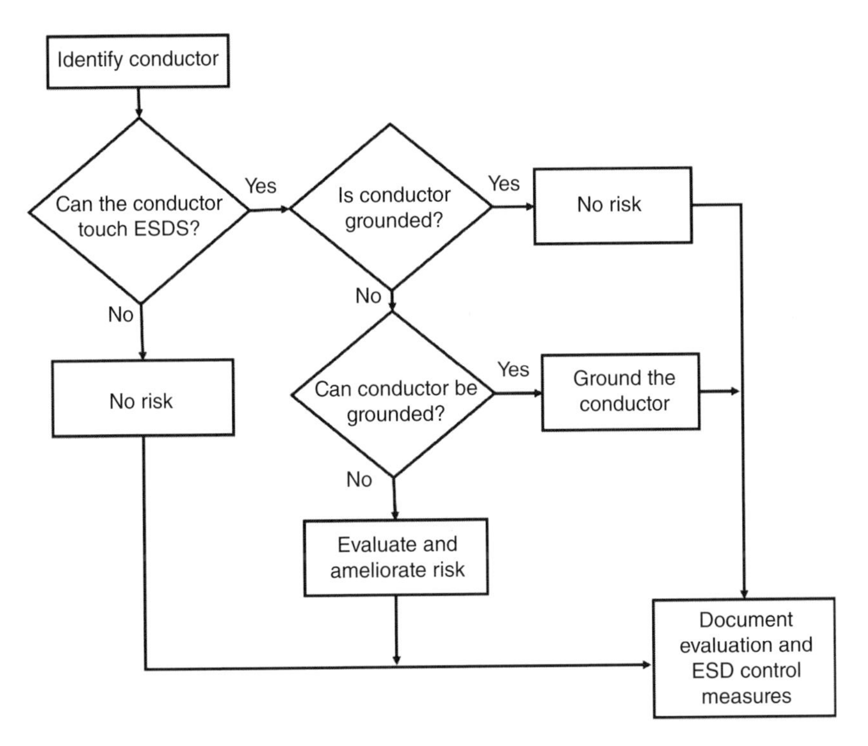

Figure 9.1 Dealing with isolated (ungrounded) conductors.

ESDS device and the conductor must be less than ±35 V. Measuring this is easier said than done. It requires two measurements.

- The voltage range and polarity on the ESDS devices must be measured under working conditions.
- The voltage range and polarity on the conductor must be measured under working conditions.
- The range in difference between the two voltages must be calculated.

Measurement of voltage on the ESDS device and conductor under working conditions may be far from easy in practice, especially under working conditions in a fast-moving automated process. If the ESDS device or the conductor is small, contact or noncontact electrostatic voltmeters capable of measurement on small objects must be used (Steinman 2010, 2012).

One way to bring the voltage difference between a conductor and an ESDS device to a low level is to place them in the ion stream from an effective ionizer. Given sufficient time and low charge generation processes, each will come to the same voltage - the offset voltage of the ionizer. This may not work well, however, in a fast-moving process where sufficient time may not be available or if charge generation processes are greater than the rate of neutralization that the ionizer can achieve in the installation.

If the ESD concern is restricted to charged device ESD (i.e. an otherwise unconnected ESDS device makes contact with the conductor) and if the conductor has surface resistance >10 kΩ, the ESD risk is likely to be negligible (see Section 9.3.4).

9.3.4 Charged Device ESD Risks

Charged device ESD risks are often the main risks occurring in automated processes and in any situation where a charged device can contact a highly conductive material (e.g. metal). Charged device ESD is a "one-pin" ESD risk that can occur by the device pin contacting a low-resistance conductor. Contacting a highly resistive conductor gives a reduced risk of damage because the ESD current is reduced by the resistance of the material at the point of contact. For a simple analysis, an assumption can be made that if the voltage difference between the device and the conductor is less than the CDM withstand voltage, ESD damage is unlikely to occur.

In this case, a visual inspection, observing the path taken by ESDS devices through the system, can be undertaken to detect the locations where contact with metal or other low-resistance conductors may occur. At these contact points, charged device ESD risk is apparent if the voltage difference between the ESDS devices and the contacted metal could exceed the ESDS devices CDM withstand voltage.

Charged device ESD could also be a risk in a manual process if the ESDS device contacts a low-resistance material. This is the main reason why intimate ESD packaging and bench mats should usually be made from materials having surface resistance greater than about 10 kΩ. When ESDS devices can contact materials having surface resistance <10 kΩ, charged device ESD risks should be evaluated.

It is often not easy to determine the voltage difference between the device and contacting material, particularly in an automated process during operating conditions. If one of the conductors is grounded, then at least the voltage of this is known to be zero. If the conductor is not grounded, grounding the conductor does not prevent charged device ESD as it is the charged device itself that is the ESD source. Grounding the contacting conductor does simplify the analysis of risk and ensures that the conductor cannot become charged and an additional source of ESD risk (see Section 9.3.3).

In practice, it is the peak current in the discharge that normally causes charged device damage, and the risk of this can be reduced by increasing the resistance of the material contacting the ESDS devices. So, charged device ESD risk can be reduced either by maintaining the voltage difference less than the CDM withstand value or by increasing the contacting material resistance, or preferably both. The charged device risk is greatly reduced by contact with a sufficiently resistive material. The resistance of the contacting material reduces the ESD peak current and absorbs some of the energy of the discharge current. ESD S20.20 and IEC 61340-5-1 recommend a minimum contacting material surface point-to-point resistance of 10 kΩ for avoiding charged device ESD risk.

Adding resistance into the ground path of a low resistance conductor, for example by adding a discrete resistor, does not reduce charged device ESD risk as the discharge is into local capacitance of the conductor (Wallash 2007). Resistance in the ground path does not act to limit the ESD current into this capacitance (Figure 9.2) The device capacitance can be discharged with the peak current limited only by the impedance of the discharge (e.g. spark) and the conductive material.

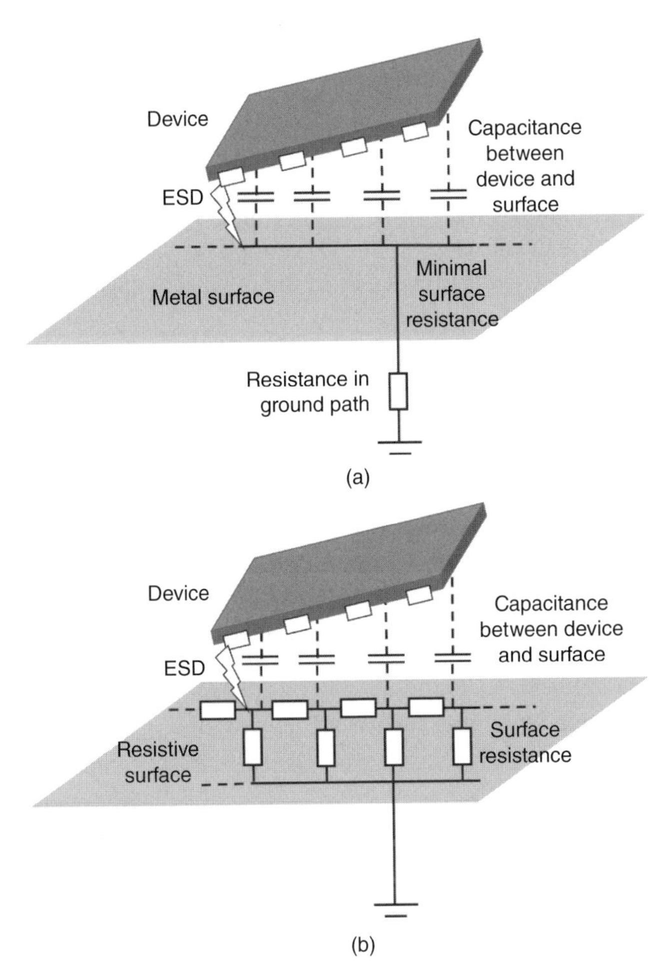

Figure 9.2 Charged device ESD to metal (above) or resistive (below) surface. Adding resistance to a ground path from a contacting conductor does not limit the charged device ESD current as using a resistive surface does.

Evaluation of the voltage likely to be produced on a device in a production process is not easy (see Section 11.9.4). Voltages can be produced on the device by triboelectrification during handling or by induction due to nearby electrostatic fields. If it is to be measured, it must be done in a way that minimizes the effect of the measurement equipment on the device voltage. The measurement method must also distinguish the voltage on the device from any other electrostatic fields and voltages in the region. This requires use of a low capacitance very high input impedance contacting or noncontacting electrostatic voltmeter capable of accurate measurement of voltage on small items.

If a measurement of the device voltage is successfully made, it remains a challenge to ensure that this represents the maximum likely during the full range of possible process conditions including low atmospheric relative humidity.

If contact with an isolated conductor occurs, then the voltage arising on the conductor must be evaluated in the same way as for the device. The range of possible voltage difference between the device and the conductor must then be calculated.

If reduction of the voltage produced on the device by triboelectrification is needed, then the charging effect due to contact with materials leading to that process point must be evaluated. Careful change of the materials that contact the device may be one way of reducing triboelectrification. Against this, triboelectrification is a variable phenomenon, and attempts to use it in ESD control are often unreliable.

Evaluating and reducing voltages induced on the device may be simpler. The electrostatic fields occurring due to voltages on materials close to the device must be evaluated. The voltages induced on the device by these fields cannot exceed the original voltage of the field source. Therefore, if the voltages produced are less than the device CDM withstand voltage, they can be considered insignificant. As with measurement of the device voltage, it remains a challenge to ensure that this represents the maximum likely during the full range of possible process conditions including low atmospheric relative humidity.

If the voltage is greater than the CDM ESD withstand voltage, then its effect can be reduced or eliminated by measures such as

- For a charged insulator, it may be possible to replace the insulator with a grounded conductor.
- Increasing the separation distance between the voltage source and the device path.
- Adding an electrostatic field shield between the voltage source and the device.

9.3.5 Damage to Voltage-Sensitive Structures Such as a Capacitor or a MOSFET Gate

Some components such as capacitors and MOSFET gates can be damaged by injection of sufficient charge to raise the voltage to a breakdown voltage value. (see Section 3.4.3.) ESD risk to these components can be evaluated in terms of the charge that can be transferred to the ESDS devices by triboelectrification or in an ESD event (Paasi et al. 2003, 2006; Smallwood and Paasi 2003; International Rectifier AN-986 2004). This can happen by various means including the following:

- Charging by contact with another material
- Contact with a charged person or conductor
- Field-induced voltage differences
- Charging by ion influx from an unbalanced ionizer output

The parts of an ESDS device most at risk from this are likely to be high-impedance lines that have low capacitance and low-voltage breakdown. This result in sensitivity to damage from electrostatic fields even where no contact is made with the ESDS device.

Paasi et al. (2006) showed that for a MOSFET, the ESD risk threshold was adequately described by a charge threshold, whereas the HBM voltage value was inadequate as an indicator of ESD risk. The damage risk occurred when the induced charge could elevate the voltage to the gate breakdown value. When the MOSFET was mounted on a PCB, the gate breakdown voltage could replace the charge threshold as a risk indicator. The real ESD risk is of course modified by the circumstances and the probability of making contact to the MOSFET in a way that could cause ESD damage. ESD risk did not arise directly from the electrostatic field strength but rather from charge induced on the MOSFET or PCB. The field threshold for damage to a device on a PCB was lower than for an unmounted component because the PCB had greater area exposed to the field and greater induced charge for the same field.

9.3.6 Evaluating ESD Risk from Electrostatic Fields

ESD risk is in many cases a result of the electrostatic field from charged materials such as essential insulators but can also be from equipment such as old types of cathode ray tube (CRT) display equipment screen or high-voltage cables. Nonessential field sources should be dealt with by removing them from the vicinity of the ESDS device.

The risk often arises because voltage differences arise between the ESDS device and another conductor in an electrostatic field. If the ESDS device comes close enough to or touches another conductor at a different voltage, ESD will occur between them. The conductor may or may not be grounded – ESD will occur in either case.

The classification of insulating items as essential or nonessential is a matter of opinion. The same item may be considered essential in one process and nonessential in another. For example, it may be easy to remove papers from one workstation process, but in another it may be difficult to proceed without them if they must be updated or signed on completion of process steps.

In most cases, the risk from insulators can be assessed by a simple process of evaluation such as the procedure shown in Figure 9.3.

Most ESDS devices are not inherently sensitive to damage directly from electrostatic fields. The ESD risk is usually significant only if there are significant electrostatic fields and there is the possibility of contact between the ESDS device and another conductor within the field. If there is no contact with the ESDS device within the field, it may not be necessary to control the field. If the insulator is sufficiently far from the ESDS device, the electrostatic field arising at the position of the ESDS device may be negligible. (Take care that the ESDS device and risk of contact are not likely to be moved into a position where the field may be significant.) If the insulator is not likely to be handled or moved or become charged, the risk of a field arising may be negligible.

Electrostatic charging of the item should preferably be measured under worst-case conditions, which usually means under low humidity (<30% rh) ambient atmosphere. In practice, measurements may have to be done under ambient atmospheric conditions, as a humidity-controlled facility is usually not available. Nevertheless, an initial evaluation done under ambient condition at higher humidity may give a useful first evaluation and should then be repeated when the weather conditions give lower humidity.

The question then arises – what level of charging can be considered negligible? Unfortunately, this may not be easy to answer and depends on the withstand voltages of the ESDS device being handled and other factors. For example, if it is charged device damage to the ESDS device that is of concern, if the voltages produced on the device are lower than the CDM withstand of the ESDS device, they can be considered negligible. The voltage induced on a conductor can never exceed the voltage of the electrostatic field source.

In practice, higher voltages than the ESD withstand voltage may also be negligible, but evaluation of this is more difficult. Charged device ESD damage is caused by peak ESD current and not the source voltage. For ESD between an ESDS device and a conductor, the peak ESD current is determined by the total impedance of the discharge circuit through

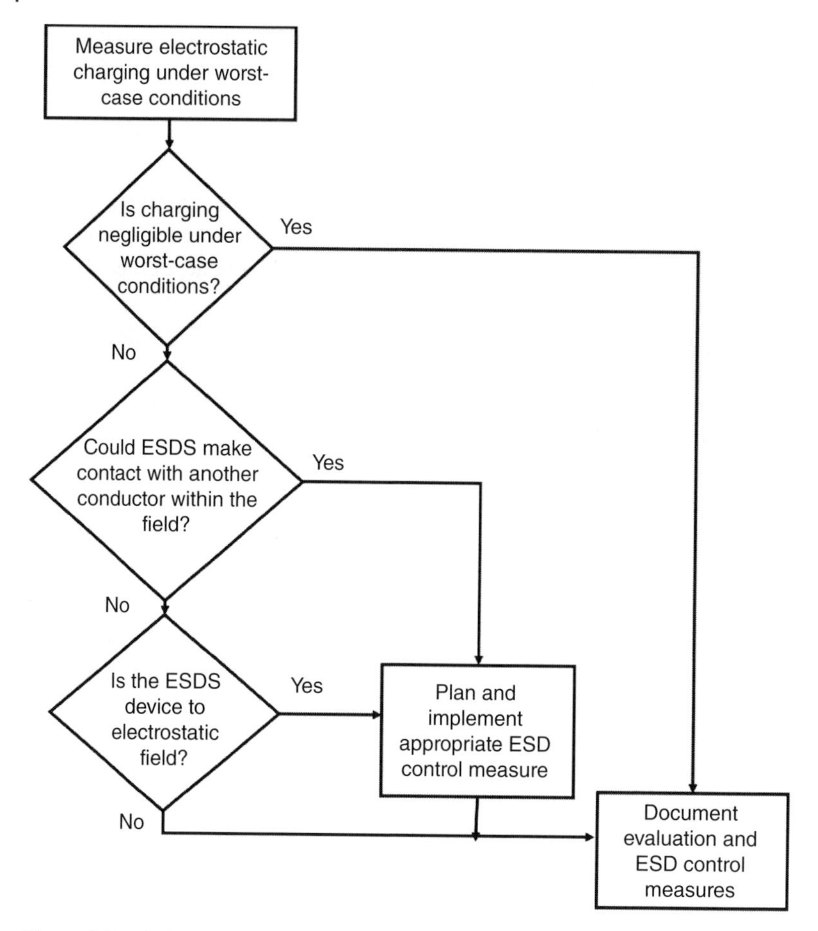

Figure 9.3 A simple electrostatic field risk evaluation.

any spark and the impedance of the contacting material at the point of contact as well as voltage difference between the ESDS device and conductor. The impedance of this circuit typically changes with material characteristics and ESDS device position and orientation.

Standards may also give requirements that can be used to evaluate fields from charged insulators. For example, the IEC 61340-5-1:2016 standard gives requirements that the electrostatic field at the position of the ESDS device must be $<5\,\text{kVm}^{-1}$. In IEC 61340-5-1:2016 and ESD S20.20-2014, insulators charged to $>125\,\text{V}$ must be kept at least 2.5 cm from the ESDS device, and if charged to $>2\,\text{kV}$ must be kept $>30\,\text{cm}$ from the ESDS device. If these conditions are fulfilled, the electrostatic fields and voltages can be considered negligible for the purposes of these standards. Nevertheless, these limits may not be sufficiently low or well specified for safe handling of some types of ESDS devices (Stadler et al. 2018). This is a subject of research and discussion among experts at the time of writing.

9.3.6.1 Risk to ESDS Devices Due to Electrostatic Fields

Swenson investigated the susceptibility of a voltage-sensitive component to damage from an electrostatic field and helped establish field limits for use in ESD control (Swenson

2012). He used a 14-pin DIL package positive metal-oxide semiconductor (PMOS) device (Siliconix SM110CJ with 200 V HBM and 125 V CDM ESD withstand voltage) and a discrete MOSFET (Motorola 3 N157 with 200 V HBM and 150 V CDM ESD withstand voltage) as his test devices. He mounted these in an electrostatic field created by charged plates of various sizes charged by momentary contact with a preset high voltage. Devices were placed on a glass plate or a grounded metal plate. A grounded metal-wire probe was touched to a device pin while within the field. The discharge current waveform was recorded. Swenson then produced charts showing the damage rates experienced by devices at different applied voltage (electrostatic field) and separation distance from the ESDS devices.

With the SM110C device, Swenson found that the field level required for damage was greater when the device was placed on a ground plane compared to when it was on a glass plate. With the device on the glass plate, it is likely that the field coupled to the device is intensified. With the ground plane present, some of the field is coupled to the plane, the field at the device is reduced, and voltage required for damage is increased. Differences were also observed with the 3N157 device. Swenson noted that damage actually occurred when the devices were touched by the ground lead, creating a field-induced charged device ESD.

Swenson went on to perform experiments with a uniformly charged insulating plastic plate. He found that with the SM110C device on a ground plane, direct contact between the ESDS device and the charged plastic plate did not damage the device until the charged insulator voltage was 18 kV. When the device pin was grounded in the presence of the field, damage occurred at 10 kV. (In practice, the surface voltage and surrounding field will be altered by the presence of conductors such as the ground plane and ESDS device. The insulator voltage was measured using a field meter at 2.5 cm distance. (The measured insulator surface voltage would not have been the same as at the time of damage occurring.)

In experiments with the 3N157, again direct contact between the ESDS device and the charged plastic plate did not damage the device until the charged insulator voltage was 18 kV. Touching a device pin with a grounded probe resulted in damage with the insulator charged to only 2 kV.

Swenson's experiments showed that an ESDS device, if grounded via a metal contact in the presence of an electrostatic field, experiences a potentially damaging ESD. The device becomes charged in the process and, if the field is removed, can subsequently experience another potentially damaging ESD. The risk to the ESDS device depends on the size of the electrostatic field source and its distance from the ESDS device, as well as the charged surface voltage. Low voltages close to the device can be of concern where very sensitive devices are handled. It is the act of grounding, rather than the field, that caused the damaging ESD.

Direct contact with an insulator did not result in damage up to high charge levels. Swenson's results show that proximity or contact between a charged insulator and ESDS device is not a significant ESD risk in most circumstances, providing the ESDS device does not make contact with a conductor in the presence of the field. Small charged objects provide lower risk than large objects – Swenson found damage occurred to both devices when they were grounded in the presence of plates >4 cm × 4 cm size charged to 500–1000 V. The presence of a ground plane nearby increased the field required for damage.

Swenson's results supported the strategy of removing nonessential insulators from the vicinity of the ESDS device to reduce ESD risks. They also confirm that field strength is an indicator of these risks where contact between the ESDS device and a grounded conductor occurs in the presence of the field.

It is arguably the CDM ESD withstand voltage of the devices that is of most relevance to the ESD risk in Swenson's experiments. The voltages required for damage in the experiment (>500 V) were considerably greater than the CDM ESD withstand voltage of the devices (125 and 150 V).

9.3.7 Troubleshooting

Where failures have been experienced and ESD damage is suspected, the first action should be to check that normal ESD control processes are in place and are operating correctly. Rectify any deviations from the ESD control program and evaluate whether they could cause the damage.

Check that personnel handle the ESDS device using the required procedures and personal grounding equipment. Human nature sometimes causes personnel to invent more quick and convenient procedures that may not be so effective for ESD control, especially when under pressure to work fast.

If possible, obtain ESD withstand data for the damaged components to help identify possible sensitivity to likely ESD sources. If possible, failure analyze the damaged components to establish, or rule out, ESD as a possible cause. Semiconductor manufacturers will sometimes failure analyze components on behalf of their customers.

If atmospheric humidity data is available, check whether failure rates correlate with low atmospheric humidity. If this correlation exists, it can be a strong indicator that ESD is a likely cause of damage and ESD control is insufficient. Low humidity can occur in areas where strong heat sources are present, such as near ovens or near the cooling fan exhausts of equipment or computers. Local low humidity conditions can occur near heat sources even in air-conditioned rooms where humidity is controlled.

If there is no obvious deviation from the ESD program that could be responsible for the failures, more detailed evaluation of the process may be required, including following the path of the ESDS device through the process. Try to narrow down the problem to part of the process. For example, if failed products pass tests at one stage and then fail at a later stage, it is likely that the problem has occurred in the process steps between the two test points.

Critically examine the process for stages at which the ESDS device contacts other conductors. In a manual process, this should include handling by personnel. These contact points are points at which ESD is likely to occur. Processes in which ESDS devices may contact metal or highly conductive items should be examined carefully for ESD risks. This is discussed further in Section 9.4.

One area that is often overlooked is where an ESDS device is placed on a test jig that makes contact prior to test. If the ESDS device becomes charged before contact or contact is made in the presence of an electrostatic field, ESD can occur. Test jigs often contain essential insulating parts that could charge and give rise to an electrostatic field during use. I have also often found jig covers made from high-charging materials – these could induce voltages on ESDS devices immediately prior to contact with test jig contacts. The possibility that the test jig contacts and wiring might be charged or energized should also be checked.

Another often overlooked area is connection of cables to an ESDS device. ESD can occur if the cable is charged, if the ESDS device is charged, or if the connection is made in the presence of an electrostatic field. Cables can be highly charged as a result of being supplied packaged within a polythene bag or taken off a reel. The presence of charged packaging can also induce high voltage on cables, modules, or assemblies nearby.

It is often assumed that a potted module is immune to ESD. A module can have its exterior highly charged by contact with, or storage within, insulating packaging. The surface charge on the module can then induce high voltage on internal conductors that can lead to ESD when connected to cables or other conductors.

One area that can be problematic is handling or testing ESDS devices where high voltages are present, e.g. test process. Safety considerations often dictate that personal grounding or other standard ESD control measures cannot be used. Personal protective equipment such as high-voltage protection gloves or footwear may be demanded.

9.4 Evaluating ESD Protection Needs

9.4.1 Standard ESD Control Precautions Do Not Necessarily Address all ESD Risks

Standard ESD control precautions applying ESD control equipment by rote address most well-known ESD risks, but they do not necessarily address all ESD risks. Every step of processes where ESDS devices are handled should be scrutinized for possible ESD risks (Gaertner 2007). This includes any automated processes. Any possible risks found should be evaluated. Risks in general occur where the ESDS device is brought into contact with other conductors. Wherever this happens, the following questions should be asked:

- Is the conductor grounded? If not, what would be the ESD risk if it became charged?
- Could the ESDS device become charged before touching the conductor? Could the voltage difference between the conductor and the ESDS device result in ESD risk?
- What is the resistance of the conductor? Is it high enough to prevent ESD risk?

ESDS devices or isolated conductors can attain high voltage by two means. First, they can become charged by contact with other materials (triboelectrification). Second, they can attain high voltage under the influence of an electrostatic field.

Insulating materials are often the main source of electrostatic fields when they become charged by contact with other materials. Nonessential insulators are therefore removed from the vicinity where unprotected ESDS devices are handled, or even from the EPA. Many insulators are, however, an essential part of the product or process. These essential insulators cannot be removed from the vicinity of the unprotected ESDS devices. The ESD risk associated with the presence of the insulator must be evaluated.

If an ESDS devices is exposed to an electrostatic field but does not contact another conductor, there is unlikely to be a risk of ESD damage (Gaertner 2007). Few ESDS devices are susceptible to damage from electrostatic fields alone. An exception may be where ESDS devices contain voltage-sensitive devices such as MOSFETs with high-impedance circuit nodes (Smallwood 2019). So, the presence of an electrostatic field from a charged insulator does not necessarily amount to an ESD risk. An ESDS device can enter a field and exit without damage. If, however, contact is made with another conductor while within the electrostatic field, ESD will occur. An ESDS device that enters the field uncharged will exit the field uncharged unless charge has been lost or gained by ESD, by conduction through contact with other materials or via unbalanced ionization.

Triboelectric charging of ESDS devices can happen whenever the ESDS device contacts other materials. The amount of charge generated depend on the material that contacts the ESDS device and many other factors and conditions (e.g. humidity and rubbing action). It is often thought that triboelectrification occurs with contact only by insulating materials. This is not correct as any material contacts give triboelectrification, although if a conductor is grounded, the separated charge escapes from the conductor without any voltage developing. Insulating parts of an ESDS device such as PCB substrate, coatings, or potting compound can be charged by contact with conductors, and the charge developed on the insulator is not removed by contact with the conductor. In some cases, a greater level of charging has been found for contact with static dissipative materials than with insulators or lower-resistance conductors (Viheriäkoski et al. 2012). The voltage developed on the charged item depends on the amount of charge and its capacitance. Capacitance is in turn dependent on the proximity of other materials and conductors.

The risk of damage occurring when an ESDS device at high voltage contacts a conductor depends on the part of the ESDS device that is subjected to ESD, the magnitude (energy, current, or other parameter related to possible damage) of the ESD, and the sensitivity of the ESDS device to the form of ESD that occurs. Unfortunately, it is difficult to predict the susceptibility of ESDS devices to a real ESD event even if HBM, MM, or CDM ESD withstand voltage data is available.

Common ways in which an ESDS devices could attain high voltage include

- Triboelectrification during handling by operators wearing gloves
- Triboelectrification during transport by suction grip
- Triboelectrification during contact with a support wheel or conveyor
- Induced voltages due to proximity to charged insulators, machine parts, computer screens, clothing, gloves, or another electrostatic field source
- Triboelectrification during removal of a label, masking material, or adhesive item
- Triboelectrification during brushing to remove debris
- Induced voltages due to assembly into a charged insulative housing
- Triboelectrification during rubbing or removal adhesive tape or packaging from a plastic housing or potted component

Common ways in which conductors often contact ESDS devices include

- Touching a metal tool or machine part
- Placing of a component on a PCB track
- A PCB track coming against an end stop in a machine
- A PCB track contacting "bed of nails" pogo pin or support
- Contact of a cable with a terminal on the ESDS device

If an ESDS device is found to attain high voltage and contact another conductor while in this state, an ESD risk is clear. Similarly, a risk is clear if an isolated (ungrounded) conductor is found to attain high voltage and contacts an ESDS device while in this state. To avoid these risks, a means must be found of reducing voltage difference between the conductor and the

ESDS device before contact. A possible alternative is to replace the conductor with a static dissipative material to reduce the ESD current occurring on contact.

It is usual, if possible, to ground any conductor that contacts the ESDS device. This eliminates the risk that the conductor will attain high voltage due to triboelectric or induction charging. It does not, however, eliminate the possibility that the ESDS device may attain high voltage due to triboelectrification or induction. If this occurs, charged device ESD can occur on contact.

Understanding the process and evaluating the associated risks can be particularly difficult for automated processes. A working automated process is usually fast and inaccessible. Seeing the steps can be difficult and measuring ESD-related parameters during live operations is often impossible. Some common practices used in automated equipment are discussed in Chapter 5.

9.4.2 Evaluating Return on Investment for ESD Protection Measures

It can be useful to estimate return on investment (ROI) for individual ESD control measures or the whole ESD control program (see Section 9.5.4). This can be done to help decide which particular ESD control measures should be prioritized or may be omitted.

9.4.3 What Is the Maximum Acceptable Resistance to Ground?

In Section 4.5 we looked at a simple model of electrostatic charge build-up and definitions of insulator. This model also allows us to evaluate the upper limit of resistance to ground R_g in many circumstances. This can be looked at in two ways.

- In a quasicontinuous charging process, for a given charging current, what is the allowable voltage build-up?
- How long would it take a stored charge to dissipate?

In practice, we do not have to do this type of evaluation if the circumstance is covered by ESD control standards. There may, however, be circumstances not covered by standards or in which the requirements of standards need to be questioned for some practical reason.

9.4.3.1 Charging Current in a Quasicontinuous Process

In a quasicontinuous charging process, the charging current I can sometimes be measured using a picoammeter or electrometer. This might be useful, for example, in an automated process where grounding of some moving machine parts may be difficult.

Typically, measurement of the charging current would be attempted in worst-case (usually dry atmospheric humidity) conditions where possible. If a maximum voltage build-up V_{max} can be specified, then Ohm's Law can be used to specify a maximum resistance to ground R_{gmax}.

$$V_{\max} = IR_{gmax}$$

If, for example, a maximum charging current on a part is found to be 10 nA (10^{-9} A) and the maximum voltage acceptable is specified as 10 V, the maximum resistance to ground is $10/10^{-8} = 10^{9}\ \Omega$ (1 GΩ).

9.4.3.2 Maximum Decay Time

In some cases, it may be that a conductor with measurable capacitance C has been identified as a possible ESD risk and must be grounded to prevent charging. Often, as a general guide, it can be assumed that if there are no continuous charging processes present, a charge decay time τ less than a second or so would be adequate to ensure this. In a fast-moving automated process, a faster decay time might be desirable. The maximum resistance to ground R_{gmax} can then be specified as

$$R_{gmax}C = \tau$$

An example might be the resistance through the handle of a handheld tool. If the capacitance from bit to hand through the handle is measured to be 20 pF and a decay time of one second is deemed to be adequate, the maximum acceptable resistance through the handle is $1/10^{-11} = 10^{11}\ \Omega$ (100 GΩ).

In practice, it might be convenient to set a lower resistance that is easier and more convenient to measure (see Section 9.4.5).

9.4.4 Should There Be a Minimum Resistance to Ground?

There are usually possible two reasons that a minimum resistance to ground might be specified.

- Protection of personnel against electric shocks in a fault condition with high voltages present (see Section 4.7.6)
- Incorporation of inherent resistance in a material that contacts an ESDS device, for reduction of charged device ESD risk (see Section 4.7.5)

If neither of these is necessary, minimum resistance to ground need not be specified.

9.4.5 ESD from Charged Tools

A common risk in manual assembly is that of ESD due to charged tools. Many ordinary hand tools have a metal part (e.g. cutter, plier, or screwdriver bit) that is mounted in an insulating handle. This is an isolated conductor that could attain high voltage by induction or triboelectrification. If this metal part contacts an unprotected ESDS device during a process, a potentially damaging ESD could occur.

To avoid this risk, the insulating tool handle can be replaced by a conducting material, usually having resistance in the range 1 MΩ–100 GΩ. The metal bit is then connected to the grounded operator's hand via the conducting handle material.

In some cases, where charged device ESD risk is a concern, the tool bit may be replaced with a high-resistance material to reduce charged device peak ESD current levels below the likely damage threshold.

Grounding of the tool is achieved unless the operator is wearing insulating gloves. If they do so, the tool is isolated from the grounded hand by the glove, and contact with the glove is likely to charge the tool to some voltage. So, where ESD control tools are used, any gloves used must also be conducting through the grip area (see Section 9.4.6).

In practice, it might be convenient to set a lower resistance that is easier and more convenient to measure. For example, setting an upper limit of 20 MΩ might make the tool more easily verifiable using a lower specification resistance meter or even a wrist strap checker.

9.4.6 Use of Gloves or Finger Cots

Some processes require the operator to wear gloves to protect their hands or the product being handled. Contact with the glove material handled causes triboelectrification of the items handled.

If the glove is a conducting material, the charge retained on the conductors of the ESDS devices can be reduced via grounding via the conducting glove and operator's body. The resistance through the glove is usually chosen in the range 1 MΩ–1 GΩ, although up to 100 GΩ may be acceptable in some cases. Any charge retained on insulating parts of the ESDS device is not removed, but it is usually the voltage on the conductors of the ESDS device that is the critical issue in charged device ESD control.

9.4.7 Charged Cable ESD

Long cables are often supplied on reels, and short cables are often supplied in insulating plastic packaging. The insulating sheaths of the cables and the core conductors can easily become charged by triboelectrification between the core and sheath and between the sheath and packaging or other sheath areas. An additional voltage is often induced on a core due to proximity with the charged insulating materials. If a cable is removed from packaging, its core voltage can rise to several kilovolts (see Section 2.6.6).

ESD can occur when the cable is connected to an ESDS devices. Damage could occur if the circuitry connected to the ESDS device connector has insufficient immunity to ESD damage.

9.4.8 Charged Board ESD

PCBs usually have some accumulated voltage during manual or automated production processes due to triboelectrification or induced by nearby electrostatic fields. ESD occurs if a conductor of the PCB contacts another conductor during handling. This can happen by contact with a machine part such as end stop, metal PCB support, or "bed of nails" pogo pin contact.

It can be difficult to prevent this type of ESD, although charge on a PCB can be neutralized using an ionizer. The ESD peak current, and potentially damaging effect, can be limited by ensuring that materials that contact the ESDS device have high resistance (e.g. 100 kΩ–100 GΩ) rather than being low resistance or metallic.

9.4.9 Charged Module or Assembly ESD

Some types of system components have an ESDS PCB contained within a potted block or plastic housing. The component electronics may have only a flying lead or connector for connection to the external wiring looms or system components. In other cases, an ESDS

PCB may be contained within a subassembly isolated from the main assembly. In each of these cases, the ESDS PCB can achieve high voltage by triboelectrification or induction due to nearby electrostatic fields. ESD can occur when the ESDS PCB is connected to another system component. Damage can occur if the connection and ESD occurs to a susceptible part of the circuit.

It can often be difficult to prevent this type of problem unless the system components are designed to avoid it. The risk of damage can often be reduced by connecting first to a 0 V, ground, or power connection before I/O connections are made.

When packaging for transport modules that have flying leads, the ESD protective packaging should be specified to prevent accidental contact between the flying leads and external items.

9.5 Evaluation of Cost Effectiveness of the ESD Control Program

9.5.1 The Cost of an Inadequate ESD Control Program

In an interview with Halperin, Brandt (2003) reported that independent consultants and corporate studies have found that ESD losses can be as high as 10% of annual revenues with an estimated average negative impact of 6.5% of revenues. Based on 1997–2001 numbers, this represented estimated losses of $84 billion per annum to international electronics industry. While this is difficult to verify it represents significant losses, these are not just due to material costs, which are often the smallest part of the ESD impact, but include costs such as rework, warranty, field service, and customer service. Customers may also experience costs such as lost productivity due to failures during operation. The total cost of even a small ESDS device loss may be significant.

Ever since the recognition of thc nccd for ESD control, engineers suspecting ESD damage have asked two questions (Halperin 1986).

- How do I know static-related problems are affecting our operations, and to what degree?
- How do I define static impact in a way that attracts management attention and support?

Most company managers have to identify priorities for their resources. They respond by releasing resources and action to clearly defined problem areas that have specific causes and measureable value impact, with a probable return on investment. Lack of quantified information on cost or impact, and value of investment, can lead to inadequate investment in the issue. Halperin went on to discuss ways of estimating the cost of ESD to the organization and recommend how to present this to management to encourage their support.

It might seem at first sight that the cost of the lack of adequate ESD control should be easily determined, as the cost of product failed due to ESD within the organization and at the customer's site. Unfortunately, this is not so simple for various reasons.

One problem is that few organizations evaluate failures to the level at which ESD failures could be identified. This failure analysis can be expensive and time-consuming, often taking several day's work. The outcome may be inconclusive. ESD failures often are difficult to distinguish from electrical overstress (EOS) failures, and EOS failures could sometimes be caused by earlier ESD.

The costs of an inadequate ESD control program can include (Smallwood et al. 2014)

- Repair and replacement of ESDS devices product failures
- Dealing with product unreliability or drift in characteristics
- Failure analysis
- Purchase of materials that have incorrect properties for ESD control
- Production delays
- Need to overstock commonly failing components
- Expenditure on unnecessary or ineffective control materials or equipment
- Dealing with customers failures occurring in the field
- Dealing with disputes with customers about their perception of adequacy of ESD control in the facility
- Effects on product and company reputation, and sales

One way to estimate component losses is by throughput evaluation (Halperin 1986). This includes the following steps:

1. Identify the ESDS devices and determine the discrepancy between the volume purchased and the volume used in production.
2. Analyze ESDS usage, including average inventory levels and locations, requisitioning departments, purchase volume, and unit cost.
3. Define burden costs associate with ESDS devices and assemblies.
4. Evaluate the overall impact of ESD damage.

One approach is to identify the finished products produced during a particular period that contain ESDS components or assemblies, especially those that contain very sensitive (low ESD withstand voltage) devices. The number of product produced is usually easily found from production statistics, and the number of devices or assemblies used in the product can be calculated. Sometimes a difference between planned and realized product numbers is evident, which can itself indicate that something was going on that might be worth investigation, such as excessive rework, shortage of parts, or field problems.

If the ESD withstand data for the ESDS devices can be obtained, then the lowest ESD withstand devices are the ones most likely to be worth further investigation. Unfortunately, this data is often difficult to obtain as it is not always published on device data sheets.

For the devices selected for further investigation, inventory and purchasing records can be obtained. The actual number of ESDS items purchased in the period can be obtained, including the starting and final inventory and the cost of the ESDS devices. It can also be useful to find details of the locations or departments at which the ESDS devices are used.

Analysis of these data and the production figures may reveal a discrepancy between the number of devices leaving in the product and the number purchased and remaining in storage. An excessive difference may indicate, for example, rework or field service consumption due to failures. Of course, this would not in itself confirm ESD as the failure cause in all cases, but identification of any failure mode is usually a useful outcome. The analysis may give useful indication of where to focus further resources.

Knowledge of the number of failed devices and their cost allows calculation of failed component cost, but this represents only a part of the cost of the failures. If the number of reworked items or field or other failures can be identified, the associated costs can be estimated. These may include labor, facility, power, and other expenditure. Average costs may be easier to estimate than real costs per individual failed item. Halperin (1986) described preparation of the data in the form of a spreadsheet table, listing the sensitive parts, and for each part the ESD withstand voltage, difference between numbers sourced and those used in final in product, cost of each item, estimated "burden" of associated costs, and total lost cost of the components and associated burden. The magnitude of total cost per item give a means of ranking for further work, and the overall total cost for all items gives a view of the possible cost of failures.

The cost of the failure is likely to vary with the production stage at which the failure is found, including field failures. The cost of failure typically increases as the product advances through the production process. Field failures are usually the costliest of all. With some markets and product types (e.g. satellites and aerospace), the cost of a field failure, in economic and other terms, is extremely high and could include equipment downtime, loss of the product, and even threat to life or property. Some product may be irreplaceable or unserviceable. This consideration in itself may justify considerable care in ESD control during product manufacture, storage, transport, and handling.

Halperin (1986) provided an example of this type of analysis. An interesting case study was also later provided by Helling (1996) (see Section 9.5.4).

While it may be desirable to confirm ESD damage with failure analysis of the failed components, few organizations do this in practice. One reason may be that failure analysis of components can be a time-consuming process requiring several person-days specialist effort and equipment. The outcome is not always conclusive in identification of ESD failures. In particular, ESD damage and electrical overstress damage often show similarities. The situation can be worsened if adequate ESD control is not maintained at all stages of handling including between discovery of failure and into failure analysis, as ESD could still damage an already failed component!

Dangelmayer (1999) reported several interesting case studies that formed a convincing demonstration of the economic benefit of ESD control. These used several different approaches to establish the source or presence of ESD failures and benefit of ESD controls, including the following:

- Use of failure analysis to identify ESD failures
- Correlation of ESD control deviations and ESD losses
- Comparison of failure rates of batches of product assembled with, and without, ESD control measures
- Reproduction of failure of components using simulated ESD
- Testing of devices before and after operations to determine the source of ESD damage

9.5.2 The Benefit Arising from of the ESD Control Program

It can be difficult to know with certainty the cost of ESD failures to the business. It can be much easier to establish at least a standard ESD control program addressing the usual ESD risks. Nevertheless, the cost of ESD failures can be considerable and has been evaluated in some studies (e.g. Helling 1996; Halperin 1986). One problem is that failure analysis to a level that can positively identify ESD failures can be time-consuming and therefore expensive. It can be difficult to distinguish between failures due to EOS and ESD (Lin et al. 2014).

One common reason for my clients to first contact me for help has been to evaluate their ESD control after a customer audit has declared it inadequate. The customer is not always right in terms of their perception of control of real ESD risks. They do, however, sometimes opt to take their business elsewhere due to perceived inadequate of ESD control in the organization's facility. A dispute with a customer over this costs time and resources and is not good for customer relations, even if their evaluation is of ESD control is incorrect.

Implementation of a good ESD control program, with good documentation of ESD control processes and compliance with a standard, can do much to convince a customer that adequate care is taken and ESD control is effective. For some customers, compliance with an ESD control standard can be a prerequisite to doing business with a supplier organization. In this situation, the organization's positive attitude to ESD control can even become a useful marketing benefit.

9.5.3 Evaluation of the Cost of an ESD Control Program

The costs of implementing an ESD control program includes aspects such as

- Identification and analysis of ESD related failures during production
- Identification of process stages at which failures are found and possible ESD sources
- Estimation of cost of failure and rework at each stage
- Analysis of failures occurring at the customer site
- Estimation of cost of ESD control
- Setup and maintenance of EPAs and acquisition of ESD control equipment
- Documentation of the ESD control process
- Compliance verification
- ESD training
- Maintenance and replacement of ESD control equipment

Some of these are easier to estimate than others. Some costs are likely to vary considerably with product type and market characteristics.

One question that is usually worth considering is, what would be the cost of an ESD failure occurring at the customer's site? The answer to this will help put into context the importance of ESD control for the organization. A throwaway product in a consumer market might indicate that minimal ESD control sufficient to maintain acceptable failure rates is adequate. In a high-reliability high-value product market, the cost of a single failure, and associated costs like downtime or safety issues, can justify considerable investment. Some products lines require failure levels in the parts per million range. In a market such as satellite manufacture, a single failure may be intolerable.

9.5.4 ROI in ESD Control

Effective investment in ESD control should lead to a worthwhile return. According to Halperin (Brandt 2003), a properly implemented ESD control program can have a return on investment exceeding five to one within six months. Unfortunately, there are few published research studies on this aspect of ESD control program evaluation. An early example is Downing (1983). An ROI estimate can also be made for an individual ESD control measure. Gumkowski and Levit (2013) examined the use of air ionization and compared different types of ionizers in a semiconductor manufacturing process.

Helling (1996) found that internal studies had shown that a failure to observe an ESD control measure in their facility typically resulted in about 1% of ESDS devices being stressed by ESD. About 10% of stressed devices resulted in a defect or failure. He therefore assumed an ESD failure rate of 0.1% per ESD control fault. He calculated on this basis failure rates for five production lines. He added the cost of repair of ESD failures. He found that 60% of failures were found at PCB test, and 30% at system test. Ten percent of failures were found at the customers site. His calculations were compared over two business years. The repair costs at each stage was calculated as follows:

$$\text{Stage repair cost} = \text{number of product} \times \%\text{failures} \times \text{ESD failure rate } (0.1\%) \\ \times \text{ repair cost per item}$$

An example summary of some of Helling's results for a manual process handling 80 000 products per year is given in Table 9.1. Two interesting things can be seen from this example. First, the cost of failures increases as the failure is discovered at later stages in production. Second, the most expensive cost is that of failures at the customers site. These characteristics are likely to be typical of most cost of product failure profiles.

Helling estimated the repair costs for other processes in the same way and was then able to add these to estimate the cost of ESD damage related repairs to the business. He then estimated the cost of ESD control measures required for an example facility in terms of packaging, ESD control equipment, training, audit, and other items for each business year under consideration. The ROI could then be calculated as the cost/benefit ratio of expenditure on ESD control compared to anticipated saving on ESD failures. For the two business years considered he found ROI values of 3:1 and 11:1, respectively. He commented that ESD protection faults led to the product being more expensive than necessary, and this alone, aside from the loss of customer reputation that results from defects and failures, cost the business more than ESD protection measures.

Other workers have reported high return on investment figures. Danglemayer (1999) reported in some of his case studies ROI as high as 185% and even 950%. In another case study, an investment of $1000 resulted in a saving of $6 000 000. He concluded that using ESD controls can lower operating costs with a ROI of up to 1000% while simultaneously enhancing product quality and reliability.

Table 9.1 Costs of repair of ESD damage in a manual process (Helling 1996).

Stage	Failures found at this stage (%)	ESD failure rate	Stage item repair cost (DM)	Total stage repair cost (DM)
PCB test	60	0.1%	100	4800
Final test	30		400	9600
Customer's site	10		2000	16 000
			Total cost to business	30 400

9.5.5 Optimizing an ESD Control Program

Establishing and maintaining a good ESD control program requires investment in time and resources. A key to achieving a good return on investment is acting through knowledge and understanding. Each action should be taken and ESD control measures should be applied for a reason addressing a need. The need may be to address an ESD risk or improve the effectiveness or efficiency of the ESD control process or improve value (e.g. in customer perceptions or compliance with a standard). Improving cost effectiveness of an ESD control program is likely to require balancing trade-offs between costs, for example in equipment purchases, documentation time, training, and compliance verification (Smallwood et al. 2014).

ESD controls applied without knowledge and understanding can often be unnecessarily expensive or make little real contribution to ESD control. ESD control equipment often works as part of a system with other equipment (see Section 4.7.8.2). If the system is not well understood, another part may be omitted or incorrectly specified. The result may be that the system may be compromised or even rendered ineffective.

Investment is often thought of in terms of ESD control equipment and materials, but it also needs resources such as the following:

- Planning, development, and implementation time
- Documentation time
- Selection and qualification of ESD control equipment, packaging, and materials
- Purchase, installation, and commissioning of EPA equipment and materials
- Development and regular provision of an ESD training program
- Development and regular execution of a compliance verification program
- Product failure detection, tracking, and analysis

All of these costs are really investments that should together provide a return on the investment. ESD control applied without knowledge and understanding is akin to financial investment applied without knowledge and understanding. Either can result in expenditure with little return. Furthermore, the investment in ESD control should be appropriate to the value of the product, market needs, and consequences of an ESD-related failure. A low-cost throwaway consumer product with unimportant consequence of failure may merit minimal ESD control investment. A high-cost high-reliability product with a high cost of failure would merit larger investment.

Many of the costs are inter-related. For example, a complex ESD control program using a variety of control techniques and equipment and different implementation in different areas is also likely to have more complex training and compliance verification needing extensive documentation. Each variation on ESD control equipment or procedure needs documentation, training, and compliance verification.

Implementation of different ESD control measures in different areas can sometimes lead to equipment savings. It may also lead to confusion in workers who enter the different areas, and problems with maintaining compliance. It can also lead to conflict with customers who audit the areas, especially if unconventional ESD control measures are used or common standard ESD controls are omitted, as they may believe the implemented control measures are inadequate or incorrect.

In contrast, a simple standardized ESD control program implementing the same control measures in different areas may lead to reduced training and compliance verification needs and costs and help prevent operator confusion. It may also reduce the risk and cost of conflict with customers or auditors.

Against this, equipping EPAs with the same equipment, whether it is necessary for ESD risk control or not, is likely to increase equipment and compliance verification costs. Any additional equipment must of course be regularly tested and verified.

When optimizing an ESD control program, it is essential to analyze and understand the ESD risks that are present in the processes and facilities. Without this, ESD controls may be implemented that are not necessary or risks may be present that are not addressed. ESD risks are often not obvious or well understood. Adequate skills and experience are required for successful analysis of these. Implementation of a standard will address the standard well-known ESD risks but may not cover more unusual ESD risks (Gaertner 2007).

With a little knowledge and understanding, the cost of ESD control can be high and the effectiveness low (Smallwood et al. 2014). This is perhaps most likely to occur when the low ESD withstand voltage devices are handled. With a high level of knowledge and understanding, the cost of ESD control can be reduced and the effectiveness maximized. Investment in high-level training in the principles and practice of ESD control for the ESD coordinator and other personnel working on development of the ESD control program can be worthwhile if not essential. Strategies for optimizing the ESD control program are further discussed in Section 10.7.

9.6 Evaluation of Compliance of an ESD Control Program with a Standard

9.6.1 Two Steps to Compliance Evaluation

There are arguably two steps to evaluating compliance of an ESD control program with a standard. First, the ESD program documentation must be evaluated for compliance with the standard. The second step is to evaluate the ESD program in practice for compliance with the ESD program documents. If both steps find compliance, then compliance with the standard is achieved.

9.6.2 Using Checklists to Evaluate Compliance of Documentation with a Standard

Compliance of an ESD control program documentation with a standard can be evaluated with the aid of a check list of the requirements of the standard. The check list is compiled by going through the standard in detail, noting every requirement.

The following tables are an example of a check list compiled in this way, with reference to the IEC 61340-5-1:2016 standard. For compliance, all the items in the tables should be addressed by the ESD program plans. The compliance evaluation response can be in terms such as “satisfactory,” “major non-compliance,” “minor non-compliance,” or “not applicable” to suit the user.

The IEC 61340-5-1 states under the heading "ESD Coordinator" that "A person shall be assigned by the organization with the responsibility for implementing the requirements of this standard including establishing, documenting, maintaining, and verifying the compliance of the program." This translates into the checklist of Table 9.2.

Table 9.3 gives a checklist of the broad requirements of the ESD Control Program Plan. The detail of compliance of the Compliance Verification Plan is covered in Table 9.7. As an example, the Compliance Verification Plan might be counted as "defined" if some elements of compliance verification are covered in some way, e.g. testing of wrist straps. That does not mean it is necessarily considered adequate – further requirements are tested in Table 9.7. Table 9.4 covers the requirements for "tailoring" given by the standard. This does not mean that some tailoring must be specified – if none is needed, none needs be documented.

Table 9.2 ESD Coordinator duties.

Requirement	Compliance evaluation	Notes
The organization must assign an ESD coordinator		
Stated responsibilities of ESD coordinator include		
Establishing the program		
Implementing the requirements of the standard		
Documenting the program		
Maintaining the program		
Verifying the program		

Table 9.3 ESD control program plan – topics covered.

Requirement	Compliance evaluation	Notes
ESD training plan is defined		
Product qualification plan is defined		
Compliance Verification plan is defined		
Grounding/bonding systems are defined		
Personnel grounding is defined		
EPA requirements are defined		
Packaging systems are defined		
Marking requirements are defined		
The ESD Control Program Plan is applied to all relevant aspects of the organization's work		

Table 9.4 Tailoring.

Requirement	Compliance evaluation	Notes
Evaluation of applicability of each requirement is adequate.		
Documentation of tailoring decisions is adequate.		
Tailoring statements cover situations where limits of the standard are exceeded.		

Table 9.5 Content of the ESD training plan.

Requirement	Compliance evaluation	Notes
Define all personnel that are required to have training.		
Initial awareness and prevention training to be provided before personnel handle ESDS devices.		
Recurrent training is required.		
The type of training is defined for all relevant personnel.		
Frequency of training is defined for all relevant personnel.		
There is a requirement for maintaining training records.		
The location where records are kept is adequately documented.		
The training methods used are adequately documented.		
The methods used to ensure comprehension and training adequacy are documented.		

Table 9.5 covers the required content of the ESD Training Plan in detail, and Table 9.6 covers the detailed content of the ESD Control Product Qualification Plan. Table 9.7 covers in more detail the required content of the Compliance Verification Plan.

Table 9.8 concerns the definition of grounding and bonding systems. Suitable grounding methods must be defined, but not all the methods in this table must be used in all facilities. Usually only one will be sufficient. For example, many facilities will use the electrical protective earth as ESD earth (ground).

Table 9.6 ESD control product qualification plan.

Requirement	Compliance evaluation	Notes
All ESD control items selected for use are qualified in some specified way.		
The technical requirements to be verified are adequately defined.		
The pass criteria are adequately defined.		
Any test methods used are adequately defined and documented.		
ESD control items not listed in the standard but considered to be part of the ESD control program are qualified		

Table 9.7 Compliance verification plan.

Requirement	Compliance evaluation	Notes
The technical requirements to be verified are adequately defined.		
The pass criteria are adequately defined.		
The frequency of verifications is adequately defined.		
All test methods used are adequately defined and documented.		
If test methods are nonstandard, correlation with standard measurements is adequately documented.		
Test methods used to verify items not covered in the standard are documented with corresponding test limits.		
Compliance verification records are required to be established.		
Compliance verification records are required to be maintained.		
Test equipment is capable of making the required measurements		

Table 9.8 Grounding and bonding systems.

Requirement	Compliance evaluation	Notes
It is required that all conductive and dissipative equipment are connected to ground or each other.		
At least one of the following are defined as "ground."		
Grounding via electrical protective earth		
Functional ground		
Equipotential bonding		

Table 9.9 Personal grounding.

Requirement	Compliance evaluation	Notes
All personnel are required to be grounded.		
When personnel are seated, they are required to wear a wrist strap.		
Personnel using footwear-flooring grounding must wear footwear on both feet.		
Resistance to ground and body voltage criteria are both addressed.		

Table 9.9 concerns personal grounding. The detailed pass criteria given in current standards 61340-5-1:2016 and ESD S20.20-2014 are given in Chapter 6, in this case in Table 6.5. Table 9.10 gives general requirements for EPAs. Requirements for EPA equipment such as floors, bench mats, and chairs are given in in Chapter 6, in this case in Table 6.6.

Table 9.11 covers the requirements for ESD protective packaging. The classification of packaging materials and requirements of IEC 61340-5-3:2015 are given in in Chapter 6, in this case in Tables 6.10 and 6.11. Table 9.12 covers the requirements for marking for ESD control purposes.

9.6.3 Evaluation of Compliance of a Facility with the ESD Control Program

The task of evaluation of compliance of a facility with its ESD Control Program Plan should be specified in the Compliance Verification Plan part of the documentation. Check lists can also be used to aid this process. Every requirement of the ESD Control Program Plan should

Table 9.10 General requirements for EPAs.

Requirement	Compliance evaluation	Notes
Handling of unprotected ESDS devices must always be within an EPA.		
The boundaries of the EPA must be clearly identified.		
Access to the EPA is limited to trained personnel or personnel escorted by trained personnel.		
All nonessential insulators must be removed from locations where unprotected ESDS devices are handled.		
ESD threat evaluated such that field $<5\,\text{kV}\,\text{m}^{-1}$.		
ESD threat evaluated such that 2 kV potentials are kept >30 cm from ESDS devices.		
ESD threat evaluated such that 125 V potentials are kept >2.5 cm from ESDS devices.		
Where fields or potentials exceed limits, method of mitigating the ESD risk is defined.		
Where conductors come into contact with ESDS devices and cannot be grounded, the voltage difference between the conductor and ESDS devices is reduced below 35 V.		

Table 9.11 Packaging.

Requirement	Compliance evaluation	Notes
Use of packaging in accordance with customer contracts, purchase orders, drawing, or other documentation is required where it is defined.		
(61340-5-1) Packaging requirements must be defined for ESDS devices not covered by customer contracts or documentation, in accordance with the packaging standards.		
(61340-5-1) Packaging where required is defined for all material movement within EPAs, between EPAs, between job sites, field service operations, and to the customer.		

Table 9.12 Marking for ESD control purposes.

Requirement	Compliance evaluation	Notes
Marking according to customer contracts, purchase orders, drawing, or other documentation is required where it is defined.		
Evaluation of need for marking is required where marking is not defined by customer contract, purchase orders, drawing, or other documentation.		
If marking is required, marking requirements must be documented in ESD Control Program Plan.		

be tested in the Compliance Verification Plan and hence in evaluation of compliance of the facility with the plan.

9.6.4 Common Problems

External auditors often audit and comment on an ESD Program based on what is practiced in their own facility. This is a mistake – an organization's facility and practices should always be audited against its own documented ESD Control Program Plans. So, if an organization has a well-documented ESD Control Program Plan that is compliant with a standard and complies with the plans in practice, this can go a long way in preventing adverse comments from customers who are used to doing things in a different way.

Tailoring allows the organization to do things in a different way to that required by the standards. But, if the tailored practice is not adequately documented, it represents a non-compliance. Good documentation of a tailored ESD control measured, with documentation of the technical rationale and any tests supporting its implementation, is necessary to defend the user against accusations of non-compliance. The detailed documentation may be in a separate report cross referenced by the ESD Control Program Plan.

References

Bellmore D. G. (2004) Paper 4A.6. Characterizing Automated Handling Equipment Using Discharge Current Measurements. In: *Proc EOS/ESD Symp. EOS-26*. Rome, NY, EOS/ESD Association Inc.

Brandt M T. (2003) What does ESD really cost? Available from: http://circuitsassembly.com/cms/images/stories/pdf/0306/0306esd.pdf [Accessed 6th March 2019]

Dangelmayer, T. (1999). *ESD Program Management*, 2e. Clewer. ISBN: 0-412-13671-6.

Downing, M.H. (1983). ESD Control Implementation and Cost Avoidance Analysis. In: *Proc EOS/ESD Symposium EOS-5*, 6–11. Rome, NY: EOS/ESD Association Inc.

EOS/ESD Association Inc. (2014). ANSI/ESD S20.20–2014. ESD Association Standard for the Development of an Electrostatic Discharge Control Program for – Protection of Electrical and Electronic Parts, Assemblies and Equipment (excluding Electrically Initiated Explosive Devices). Rome, NY, EOS/ESD Association Inc.

EOS/ESD Association Inc. (2015). ESD TR17.0–01-15. *Technical Report for ESD Process Assessment Methodologies in Electronic Production Lines – Best Practices used in Industry*. Rome, NY, EOS/ESD Association Inc.

EOS/ESD Association Inc. (2016a) ESD SP10.1–2016. *Standard practice for protection of Electrostatic Discharge Susceptible Items–Automated handling Equipment (AHE)*. Rome, NY, EOS/ESD Association Inc.

EOS/ESD Association Inc. (2016b) ESD Association Electrostatic Discharge (ESD) Technology roadmap – revised 2016. Available from: https://www.esda.org/assets/Uploads/docs/2016ESDATechnologyRoadmap.pdf [Accessed: 10th May 2017]

Gaertner R. (2007) Do We Expect ESD-failures in an EPA Designed According to International Standards? The Need for a Process Related Risk Analysis. Proc. EOS/ESD Symposium EOS-29 Paper 3B.1 pp. 192–197

Gaertner R., Stadler W. (2012) Paper 3B.5. Is there a Correlation Between ESD Qualification Values and the Voltages Measured in the Field? In: *Proc. EOS/ESD Symposium EOS-34*. Rome, NY, EOS/ESD Association Inc.

Gumkowski G., Levit L. (2013) EOS-35 Paper 7B4. A New Look at the Financial Impact of Air Ionization. *In: Proc. EOS/ESD Symposium*. Rome, NY, EOS/ESD Association Inc.

Halperin, S.A. (1986). Estimating ESD losses in the complex organisation. In: *Proc. EOS/ESD Symp. EOS-8*, 12–18. Rome, NY: EOS/ESD Association Inc.

Halperin S. A., Gibson R, Kinnear J. (2008) EOS-30 2B-2. Process Capability & Transitional Analysis. In: *Proc. EOS/ESD Symp*. Rome, NY, EOS/ESD Association Inc.

Helling, K. (1996). ESD protection measures – Return on investment calculation and case study. In: *Proc. EOS/ESD Symp. EOS-18*, 130–144. Rome, NY: EOS/ESD Association Inc.

Industry Council on ESD Target Levels (2010) White paper 2: A case for lowering component level CDM ESD specifications and requirements. Rev. 2.0. http://www.esdindustrycouncil.org/ic/en/documents/6-white-paper-2-a-case-for-lowering-component-level-cdm-esd-specifications-and-requirements [Accessed: 10th May 2017]

Industry Council on ESD Target Levels (2011) White paper 1: A case for lowering component level HBM/MM ESD specifications and requirements. Rev. 3.0. Available from: http://www.esdindustrycouncil.org/ic/en/documents/37-white-paper-1-a-case-for-lowering-component-level-hbm-mm-esd-specifications-and-requirements-pdf [Accessed: 10th May 2017]

International Electrotechnical Commission. (2015) IEC 61340–5-3:2015. *Electrostatics. Protection of electronic devices from electrostatic phenomena. Properties and requirements classifications for packaging intended for electrostatic discharge sensitive devices*. Geneva, IEC.

International Electrotechnical Commission (2016) IEC 61340–5-1: 2016. *Electrostatics – Part 5–1: Protection of electronic devices from electrostatic phenomena - General requirements*. Geneva, IEC.

International Rectifier. (2004). ESD Testing of MOS Gated Power Transistors. AN-986. http://www.infineon.com/dgdl/an-986.pdf?fileId=5546d462533600a40153559f9f3a1243 [Accessed: 10th May 2017]

Jacob P., Gärtner R., Gieser H., Helling K., Pfeifle R., Thiemann U., Wulfert F., Rothkirch W. (2012) EOS-34. Paper 3B.8. ESD risk evaluation of automated semiconductor process equipment – A new guideline of the German ESD Forum e.V. In: *Proc. EOS/ESD Symp.* Rome, NY, EOS/ESD Association Inc.

Lin N., Liang Y., Wang P. (2014) DOI: 10.1109/ICEPT.2014.6922951. Evolution of ESD process capability in future electronics industry. In: *Proc. 15th Int. Conf. Electronic Packaging Tech.* Bristol, England, IOP Publishing.

Paasi, J., Smallwood, J., and Salmela, H. (2003). EOS-25 Paper 2B4. New Methods for the Assessment of ESD Threats to Electronic Components. In: *Proc. EOS/ESD Symp*, 151–160. Rome, NY: EOS/ESD Association Inc.

Paasi, J., Salmela, H., and Smallwood, J.M. (2006). Electrostatic field limits and charge threshold for field-induced damage to voltage susceptible devices. *J. Electrostat.* 64: 128–136.

Smallwood J M. (2019) Can ElectroStatic Discharge Sensitive electronic devices be damaged by electrostatic fields? In: *Proc. Electrostatics 2019. J. Phys. Conf. Se. Vol. 1322 01 2015* Available from: https://iopscience.iop.org/article/10.1088/1742-6596/1322/1/012015/pdf [Accessed Oct. 2019]

Smallwood J., Paasi J., (2003) Assessment of ESD threats to electronic devices, VTT Research Report No BTUO45-031160

Smallwood J., Taminnen P., Viheriaekoski T. (2014) EOS-36. Paper 1B.1. Optimizing investment in ESD Control. In: *Proc. EOS/ESD Symp.* Rome, NY, EOS/ESD Association Inc.

Stadler W., Niemesheim J., Seidl S., Gaertner R., Viheriaekoski T. (2018) EOS-40 Paper 1B.4.The Risks of Electric Fields for ESD Sensitive Devices. In: *Proc. EOS/ESD Symp.* Rome, NY, EOS/ESD Association Inc.

Steinman A. (2010) EOS-32 Paper 3B3. Measurements to Establish Process ESD Compatibility. In: *Proc. EOS/ESD Symp.* Rome, NY, EOS/ESD Association Inc.

Steinman A. (2012) EOS-34 Paper 2B.4. Process ESD Capability Measurements. In: *Proc. EOS/ESD Symp.* Rome, NY, EOS/ESD Association Inc.

Swenson D. E. (2012) EOS-34 paper 3B.6. Electrical fields: What to worry about? In: *Proc. EOS/ESD Symp.* Rome, NY, EOS/ESD Association Inc.

Tamminen, P. and Viheriäkoski, T. (2007). EOS-29 Paper 3B3. Characterization of ESD Risks in an Assembly Process by Using Component-Level CDM Withstand Voltage. In: *Proc. EOS/ESD Symp*, 202–211. Rome, NY: EOS/ESD Association Inc.

Viheriäkoski T, Ristikangas P, Hillberg J, Svanström H, Peltoniemi T. (2012). Paper 3B.2. Triboelectrification of static dissipative materials. In: *Proc. EOS/ESD Symp.* Rome, NY, EOS/ESD Association Inc.

Wallash, A. (2007). EOS-29 Paper 2B8. A Study of "Soft Grounding" of Tools for ESD/EOS/EMI Control. In: *Proc. EOS/ESD Symp*, 152–157. Rome, NY: EOS/ESD Association Inc.

10

How to Develop an ESD Control Program

10.1 What Do We Need for a Successful ESD Control Program?

10.1.1 The ESD Control Strategy

An electrostatic discharge–sensitive (ESDS) device needs protection from potentially damaging electrostatic discharge (ESD) at all times. This can be done by adopting a simple overall ESD control strategy.

- Unprotected ESDS devices are handled only in an area where ESD risks are controlled to an acceptable level. This area is usually called an *electrostatic discharge protected area* (EPA).
- Outside EPAs, in unprotected areas (UPAs), the ESDS device is always protected within ESD protective packaging.

If this strategy is implemented and the ESD packaging and control measures used within the EPA are sufficient and appropriate, it's likely that the ESD program will be effective. Notice that this strategy immediately implies that an ESDS device is not removed from its ESD protective packaging unless it is within an EPA.

This chapter looks at how the ESD control program can be developed and documented according to the International Electrotechnical Commission (IEC) 61340-5-1: 2016 standard and determine the control measures that will be used in the EPA. ESD control equipment is discussed in more detail in Chapter 7. ESD protective packaging is discussed in Chapter 8. ESD measurements are covered in Chapter 11 and ESD training in Chapter 12.

10.1.2 How to Develop an ESD Control Program

Dangelmeyer (1990), based on his experience at AT&T, commented that developing, implementing, and managing a successful ESD program requires a system approach from product design to customer acceptance. A well-managed program can be much more effective than one well stocked with expensive supplies. He found that 12 critical factors form the basis of a successful ESD control program.

1. An effective implementation plan
2. Management commitment

The ESD Control Program Handbook, First Edition. Jeremy M Smallwood.

3. A full-time ESD coordinator to serve as a consultant and oversee the plan
4. An active committee to help implement the ESD program
5. Realistic requirements
6. Training for measurable goals
7. Auditing using scientific measures
8. ESD test facilities to qualify and test ESD control equipment
9. A communication program to keep people aware of the ESD issue and demonstrate progress
10. Systematic planning
11. Human factor engineering to take care of employee needs and render human error unlikely
12. Continuous improvement

Effective implementation of these factors will produce an effective and cost-effective ESD control program, and continuous improvement sustains the success. Dangelmeyer states that failure to continuously improve the process will translate into complacency and deterioration.

Effective ESD protection requires the following elements:

- Identification of ESDS device.
- Understanding in which processes and facilities they must be exposed and handled in an unprotected state. These areas will be EPAs. All other areas will be regarded as UPAs.
- Ensuring that within EPAs, ESD risks are adequately controlled. This is done by specification of appropriate ESD control equipment (e.g. wrist straps, bench mats) as well as removal of items and equipment that do not comply with ESD control requirements or could cause ESD risk (e.g. ordinary packaging materials).
- Ensuring that when outside EPAs, ESDS devices are protected from ESD damage (e.g. by enclosure within equipment or ESD protective packaging).
- Training of personnel in ESD-related matters.
- Regular checks and tests (Compliance Verification) to ensure that equipment is working and procedures are being observed correctly.
- Documentation of ESD control, training, and compliance verification procedures and requirements.

To specify and use appropriate and effective ESD controls, it is important to understand ESD control technology and the role of different items of ESD control equipment. Specification of equipment without this understanding can lead to an overly expensive and yet poorly effective ESD control program (Smallwood et al. 2014). The equipment specified may be unnecessary or may fail to function as a system in the absence of other items.

10.1.3 Safety and ESD Control

Safety considerations and local safety regulations must always be given top priority over ESD control.

One area in which apparent contention between safety and ESD control requirements often occurs is in handling ESDS devices that are in a powered state with high voltage present or that contain batteries that could give electric shock risk to personnel. Another area is where processes require the operator to use personal protective equipment against the effects of chemicals, high temperatures, or other risks.

Often, by careful evaluation of the risks and process steps, ways of working can be found that both maximize safety and minimize ESD risks. As an example, handling an unprotected ESDS device may be required while setting up for live testing with high voltages present. Using personal grounding with live voltages present could give electric shock risks. If it can be arranged that the handling of the ESDS device can be done before the high voltages are present, the safety risk is removed. Not touching an ESDS device when live voltages are present at the same time prevents ESD risk from charged personnel and prevents electric shock to the personnel! If it is necessary to work on the ESDS device in a live state, precautions such as use of insulating rubber gloves may be required to protect the person against electric shock risks. The same gloves may give some protection to the ESDS device against human body ESD at body voltages up to the breakdown voltage of the gloves. (If the person's electrostatic body voltage could exceed the breakdown voltage of the glove, an ESD risk could result, but more importantly the protection provided by the glove against shock risk is compromised.) When using rubber gloves in this way, an ESD risk could arise from electrostatic fields due to charging of the glove material. This is usually a lower risk but should be evaluated and if necessary countered in some way.

In practice, the effect of electric shock depends on the electrical current flowing through the body, rather than the source voltage present. The effects were summarized by Dalziel (1972) and become more hazardous as current flow level and duration of the shock increases. The sensitivity of the body to shock also depends on the body parts that are subjected to the current flow, and whether the source is direct current (DC) or alternating current (AC), and the frequency of AC current. Women are typically susceptible at lower current levels than men, and individuals vary greatly in their sensitivity. Other conditions such as presence of moisture can be influential. As current is increased, the effects start with perception of the current (tingling or warmth) and can include at higher levels painful shocks, stopping of breathing, inability to let go, burns, ventricular fibrillation (cessation of heart action), and immediate or delayed death.

Where safety guidelines or regulations do not cover a situation, the guidelines in (Table 10.1) are often given in electrical textbooks, e.g. Nave and Nave 1985. Safety can be evaluated by comparing possible electrical currents with these thresholds.

Table 10.1 Electrical current physiological effects for AC 60 Hz current.

Current level (mA)	Physiological effect
1	Threshold of perception
5	Maximum harmless current
10–20	Sustained muscular contraction "let go current"
50	Pain, possible fainting, exhaustion, heart, or respiratory effects
>100	Ventricular fibrillation, death possible

10.2 The EPA

10.2.1 Where Do I Need an EPA?

A major contribution to ESD protection may be made by minimizing handling of ESDS devices in an unprotected state. Where handling is necessary, some form of ESD control is required – meaning the ESD protected areas.

Any area in which unprotected ESDS devices may be handled in an unprotected state needs to be an EPA in some form. That does not necessarily mean it needs to have all the possible range of ESD control measures and equipment. To be effective, the EPA must have the following features:

- A known boundary in which ESD control measures are applied
- ESD control measures addressing the significant ESD risks to an acceptable level

The equipment and practices are discussed in more detail in the following sections.

The first step in determining ESD hazards is to identify where ESDS components, printed circuit boards (PCBs), or other items (ESDS device) are handled in an unprotected state in the facility. Many semiconductor devices and some passive devices are ESD susceptible. In general PCBs, modules and similar assemblies should be considered ESD susceptible if they contain ESDS devices and are not protected from ESD by being fully contained within an enclosure or some other factor. Even PCBs or modules that are fully contained within an enclosure may have some susceptibility to ESD that may occur via a connector or flying leads.

10.2.2 Boundaries and Signage

A clear boundary, obvious to all personnel approaching the entrance to the EPA, is necessary. If this is not present, personnel will not know whether they are inside or outside the EPA and where the ESD control measures are to be applied (a simple control measure is to keep untrained personnel out of the EPA and reduce handling of unprotected ESDS devices to a minimum).

Modern ESD standards require that the EPA is marked by signage visible to personnel before they enter the EPA.

EPA boundaries should be carefully chosen to include all areas where ESD control is required (because unprotected ESDS devices are to be handled) but preferably exclude areas or processes where ESD control is not required or not possible. Including inappropriate processes within an EPA can lead to difficult noncompliance issues and confusion over acceptable ESD practices. Nevertheless, EPAs and UPAs should be designed for operator convenience, as convenient practices will assist compliance and inconvenience can encourage noncompliance.

10.3 What Are the Sources of ESD Risk in the EPA?

The next step is to identify the potential ESD sources in each process. The most common of these are

- A charged person touching the ESDS device
- A charged metal or conductive object, tool, or another item touching the ESDS device
- The ESDS device becoming charged and touching a conductive item (e.g. metal part or equipment)

Electrostatic fields are not usually in themselves damaging (with a few exceptions). However, electrostatic fields help set up the conditions in which ESD can occur because any isolated conductor (e.g. metal parts or the device itself) within the electrostatic field will attain a voltage. If two conductors touch (or become sufficiently close to each other) within an electrostatic field and at least one is isolated (i.e. not grounded), then ESD will occur between them due to voltage induced on the isolated conductor.

A particularly damaging form of ESD can occur when an ESDS device contacts a high-conductivity (low-resistance, e.g. metal) item. An ESD event occurring in this circumstance can have a short-duration high discharge current due to the low resistance and inductance of the discharge circuit between them. The susceptibility of the ESDS device to this type of event is characterized by its charged device model (CDM) withstand voltage. This type of damage can often be avoided by ensuring that, where possible, the device instead contacts a higher resistance ($>10^4\ \Omega$) material.

10.4 How to Determine Appropriate ESD Measures

10.4.1 ESD Control Principles

The first step in determining the appropriate ESD control measures is to define the boundary of each EPA. These must then be marked and signed so that personnel can easily identify which areas are EPA and which are uncontrolled areas. Within the EPAs, the ESD control measures can then be determined.

- Personnel handling ESDS devices must be grounded so that they cannot be at a high enough voltage to damage the ESDS device that they touch. This is normally accepted to mean that the body voltage on personnel must not exceed the human body model (HBM) ESD withstand voltage of the most sensitive ESDS device. For an ESD program handling 100 V HBM devices, the maximum voltage allowed on a person handling ESDS devices must be less than 100 V.
- Any metal or conductive items that contact ESDS devices must be grounded where possible to ensure that they are not charged.
- Sources of high electrostatic fields (e.g. charged insulators or equipment that generates external electrostatic fields) must be kept far enough away from ESDS devices not to risk inducing high voltages on the device (or on isolated conductors that may contact the ESDS device).

An ESDS device will often be in an isolated (nongrounded) state and must make contact with metal objects or other components during the process. ESDS devices are not normally grounded as an ESD protection measure. Because of this, it will normally have some level of residual voltage on it. If the ESDS device is susceptible to charged device ESD and the voltage difference is sufficient, a risk of damage could occur, so the act of grounding an ESDS device to a low-resistance conductor often carries a charged device ESD risk. This risk can be minimized by contacting the ESDS device through a resistive material (see Section 10.3) rather than a low-resistance material such as a metal. Where ESDS devices must contact conductive items (including other devices, tools, or PCBs), the ESD risks should be evaluated, and ESD prevention measures may be required e.g. to reduce the voltages possible on the ESDS device and conductors or increase the resistivity of the conductor.

The ESD control measures required within the EPA will be selected to address the ESD risks found. In a typical EPA with manual processes, these will include

- Grounding of personnel handling ESDS devices either by wrist strap or by ESD control footwear and flooring. (Occasionally other means e.g. personal grounding garments may be used.) Suitable ground points for wrist straps must be provided.
- Where personnel are to be seated, ESD control seating is used, and personal grounding via wrist straps is used.
- Specification of grounded ESD control materials for surfaces on which unprotected ESDS devices could be placed, e.g. workstations and carts.
- Insulating materials are evaluated to decide whether they are essential to the process or not.
  - Nonessential insulators should be kept well away from unprotected ESDS devices. The ESD program may specify that they are excluded from the EPA.
  - Insulators that are essential to the process must be evaluated to determine whether any ESD risk arises in their use. If an ESD risk is found, some means of ameliorating it must be devised. Often this may be done by using an ionizer.
- Ionizers may be specified for control of voltages on essential insulators or nongroundable isolated conductors
- ESD control packaging may be selected to protect ESDS devices when outside the EPA or to prevent ESD risks arising from packaging materials used inside the EPA.
- ESD control garments may be specified to reduce the risk of electrostatic fields arising from clothing worn by operators.

All the equipment and processes in the EPA should be designed with these requirements in mind. Well-designed common ESD control equipment normally fulfills these requirements.

Other control items such as ESD control tools may also be specified. These control measures are further discussed in Section 10.5.13.4.

Sometimes ESD risks are identified that are not addressed by the usual ESD control measures and equipment. In this case, special ESD control measures and precautions must be devised.

Operator safety should always be considered when developing an ESD program (see Section 10.1.3).

10.4.2 Select Convenient Ways of Working

If ESD control procedures are inconvenient, it's likely that they may not be properly and consistently followed, especially when in a hurry or under pressure. If, on the other hand, they provide the easiest and most convenient way of working, it is likely they will be consistently followed.

As an example, in a Stores area it may be sufficient to provide a simple single workstation for occasional inspecting or counting incoming ESDS products. If the operator does not need to sit and is highly mobile in their work, it may be most convenient to provide them with ESD control footwear and a local ESD control floor mat around the workstation for personal grounding. With no ESD control seat in use and footwear and flooring selected to give adequate grounding, no wrist strap is required. Merely walking onto the ESD control mat establishes the connection to ground.

If instead the operator were required to connect a wrist strap for personal grounding as they enter and leave the workstation, they may occasionally forget in the heat of busy activity.

In an otherwise similar facility, it could be that the operator needs to sit for their work for longer periods at the workstation. In this case, an ESD control seat should be provided. The ESD control floor must now withstand the repeated rolling action of the seat, and the seat can be selected to be conveniently grounded through its wheels and the floor. The ESD floor area must be large enough to ensure that the operator and the seat remain on the mat during use. A wrist strap must now be provided for effective grounding of the person during seated work, as grounding through footwear and flooring is not normally relied up for seated operations.

10.5 Documentation of ESD Procedures

10.5.1 What Should the Documentation Cover?

For long-term effective ESD control, four aspects should be documented.

- The ESD Control Program Plan, specifying the equipment and other ESD control measures to be used.
- An ESD Training Plan, specifying necessary ESD-related training requirements.
- An ESD Control Product Qualification Plan, specifying the criteria and requirements by which ESD control equipment and products will be selected for use in the ESD control program.
- A Compliance Verification Plan, specifying the checks and test program to be used.

When writing the ESD control program, bear in mind that there are usually at least two audiences who must clearly understand the documents. First, the personnel who must work with the documents, using them to maintain and implement the ESD control program, must understand them.

Second, auditors (first, second, or third party) may need to audit the ESD control program and check that what is happening in practice reflects the documented ESD control requirements. These auditors may need also to check that the documents are compliant with the selected standard. It can save a lot of time and effort for auditors if the documentation is laid out in a way that quickly shows that these requirements are met. If parts of the ESD control program are documented in separate documents or procedures or are provided in a different form (e.g. paper or intranet-based documents), suitable cross-referencing can be used to make this clear.

10.5.2 Writing an ESD Control Program Plan That Is Compliant with a Standard

Often it is advantageous to develop an ESD control program that is compliant with an ESD control standard such as IEC 61340-5-1 or American National Standards (ANSI)/ESD S20.20 (EOS/ESD Association Inc 2014). While compliance with an ESD control standard is not a necessary part of effective ESD control, it can be an effective way of demonstrating to customers that ESD control is taken seriously and has reached a certain level of benchmarked professionalism. A summary of the main requirements of the 61340-5-1 and S20.20 standards (which are almost identical) is given in Chapter 6. New versions of these standards will be published from time to time. These may have some differences from the cited versions but are expected to remain broadly similar. A copy of the chosen standard should be kept for reference.

When working to a standard, it is usually worth selecting the standard that is most likely to be recognized by the market or geographical area in which the products will be marketed. It is worth noting that it is easily possible to write an ESD control program that complies with both 61340-5-1 and 20.20 standards.

One way to commence documentation that is to be compliant with a standard is to first compile a checklist of requirements of the standard.

Using the same or similar section headings as given in the standard can give a compliant document structure that is also easily seen by an auditor to address the requirements of the standard. Table 10.2 gives an example of a three-level heading structure based on the headings of IEC 61340-5-1: 2016. The headings can be modified as wished to add clarity for the intended readers.

Table 10.2 An example of a document heading structure derived from the headings of IEC 61340-5-1: 2016.

Introduction
Scope

Terms and definitions
Personal safety

ESD control program
- ESD control program requirements
- ESD coordinator
- Tailoring ESD control requirements

ESD training plan

ESD control product qualification

Compliance verification

ESD control program technical requirements
- ESD ground
- Personal grounding
- ESD protected areas (EPAs)
  - Handling ESDS device and access to the EPA
  - Insulators
  - Isolated conductors
  - ESD control equipment
- ESD protective packaging
- Marking of ESD related items

References

This was used to produce the example of Appendix A.

Once the heading structure is in place, the text under each heading can be added according to the requirements for ESD control in the facility and processes, in compliance with the requirements of the standard. In some cases, it can be appropriate to include text based on that of the standard being followed to ensure compliance with the standard. The following sections give an example of how the basis of an ESD program based on IEC 61340-5-1: 2016 can be drafted from the heading structure. An outline of the content of each heading is given. The requirements of 61340-5-1: 2016 are discussed in more depth in Chapter 6. Once the document content is drafted, the headings can, if necessary, be reorganized to better reflect the content and for clarity.

The ESD program documents should be written to conform to the organization's internal quality system procedures.

10.5.3 Introduction Section

An introduction section can include anything that will be useful to the reader in introducing the document and ESD control program. A brief description of typical ESDS handled in the facility, and the main ESD risks, can be useful.

10.5.4 Scope

The section summarizes the scope of the ESD program. This should specify the areas and activities in which ESDS devices are handled and will be covered by the ESD program. It should cover all applicable parts of the organization's work. The standard lists "activities that: manufacture, process, assemble, install, package, label, service, test, inspect, transport, or otherwise handle" ESDS devices.

The scope should also specify the range of ESD susceptibility of the components covered by the ESD program. The default range from the standard gives "electrical or electronic parts, assemblies, and equipment with withstand voltages greater than or equal to 100 V HBM, 200 V CDM, and 35 V for isolated conductors." If the devices handled are limited to an ESD withstand voltage range higher than this default, the default can normally be accepted. If devices of lower ESD withstand voltage are handled, a withstand voltage range should be specified that includes the lowest withstand voltage device to be handled. For example, if 60 V HBM or 180 V CDM devices are handled, the ESD withstand voltage range should be specified to include these, e.g. greater than or equal to 50 V HBM and 150 V CDM.

10.5.5 Terms and Definitions

The terms and definitions used in the adopted standard should be used. Using different definitions for terms specified in the standards can lead to great confusion. It can also lead to, for example, purchase of incorrectly specified equipment if the term is used in equipment specification. Where terms are not defined in the standards, they should be clearly and carefully defined in this section.

10.5.6 Personal Safety

It is always worth stating clearly that compliance with local safety laws and regulations is essential. These requirements take precedence over standard ESD requirements. Specific safety issues could be covered here.

10.5.7 ESD Control Program

10.5.7.1 ESD Coordinator

61340-5-1 and S20.20 require that an ESD coordinator is assigned with responsibility for the ESD control program. 61340-5-1 lists these responsibilities as "implementing the requirements of this standard including establishing, documenting, maintaining, and verifying the compliance of the program." It could be easiest just to quote these requirements. Having an ESD coordinator is a good way of ensuring that there is a recognized point of contact for all personnel who need help in compliance with ESD control matters.

It's not necessary to name the person here – doing so could just make it necessary to update the document if they leave the role. Just make it necessary to appoint one and define their role. The ESD Coordinator will normally have to specify ESD control equipment, procedures, and training and so should have a good understanding of ESD control practice. An allowance should be made for specialist training and update for the role as well as time and resources necessary for the role.

10.5.7.2 Tailoring ESD Control Requirements

An ESD program may, or may not, need to tailor the requirements of the ESD program so that they lie outside the requirements of the standard. Either way, it is worth specifying in the ESD Control Program Plan that the process is allowed and how it is to be done and documented.

Tailoring of ESD control equipment requirements is also discussed in Section 10.5.13.4.

10.5.8 ESD Control Program Plan

In 61340-5-1:2016 this heading gives the requirement that an ESD control program is written and that it must address ESD training, product qualification, compliance verification, grounding/bonding systems, personnel grounding, EPA requirements, packaging systems, and marking. It also notes that the plan is the principal document for implementing and verifying the program, the program must conform to internal quality system requirements and apply to all applicable facets of the organization's work. Providing all these are addressed in the ESD Control Program Plan, this heading is not needed within it.

10.5.9 ESD Training Plan

The ESD Training Plan should define the personnel who need ESD training, the type of the training they need, and when it is to be given. It should define the need for training records, and where they are to be kept. Tests used to evaluate the effectiveness of training should also be covered.

One way of doing this is to develop a training matrix that lists the types of personnel that require training in the left column of a matrix table and the types of training in columns across the page. The training appropriate to each type of trainee can then be indicated in each cell of the table (see Table 12.1).

ESD training is discussed in more detail in Chapter 12.

10.5.10 ESD Control Product Qualification

The ESD control products that are to be used within the ESD Control program must be qualified to demonstrate that they will do the job intended, for the duration of their required working life. The way that this is done should be documented under this heading. The standards give requirements for some basic tests for the standard ESD control equipment.

The tests and pass criteria required will vary with the type of product being qualified (see Chapter 7). In the simplest cases, the tests and pass criteria may replicate the measurements used in compliance verification. For example, qualification of a wrist band or ground cord would be based on a simple end-to-end resistance measurement. The same measurement could be made in compliance verification, although in practice a combined system test is used of the wrist band and cord as it is worn by the operator.

In other cases, additional tests are made in qualification that are not repeated in compliance verification. For example, a workstation bench mat would be qualified by measurement of the resistance from the surface to a groundable point (earth bonding point). This measurement is related to the compliance verification system test of bench surface resistance to ground. But we often also want to know what is the surface resistance characteristic of the bench mat material, as too low resistance could give a charged device ESD risk. So, in selecting and qualifying a bench mat, we do a point-to-point resistance measurement. We do not normally repeat this in compliance verification tests, as material resistance changes are adequately tested in the resistance to ground system test.

Qualification of ESD control products can be evaluated based on data sheet values (if you trust them), in-house tests, or third-party evaluations. Taking the trouble to test things in-house can be useful in eliminating products that do not behave as the data sheets specify and in understanding how well (or not!) ESD control products work (Smallwood et al. 2014).

10.5.11 Compliance Verification Plan

This heading must define the routine tests that will be made to check compliance of the ESD control products and equipment used in the ESD control program. Everything that is specified must be tested. The test methods used, the pass criteria, and how often the tests will be made must be documented. If there is ESD control equipment not specified by the standard, compliance verification test methods, pass criteria, and frequency of testing must also be defined for these. One way to conveniently specify all these things is in the form of a table, such as the examples of Tables 10.3 and 10.4.

The standards give reference to test method standards that must usually be applied for testing ESD control equipment. Normally these will be documented by the user as simple test procedures that can easily be followed in compliance verification. Chapter 11 gives examples of these test methods. Often the test procedures are in other documents – these should be referenced in the Compliance Verification Plan.

If the user wants to use test methods that are different from those specified by the standard, it must be shown that the results of the test method used correlate with the standard test methods. A reference should be included here to the test reports that demonstrate correlation.

The need to keep compliance verification records must be defined.

Table 10.3 An example of specification of requirements for personal grounding equipment with test method and test frequency.

ESD control item	Test method	Pass criteria	Test frequency
Wrist strap, while worn, resistance from body to groundable point	TM1	<35 MΩ and >750 kΩ indicated by green light on tester	On each entry to EPA
ESD control footwear, while worn, hand to metal foot plate	TM2	<10 MΩ and >750 kΩ indicated by green light on tester	On each entry to EPA
ESD control footwear worn while standing on ESD control floor, hand to ground.	TM3	<1 GΩ	6 monthly
Body voltage generated while wearing ESD control footwear and walking on ESD control floor	TM4	100 V peak body voltage	6 monthly

Table 10.4 An example of EPA equipment pass requirements, test methods and test frequency summarized as a table.

ESD control item	Test method	Pass criterion	Test frequency
Workstation or rack surface	TM5	$R_g < 1\,\text{G}\Omega$	6 monthly
Floor	TM5	$R_g < 1\,\text{G}\Omega$	6 monthly
Seat	TM5	$R_g < 1\,\text{G}\Omega$	6 monthly
Wrist strap earth bonding point (temporary)	TM6	$R_g < 5\,\text{M}\Omega$	Daily before use
Wrist strap earth bonding point (permanent)	TM6	$R_g < 5\,\text{M}\Omega$	Monthly
ESD garment	TM7	$R_{p\text{-}p} < 100\,\text{G}\Omega$	6 monthly
ESD tool	TM8	$R_g < 1\,\text{G}\Omega$	6 monthly
Ionizer decay time	TM9	<5 s	6 monthly
Ionizer offset voltage	TM8	±35 V	6 monthly

10.5.12 ESD Program Technical Requirements

10.5.12.1 Documenting Technical Requirements

In the ESD program document, these requirements could be specified in any one of several places (e.g. in this section or under product qualification and compliance verification). It is probably advisable not to repeat the requirements in more than one place. If they are later changed, one of the places they are specified could be missed, and this could lead to conflicting or ambiguous requirements being specified in the document.

10.5.12.2 ESD Ground

The ESD control grounding method that is to be used (i.e. mains electricity protective earth, functional earth, or equipotential bonding point) must be specified so that it is easily recognized by the user. All ESD control items will be connected to a common ground point using this method. Grounding should comply with requirements of the National Electrical Code. Any requirements of this that apply to the ESD ground can be specified or referenced under this heading, if appropriate.

It is worth specifying that all conducting items and materials that contact ESDS devices must be connected to this common ground point if possible. Any conductor that contacts ESDS devices but for some reason cannot be electrically connected to ground must be treated as an isolated conductor.

It is not good practice to have two different ground points in an EPA that are not connected together as they could be at different voltages. It follows that if mains electricity protective earth is present in the EPA, this should be used as, or connected to, ESD ground if possible.

10.5.12.3 Personal Grounding

Personal Grounding Equipment

Personal grounding is one of the key control measures of an ESD control program. 61340-5-1:2016 gives some basic requirements that are worth stating in the ESD program document.

- All personnel must be grounded according to the requirements of the ESD program when handling ESDS devices.
- Seated personnel must be grounded via a wrist strap.
- Standing personnel can be grounded via a wrist strap or via footwear and flooring.
- When grounding via footwear and flooring is used, ESD control footwear must be worn on both feet. The resistance from the person's body to ground through the system of person, footwear, and flooring must be less than 1 GΩ *and* the maximum voltage generated on the body (measured using a walk test) must be less than 100 V (for an ESD program protecting devices down to 100 V HBM).

The main requirement is that personnel handling ESDS devices should be grounded. Whether wrist straps or footwear-flooring grounding, or both, are used is optional and should be defined by the ESD control program. An exception is the case of seated personnel handling ESDS device, where current ESD standards require that a wrist strap must be used. Note that ESD seating is not commonly specified in such a way that reliable grounding of personnel can be achieved and should not normally be considered a method of personnel grounding (See Section 7.9.8).

Other methods of grounding have been used in some ESD control programs, for example grounding the person via specially designed groundable ESD control garments or seats. Use of nonstandard grounding methods such as this under a current standard ESD control program would require tailoring, with documentation of the rationale for the methods and evidence that it achieves the required personal grounding performance.

It is necessary to specify the limits of the measurable parameters for grounding equipment used in the ESD program. 61340-5-1:2016 gives a table (table 2 in the standard) that specifies these for product qualification and compliance verification.

It can be helpful to specify the requirements, test methods, and frequency of testing in a single table such as the example of Table 10.3. The test methods TM1–TM4 must be written up and referenced. The pass criteria and test frequency should be specified after consideration of the standard requirements and appropriateness of the facility and process. In this example, the resistance of footwear, while worn by personnel, measured between the hand and a metal plate under the foot, is specified to be <10 MΩ, which is much less than the requirements of the standard. As it is within the requirements of the standard, no tailoring explanation or justification is required.

As a key ESD control measure, it is wise to test personal grounding (wrist straps or footwear) before use on each day of use. Some organizations require personnel to test these at the beginning of each work shift or on each entry into an EPA. It is good practice to keep records, such as a simple log (e.g. signed list) of this test. A failed wrist strap should always be removed from use and repaired or replaced.

Where wrist straps are used, wrist strap grounding points should be regularly checked for connection to ground. Where these are of a temporary nature (e.g. via connection to a ground plug in a mains socket), tests should be more frequent due to the risk that accidental disconnection of the ground path may occur. In the standard, testing of wrist strap bonding points is addressed under "EPA Equipment."

Occasional Verification of ESD Control Footwear Worn on ESD Control Flooring

The resistance from the operator's body to ground through footwear and flooring and the body voltage generated while walking are dependent on the type of footwear and flooring used. These parameters are primarily confirmed during product qualification. Thereafter, only the combinations of footwear and flooring checked in product qualification should be used in practice. If a new combination is used, it should be qualified before use.

Nevertheless, these parameters can be affected by floor or footwear contamination, humidity, and other factors. So, it is wise to check them occasionally to make sure nothing has changed to compromise their effectiveness.

Personal Grounding Testers

Perhaps an obvious point, although often missed, is that personal grounding (wrist strap or footwear) testers must be specified to test according to the parameter limits specified in the ESD Control Program Plan. In practice, it may be easiest to select suitable personal grounding testers and then specify the parameter limits that they use, in the ESD Control Program Plan. Many personal grounding testers use upper parameter limits specified in the standards. Most also specify minimum limits, but some may not do so. Minimum limits are not specified in the standards but are user defined, usually based on a minimum resistance from body to ground for reduction of electrical shock risk if a live conductor is accidentally touched. Minimum values selected are often around 750 kΩ for up to 250 VAC and 500 VDC power systems.

10.5.13 ESD Protected Areas

10.5.13.1 Handling ESDS Devices and Access to the EPA

First, make it clear that handling unprotected ESDS devices must be done only in an EPA. This applies equally to ESDS devices that are protected by ESD protective packaging or those that would normally be enclosed e.g. within equipment. It is worth stating explicitly that an ESDS device must not be taken out of ESD protective packaging unless it is within an EPA.

The boundaries of the EPA must be clearly identified and noticeable to personnel entering or leaving the EPA. The way that the organization does this should be defined and described here. Examples of typical signage should be shown.

The 61340-5-1: 2016 requires that "Access to the EPA shall be limited to personnel who have completed ESD training. Untrained individuals shall be escorted by trained personnel while in an EPA." A clear statement like this should be included here, or at some other appropriate point in the ESD Control Program Plan.

10.5.13.2 Insulators

This section should be used to explain and define identification and treatment of essential and nonessential insulators used in the EPA. A similar approach can be taken to dealing with other known electrostatic field sources.

Essential insulators are those that are required to be present as part of the product or process, without which the process cannot be completed. Nonessential insulators are those without which the process can be completed. The most common source of high electrostatics fields in the workplace include insulators used in plastic (non-ESD protective) packaging, documentation (file dividers and covers), equipment housings and exposed construction, and other components used in the product. While electrostatic fields do not in general cause direct damage to ESDS devices, they do lead to induced voltages on ungrounded conductors and lead to the conditions in which ESD can arise if the conductors are, or touch, ESDS devices. Contact between the ESDS device and a metal item while within an electrostatic field can give a charged device ESD. It may be worth explaining this risk in this section or somewhere else in the ESD control program such as the introduction.

There are essentially three ways of dealing with nonessential insulators in an EPA. The organization should decide which approach they are going to take in each case and write the ESD control program accordingly.

The first approach is to replace the insulator with a noninsulating (conducting) material and ground it.

The second approach is to remove or eliminate the nonessential insulator from the EPA. This approach is simple and clear and often easy to manage and train personnel to do. In many cases, ESD control materials (static dissipative or conductive) can replace insulators for use in the EPA. In some facilities and processes, however, it could give great inconvenience for various reasons. Inconvenience usually leads to difficulty in compliance.

A third approach is to keep nonessential insulators a minimum distance from workstations where ESDS devices are handled. This approach can sometimes be used in convenient practices. (An example might be to have sign-off papers or computer equipment at a workstation next to, but sufficiently far from, the workstation area at which work is done on the ESDS device.) The disadvantage can be that effective procedures can be less clear to EPA personnel and may require greater training for reliable implementation. Insulators or ESDS can easily be transported from a position where they are allowed to a position where they might cause ESD risk by personnel who are unwary or lack sufficient understanding of the ESD control procedures.

The requirements for electrostatic field or voltage limits should be stated in the ESD Control Program Plan. These limits can be taken from the standard, or other limits can be defined. If the electrostatic field and voltage limits defined in the ESD Control Program Plan are outside the limits given in the standard, this requires justification and documentation as tailoring. This section should also define the actions to be taken if these limits are found to be exceeded.

Papers and cardboard are highly variable materials that can vary by orders of magnitude in their electrical characteristics between grades and with humidity changes (and the weather!). So, it is considered good practice to keep these materials away from ESDS devices. 61340-5-1: 2016 requires that "All non-essential insulators and items (plastics and paper), such as coffee cups, food wrappers, and personal items shall be removed from the workstation or any operation where unprotected ESDS are handled." A clear statement such as this should be included in this section of the ESD Control Program Plan.

Pink polythene material used for ESD control relies on the presence of humidity for its low-charging properties. It is often used within an EPA as a replacement for polythene packaging e.g. for enclosing non-ESDS device items or documents. This material becomes ineffective and can act like an insulator at low ambient humidity (<30% rh), charging highly.

There are other common items that could occur in practice and must be controlled because they generate electrostatic fields, such as cathode ray tube (CRT) displays. These should be identified, and how they will be treated should be defined.

Essential insulators cannot, by definition, be removed from the process in which they are used. A strategy for dealing with these is shown in Figure 10.1. The first task is to determine whether the insulator provides a significant ESD risk to the ESDS device in the process. This can be evaluated by determining whether it can charge up and give a significant electrostatic field at the possible position of the ESDS device. If not, it is necessary only to document the evaluation.

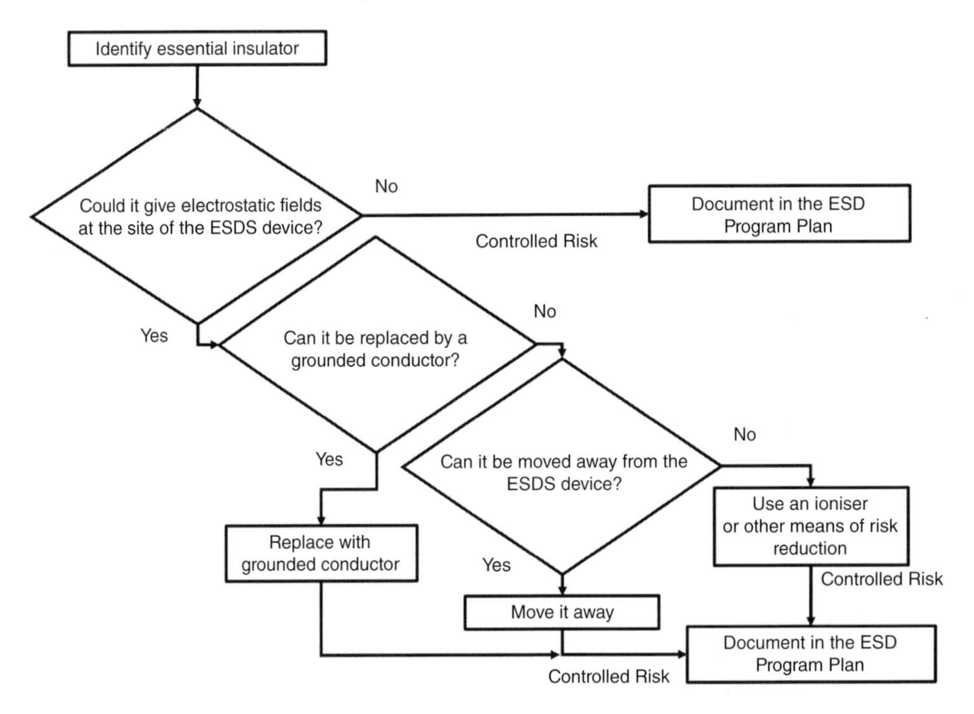

Figure 10.1 A strategy for dealing with essential insulators.

If the essential insulator can provide significant electrostatic fields at the position of the ESDS device, it remains to decide what to do with it. Sometimes they can be replaced with static dissipative materials and grounded. An example might be part of a test or assembly jig that does not need to be made of insulating material. Sometimes, although the insulator must remain nearby, it can be moved sufficiently far away from the position where ESDS devices are handled to reduce the field at the position of the ESDS devices to a low level. If neither of these techniques can be used, an ionizer or some other means of reducing the electrostatic field must be found. Whichever technique is used should be documented in the ESD Control Program Plan.

10.5.13.3 Isolated Conductors

The 61340-5-1:2016 standards states that "All items that come into contact with ESDS device and are capable of conducting electricity shall be connected to ground or electrically bonded in order to eliminate differences in potential" (Section 5.3.2 of the standard).

The standard also says that "if a conductor that comes into contact with an ESDS device item cannot be grounded or equipotential bonded together, then the process shall ensure that the difference in potential between the conductor and the contact of the ESDS device item shall be less than 35 V." It is worth including this statement, or an equivalent, in the ESD control program.

In practice, it is easy to say and difficult to do this. If at all possible, all conductors that contact the ESDS device should be grounded per Section 5.3.2 of the standard. One way of bringing the ESDS device and an isolated (ungrounded) conductor to nearly the same voltage is to use an ionizer to neutralize the charge on them to the same level. It is, however, difficult to measure and verify that this control is working as intended, especially for small items and automated processes.

The standard is not clear as to what should be considered an isolated conductor. Perhaps a good, although deliberately vague, definition of an isolated conductor is "a conductor on which electrostatic charge can build up, during normal operations, to a level where it could cause ESD risk on contact with ESDS." This definition implies that some evaluation is required to know whether the charge can build up on the conductor during normal operations or not.

Perhaps the easiest way to specify this would be to specify a maximum resistance from the conductor to ground. If the resistance from the conductor to ground is greater than the limit, the conductor is considered isolated. While the standard does not specify such a limit, from other specifications in the standard, resistance to ground below 1 GΩ would usually be considered adequately grounded. For small items and where charge generation or voltage changes due to capacitance changes are not an issue, up to 100 GΩ might be acceptable resistance to ground. Above 100 GΩ, resistance to ground will almost certainly be considered isolated unless it can be shown that the voltage on the item is reliable kept at zero during operation.

If an isolated conductor is found necessary, as it cannot easily be grounded and charge is found to build up on it during normal operation, it is likely to be necessary to specify control measures on a case-by-case basis. A strategy for dealing with ungrounded conductors is given in Figure 10.2. If the ungrounded conductor cannot touch an ESDS device, then there is probably no ESD risk. In this case, it is enough to document this evaluation.

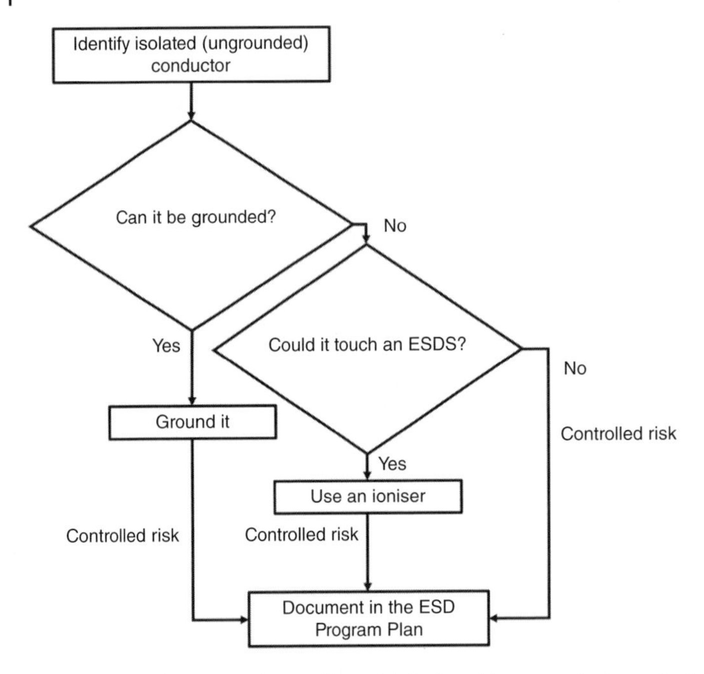

Figure 10.2 A strategy for dealing with isolated (ungrounded) conductors.

If the isolated conductor could touch an ESDS device, one way to reduce the voltage on it is to use an ionizer. If an ESDS device and isolated conductor spend sufficient time close together in the ion stream from an ionizer, they are both likely to reach a similar voltage near the offset voltage of the ionizer. The time taken to achieve this will depend on the ionizer and process circumstances. To show that this has been successful, it would be necessary to measure the voltages on the ESDS device and the conductor with an electrostatic voltmeter and compare them. Measurement of these voltages would require an electrostatic voltmeter capable of measuring voltage on small items. This is not an easy task, especially in a fast-moving automated process.

10.5.13.4 ESD Control Equipment

How to Specify ESD Control Equipment

The ESD control equipment and materials used in EPAs must be specified both for qualification and compliance verification. This means that test methods and pass criteria must be specified. The 61340-5-1:2016 standard specifies basic limits for work surfaces, storage racks and carts, wrist strap bonding points, flooring, ionizers, seats, and static control garments (see Chapter 6). It is not essential that all the items listed in the standard are used in an EPA, but any equipment used should be specified to comply with the requirements of the standards. For example, an ESD control program may not specify ESD control garments. Or, if personnel handling ESDS devices are always grounded via wrist straps, it may not be necessary to specify ESD control footwear. However, if these items are specified, they should be specified according to the standard used or covered by tailoring of the standard requirements.

Requirements for other items not specified in the standard may also need to be specified, e.g. gloves, finger cots, tools, or soldering irons. Test methods and pass criteria will also need to be specified for these items.

Sometimes it is desirable to specify a limit that is different from that given in the standard. If this limit is within the range given by the standard, it does not represent tailoring of the ESD program. For example, the organization may have selected a floor material that when installed has a resistance to ground below 10 MΩ. In this case, 10 MΩ could be specified as the upper limit for compliance verification to pick up changes in material performance or due to surface contamination.

In contrast, if the proposed limit is outside the range given in the standard, this should be considered a tailored requirement. As an example, many organizations that established an ESD control program under earlier versions of 61340-5-1 would have accepted seats with resistance to ground up to the previously specified limit of 10 GΩ. In the 2016 version, the limit for seats was dropped to 1 GΩ. For compliance with 61340-5-1:2016, the organization would have the choice of replacing all the seats according to the new requirements or including a tailored limit of 10 GΩ in their ESD control program. If they choose the tailored option, the decision should be supported by some tests demonstrating that no additional risk is provided by this specification.

Don't use words like *conductive*, *antistatic*, or *dissipative* to specify equipment unless they are defined for that product in a standard that is referenced as part of the specification. Contrary to popular usage, these words do not have generally standardized meanings for many products. Even worse, they can mean different things for the same product in different industry areas and for different products. So, unsuitable products may be purchased if using these words as the specification.

Instead, all products are best specified using verifiable parameters (e.g. resistance), measured using standard test methods. It is important to specify the test method used, because different test methods may give different measurement results. Common measurable parameters include resistance to ground or to a groundable point for many types of ESD control equipment, resistance point to point on a surface, material or garment, surface or volume resistance for packaging, and decay time and offset voltage for ionizers.

An example of ESD control equipment summarized with test method, pass criteria, and test frequency is given in Table 10.4. The pass criteria and test frequency should be specified after consideration of the standard requirements and the needs of the facility and processes present. The test methods TM5–TM9 must be written up and referenced. These should be based on those given in IEC 61340-5-4 (International Electrotechnical Commission 2019) or ESD TR53-01-15 (EOS/ESD Association Inc 2015; see Chapter 11).

Bench Mats and Other Surfaces on Which ESDS Devices Are Placed

Any surface on which unprotected ESDS device could be placed, for example storage racks and cart shelves, should be subject to the same requirements as work surfaces. If ESD control packaging materials are used in this way, they should also be specified as per work surfaces.

Compliance verification of these surfaces is normally specified and measured as resistance between the surface and ESD ground (R_g) (see Section 11.8.1).

ESD Control Flooring

ESD control flooring can be a useful way of establishing an EPA containing multiple workstations. Without the EPA floor, each workstation is effectively a stand-alone EPA with UPAs in between. ESD risks must be considered when transporting ESDS devices through these UPAs in between workstations, and ESD packaging is likely to be required. Using ESD flooring, the necessity of this packaging can be eliminated.

The function of ESD flooring is to provide convenient grounding of personnel wearing ESD footwear or using seats, carts, racks, and any other items standing on the floor.

The resistance of ESD control flooring should be chosen according to its purpose. The standards specify a maximum resistance to ground of 1000 MΩ. Organizations often have, or source, a floor with much lower resistance than this limit. A lower-resistance floor is more likely to give reliable low body voltage on personnel using suitable footwear or may be required to help give low system resistance to ground for some items.

If a floor has a considerably lower resistance to ground than required by the standard, it may be desirable to set the upper limit for flooring resistance to ground in the ESD Control Program Plan at a level just above the maximum expected in practice for the installed floor. Compliance verification measurements will then flag up rising floor resistance due to contamination or other factors.

Compliance verification of floors is normally specified and measured as resistance between the floor surface and ESD ground (R_g) (see Section 11.8.1).

Seating

If seating is needed at a workstation where ESDS devices are handled, then it must be specified for ESD control according to the standard requirements. Seats are normally most conveniently grounded through an EPA floor, and so an ESD control floor must usually also be present. In the absence of an ESD control floor, it is sometimes possible to ground the seat through a ground cord, but this can be inconvenient and may easily become disconnected. The requirement for seating should be thoroughly investigated before adding to the ESD Control Program Plan.

Compliance verification of seats is normally specified and measured as resistance between the seat surface and ESD ground (R_g) (see Section 11.8.1).

Tools

When handling very sensitive ESDS devices, any tools that could contact ESDS devices (e.g. cutters or pliers) could provide a risk of ESD damage. This can be addressed by specifying ESD tools in which the parts that contact ESDS devices are grounded through the user's body.

Compliance verification of tools can be specified and measured in various ways. There is no standard test specified in current standards. One method is to specify the resistance between the tool contact tip and ESD ground (R_g) when the tool is held by the operator (see Section 11.9.5).

Another way is to specify the maximum charge decay time measured using a charged plate monitor (CPM) (see Section 11.9.8.1).

Gloves and Finger Cots

Gloves and finger cots may need to be specified for ESD control, especially when used to handle ESDS devices or ESD control tools. Compliance verification of gloves and finger cots can be specified and measured in various ways. There is no standard test for these items specified in current standards, although annex A of S20.20-2014 points to ANSI/ESD SP15.1 for testing gloves and finger cots.

One method is to specify the resistance between the glove contact area and ESD ground (R_g) when the glove is worn by the operator (see Section 11.9.7). Another way is to specify the maximum charge decay time measured using a CPM (see Section 11.9.8.2). The latter method can also indicate whether the CPM plate is significantly charged by contact with the glove material.

Ionizers

Ionizers may be needed if there are essential insulators or isolated conductors present that evaluation has shown can charge up and cause ESD risk (see Section 4.6).

Ionizers are specified for compliance verification in terms of charge decay time and offset voltage, measured using a CPM (see Section 11.8.8).

ESD Control Garments (Coats)

It is not always necessary to use ESD control garments in every ESD control program. Those that choose to use ESD garments are often those that have heightened concern over ESD damage, for example those involved in handling highly sensitive ESDS devics or producing items for a high reliability market. Some ESD coordinators choose to use ESD control garments to help give a favorable impression to visitors regarding the care taken in the facility, and wearing a special garment in the EPA can help reinforce EPA discipline.

If an ESD control garment is required by the ESD program, it should be selected for compliance with the standard. These are usually specified in terms of point-to-point resistance ($R_{p\text{-}p}$) of the garment material (see Section 11.8.2.2) or the resistance to a groundable point (R_{gp}) for a groundable static control garment (see Chapter 6). User-defined test methods and pass criteria can also be used.

Some types of ESD control garments are available that do not easily pass the standard tests because they are not made from resistive materials. If these are to be used in a standard (IEC 61340-5-1 or ANSI/ESD S20.20) ESD control program, then tailoring will need to be used to give the rationale and justification for their use (see Section 6.5.6). Suitable test methods and pass criterial will need to be defined to qualify and verify these garments.

10.5.14 ESD Protective Packaging

The 61340-5-1:2016 standard states first and foremost that "ESD protective packaging and package marking shall be in accordance with customer contracts, purchase orders, drawing, or other documentation." For organizations that use packaging agreed with customers, a statement like this should be included in the ESD control program.

Where customers or other documentation does not define ESD protective packaging, ESD protective packaging requirements must be defined in the ESD Control Program Plan for wherever packaging is needed, according to IEC 61340-5-3. In situations where packaging is not needed, it need not be specified.

For otherwise unprotected ESDS devices taken outside the EPA, including packaging used for sending ESDS devices to the customer, ESD protective packaging must be defined that gives adequate protection against the ESD risks likely to be met in an unprotected area.

Packaging used within the EPA must also be specified according to IEC 61340-5-3.

ESD protective packaging is normally specified in terms of the surface resistance (R_s; see Section 11.8.4), the volume resistance (R_v; see Section 11.8.5), and for ESD shielding bags, the energy determined in an ESD shielding test. The latter is not normally a test that many users will do themselves due to its complexity. For packaging systems other than bags, ESD shielding capability is evaluated as the presence of a barrier against ESD current being conducted to the inside of the package (see Section 11.8.6).

10.5.15 Marking of ESD-Related Items

The 61340-5-1:2016 standard states that "ESDS, system or packaging marking shall be in accordance with customer contracts, purchase orders, drawing, or other documentation." For organizations that have marking of ESD control–related items agreed with customers, a statement like this should be included in the ESD control program.

Organizations that do not have marking of ESD control–related items agreed with customers should also consider whether marking of items for ESD control purposes would be of benefit or is in practice happening in their ESD control program. If there is a need or if markings are used in practice, then this should be documented. Common examples are marking of packaging and ESD control equipment. Some organizations also decide to mark documents related to ESDS devices (e.g. data sheets or component lists) or even the ESDS devices themselves (e.g. PCBs) to assist identification.

10.5.16 References

It is normally useful to list the references used in preparing the document, including the standards used and their user guides, and any in-house procedures, reports, test procedures, or other documents specified.

10.6 Evaluating ESD Protection Needs

Standard ESD control precautions applying ESD control equipment by rote address most common and well-known ESD risks, but they do not necessarily address all ESD risks. Every step of processes where ESDS devices are handled should be scrutinized for possible ESD risks. This is discussed in Chapter 9.

10.7 Optimizing the ESD Control Program

10.7.1 Costs and Benefits of ESD Control

Modern ESD standards allow the flexibility to optimize an ESD program for maximum effectiveness while providing only the equipment and control practices necessary for ESD protection. It is possible to use nonstandard or tailored ESD control methods in EPAs where ESDS devices are handled by less common processes. The choice of ESD control measures used governs the cost of ESD control measures used in the EPAs and ESD protective packaging used outside those areas. Documentation, training, and compliance verification programs are also essential parts of ESD control. Trade-offs can be made between these costs in order to optimize the ESD control program, although a high level of expertise may be needed to do this successfully (Smallwood et al. 2014).

The first and most obvious benefit of an ESD control program is to protect ESDS devices during processes and maintain the level of ESD damage at an acceptably low level. There are also other benefits that could be part of the rationale for ESD control. Maintenance of an ESD control program to a recognized international standard could be an important part of the organization's quality program. It may even be essential to prequalification of the organization as a provider of product to some customers. If a customer audits the organization, a favorable impression of ESD control (whether the customer has a good understanding of the topic or not) can be a vital step in the path to obtaining an order. An adverse impression and disagreements over the adequacy of ESD control can provide a significant block to customer acceptance.

The main cost of inadequate ESD control is often thought to be damage to ESDS devices handled in the organization's processes. This cost, for many organizations, is far from clear or quantifiable as failure analysis often is not done to the level that ESD failures are detected. Other costs can also be significant but are often not considered. These can include product failures, unreliability or drift in characteristics, additional test and rework expenditure, delays to shipments, or need to overstock frequently failing items. Some of the most expensive costs can be those related to customer service – cost of reaction to customer complaints and failures at the customer's site, impaired product or company reputation, and a possible result in reduction of sales.

The real cost/benefit balance of an ESD control program depends on the time and resources spent on all the aspects of ESD control. Optimizing the ESD control program should consider all the desired benefits and areas of cost and determine an appropriate balance between them. This means that trade-offs between investment in the different aspects are possible and probably necessary.

10.7.2 Strategies for Optimization

10.7.2.1 Minimization of the EPA

A major contribution to ESD protection may be made by minimizing handling of ESDS devices in an unprotected state. As well as minimizing the opportunity for ESD to occur, it has other advantages. For example, a smaller EPA usually requires less equipment. This also leads to reduced overhead of compliance verification testing overhead.

10.7.2.2 Choice of EPA Boundary

Including too many parts of a process in an EPA can result in processes that are incompatible with ESD control being included in the EPA. This can then make it difficult or impossible to comply with the ESD control program requirements.

Nevertheless, EPAs must be defined for all activities in which unprotected ESDS devices are handled.

So, it is worth carefully considering where the EPA boundary should be placed in order to enclose all necessary activities while minimizing the EPA size and excluding unnecessary and incompatible activities.

10.7.2.3 Minimizing Variation of the ESD Control Program

Variations are often specified in ESD control measures to minimize ESD control equipment required in some areas. It is perfectly possible to specify considerable variation even in an ESD control program compliant with a standard.

It can, however, be advantageous to minimize the variation for various reasons. The impact on documentation, training, and compliance verification should be considered. Moreover, customers viewing an ESD program are inclined to evaluate them against their own idea of what a good ESD control program should look like. If they see that elements are missing or different, they may jump to a conclusion that ESD control is insufficient. Disputes or unwillingness to place work with the organization can result.

Having variations on ESD control measures used in different EPAs can result in an increased need for ESD training, as personnel working in each area must be trained in the specific requirements for each area. Personnel who work in several different areas may get confused between differing requirements used in each area, and this can lead to unintended noncompliances.

Variation of required ESD control equipment can reduce compliance verification overheads if the amount of equipment needing to be verified can be reduced in some areas.

If, however, the same equipment is specified in different areas but has different pass criteria, confusion can arise with equipment meeting the wrong criteria being introduced into an area. It can be difficult to notice this type of problem until picked up through subsequent compliance verification tests.

It is inadvisable to expect workers to make decisions or change ESD control behavior for different ESDS devices, e.g. of different sensitivity, or for non-ESD-sensitive items. Confusion, wrong decisions and poor compliance habits may result. It is probably better in most cases to design and implement an ESD program according to the requirements for handling the most sensitive ESDS devices and apply this generally to simplify training and compliance verification and reduce likely noncompliances.

10.7.2.4 Standard or Tailored ESD Control

Standard ESD control measures provide a generally recognized set of controls that are needed and control the most common ESD risks in most cases. It is relatively easy to implement an ESD control program with standard ESD controls with relatively little expertise.

With additional effort and evaluation, it may be possible to devise more cost-efficient, tailored, nonstandard controls or to leave out some controls found unnecessary. This requires greater expertise for successful evaluation and specification of effective tailored controls. With inadequate expertise, there is a risk that the specified ESD controls may be inadequate.

This approach is also likely to increase the overheads of documentation and training and the variation of ESD controls specified in different areas. Visiting customers may be more inclined to jump to conclusions that ESD controls are inadequate.

10.7.2.5 Design for Convenience

If ESD control measures are designed to be convenient, it is likely they will be followed. If inconvenient, it is more likely they will not, and noncompliances will occur.

Design for convenience may not give the lowest equipment, documentation, training, or compliance verification costs, but it may give a benefit in lower noncompliance levels and associated overheads.

10.7.2.6 Who Audits the ESD Control Program?

It is possible to delegate the task of auditing the ESD program to any personnel with acceptable expertise. Some organizations use internal personnel, whereas others delegate the task to second- or third-party auditors or even to suppliers of ESD control equipment.

Sometimes delegation of audit can result in a conflict of interest. For example, using a supplier to audit may result in a conflict of interest with their desire to sell their ESD control products.

Against this, using internal auditors may result in training and compliance verification requirements that are difficult for a small organization to resource. Under this pressure, compliance verification may become under resourced and neglected or ineffective.

10.8 Considerations for Specific Areas of the Facility

10.8.1 The Varying ESD Control Requirements of Different Areas

Many facilities include typical stages such as Goods In, Storage, Kitting, Manufacturing, Test, and Dispatch. Some organizations also have areas such as Research & Development. Some of these areas (e.g. Manufacturing) are easily controlled with formal EPAs. Other activities (e.g. Kitting) may not obviously need an EPA but nevertheless have a large effect on compliance within the EPAs they work with. Still others (e.g. Goods In, Stores, Dispatch) may have aspects requiring both EPA and unprotected areas working together. Research & Development areas also often have mixed requirements for ESD control and unprotected areas and may be difficult to organize as formal EPAs for various reasons, including acceptance of ESD control measures by the personnel working in the EPAs.

10.8.2 Goods In and Stores

An essential part of the Goods In function is to remove any packaging that is provided for protection during transport rather than ESD control ("secondary packaging"). Removal of secondary packaging should stop as soon as ESD protective packaging is encountered and before any ESDS devices are exposed, allowing ESD protective packaging to remain unopened. In ESD protective packaging (e.g. PCBs in bags), a symbol should be evident on the packaging to identify its ESD protective function (see Section 8.10). This packaging should not be opened outside an EPA. On the other hand, secondary packaging should not be taken into an EPA.

ESDS device items that are protected against ESD damage (e.g. enclosed within equipment or ESD protective packaging) can be stored under ordinary unprotected conditions. Items that have exposed ESDS devices should be stored under EPA conditions.

If inspection or test of ESDS devices is required, an EPA should be provided for the purpose. It is preferable not to use this for any non-EPA activity, as this can be a means of introducing noncompliant materials and lax EPA procedures. An EPA bench used for other purposes should be checked for compliance and recommissioned before use as an EPA.

Goods In and Stores areas do not necessarily need an EPA. They will need one only if ESD protective packaging must be opened and the ESDS device handled (e.g. for counting or inspection) in an unprotected state. If possible, it can be preferable to avoid having this sort of activity and therefore avoid the necessity of having an EPA. If necessary, the EPA can be as simple as a single carefully specified workstation.

There is a considerable risk that EPA workstations in Goods In and Stores areas become contaminated with secondary packaging and other insulating materials. To minimize this risk, EPA workstations should not be used for non-EPA activities. Personnel using these facilities may need to be carefully trained in avoiding non-compliances and ESD risks.

Siting of EPA workstations should be carefully chosen to be well away from areas where high-level static generators such as secondary packaging are present. As an example, an EPA workstation that backs onto a storage rack in an uncontrolled area could be subject to considerable ESD risk from the static generators stored on the rack.

10.8.3 Kitting

Kitting activities may or may not require use of an EPA for handling unprotected ESDS devices. Nevertheless, any kit destined for use in an EPA will be a source of noncompliant packaging or other materials and items unless carefully specified to avoid this. Tote boxes holding kits for transport into an EPA must be specified as compliant with EPA entry. Items included within the kit, including mechanical, consumable, or non-ESDS device components, must be supplied in suitable ESD protective packaging materials. Insulating materials and items must be minimized or eliminated. If not eliminated, how they are used and where they occur in the EPA must be carefully controlled to avoid ESD risks to ESDS devices. If noncompliant items are included in a kit, they will cause noncompliance in the EPA to which they are delivered.

10.8.4 Dispatch

Dispatch is another area that may or may not need EPAs as well as unprotected areas. Dispatch areas commonly have large amounts of secondary packaging materials including insulating packing tapes and other high-level static generators. Many of the considerations for these areas are like Stores and Goods In areas.

Another consideration is that high levels of paper or cardboard dust, fibers, and larger particles can be generated during handling of packaging materials. These can cause contamination problems for some types of products handled in the vicinity.

10.8.5 Test

Test is probably the area most likely to require some tailoring of ESD control measures due to safety considerations if high voltages and manual handling are present. This often means that personal grounding requirements must be carefully thought through to maximize safety and minimize ESD risk.

10.8.6 Research & Development

Research & Development areas are perhaps notorious for contention between the desire for ESD control and the easy freedom of engineers to move between prototyping and office desk or computer simulation areas. Some organizations take the position that prototypes that are not shipped to customers are not required to be handled under ESD control conditions. This, however, neglects the fact that damage to ESDS devices by ESD can lead to considerable wasted time in debugging and reworking circuitry, and ESDS devices that may have their properties changed by ESD could result in a prototype working differently from a final produced design. Moreover, for some organizations, each product is effectively a one-off product produced by a prototyping process.

It is often possible to successfully specify ESD control for Research & Development areas that will address these problems. The ESD control design of these areas will often be very different from a production facility. The fact that highly technically competent engineers are using the facility can be an advantage providing they “buy in” to the need for ESD control and it is designed to be convenient for them to use. If not, it will probably be difficult to get them to comply with the ESD control requirements.

10.9 Update and Improvement

ESD control programs are often mistakenly treated as a process that, once specified, can be left without change for an extended period. In practice, here are several reasons why an ESD program should be regularly reviewed.

First, auditing reveals deviation patterns that can often be addressed by updating the ESD control program to be easier to use. Snow and Dangelmeyer (1994) gave an example where foot straps were often incorrectly worn. Replacing them with ESD control shoes eliminated an important source of noncompliances. Sometimes, ESDS device failures reveal weak ESD control that must be addressed by changes in the ESD control program.

Second, processes and facilities may change over time, and new or modified ESD control measures may need to be implemented for them.

Third, the types of ESDS devices often may change and may require modification to the ESD control measures used. Introduction of new, more sensitive ESDS devices may introduce special challenges.

Finally, update of ESD control standards may result in necessary changes to the ESD control program to maintain compliance with the standard.

For all these reasons, it is advisable to build in a program of regular update and continual improvement.

References

Dalziel (1972). Electrical shock hazard. *IEEE Spectrum*: 41–50.

Dangelmeyer, G.T. (1990). *ESD Program Management*. Van Nostrand Reinhold.

EOS/ESD Association Inc. (2014). ANSI/ESD S20.20-2014. *ESD Association Standard for the Development of an Electrostatic Discharge Control Program for – Protection of Electrical and Electronic Parts, Assemblies and Equipment (excluding Electrically Initiated Explosive Devices)*. Rome, NY, EOS/ESD Association Inc.

EOS/ESD Association Inc. (2015). ESD TR53-01-15. *Technical Report for the Protection of Electrostatic Discharge Susceptible Items – Compliance Verification of ESD Protective Equipment and Materials*. Rome, NY, EOS/ESD Association Inc.

International Electrotechnical Commission (2016) IEC 61340-5-1: 2016. *Electrostatics – Part 5-1: Protection of electronic devices from electrostatic phenomena - General requirements*. Geneva, IEC.

International Electrotechnical Commission. (2019) IEC TR 61340-5-4:2019. *Electrostatics - Part 5-4: Protection of electronic devices from electrostatic phenomena – Compliance Verification*. Geneva, IEC.

Nave, C.R. and Nave, B.C. (1985). *Physics for the Health Sciences*, 3e. W B Saunders. ISBN: 0 7216 1309 8.

Smallwood, J., Taminnen, P., and Viheriaekoski, T. (2014). Paper 1B.1. Optimizing investment in ESD control. In: *Proc. EOS/ESD Symp EOS-36*. Rome, NY: EOS/ESD Association Inc.

Snow, L. and Dangelmeyer, G.T. (1994). A successful ESD training program. In: *Proc. EOS/ESD Symp. EOS-16*. Rome, NY: EOS/ESD Association Inc. pp. 94-–94-12.

11

ESD Measurements

11.1 Introduction

All materials and equipment used for electrostatic discharge (ESD) control are designed and made to have certain properties that enable them to function. These commonly are based on the principles of replacing exposed insulating materials with noninsulating materials and establishing a ground path for any charge generated on the equipment or material to dissipate to electrostatic discharge protected area (EPA) ground. In addition, charge on essential insulators may need to be neutralized using an ionizer or ESD risks controlled by other means.

Many of the facilities equipment and materials needed for use in ESD prevention are specified in various standards worldwide. These standards establish performance criteria for compliance, and test methods for measuring this performance. The two main standards discussed in this book are IEC 61340-5-1 and ESD Association ANSI/ESD S20.20 (EOS/ESD Association Inc. (2014a)). IEC 61340-5-1 has also been adopted as a national standard in many countries, in Europe becoming the European Norm EN61340-5-1. Individual countries may have their own versions, and, in the United Kingdom, this is BS EN 61340-5-1 (British Standards Institute. (2016)).

This chapter explains how to make basic measurements for use in the assessment of equipment for compliance with these standards. It is a guide to the main measurements that ESD coordinators will wish to make in their facilities, giving some practical guidance and "work-arounds" where necessary, and giving some information on the more unusual tests and nonstandard tests that can be used where standard tests are not specified or are unsuitable.

For clarity, the 61340-5 series terminology is used, although the Electrostatic Discharge Association (ESDA) series test methods are in many cases nearly identical and are cross-. In some cases, differences with the ESDA standards are noted.

The ESD Control Program Handbook, First Edition. Jeremy M Smallwood.

11.2 Standard Measurements

61340-5-1 and 20.20 use a variety of basic test methods to assess the performance of ESD prevention equipment and materials. These include

- Point-to-point resistance
- Resistance to ground or groundable point
- Surface and volume resistance of ESD protective packaging
- End-to-end resistance of a ground cord
- Personal grounding resistance tests
- Electrostatic fields and voltages
- Walk test of footwear and flooring
- ESD shielding test of shielding bags

A "groundable point" can be, for example, a stud on a bench or floor mat that is intended to be connected to ESD ground in the final installation.

Table 11.1 Summary of test methods and their application in 61340-5-1, and corresponding ESD Association standards.

Item tested	Test method standard	
	IEC	**ESDA**
Work surfaces, racks	61340-2-3	STM4.1
Floor	61340-4-1	STM7.1
Seating	61340-2-3	STM12.1
Ionizers	61340-4-7	STM3.1
Hand tools	None	None
Garments	61340-4-9	STM 2.1
Wrist straps	61340-4-6	S1.1
Footwear – shoes	61340-4-3	STM9.1
Foot grounders	None	SP9.2
Footwear and flooring resistance measurement	61340-4-5	STM97.1
Footwear and flooring walk test	61340-4-5	STM97.2
ESD control packaging surface resistance – concentric ring	61340-2-3	STM11.11
ESD control packaging volume resistance – concentric ring	61340-2-3	STM11.12
ESD control packaging surface resistance – two-point resistance measurement.	61340-2-3	STM11.13
ESD shielding bags	61340-4-8	STM11.31
Resistance of gloves and finger cots	None	SP15.1
Charge decay of tools, gloves, and finger cots	61340-2-1	None
Electrical Soldering/Desoldering Hand Tools	None	STM 13.1
Compliance Verification test methods – various	TR 61340-5-4	TR53-01

Table 11.2 Other IEC test methods that can be used in ESD protective equipment, materials, and packaging evaluation.

IEC standard	IEC title	Corresponding ESDA standard
61340-2-1	Ability of materials and products to dissipate static electric charge	None
61340-2-2	Measurement of chargeability	ESD ADV 11.2

The test methods will be described and demonstrated in the following sections. Simple and clear test procedures for each measurement are given, which could be used to form the basis of a company test procedure.

The test method standards available from IEC and ESDA at the time of writing are listed in Tables 11.1 and 11.2. Many of the test methods in the IEC system have corresponding test methods in the ESDA system, and vice versa, and these are indicated. It should be noted that "correspondence" does not mean "equivalence" – there may be differences between these documents. In some cases, the correspondence is with only part of the corresponding document, or there may be other differences. As always, when aiming to comply with a standard, it is necessary to refer to that standard for full and up-to-date details.

11.3 Product Qualification or Compliance Verification?

It is recognized that there are two main reasons for making measurements on equipment and materials used for ESD control – product qualification and compliance verification.

Product qualification means the testing of ESD control products or materials to determine whether they are suitable for use within the ESD control program. Compliance verification refers to the tests routinely used for checking the equipment or materials are still working as specified, during their working life. In some cases, the same test methods may be used for both purposes.

11.3.1 Measurement Methods for Product Qualification

During product qualification test, the equipment or material will often not be tested in its working environment. The tests used are typically designed to establish its basic characteristics and ensure that it will fulfill its function under the expected variation of operating conditions over its lifetime.

For this reason, items will usually be tested in the laboratory under dry atmospheric humidity conditions as this represents the most challenging atmospheric conditions for operation. Many test methods from the 61340-x series specify 12 ± 3% rh 23 ± 3 °C, although some specify other test conditions or leave it to the user to select suitable conditions.

The tests used may include tests under simulated working conditions and may include the same test as is used for compliance verification.

11.3.1.1 Product Qualification of a Bench Mat

As an example, it may be required to test a bench mat type before approving it for purchase and fitting on workstations in a facility. The objectives may be to

- Establish the basic material performance
- Ensure it will remain within limits even at low ambient humidity
- Give an idea of variation occurring between specimens of the item

IEC 61340-5-1:2016c Table 3 specifies that the installed resistance to ground shall be $<10^9\ \Omega$ (1 GΩ). However, many ESD Control Program Plans will also require a minimum point-to-point resistance of the surface, as it is known that placing a charged component on a low-resistance surface carries a risk of charged device ESD damage. Typically, the user may select a lower limit of point-to-point resistance of, for example, $10^4\ \Omega$ (10 kΩ).

So, the standard specifies the following as product qualification tests:

- Resistance to groundable point (with an upper limit of $10^9\ \Omega$ [1 GΩ]).
- Point-to-point resistance of the surface (with an upper limit of $10^9\ \Omega$ [1 GΩ]).
- The test method standard specified is IEC 61340-2-3 (International Electrotechnical Commission 2016a).

So, the user specifies these two tests to be performed under controlled atmosphere laboratory conditions of $12 \pm 3\%$ rh $23 \pm 3\,°C$, as this will be representative of their worst-case expected dry air conditions. Three specimens are tested after the specified conditioning time (as required by 61340-2-3) as this gives an idea of the variation between specimens.

11.3.1.2 Example: Product Qualification of a Footwear and Floor Combination

A more complicated example is that of specification of a combination of footwear and flooring to give the performance required by the 61340-5-1:2016 standard. This allows a combination to be used that gives a relatively high resistance from the footwear wearer's body to ground of $10^9\ \Omega$ (1 GΩ), providing the body voltage does not rise above 100 V. IEC 61340-5-1:2016c Table 2 specifies that for product qualification, the test method standard is 61340-4-5 (International Electrotechnical Commission 2004). This gives the following:

- A test method for determining the resistance from the wearer's body to ground via footwear and flooring in combination (limit of resistance to ground $10^9\ \Omega$ [1 GΩ]).
- A "walking test" giving the voltage generated on the body (upper limit 100 V).

Note that these tests require the footwear and the flooring to be specified and tested in combination, as intended to be used in practice. This is because the combined performance of footwear with flooring is often dominated by factors such as contact resistance and charge generated between the shoe sole and the floor surface.

In addition to these tests, it is wise to test that the footwear when worn will pass the usual compliance verification test (resistance from body via footwear to metal plate).

11.3.2 Measurement Methods for Compliance Verification

Compliance verification tests tend to be simple tests specified to check that the equipment or materials of the EPA are still working according to their intended specification. They are often a simple system test of the equipment under operational conditions. A major consideration in developing compliance verification tests is that these will need to be done on a regular basis and test many items during an audit. A quick and easy but effective test is required to minimize the time and effort required. Compliance verification tests are done under the ambient atmospheric conditions in the EPA. It is good practice to make a note of these during the test.

The 61340-5-1 and S20.20 standards have related documents IEC 61340-5-4 (International Electrotechnical Commission 2019) and ESD TR53-01 (EOS/ESD Association Inc. (2018a)) that specify tests adapted from other standards specifically for compliance verification testing.

11.3.2.1 Example: Compliance Verification Test of a Bench Mat

Compliance verification test of a bench mat in use in the EPA is specified as a simple resistance to ground measurement (Section 11.8.1) from the surface of the mat to the EPA earth. In this way, the mat and its grounding system are tested in one quick measurement.

11.3.2.2 Example: Compliance Verification of Footwear and Flooring

Although a footwear and flooring combination may be specified using tests in combination (see Section 11.3.1.2), it is usual to use simple separate compliance verification tests of the footwear (while worn by personnel) and flooring.

The footwear resistance (tested while worn) is specified as a measurement of the resistance from the wearer's body to a metal contact plate under the foot (see Section 11.8.3.3). This is normally done using a proprietary tester. The floor is tested separately using a periodic test of resistance from the floor surface to ground (see Section 11.8.1.1).

11.4 Environmental Conditions

Compliance verification tests are normally made under normal operational ambient temperature and humidity. If the conditions are variable, it is useful to make occasional measurements, when possible, under the lowest-humidity conditions. These will often be conditions under which maximum electrostatic charging will occur.

It is good practice to measure and record these at the time of the test. Humidity is especially important when high resistance (>1 GΩ) materials are measured. A change in humidity, particularly below 30% rh, can have a large effect on the result with increased decay time and higher resistance measured at lower humidity.

For ESD control product qualification tests, the items are preconditioned and measured in the laboratory under set low humidity and temperature conditions. Dry conditions usually give the worst-case and highest resistance measurement results. Many standards specify 12% rh 23 °C as the test conditions.

11.5 Summary of the Standard Test Methods and Their Applications

Several standard test methods are called up by IEC 61340-5-1:2016c in Tables 1–3 within the standard. Table 11.1 summaries these standards and their corresponding standards in the 20.20 system. Some gaps in the standard test methods are also shown. Table 11.2 gives some additional test method standards that can be useful in the evaluation of ESD protective equipment, materials, and packaging.

11.6 Measurement Equipment

11.6.1 Choosing a Resistance Meter for High-Resistance Measurements

Many measurements made on ESD control equipment and materials involve high-resistance measurements. A suitable high-resistance meter is needed to make these measurements according to the standard requirements.

Two important points to consider when buying a high-resistance meter are the test voltage options and the resistance range. For use with 61340-5-1, the equipment must be capable of measurement at the standard test voltages of 10 and 100 V (Table 11.3). The resistance meter should have the capability to measure resistance from 10 times less than the lowest resistance limit required to be measured to at least 10 times the greatest resistance required to be measured. Many types of ESD control equipment have an upper limit of 1 GΩ specified in the standards, requiring a resistance meter that will measure up to at least 10 GΩ. ESD packaging and garments have a specified upper resistance limit of 100 GΩ (10^{11} Ω). Measuring these require a meter that can measure up to 1 TΩ (10^{12} Ω).

The resistance measurement result will usually vary with test voltage, being raised for a lower test voltage. The standard test methods used in 61340-5-1 and S20.20 typically require use of 10 V to measure resistances up to 1 MΩ. The test voltage is increased to 100 V if a resistance >1 MΩ is found at 10 V. Always test at 10 V first, and then move to 100 V if the result exceeds the limit.

Table 11.3 Test voltages used in resistance measurements required in 61340-5-1 and S20.20 standards.

Usage	Test voltage	Resistance range	Standards referenced
General resistance measurements	10 V	Up to 10 MΩ	IEC 61340-2-3
	100 V	100 kΩ to 10 GΩ	ESD STM11.11
Packaging and garments	10 V	10 kΩ to 10 MΩ	ESD STM11.12 ESD STM11.13
	100 V	100 kΩ to 1 TΩ	
Footwear and flooring system test	10 V [a]	10 kΩ to 10 MΩ	IEC 61340-4-5
The superscript 1 indicates the note applies.	100 V [a]	100 kΩ to 1 TΩ	ESD STM97.1
Wrist straps as worn	7 V to 30 V [a]	50 kΩ to 100 MΩ	IEC 61340-4-6 ESD S1.1
Footwear as worn, to metal plate	9 V to 100 V [a]	50 kΩ to 1 GΩ	IEC 61340-5-1 Annex A

a) Where measurements involve the human body, the resistance meter should be current limited to a maximum of 0.5 mA (See Section 10.1.3).

Where high voltages are used for measurement, always assess the safety issues and take appropriate safety precautions. Most handheld instruments are unlikely to source dangerous electrical currents, but check your instrument to evaluate any risks.

11.6.2 Low-Resistance Meter for Soldering Iron Grounding Test

For measurement of grounding of a soldering iron bit, a low-resistance meter is needed, capable of measurement of resistance in the range 0–100 Ω. Many multimeters have a continuity range that can make this measurement.

11.6.3 Resistance Measurement Electrodes

IEC 61340-5-1 uses as the basis of many of its resistance measurements electrodes specified in standards IEC 61340-2-3 and 61340-4-1 (International Electrotechnical Commission 2015b).

Electrodes may be found on the market that conform to a variety of current and historical standards, as well as others that do not conform to standards. Different electrode systems in general may give different results, although the differences in some cases may not be very great or significant. The 2.5 kg resistance measurement electrodes shown in Table 11.4 and Figure 11.1 compliant with IEC 61340-2-3, IEC 61340-4-1, ESD STM2.1, ESD STM4.1 (EOS/ESD Association Inc. (2017)), ESD STM7.1 (EOS/ESD Association Inc. (2013c)), and ESD STM12.1 (EOS/ESD Association Inc. (2013d)) are commonly available and suitable. Although the electrodes of ESDA 2.1, 4.1, 7.1, and 12.1 are specified as 2.27 kg, these fall within the tolerance of the 2.5 kg IEC specification. Electrodes to all these standards are for convenience referred to as *2.5 kg electrodes* in this chapter.

Table 11.4 Comparison of characteristics of 2.5 kg resistance measurement electrodes from various standards.

Characteristic	61340-4-1	61340-2-3	EN100015-1	ESDA STM2.1 ESDA STM4.1 ESDA STM7.1 ESDA STM12.1
Mass (kg)	2.5 ± 0.25 or 5.0 ± 0.25[a]	2.5 ± 0.25	2.5 ± 0.5	2.27 ± 0.06
Diameter (mm)	65 ± 5	63.5 ± 1	75	63.5 ± 0.25
Electrode surface hardness (Shore A)	60 ± 10[b]	50–70		50–70
Resistance (electrode placed on metal sheet)	<1 kΩ[b]	<1 kΩ[b]		<1 kΩ[c]

Notes:
a) According to 61340-4-1:2003, the 2.5 kg electrode is to be used for measurements on hard nonconformable surfaces. The 5 kg to be used for measurements on all other surfaces. The first is applicable to most applications under 61340-5-1.
b) Conductive rubber pad need not be used with conformable (e.g. textile) surfaces.
c) Measured at 10 V between two electrodes placed on metallic surface.

Figure 11.1 An example of 2.5 kg electrodes according to IEC 61340-2-3 and ESD STM11.11.

Organizations in Europe that have been involved in ESD control for many years often have EN100015 (CENELEC 1992) electrodes. The EN100015 electrodes may give different results due to their different diameter and lack of conductive rubber face material but may give results that correlate with the other electrodes. The lack of rubber face material means that they might not make good contact with hard surface materials. These electrodes are not recommended for making measurements according to 61340-5-1.

11.6.4 Concentric Ring Electrodes for Packaging Surface and Volume Resistance Measurement

One option for measurements of surface or volume resistance of packaging materials according to IEC 61340-5-3 (International Electrotechnical Commission 2015d) and ANSI/ESD S541 (EOS/ESD Association Inc. (2018c)) requires use of a concentric ring electrode (Table 11.5, Figures 11.2, and 11.3). This consists of an inner circular electrode and an outer ring electrode surrounding it. The *surface resistance* is measured between the inner and outer ring electrodes when the electrode is placed on a surface. For standard electrodes, the surface resistance is approximately a factor of 10 less than the material *surface resistivity*.

For volume resistance measurement, the material under test is placed on a flat metal electrode. The concentric ring electrode is placed on the material under test. The resistance is measured through the material between the flat metal electrode and the inner ring electrode. The outer ring electrode can be used as a guard ring to prevent surface leakage affecting the result.

Table 11.5 Comparison of concentric ring surface resistance measurement electrodes from various standards.

Characteristic	61340-2-3:2016	ESDA STM11.11-2015b
Mass (kg)	2.5 ± 0.25	2.27 ± 0.0567
Electrode surface hardness (Shore A)	50–70	50–70
Inner electrode diameter (mm)	30.5 ± 1	30.48 ± 0.64
Outer electrode inner diameter (mm)	57 ± 1	57.15 ± 0.64
Outer electrode width (mm)	3 ± 0.5	3.18 ± 0.25
Soft conductive material volume resistivity (Ω cm)		<10

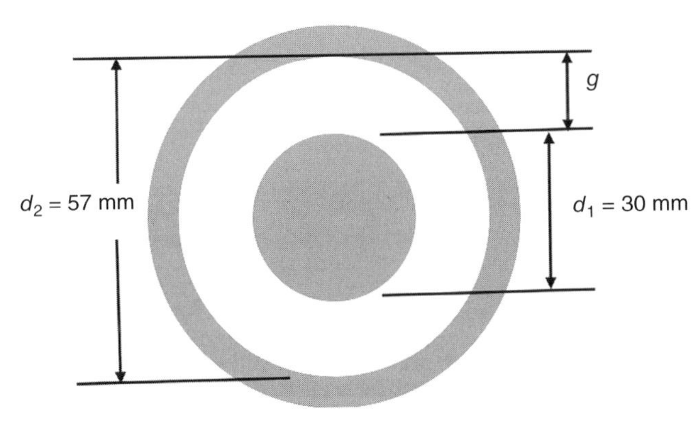

Figure 11.2 Concentric ring electrode contact area according to IEC 61340-2-3 and ESD STM11.11 and ESD STM11.12.

Figure 11.3 Examples of concentric ring electrode.

11.6.4.1 Conversion of Surface Resistance to Surface Resistivity for a Concentric Ring Electrode

Some standards quote material surface *resistivity* rather than surface *resistance*. If surface resistivity is required, the surface resistance reading must be multiplied by a factor that is related to the electrode geometry. For the concentric ring electrodes of IEC 61340-2-3 and ESD STM11.11, the factor is 10. The surface resistivity is 10 times surface resistance.

More precisely, the conversion for a concentric ring electrode is given by IEC 61340-2-3

$$\rho_s = \frac{R_s \pi (d_1 + g)}{g}$$

where ρ_s is the surface resistivity (Ω), R_s is the surface resistance (Ω), d_1 is the diameter of the inner electrode (m), and g is the gap between inner and outer electrodes (m) (Figure 11.2). For the electrodes specified in 61340-5-1, we have $g = 0.0135$ and $d_1 = 0.03$, which gives

$$\rho_s = 10.1R_s$$

For the miniature two-point probe or point-to-point resistance measurement with two 2.5 kg electrodes, the factor for conversion to surface resistivity has not been established.

11.6.4.2 Conversion of Volume Resistance to Volume Resistivity

For a homogenous sheet sample material of thickness h (m) and electrode diameter d_1 (m), the volume resistance R_v can be converted into the volume resistivity ρ_v of the material according to IEC 61340-2-3.

$$\rho_v = \pi R_v d_1^2 / 4h$$

ESD STM11.12 (EOS/ESD Association Inc. (2015c)) gives an equivalent conversion formula, where the inner electrode area A is specified to be 7.1 cm^2 and the specimen thickness t cm as

$$\rho_v = R_v A / t$$

11.6.5 Two-Point Probe for Packaging Surface Resistance Measurements

A handheld two-point probe is specified in IEC 61340-2-3 and ESD STM11.13 (EOS/ESD Association Inc. (2015d)). It consists of two 3.2 mm diameter spring loaded probes 3.2 mm apart that are held by spring action in contact with the surface being measured with a force of 4.6 ± 0.5 N (Figure 11.4). The probes are faced with conductive rubber contact material with Shore hardness 50–70. A correlation factor for the results obtained with this electrode system compared to the concentric ring probe and surface resistivity has not been determined (see Section 11.8.4).

11.6.6 Footwear Test Electrode

In testing footwear, an electrode is needed to contact the sole of the footwear. This should be a metal plate sufficiently large for the whole of the foot to be placed upon it (Figure 11.5). Some standards may specify the size of the plate to be used. It should have a connector to allow connection of test leads to other equipment.

In proprietary footwear testers, the footwear test electrode is usually incorporated into the tester equipment.

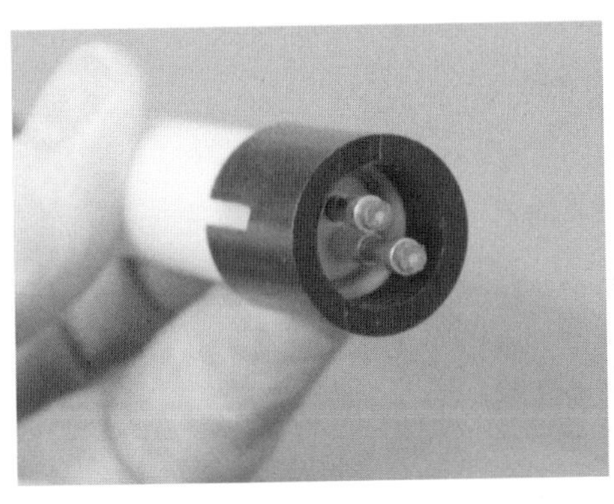

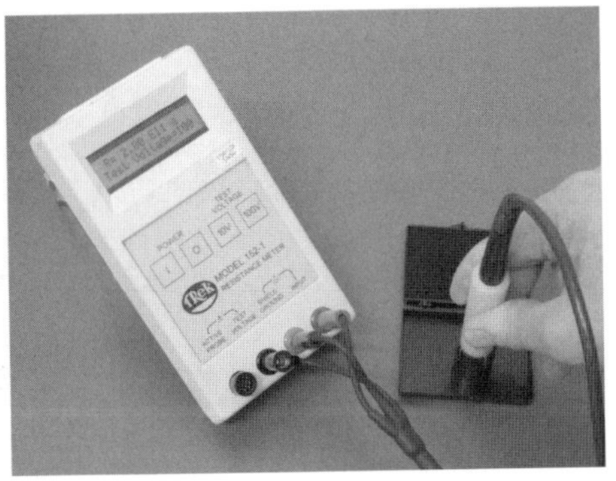

Figure 11.4 Two-point probe electrode according to IEC 61340-2-3 and ESD STM11.13.

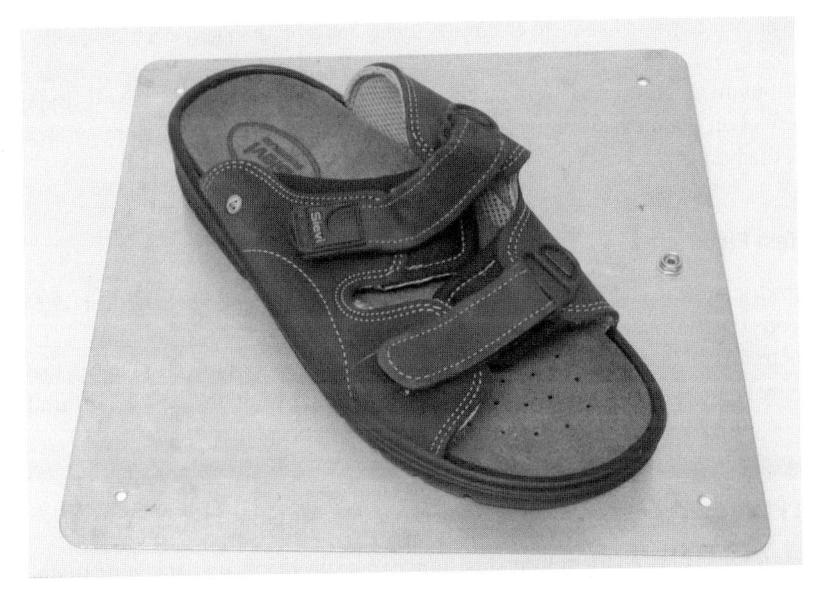

Figure 11.5 An example of a footwear test electrode.

11.6.7 Handheld Electrode

A handheld electrode is used in several types of measurement of grounding of personnel (Sections 11.8.3.2, 11.8.3.3 and 11.8.3.4). It can also be used for in-service system measurements of gloves or tools (Section 11.9.7.1, 11.9.7.3) and walk test of footwear and flooring (Section 11.8.9).

A hand touch electrode in the form of a metal plate is often built into equipment such as simple pass/fail resistance testers of wrist straps or footwear. The tool test electrode (Section 11.6.8) can also be used in this way.

A handheld electrode specified in IEC 61340-4-5, IEC 61340-4-9 (International Electrotechnical Commission 2016b), ESD STM97.1 (EOS/ESD Association Inc. (2015f)), ESD STM97.2 (EOS/ESD Association Inc. (2016c)), and ESD STM2.1 is a metal rod or tube about 25 mm diameter and at least 75 mm long that has a connector for connection of

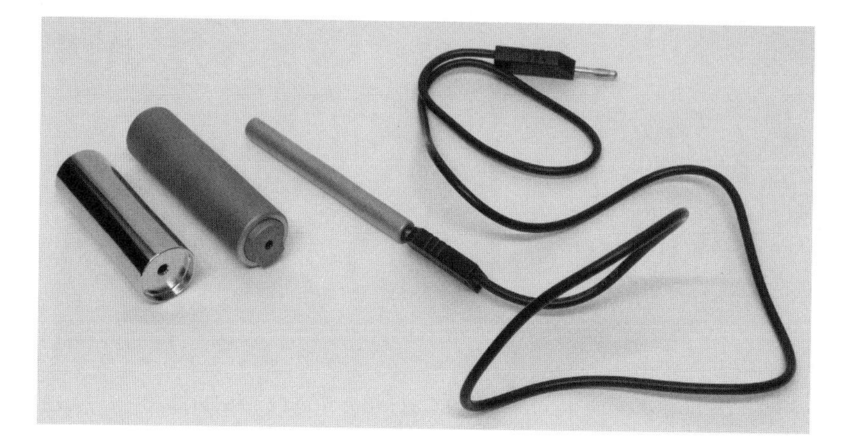

Figure 11.6 Example of standard handheld electrodes (right) and a nonstandard electrode (left).

leads to test equipment (Figure 11.6). In most cases, the handheld electrode size and shape will have little effect on the measurement results, providing there is sufficient contact area with the skin of the hand.

11.6.8 Tool Test Electrode

A tool test electrode can be used for testing the resistance to ground of tools (Section 11.9.5) and soldering irons (Section 11.9.6.2).

A tool test electrode can be a simple metal plate having a means of connection to the test equipment (Figure 11.7). The plate must be electrically isolated from any parallel ground paths. This can be done by supporting it on an insulating support plate or feet.

11.6.9 Metal Plate Electrode for Volume Resistance Measurements

A metal plate electrode is needed for volume resistance measurements on packaging materials (Section 11.8.5). The electrode must be sufficiently large to support the packaging material under test and must be made of a material such as stainless steel that does not

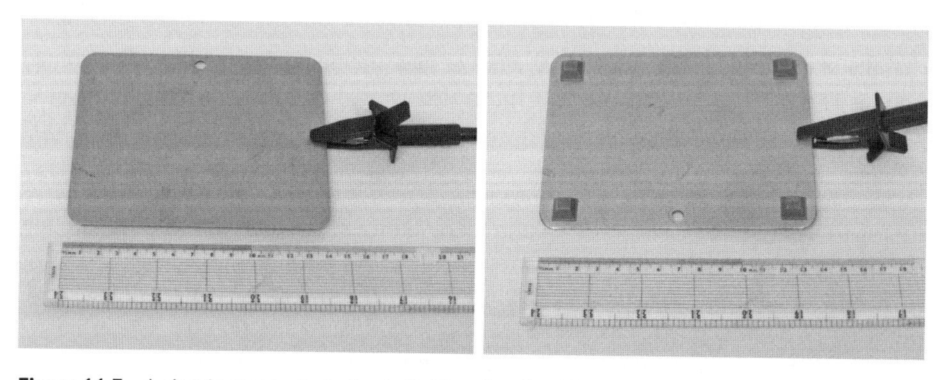

Figure 11.7 A simple tool test electrode (left) and underside (right) showing insulating feet.

oxidize. Aluminum is not suitable due to its formation of an alumina surface layer that can affect measurements.

11.6.10 Insulating Supports

Insulating supports are needed in some tests to prevent parallel conduction paths through the supporting material. The main requirement for these is that the support should be large enough to completely support and isolate the test item and at least 10 times higher resistance than expected for the item under test. So, for example, an insulating support for a packaging material that could have resistance up to over 100 GΩ (10^{11} Ω) should itself have resistance over 1 TΩ (10^{12} Ω). Insulating supports can often conveniently be made from common insulating plastic or rubber sheets. The surface and volume resistance should be verified as described in Sections 11.8.4 and 11.8.5.

11.6.11 ESD Ground Connectors

In resistance to ground tests, it is necessary to connect one terminal of the test meter with ESD ground. A suitable connector for this may depend on what ESD ground is defined. In cases where mains electrical safety earth is used as ESD ground, proprietary connectors may be used to connect to the earth pin (Figure 11.8).

11.6.12 Electrostatic Field Meters and Voltmeters

11.6.12.1 Electrostatic Field Meter-Based Instruments

Many electrostatic voltmeters are electrostatic field meters, calibrated to read voltage when positioned at a certain distance from a large flat conducting target surface. These have a wide field of view. The voltage reading is reduced for small objects and can be influenced by nearby charged surfaces or fields.

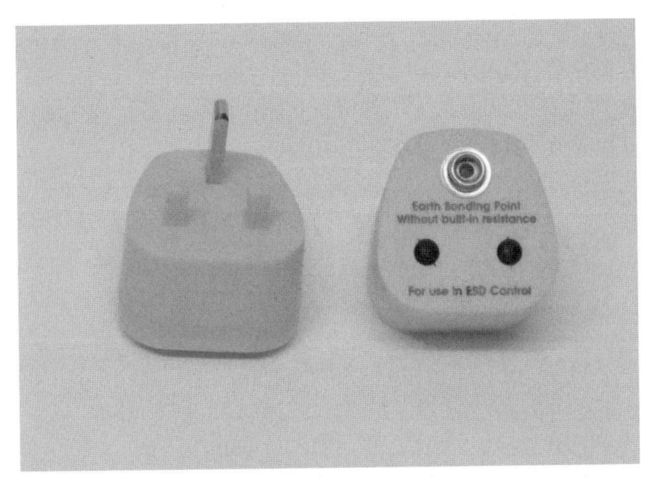

Figure 11.8 An example of a 0 Ω mains safety earth connector.

Figure 11.9 Electrostatic field meters. (left) Induction type and (center and right) two field mill type instruments.

The apparent voltage reading increases if the meter is brought closer to the surface. A correction can be made if the distance is known, but it is normally most convenient to make readings with the field meter/voltmeter held at the calibrated distance. Some meters have a means of gauging the correct distance, e.g. converging lights or pillars that are placed in contact with the surface.

These instruments operate by measuring the charge induced on a capacitive sensor plate by an electrostatic field. The instrument must be grounded correctly when measurements are made.

Some types of simple low-cost induction type instrument sense the field by charge induced on a simple metal plate (Figure 11.9, left). These tend to suffer from drift. It is important to zero these instruments before each measurement, with the sensitive aperture shielded by a grounded surface or in a zero-field region, before each measurement is made.

"Field mill" type instruments have a rotating mechanical shield interrupting the electrostatic field to the sensor (Figure 11.9, center and right). These instruments are self-zeroing and do not suffer from drift as much as the simple instruments do.

Field meters used to measure surface voltage of insulators for evaluation of risks in 61340-5-1 and 20.20 should be calibrated to read voltage at a distance of 2.5 cm (1 in.). Meters calibrated at other distances will give different voltage readings.

11.6.12.2 Noncontact Electrostatic Voltmeters

True noncontact electrostatic voltmeters are available, based for example on vibrating reed sensor technology (Figure 11.10). These must usually be held within a specified range of distance several millimeters from the surface. Within this range, the voltage reading is reasonably unaffected by distance from the surface. The instrument measures a surface voltage from a relatively small area close to the voltmeter sensor. They can be used to measure the voltage on moderately small items or areas around a centimeter dimension.

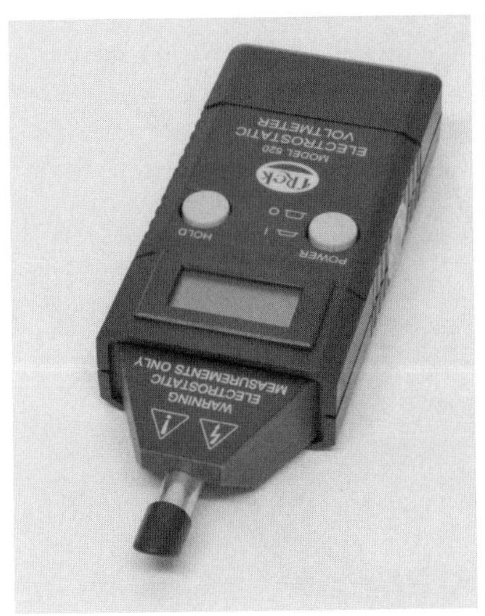

Figure 11.10 A handheld noncontact electrostatic voltmeter. The sensitive tip contains a vibrating reed sensor.

11.6.12.3 Contact Electrostatic Voltmeters

Ultra-high input resistance and ultra-low input capacitance contact voltmeters are now available (Figure 11.11). These can be used to measure voltage on small conductive or insulative items or surfaces. The sensor tip is touched to the surface of the item being measured.

11.6.13 Charge Plate Monitors (CPM)

Ionizers are tested using a charged plate monitor (CPM). A CPM can also be used for making measurements of charge decay (e.g. of tools or gloves), body voltage in a walk test, and as a means of estimating charging of objects.

A CPM has a metal plate that can be charged to a high voltage. It has an electrostatic volt meter monitoring the voltage on the plate (Figure 11.12) and a means of charging the plate to a known voltage (usually a little over 1000 V).

For ionizer performance measurements, the CPM is placed in a position simulating the normal working position of an ESD-sensitive (ESDS) board or component. After charging to an initial voltage, the time taken for the voltage on the metal plate to be neutralized by the ionizer is measured. After the voltage neutralization time, the CPM monitor will typically still show a residual reading. This is due to ionizer offset voltage, caused by ion output imbalance. This offset voltage is also an important characteristic of the ionizer.

There are many different CPMs on the market, ranging from clip-on attachments to electrostatic field meters to sophisticated dedicated instruments with built-in decay time and offset measurement capability.

To make measurements according to the ESD STM 3.1 or 61340-4-7 (International Electrotechnical Commission 2017) standards, a CPM designed according to the requirements of these standards is required. This must have a plate of size 15×15 cm that has a capacitance of 20 ± 2 pF (Figure 11.13).

Figure 11.11 A contact voltmeter. Source: D E Swenson. The sensing tip has extremely high impedance.

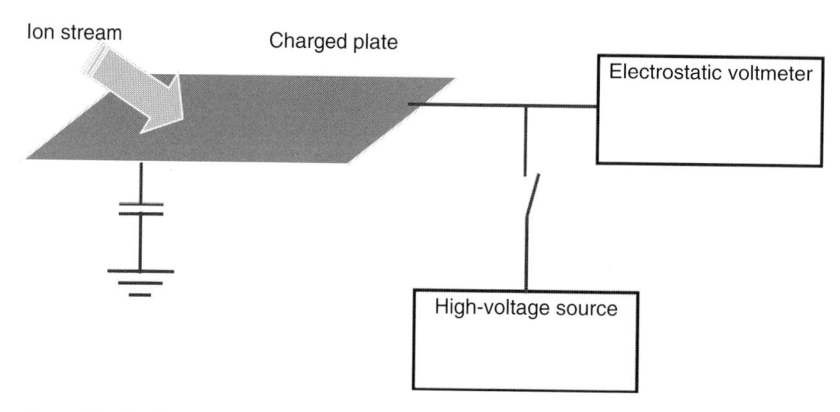

Figure 11.12 Principle of CPM operation.

Many CPMs on the market are not built according to these standards and have a smaller plate size and capacitance. They are often smaller and lighter than a standard CPM and are used for functional and comparative measurements with ionizers. These may often give different results to the standard CPM and should be correlated to a standard CPM for compliance verification measurement against standard ionizer requirements.

In practice, most organizations will use a nonstandard CPM for comparative and functionality measurements on ionizers in their working positions in a workstation. In some cases, a smaller sized nonstandard CPM plate can better represent the smaller area of a typical ESDS device.

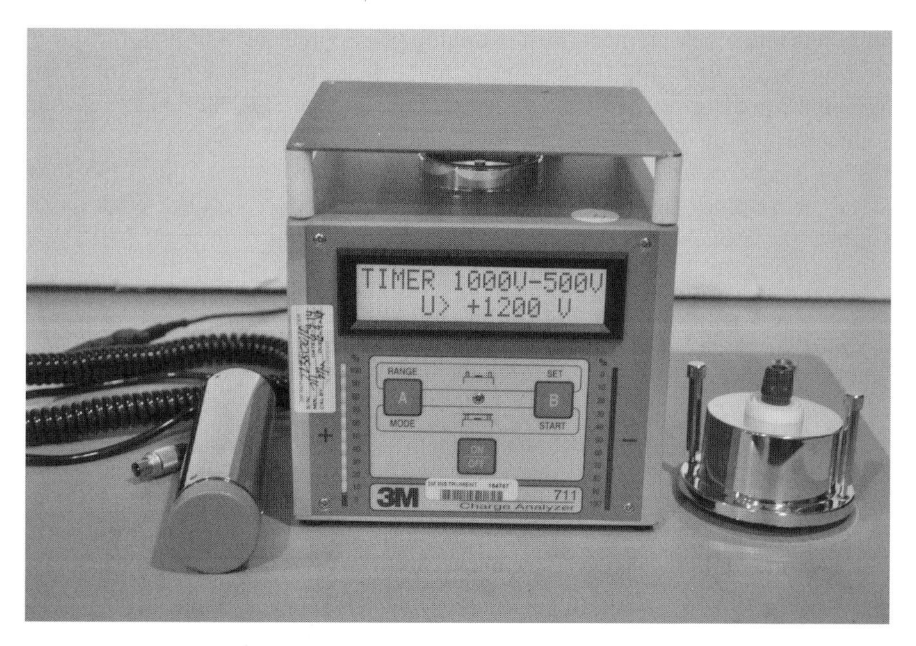

Figure 11.13 Example of a CPM built according to the IEC 61340-4-7 and ESD STM3.1 standard. Note the large 15 × 15 cm plate. Handheld electrode and voltmeter attachment accessories are also shown. Source: D E Swenson.

11.7 Common Problems with Measurements

11.7.1 Humidity

It is good practice to measure the ambient relative humidity when making electrostatic measurements. High ambient humidity can cause insulating materials to show lowered surface resistance $<10^{11}\ \Omega$ (100 GΩ). The instrument may be measuring the resistance of the water layer adsorbed on the material. If in doubt, repeat the measurement under dry (<30% rh) conditions.

11.7.2 Accidental Measurement of Parallel Paths

It is easy to accidentally "short circuit" the object you are trying to measure by a lower resistance path. One common mistake is to hold the test probes and/or object under test in the hands – and inadvertently measure the resistance of a parallel conduction path through the body!

Some objects such as garments must be placed on a highly insulating surface for test to avoid the underlying surface affecting the measurement. The surface may need to be cleaned before use – for example with a material such as isopropyl alcohol wipes, which lift off the contamination. The alcohol evaporates quickly but allows the surface to dry for a reasonable time before use.

Some test leads and probes may not be as good insulators as would be desired. If the insulating parts of these are held in the hand or placed on a dissipative surface, leakage to the hands may affect the measurement. This is usually only a problem with high-resistance measurements above about 10^{10} Ω (10 GΩ) but can be worse under high-humidity conditions. Sometimes test leads can be further isolated for use in high-resistance measurements by threading them through a clean polyethylene tube.

Sometimes it is necessary, e.g. for resistance to ground measurement, to use very long leads at least on one side of the meter. Connect the long lead on the earthy side of the meter to reduce leakage problems.

11.8 Standard Measurements Specified by IEC 61340-5-1 and ANSI/ESD S20.20

11.8.1 Resistance to Ground

Resistance to ground is a basic measurement used in many situations to evaluate the likelihood of charge dissipating quickly to ground. A 2.5 kg electrode is placed on the surface to be tested. A high-resistance meter is connected between the electrode and ESD ground (Figure 11.14).

11.8.1.1 Resistance to Ground from a Work Surface or Floor

Figure 11.15 shows measurement of resistance to ground of a work surface. The same method is used for a floor, cart (trolley), or storage rack.

Equipment required:

- One 2.5 kg resistance measurement electrode.
- Connector to ESD ground (0 Ω resistance).
- 10/100 V high-resistance meter with test leads.
- Work surface or floor to be tested.

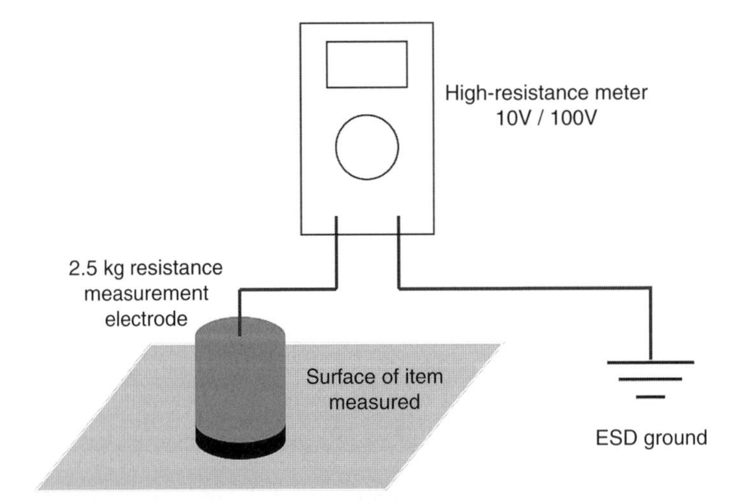

Figure 11.14 Measurement of resistance to ground.

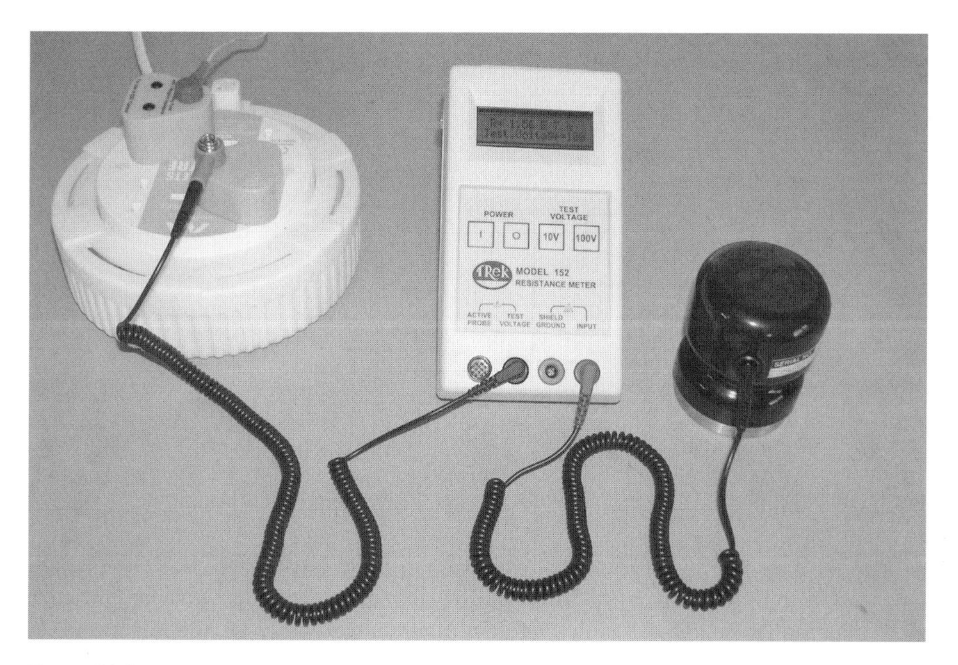

Figure 11.15 Measurement of resistance to ground from a work surface.

Procedure:

- Connect the resistance meter and place electrode on surface.
- Measure result at 10 V.
- If R > 1 MΩ, measure result again at 100 V.
- Note the result.

Common Problems

ESD ground is often, but not always, mains electrical safety earth. Make sure you have identified the correct ESD ground before making connection. The ESD ground should be specified in the ESD Control Program Plan document.

If the ESD earth is mains electrical safety earth, it is usually convenient to measure the resistance from the electrode to an ESD earth connector. Many of these have an internal 1 MΩ resistor. If present, the value of this must be subtracted from the measurement result. If the measured item has resistance very much greater than 1 MΩ, this added resistance may be neglected. To avoid confusion, it is often better to reserve ground cords and earth connectors that do not have these built-in resistors for use in ESD measurements.

If you check resistance to ground back to a convenient mains electrical earth point, don't forget to check that the earth point is really connected back to safety earth. An extension lead can be found to be unplugged or the electrical earth had somehow disconnected!

A higher than expected resistance result can be due to dust and dirt or other contamination accumulated on the surface. This is an indication that the cleaning regime may need to be improved.

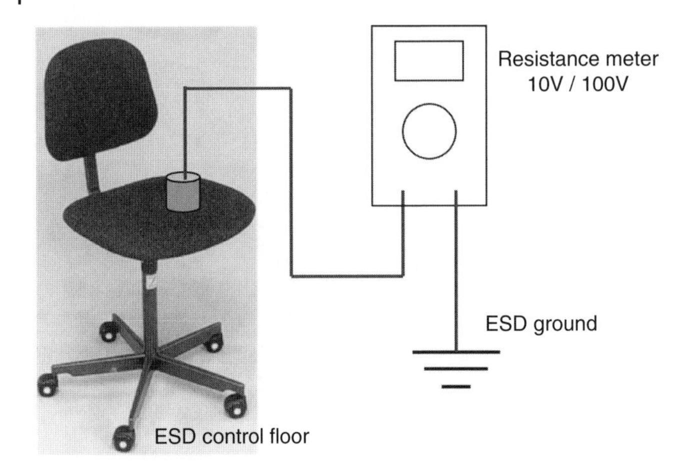

Figure 11.16 Compliance verification resistance to ground measurement of seating.

If the surface is dusty, it can be wiped with a dry cloth or paper towel before repeating the measurement. If the surface is sufficiently dirty to require cleaning with a liquid cleaning material, it must be allowed to dry before measurement; otherwise, the measurement results may be affected. Even a small amount of moisture on the surface will lower the measured resistance results.

11.8.1.2 Compliance Verification of Seating in the EPA

This simple resistance to ground measurement is used for compliance verification test of seating in the EPA including the ground path through the flooring. It is done while the seat is standing on the ESD control floor at the EPA workstation (Figure 11.16).

Equipment required:

- One 2.5 kg resistance measurement electrode.
- ESD earth connector.
- 10/100 V high-resistance meter with test leads.
- Sample chair for testing.

Procedure:

- Place electrode on chair seat.
- Connect the resistance meter between electrode and ESD earth.
- Measure result at 10 V.
- If $R > 1\,M\Omega$, measure result again at 100 V.

11.8.1.3 Qualification of Seating

In product qualification of seating, we want to measure the characteristics of the chair alone separate from the flooring system. Also, we want to know that all relevant parts of the chair such as the back and the arms are also suitable for ESD control use. So, the resistance measurement is made to a groundable point, e.g. an electrode placed under the wheel or foot

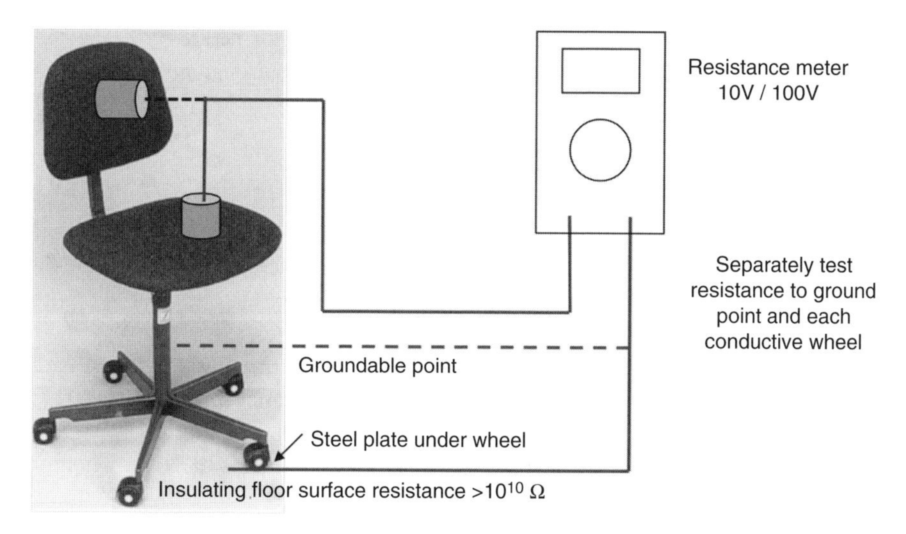

Figure 11.17 Product qualification measurement of seating.

(Figure 11.17). Resistance to several wheels or feet may be tested to see how many provide groundable points.

To isolate the chair from any conduction paths through the floor, the chair must be placed on an insulating surface that has a resistance considerably higher than the resistance expected of the chair. This is a test that is best done in a test workshop and not in the EPA.

Equipment required:

- One 2.5 kg resistance measurement electrode.
- 10/100 V high-resistance meter with test leads.
- Insulating support $>10^{10}\ \Omega$.
- Metal plate electrode (for under wheel).
- Sample chair for testing.

Procedure:

- Set the chair on the insulating support. Place metal plate electrode under a grounding wheel or foot.
- Place the resistance measurement electrode on chair seat.
- Connect the high-resistance meter.
- Measure result at 10 V.
- If $R > 1\ \text{M}\Omega$, measure result again at 100 V.

Common Problems

It is usually wished to test parts of the chair such as the seat back or arms, which may not be horizontal or have large flat areas on which an electrode can be easily balanced. Some ingenuity is required to get around these problems. If the chair has an easily connected groundable point, then the matter is simplified. The resistance from seat back to groundable point may be measured with the chair laid on its back on an insulating support plate.

If the chair has an exposed metal part of the chassis that is confirmed to be connected reliably to the groundable point, then the resistance of arms and back can be measured to this metal part in subsequent measurements.

11.8.2 Point-to-Point Resistance

Point-to-point resistance has in current practice replaced surface resistivity measurement for evaluating surfaces such as bench mats, rack and cart surfaces, and garments. The method consists of a simple resistance measurement between two electrodes placed on the surface (Figures 11.18 and 11.19).

Measurements can be made in several locations and different orientations to evaluate if there is variation or directionality in the material surface resistance characteristic. This can happen with contaminated or worn surfaces and nonhomogenous materials. Measurements should be made across surface features such as garments seams to detect whether continuity is broken by these boundaries.

11.8.2.1 Point-to-Point Resistance of Work Surface

During product qualification of a work surface, a point-to-point resistance test will give an evaluation of the resistance of the surface material (Figure 11.19). An ESD program will often give a minimum surface resistance requirement to prevent charged device ESD risks. Making several measurements in different areas and orientations of electrodes will indicate the variability of the surface resistance and whether there is any directionality. IEC 61340-2-3 specifies that the electrodes should be placed at least 250 mm apart center to center. This may not be practical in some cases (e.g. measurements on garments).

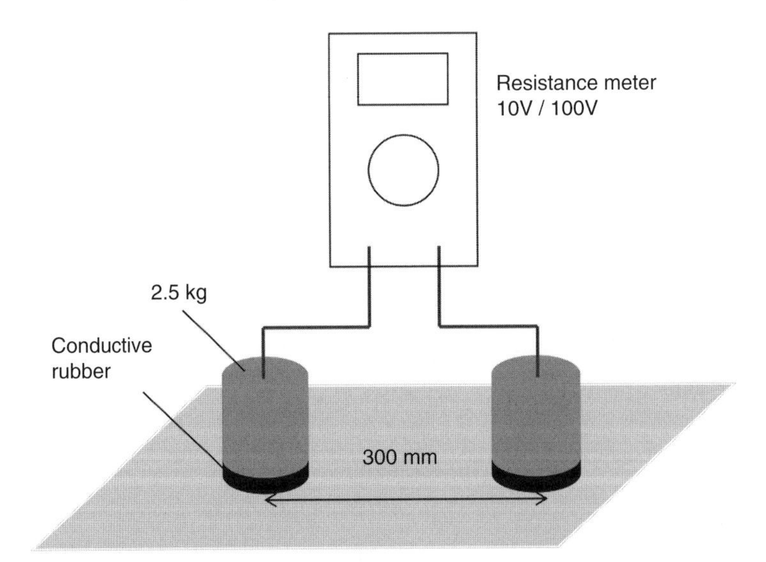

Figure 11.18 Point-to-point resistance measurement.

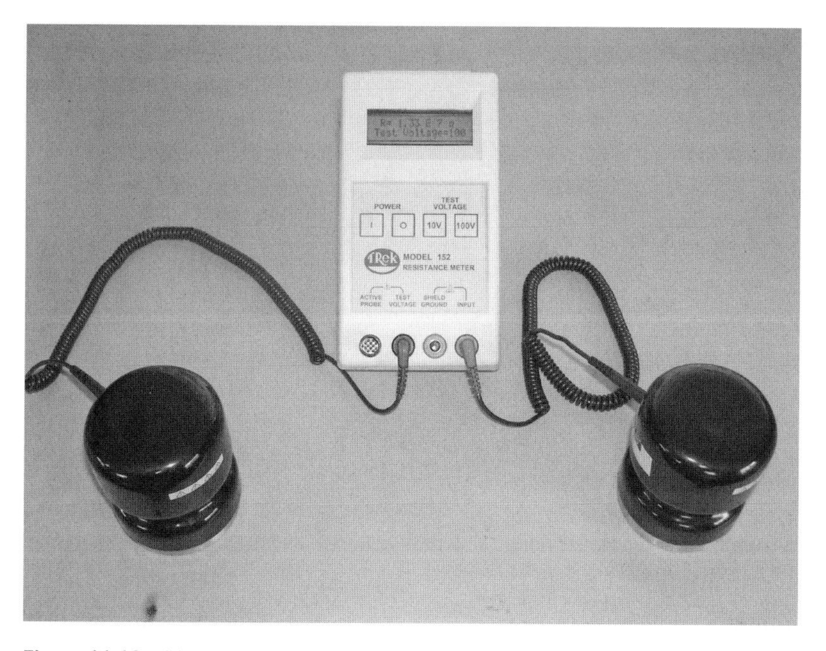

Figure 11.19 Measurement of point-to-point resistance of a work surface material.

Equipment required:

- Two 2.5 kg resistance measurement electrodes.
- 10/100 V high-resistance meter.
- Sample work surface.

Procedure:

- Place the electrodes on work surface at least 250 mm apart.
- Connect the resistance meter.
- Measure result at 10 V.
- If R > 1 MΩ, measure result again at 100 V.
- Note the result.

The measurement can be repeated in different locations and orientations to detect whether the material has significant variation with position or direction.

11.8.2.2 Point-to-Point Resistance Measurements on ESD Garments

The fabric and structure of ESD control garments can be measured using simple point-to-point measurements both for product qualification and for compliance verification. IEC 61340-4-9 and ESD STM2.1 specify tests for ESD control garments. Measurement point to point across seams verifies that the fabric conductive fibers make contact across the seams

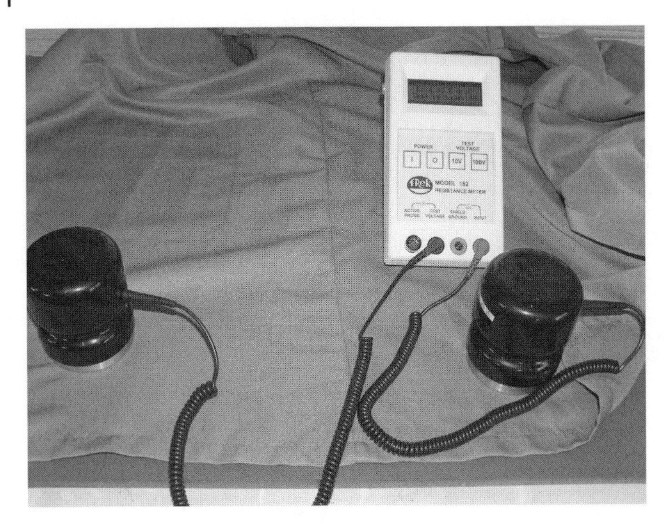

Figure 11.20 Measurement of resistance point to point of a garment across panels. The garment has been placed on a slab of insulating support material.

(Figure 11.20). Cuff-to-cuff measurements verify the connection of the garment materials via the materials and seams between the cuffs.

11.8.2.3 Simple Point-to-Point Measurement on the Garment Fabric

The garment material to be tested must be placed on an insulating support having a surface resistance at least 10 times the expected resistance of the garment material.

Some garments have a groundable point, for connection of a ground strap. In this case, a resistance to groundable point measurement should be measured to include the garment seams.

Equipment required:

- 2 off 2.5 kg resistance measurement electrodes.
- 10/100 V high-resistance meter with test leads.
- Garment specimen to be tested.
- Insulating support.

Procedure:

- Place the insulating support on a working surface. Lay out the garment on the insulating support, if possible, with a single layer of fabric.
- Place the electrodes on separate panels of the garment a convenient distance apart.
- Connect the high-resistance meter.
- Measure resistance result at 10 V.
- If $R > 1\,\mathrm{M\Omega}$, measure result again at 100 V.
- Note the result.

Making several measurements in different areas of the material and orientations of electrodes will indicate the variability of the surface resistance, and whether there is any directionality.

Common Problems

Many garment materials have high point-to-point resistance. Measurements above 10^{10} Ω (10 GΩ) tend to be unreliable, and results are dependent strongly on humidity.

If the insulating mat is not sufficiently insulating, the measurement may be affected, and the resistance result reduced. Check the point-to-point resistance of the insulating support before the test to make sure it's at least 10 times the expected resistance of the garment.

Measurement from Garment Cuff to Cuff

Cuff-to-cuff measurement tests the garment sleeve and garment body panel resistance and the connection between these as a system (Figure 11.21). Connection can be made to the inside of the sleeve to include the area that contacts the user's arms. This measurement should be done for garments that are designed to make contact with the wearer's skin at the cuffs. Alternatively, connection can be made to the outside of the cuffs (Figure 11.22).

Equipment required:

- 2 off 2.5 kg resistance measurement electrodes.
- 10/100 V high-resistance meter with test leads.
- Garment specimen to be tested.
- Insulating support.
- Insulating cuff inserts. (These can be made from sheet insulating plastic material.)

Procedure:

- Place the insulating support on a working surface. Lay out the garment on the insulating support.
- Place the electrodes on separate panels of the garment.
- Connect the high-resistance meter.

Figure 11.21 Measurement of cuff-to-cuff resistance of a garment, (above) with garment hanging or (below) with garment resting on an insulating support. Source: D E Swenson.

Figure 11.22 Contacting (left) the inside or (right) the outside of a garment cuff with a 2.5 kg electrode. The garment is placed on an insulating support. An insulating separator is used when contacting the outside of the garment sleeve.

- Measure resistance result at 10 V.
- If $R > 1\,\text{M}\Omega$, measure result again at 100 V.
- Note the result.

Measurement of Sleeve-to-Sleeve Resistance of a Garment Using Hanging Clamps

An alternative means of connecting to the garment cuffs is by hanging clamps (Figure 11.23) suspended from an insulating frame.

Equipment required:

- Two clamp electrodes suspended from an insulating frame.
- 10/100 V high-resistance meter with test leads.
- Garment specimen to be tested.

Procedure:

- Connect the clamp electrodes to the garment cuffs and suspend the garment.
- Connect the high-resistance meter to the clamp electrodes.
- Measure resistance result at 10 V.
- If $R > 1\,\text{M}\Omega$, measure result again at 100.
- Note the result.

Measurement of Resistance to Groundable Point of a Garment

For garments that have a groundable point that is connected to ESD ground, the resistance should be measured between a 2.5 kg electrode placed on the material (cuffs and panels) and the groundable point. If the cuffs are designed to ground the garment by contact with the wearer's body, the cuff body contact areas should be considered groundable points,

Figure 11.23 Measurement of cuff to cuff resistance of a garment hung from clamps. Source: D E Swenson.

and the resistance should also be measured to these. The resistance should be measured between a 2.5 kg electrode placed on the material panels and the inner surface of the cuffs (Figures 11.21 and 11.22).

11.8.3 Personal Grounding Equipment Tests

11.8.3.1 End-to-End Resistance of a Ground Cord

End-to-end resistance is a simple resistance measurement used for items such as wrist strap cords. This measurement is normally easy, providing there is a suitable means of making the connections (Figure 11.24).

Equipment required:

- 10/100 V resistance meter with appropriate leads and connectors
- Ground cord to be tested

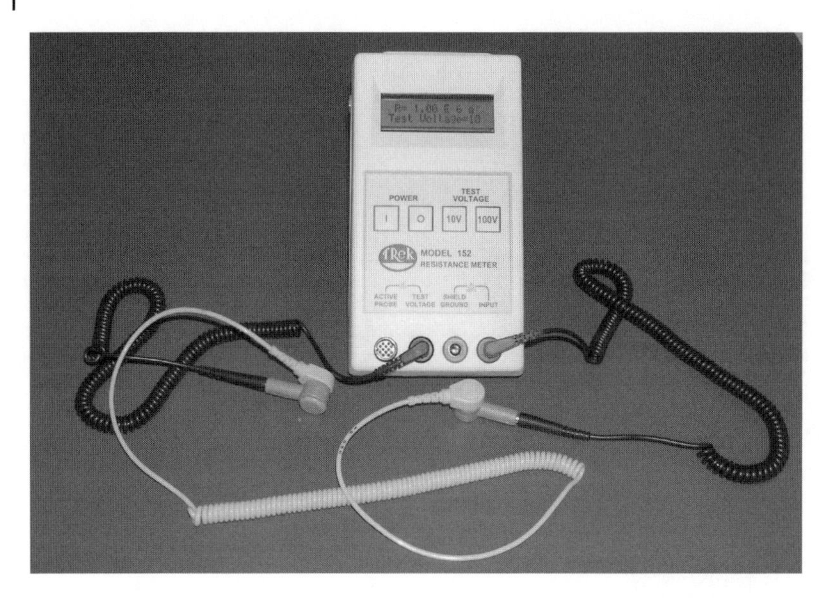

Figure 11.24 Measurement of end-to-end resistance of a wrist band cord.

Procedure:

- The wrist strap cord to be tested is connected to the resistance meter using appropriate connectors.
- Note the result.

Common Problems

Do not be tempted to hold the cord ends in contact with test leads with the fingers. If you do this, you risk connecting your body in parallel with the cord. This would then give an incorrect result. Also, make sure any connectors used do not contact a conductor such as an ESD control bench surface.

11.8.3.2 Measurement of Wrist Straps and Cords as Worn

This test is often done using a proprietary wrist strap checker giving a simple pass/fail verdict (Figure 11.25). It can also easily be done using a resistance meter (Figure 11.26).

Equipment required:

- Handheld electrode
- 10/100 V resistance meter with test leads
- Wrist strap and cord to be tested

Procedure:

- The subject to be tested should wear the wrist strap in contact with their skin.
- The wrist strap cord to be tested is connected to one terminal of the resistance meter.
- The second terminal of the resistance meter is connected to the handheld electrode.
- The subject holds the electrode. Note the result.

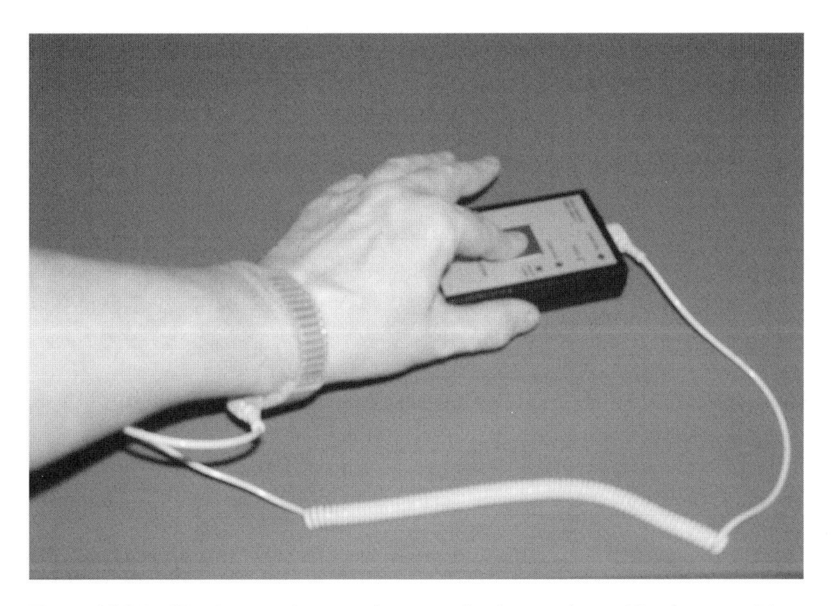

Figure 11.25 Simple proprietary wrist strap checkers make verification easy. The pass/fail limits of these checkers must correspond with the requirements of the ESD control program.

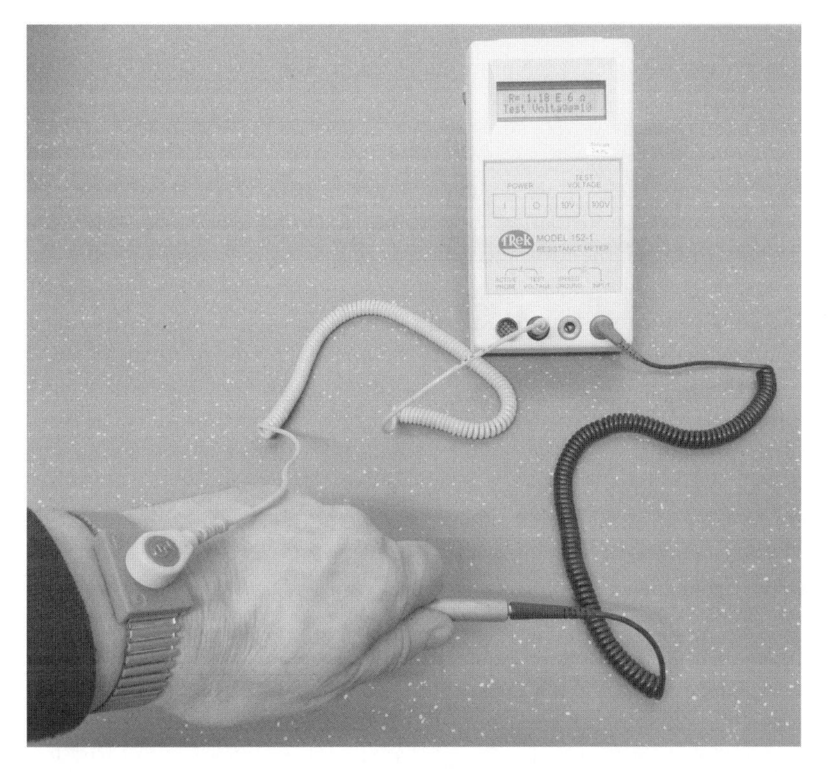

Figure 11.26 A handheld electrode used with suitable resistance meter allows checking of wrist strap as worn.

Common Problems

If a wrist strap checker is used, the pass/fail limit of the checker must be selected to correspond with the limits specified in the ESD control program.

Many wrist strap checkers also give a "low fail" if the resistance is less than a minimum value. This "low fail" limit must correspond to the lower limit specified in the ESD control program.

Note that current versions of 61340-5-1 and 20.20 do not specify a low limit for wrist strap resistance, (see Section 7.5.3). Nevertheless, many organizations will often wish to specify minimum resistance from body to ground for safety in the presence of high voltages or other reasons.

11.8.3.3 Measurement of Personnel Grounding Through Footwear to Foot Plate Electrode

This test measures the resistance from the body to a ground electrode formed by a metal plate. It is this measurement that is commonly performed by footwear test boxes found at the entrance of an EPA but can also easily be done with a resistance meter.

Equipment required:

- Handheld electrode
- 10/100 V resistance meter with test leads
- ESD footwear
- Foot plate electrode, big enough to accommodate an entire foot
- Insulating support mat

Procedure:

- The foot plate is placed on the insulating support mat and connected to the resistance meter. The second terminal of the resistance meter is connected to the hand touch electrode.
- The subject should wear the footwear to be tested.
- The subject stands with one foot on the foot plate and one foot on the insulating mat.
- The subject holds the handheld electrode.
- Note the result.
- repeat test for the other foot.

Common Problems

If a proprietary footwear checker is used, make sure that the upper and lower pass/fail limits correspond with those specified in the ESD control program.

11.8.3.4 Measurement of Resistance from Person to Ground

This measurement is useful in checking the complete installed ground path from person to ground, whether it be via wrist strap (Figure 11.27) or footwear and flooring or via a groundable garment (Figure 11.28). It is based on IEC 61340-4-5 and ANSI/ESD STM97.1. It can be done with an operator at a workstation with their personal grounding in its operating state, with little interruption to their work (Figure 11.27). It shows whether the personnel resistance to ground is within the required limits, as a system including the workstation

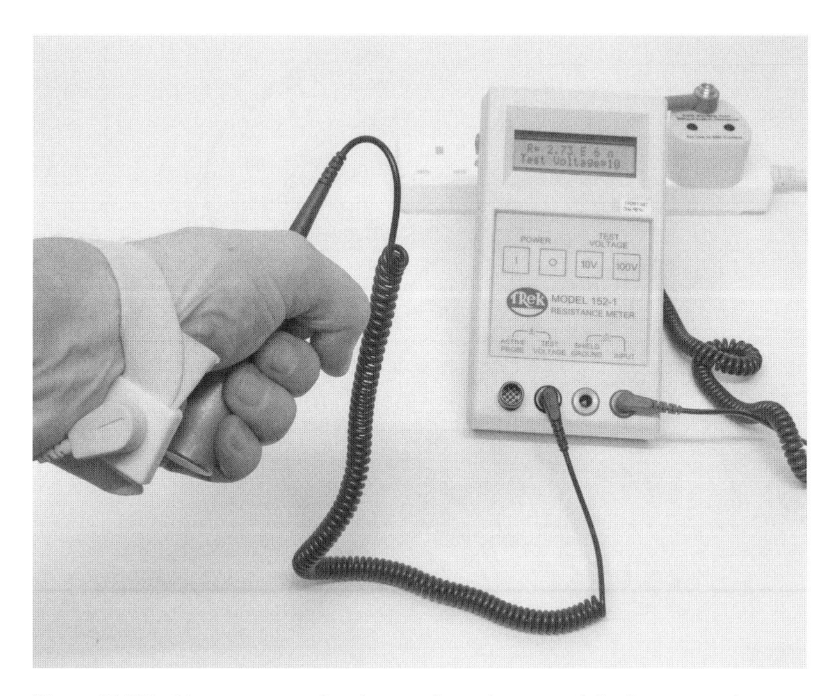

Figure 11.27 Measurement of resistance from the person's body to ground.

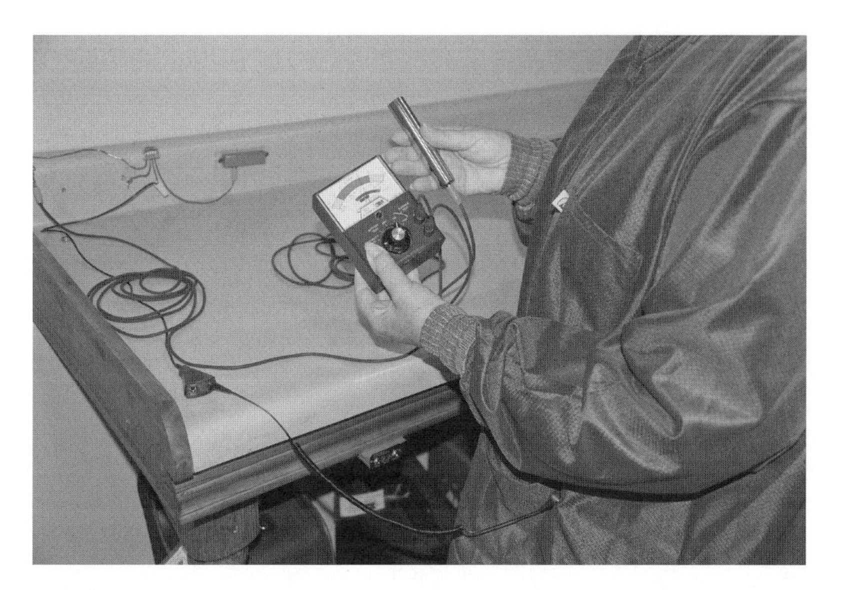

Figure 11.28 Measurement of resistance to ground when grounded via a groundable garment. Source: D E Swenson.

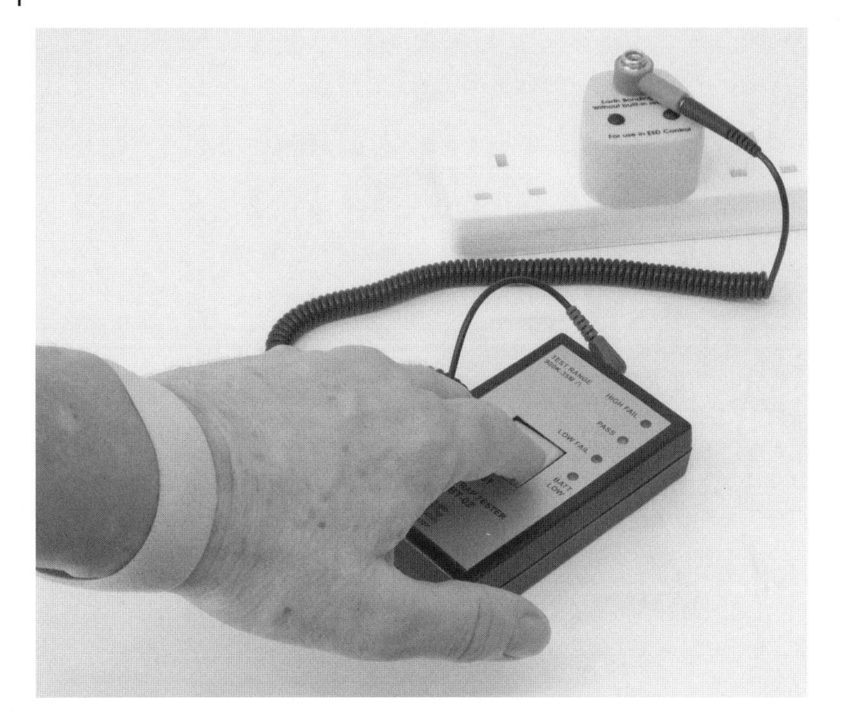

Figure 11.29 Checking the resistance from person to ground using a portable wrist strap checker.

earth bonding point or EPA floor. The test can also be done with a portable wrist strap checker (Figure 11.29) or footwear checker, providing the pass/fail thresholds correspond to the ESD control program requirements.

Equipment required:

- Handheld electrode
- 10/100 V resistance meter with test leads
- ESD ground connector (0 Ω resistance)

Procedure:

- The person to be tested should wear their personal grounding equipment (wrist strap or ESD control footwear). The wrist strap should be connected to ground in its normal working position. For footwear-flooring grounding, the person should stand on the ESD control floor to be tested.
- Connect one side of the resistance meter to ESD ground via the ESD ground connector.
- Connect the other side of the resistance meter to the handheld electrode.
- The subject holds the electrode. Note the result.

Common Problems

The test gives an easy way of checking personal grounding as a system in operation. Don't forget to check that the ESD earth point connected to the resistance meter is also connected back to ESD earth!

The test does not distinguish between individual grounding systems if more than one is in operation (e.g. simultaneous grounding through wrist strap and footwear-flooring).

Sometimes, personnel who have dry skin do not easily make good connection with wrist straps. This effect may be most evident just after the wrist strap is first put on and before a sweat moisture layer has formed between the wrist and the wrist strap band. Sometimes it can be necessary to augment the moisture layer, for example by using a skin lotion.

Contamination of the footwear sole or floor surface can lead to high-resistance results for footwear-flooring grounding. If necessary, clean these and retest. If liquid cleaners are used, make sure they are fully dried before retesting.

11.8.3.5 Resistance to Ground of an Earth Bonding Point

It is also necessary to verify the resistance to ground of the earth bonding points, e.g. on workstations.

Equipment required:

- 10/100 V resistance meter
- 0 Ω ESD ground plug
- Earth bonding point to be tested

Procedure:

- Connect one side of the resistance meter to ground via a 0 Ω ground plug.
- Connect the other side of the meter to the earth bonding point under test.
- Record the result.

Common Problems

This measurement is usually straightforward.

11.8.4 Surface Resistance of Packaging Materials

The surface resistance of ESD protective packaging material is commonly measured in three ways. Note that it is surface *resistance*, not resistivity, as it is not corrected for the electrode form.

- Using a concentric ring electrode
- Using a miniature two-pin point-to-point electrode
- Using two 2.5 kg resistance measurement electrodes in a point-to-point measurement

These three electrode systems often do not give the same results. ESD packaging materials can have highly variable surface resistance at different positions, and the electrodes respond differently to these variations. The concentric ring electrodes tend to give the lower range of resistance of the material around the circumference of the electrode (Smallwood 2017). It is not directional in response.

The two-pin point-to-point electrode tends to give a result about a factor of four greater than the concentric ring electrode, for a homogenous material of the same surface resistivity. It also is highly influenced by any small areas of high-resistance material under the point electrodes and is directional in response. So, this electrode gives greater variation in results if the material is variable in resistance. It is more likely to indicate the upper range of resistance of the material under the electrode points.

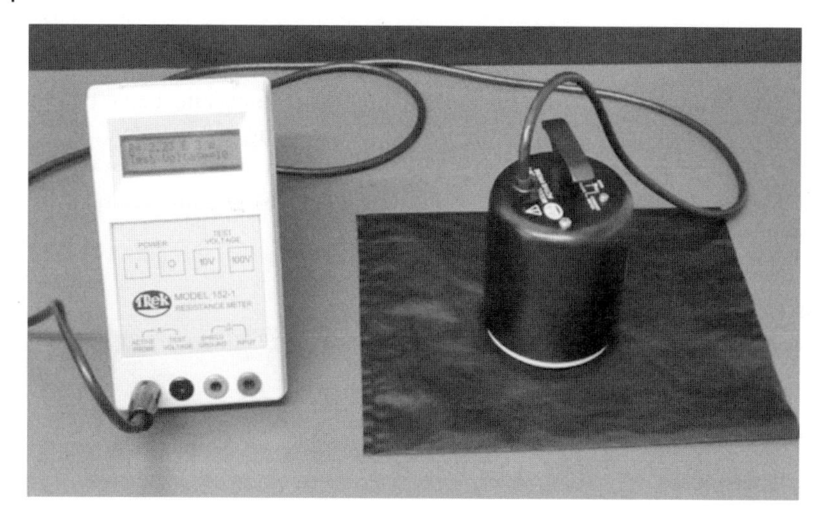

Figure 11.30 Measuring the surface resistance of an ESD protective packaging material. The sample is placed on an insulating base material.

A point-to-point measurement made with closely spaced 2.5 kg electrodes can give similar results to the concentric ring electrodes but has a directional response.

11.8.4.1 Surface Resistance of Packaging Measured Using a Concentric Ring Electrode

Surface resistance of large flat packaging surfaces can be measured using a concentric ring electrode (Figure 11.30).

Equipment required:

- 10/100 V resistance meter with test leads
- Concentric ring electrode
- ESD protective packaging material to be tested
- Insulating support

Procedure:

- Place the sample packaging upon the insulating support.
- Place the concentric ring electrode on the packaging surface.
- Connect the resistance meter to the electrode.
- Measure the resistance at 10 V.
- If $R > 1\,\mathrm{M\Omega}$, measure result again at 100 V.
- Note the result.

Common Problems

In practice, with high resistance, materials readings do not fully stabilize in a short time. In this circumstance, the reading should be taken after an appropriate electrification time, e.g. 15 seconds after application of the test voltage.

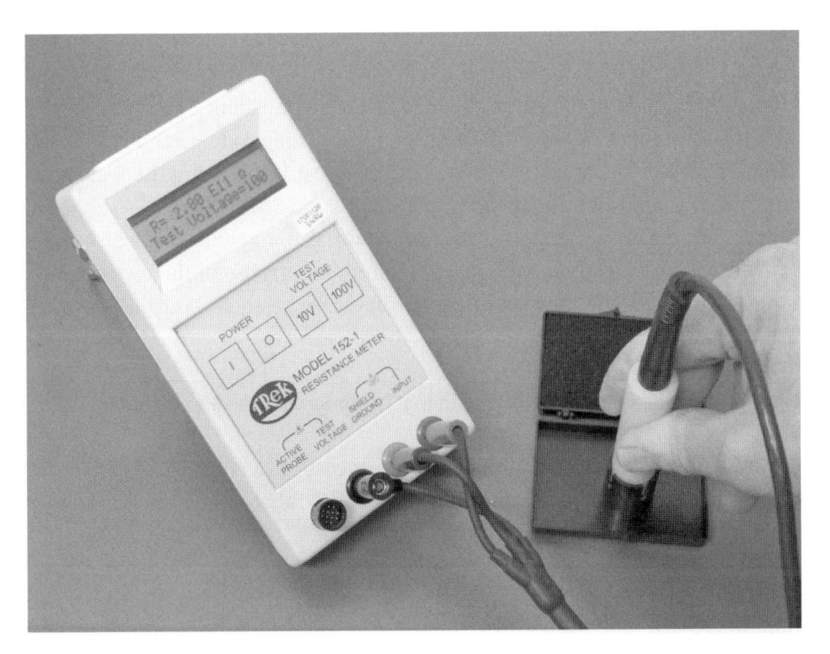

Figure 11.31 Measurement of surface resistance of small or curved ESD packaging using a two-pin probe.

The electrode needs a flat surface larger than the electrode. For profiled or curved surfaces and small items, the surface resistance is not correctly measured as the electrode contact area is reduced.

11.8.4.2 Point-to-Point Resistance of Small Packaging Items

Small or curved packaging surfaces can be measured using a two-pin surface resistance probe according to IEC 61340-2-3 or ESD STM11.13 (Figure 11.31).

Equipment required:

- Two-pin probe electrode with test leads
- 10/100 V resistance meter
- Insulating support
- ESD protective packaging to be tested

Procedure:

- Connect resistance meter to the electrode.
- Place the sample packaging upon the insulating support.
- Place the electrode in contact with the packaging surface.
- Measure result at 10 V.
- If $R > 1\ \mathrm{M\Omega}$, measure the result again at 100 V.

Common Problems

ESD protective packaging materials can have considerable variation in their surface resistance at different positions. This measurement is not directly equivalent to the concentric ring electrode or point-to-point 2.5 kg resistance measurement electrode methods and will give different results. This two-point electrode tends to give results showing the resistance of highest resistance of the material under the pins (Smallwood 2017, 2018). As such, the results may be highly variable compared to results found with the concentric ring electrode. This may be an advantage or a disadvantage depending on the purpose of the test.

For a homogenous material, the two-point probe may give results about 2.5–5 times higher than concentric ring electrodes.

11.8.4.3 Point-to-Point Resistance of Packaging Using 2.5 kg Resistance Measurement Electrodes

The surface resistance of large packaging items can be measured using 2.5 kg resistance measurement electrodes (Figure 11.32). This method is useful only on large flat packaging surfaces.

The sample under test should be placed on an insulating support. The point-to-point resistance of the support material should be at least 10 times the maximum acceptable resistance of the sample under test.

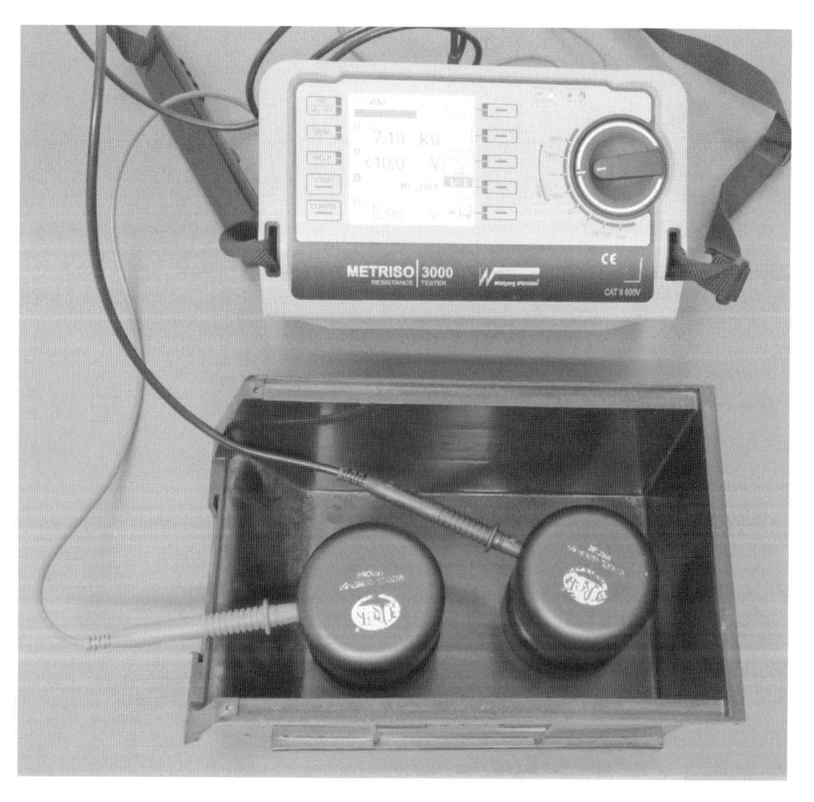

Figure 11.32 Measurement of surface resistance ESD packaging using two 2.5 kg solid resistance measurement electrodes.

The electrodes should be placed close together without risking touching each other to give results similar to the concentric ring electrode.

Equipment required:

- Two 2.5 kg resistance measurement electrodes
- 10/100 V resistance meter with test leads
- Insulating support
- ESD protective packaging to be tested

Procedure:

- Connect resistance meter to the electrodes.
- Place the sample packaging upon the insulating support.
- Place the electrodes close together on packaging surface.
- Measure result at 10 V.
- If $R > 1\ M\Omega$, measure result again at 100 V.
- Note the result.

Common Problems

This measurement is not directly equivalent to the concentric ring electrode or two-point probe methods and will give different results. The results vary with the distance between the electrodes. This arrangement is most likely to give results similar to the concentric ring electrodes when the two electrodes are separated by only a few millimeters.

11.8.5 Volume Resistance of Packaging Materials

The volume resistance of a packaging material gives the resistance from the outer surface to the inner surface. This can be measured using a concentric ring electrode. In this case, the test voltage is applied using a metal plate electrode under the packaging material (Figure 11.33). Resistance is measured to the inner electrode of a concentric ring electrode placed on top of the material under test. The outer electrode ring can be used as a grounded guard ring to eliminate surface conducted currents or can be left disconnected.

Equipment required:

- 10/100 V resistance meter with test leads
- Concentric ring electrode
- Metal plate electrode
- ESD protective packaging material to be tested
- Insulating support

Procedure:

- Place the metal plate electrode upon the insulating support.
- Place the sample packaging upon the metal plate electrode.
- Place the concentric ring electrode on the material surface to be measured.

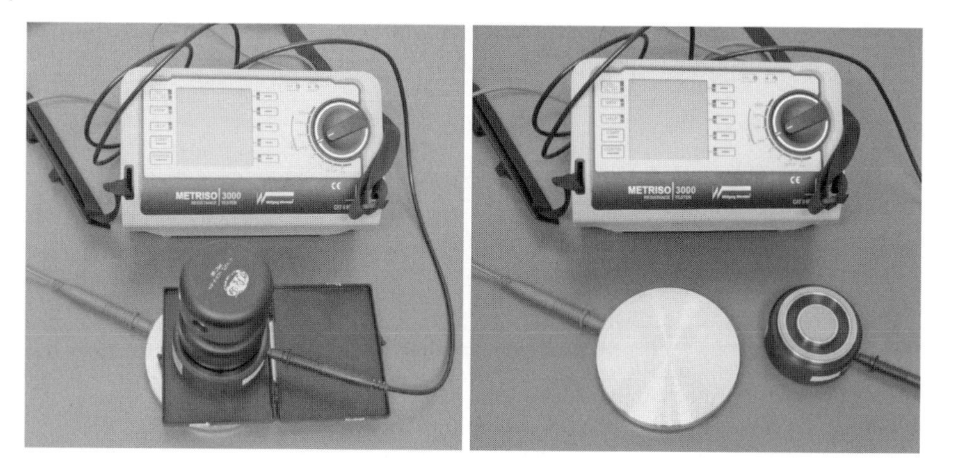

Figure 11.33 Volume resistance measurement using a concentric ring electrode (left). The electrodes are shown on the right. This concentric ring electrode requires a weight to give the correct total mass.

- Connect the resistance meter to the electrodes. The applied voltage is normally connected to the metal plate electrode, and the current sense terminal is connected to the inner ring electrode. The outer ring electrode is optionally connected to the resistance meter guard terminal, if available.
- Measure the resistance at 10 V.
- If R > 1 MΩ, measure result again at 100 V.
- Note the result.

Common Problems

In practice, with high resistance materials, the readings do not fully stabilize in a short time. In this circumstance, the reading should be taken after an appropriate electrification time, e.g. 15 seconds, after application of the test voltage.

11.8.6 ESD Shielding of Bags

Shielding packaging is defined as an enclosure that limits the passage of ESD current and energy to the device within the bag, such that the maximum energy from a 1000 V human body model (HBM) discharge is reduced to less than a limit specified in IEC 61340-5-3 or ESD S541. It is tested by application of a standard 1 kV HBM ESD to the outside of the enclosure, and measurement of the transient that appears at the position normally occupied by a sensitive component (Figure 11.34).

The shielding test is rather specialist and is normally beyond the means of ESD packaging users. It is normally performed as a type test on bags by manufacturers.

A capacitive sensor probe is placed in a central position in the bag under test. The sensor is connected via a 500 Ω wide bandwidth resistor to a current sensor probe and fast digital storage oscilloscope. The bag is placed on an earthed electrode, and an upper electrode is placed on top of the capacitive sensor outside the bag. The upper electrode is connected to a

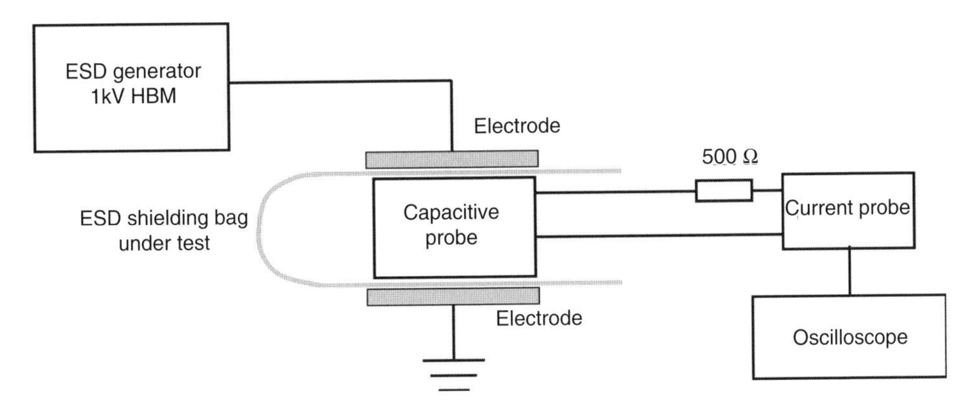

Figure 11.34 ESD bag shielding measurement.

standard HBM ESD generator. The applied waveform is similar in specification to that used to test ESD withstand voltage of components.

When an HBM ESD is applied to the upper electrode, a small impulse is detected by the capacitive electrode. A transient current flow through the resistor and current sensor and is recorded by the digitizing oscilloscope. The energy, W, in the detected transient is calculated from the digitized current samples I over n samples covering the duration the waveform.

$$W = 500 \sum_{0}^{n} I^2 dt$$

11.8.7 Evaluation of ESD Shielding of Packaging Systems

There is no measurement test method for ESD shielding of packaging systems other than bags. This is in part because such systems are highly variable in format. It would be difficult to design a test method that would be generally usable with all packaging types.

ESD shielding packaging systems can be evaluated by the procedure given in the flow chart of Figure 11.35. The surface resistance of the inner and outer packaging surfaces, and the volume resistance, can be measured using the methods of Section 11.8.4 and Section 11.8.5.

11.8.8 Measurement of Ionizer Decay Time and Offset Voltage

For ionizer qualification, ionizers can be placed in a measurement arrangement with a CPM in which the position of the CPM and ionizer are standardized. For compliance verification and measurement of ionizer performance in the workplace, the CPM should be placed in positions on the workstation representative of likely positions of the items requiring charge neutralization (Figure 11.36). Standard practices ANSI/ESD SP3.3 (EOS/ESD Association Inc. (2016a)) and ANSI/ESD SP3.4 (EOS/ESD Association Inc. (2016b)) specify that the CPM plate should face the direction of the air flow from the ionizer. In some cases, it may be more realistic to have the CPM plate in an orientation that imitates the target to be neutralized in practice.

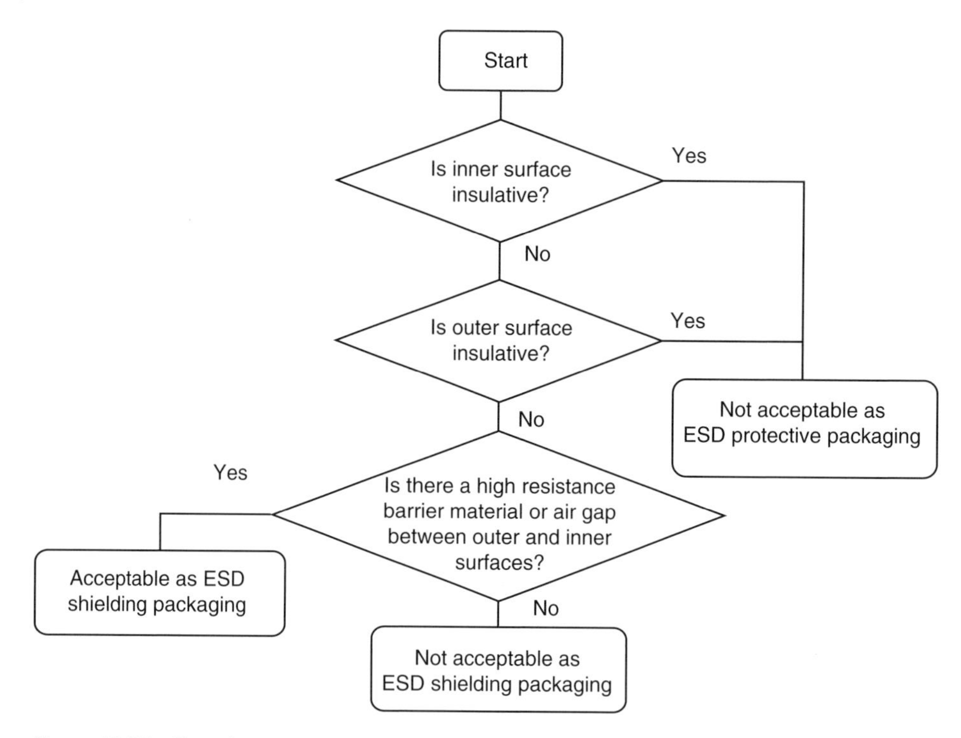

Figure 11.35 Flowchart procedure for evaluation of ESD shielding properties of packaging.

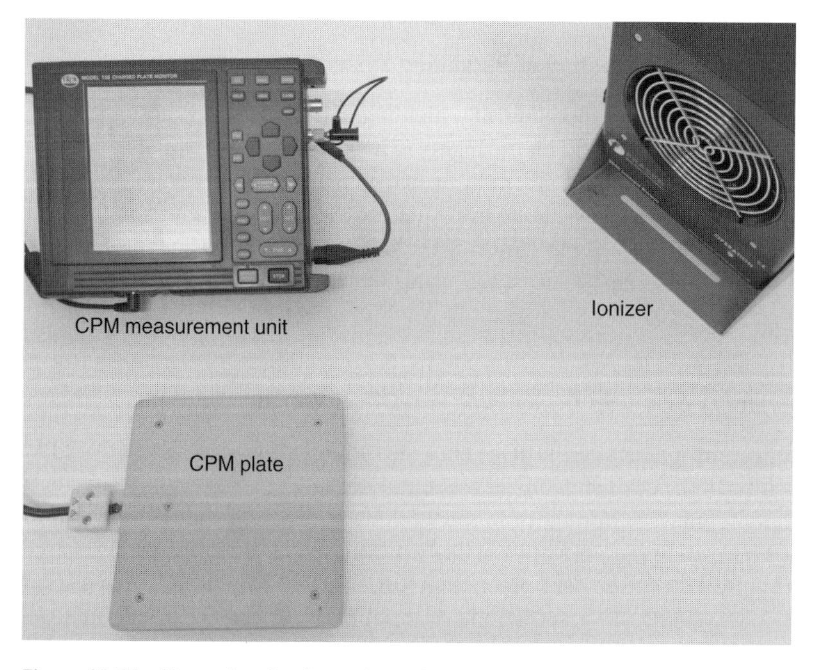

Figure 11.36 Measuring the decay time of an ionizer. The CPM plate is positioned simulating the normal position of items requiring neutralization.

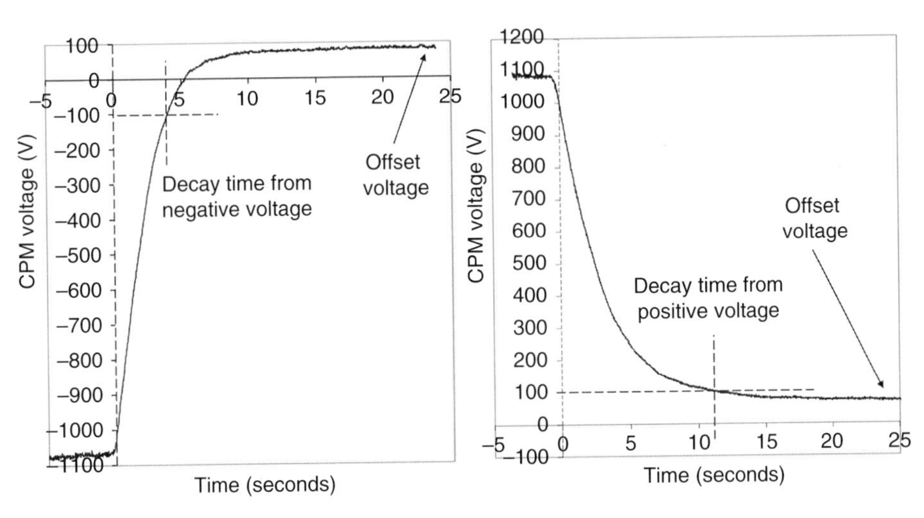

Figure 11.37 Decay time and offset voltage of an ionizer.

Standard tests require an initial CPM plate voltage of over 1000 V and measures the decay time from 1000 V to a final voltage of 100 V (Figure 11.37). The offset voltage is measured after the plate has reached a reasonably constant voltage sometime after the decay curve. Some fluctuation of the offset voltage is normal.

Equipment required:

- Charge plate monitor
- Ionizer under test

Procedure:

- Set up the ionizer in its intended working position.
- Set up the charge plate monitor in a position typical of items requiring neutralization.
- Charge the plate and observe the decay of voltage.
- Measure the decay time between the initial voltage and the final voltage required by the test.
- Note the residual "offset" voltage after it has stabilized.

11.8.8.1 Common Problems

The effectiveness of ionizers is dependent on position, drafts, and orientation of the ionizer and CPM. It may be advisable to try different CPM positions and orientations and evaluate these effects.

Standard practices SP3.3 and SP3.4 specify that the CPM plate should face the direction of the air flow from the ionizer. In many cases, it may be more realistic to have the CPM plate in a position simulating the position of a typical target to be neutralized. For example, neutralization of a printed circuit board (PCB) resting on a surface by an ionizer airstream from the side may be better simulated by a CPM plate lying parallel to the surface rather than one at right angles to the surface.

11.8.9 Walk Test of Footwear and Flooring

Body voltage measurement while walking is a test used mainly for qualification of the performance of ESD control footwear and flooring system.

Body voltage measurements can be made using a CPM or a purpose-built electrostatic body voltage measurement instrument (Figure 11.38). An ordinary multimeter or voltmeter cannot be used as the input resistance would be too low (typically 10 MΩ) effectively grounding the body through the meter. It is useful to have the CPM or body voltage measurement instrument connected to a recording system such as a computer for convenient display and logging of the results. If this is available, recordings of body voltage can be made for future reference and analysis. The subject is given a handheld electrode that connects them via a wire to the CPM plate or instrument.

Equipment required:

- CPM or body voltage measurement instrument
- Handheld electrode and long wire
- Computer or recording equipment (optional)

Procedure:

- Set up the CPM in a convenient position.
- Connect the long wire to the hand-held electrode and CPM plate.
- Ask the subject to hold the electrode and walk around. (Note: ESD STM97.1 requires a specific step pattern to be used.)
- Monitor the voltage readings and record the "peaks" of positive or negative polarity.
- Calculate the average of the five highest peaks (if required).

11.8.9.1 Common Problems

It can be difficult to record peaks and valleys without a chart recorder, computer, or some other graphical means of recording and display. Some equipment is available that has built-in decay curve timing functions.

Digital instruments often update their display every half or one second. Peaks are easily missed when monitoring these displays.

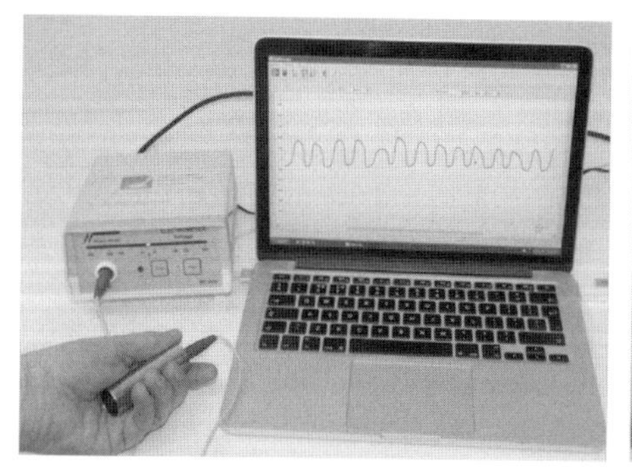

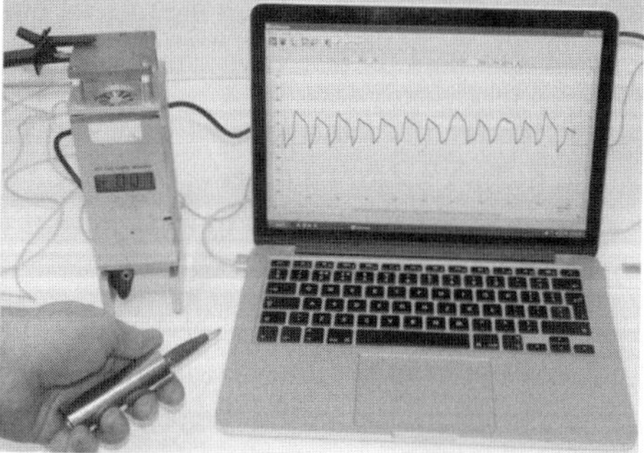

Figure 11.38 Measurement of body voltage of personnel (left) using electrostatic voltmeter instrument and (right) using CPM.

11.9 Useful Measurements Not Specified by IEC 61340-5-1 and ESD S20.20

Although they may be covered by standards, the following test methods are not specified by the current versions of the 61340-5-1 and ESD S20.20 ESD standards for measurements of equipment and materials used in ESD control. Nevertheless, they can be useful in ESD control work, especially in situations where standard measurements are not appropriate. They may also be used to give additional information about the performance of equipment or materials.

11.9.1 Electrostatic Fields and Voltages

Measurements of electrostatic fields and voltages are an important way of checking that all static measures are operating correctly and of detecting any unforeseen static sources. The most common type of instrument used is an electrostatic field meter calibrated to read voltage on large flat surfaces (Section 11.6.12.1). Other types of instrument for measurement of voltages include noncontact and contact electrostatic voltmeters (Sections 11.6.12.2 and 11.6.12.3).

11.9.2 Measurement of Electric Fields at the Position of the ESDS

An electrostatic field meter can be used to measure the electrostatic field at the likely positions of the ESDS (Figure 11.39).

Equipment:

- Electrostatic field meter
- Grounding wire (if required)
- Workstation position to be evaluated

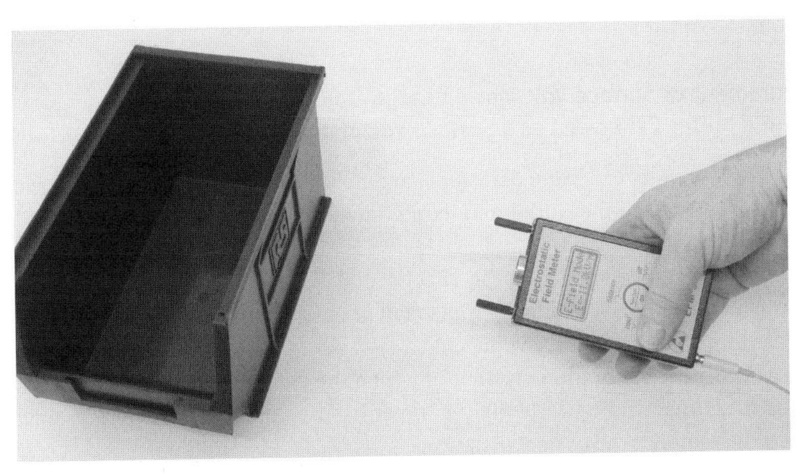

Figure 11.39 Measurement of electrostatic fields in the region where an ESDS device may be present. The field meter is held at different positions and orientations to look for electrostatic field sources and measure the strength of fields.

Procedure:

- Ground the electrostatic field meter. (Many field meters can be grounded simply by being held in the hand of a grounded person.)
- Move the field meter around the region in which the ESDS may be situated and monitor the readings.
- Note any high field readings and the positions of their sources.

When high field readings are noted, the field meter can be used to home in on the source for further evaluation.

11.9.2.1 Common Problems

Many meters do not give a direct reading of electrostatic field but are calibrated to read surface voltage of a large flat surface held at a set distance from the field meter. A meter calibrated in this way can still be used to detect electrostatic fields and determine whether they are greater or less than a required level, e.g. $5\,\mathrm{kV\,m^{-1}}$. To do this, the voltage reading that is given by a field of $5\,\mathrm{kV\,m^{-1}}$ must be known.

To a first approximation, this reading can be easily obtained by simple calculation using the calibration distance d of the field meter. The voltage reading V_E equivalent to a field E is given by

$$V_E = Ed$$

So, for example, the JCI 140 field meter shown in Figure 11.41 is calibrated to read surface voltage with the meter 10 cm from the field source. If the required field limit is $5\,\mathrm{kV\,m^{-1}}$, the corresponding meter reading is $0.1 \times 5000 = 500\,\mathrm{V}$. When the field meter is used to evaluate electrostatic fields, if the meter reading remains less than this value, the electrostatic field is less than $5\,\mathrm{kV\,m^{-1}}$.

The voltage reading equivalent to the field limit will be different for other meters of different calibration distance. For example, for a meter having a calibration distance of 2.5 cm (1 in.), the voltage reading for a $5\,\mathrm{kV\,m^{-1}}$ field would be $0.025 \times 5000 = 125\,\mathrm{V}$.

11.9.3 Measurement of Surface Voltages on Large Objects Using an Electrostatic Field Meter Calibrated as a Surface Voltmeter

Many electrostatic voltmeters or "static detectors" are electrostatic field meters that have been calibrated to read voltage measured on a large flat surface when held at a specified distance from the surface (Figures 11.40 and 11.41).

These meters give a response reflecting the effect of the charge over a wide area of the surface. While the meters will give readings from small or curved surfaces, the reading will in this case be less than the true surface voltage.

It is essential to know the calibration distance at which the voltmeter must be held from the surface being measured. In Figure 11.40, the meter has two pins that are held against the target to keep the meter at the correct distance.

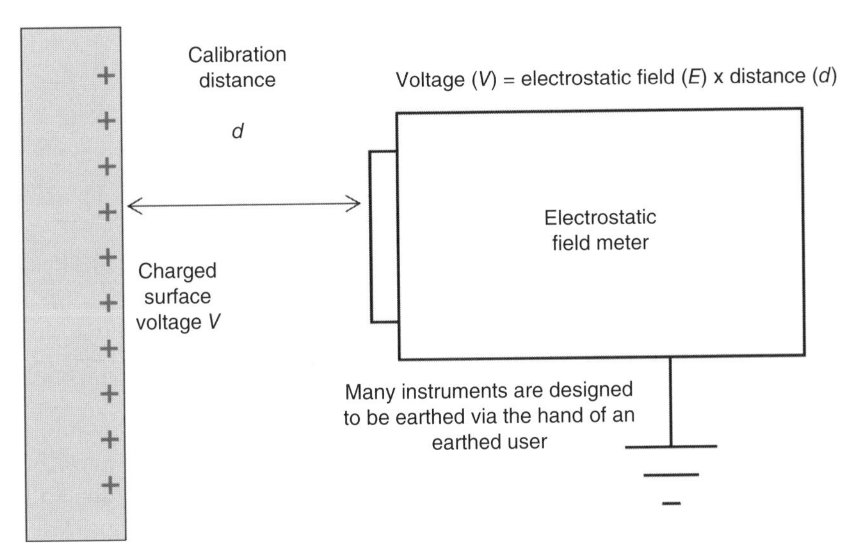

Figure 11.40 Measuring electrostatic voltages on a planar target surface using an electrostatic field meter calibrated as a voltmeter.

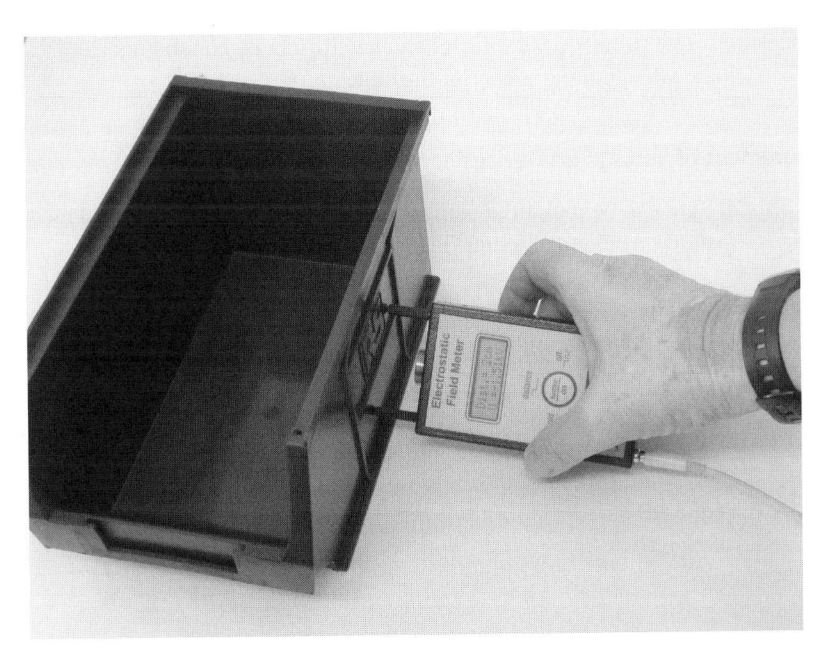

Figure 11.41 Using a field meter calibrated as a voltmeter to measure electrostatic surface voltages on a charged object.

Equipment:

- Electrostatic voltmeter
- Grounding wire (if required)
- Objects or material to be tested

Procedure:

- Ground the electrostatic voltmeter. (Many meters can be grounded simply by being held in the hand of a grounded person.)
- Hold the voltmeter the correct calibration distance from the surface to be measured.
- Take a reading.

Common Problems

The electrostatic voltmeter must be grounded, or it will not read the target voltage correctly. Any voltage on the meter is added to the voltage read on the target surface. The reading will also change if not held at the correct distance from the surface being measured. If too close, the reading will be too high.

Do not hold a sample in your hand for measurements – if it conducts electricity, any charge would be conducted to your body. If your body is not grounded, any voltage on your body could appear also on the sample.

The voltage on small objects, curved objects, insulators, or small isolated conductors is not correctly measured. The voltage on insulators and small isolated conductors changes with increased capacitance due to the presence of the meter.

11.9.4 Measurement of Voltage on Devices or Small Conductors

The voltage on a small item can be measured using a noncontact voltmeter (Figure 11.42) or ultra-high input resistance contact voltmeter (Figure 11.43).

11.9.4.1 Measurement Using a Non-contact Voltmeter

A noncontact electrostatic voltmeter can be used to measure the voltage on a small item under test.

Equipment:

- Noncontact electrostatic voltmeter
- Grounding wire (if required)
- Objects or material to be tested

Procedure:

- Ground the electrostatic voltmeter. (Many meters can be grounded simply by being held in the hand of a grounded person.)
- Hold the voltmeter within the correct calibration distance range from the surface to be measured.
- Note the reading.

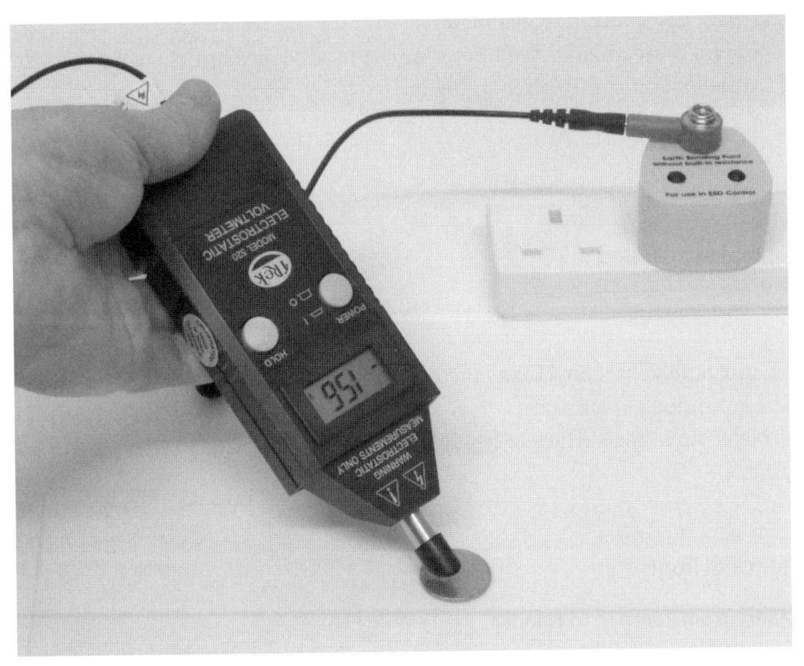

Figure 11.42 Measurement of voltage of a small item using a noncontact electrostatic voltmeter.

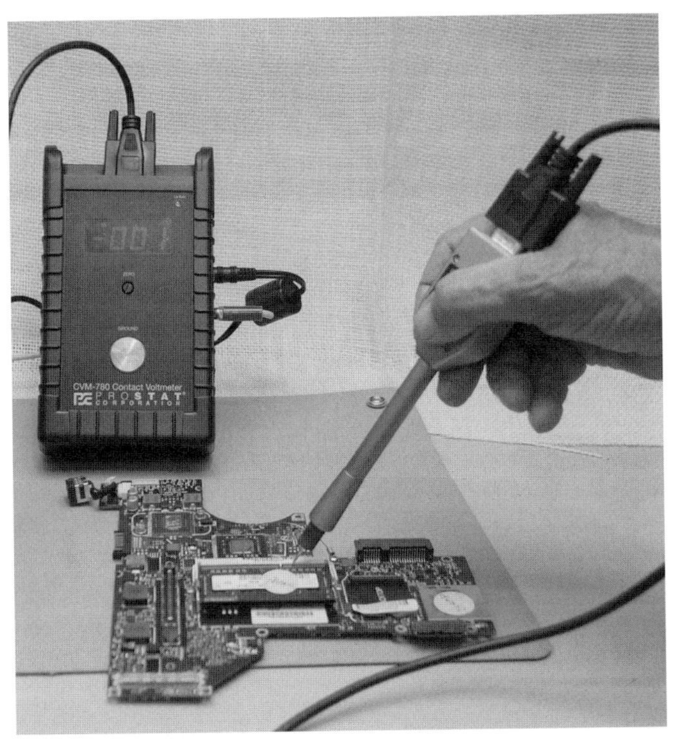

Figure 11.43 Measurement of voltage of a small item using a contact electrostatic voltmeter.
Source: DE Swenson.

11.9.4.2 Measurement Using a Contact Voltmeter

A high impedance contact electrostatic voltmeter can be used to measure the voltage on an item under test without significantly discharging it.

Equipment:

- Contact electrostatic voltmeter
- Grounding wire (if required)
- Objects or material to be tested

Procedure:

- Ground the electrostatic voltmeter. (Many meters can be grounded simply by being held in the hand of a grounded person.)
- Touch the voltmeter tip to the surface to be measured.
- Note the reading.

11.9.5 Resistance of Tools

11.9.5.1 Resistance from Tool Tip to Handle

This test measures the resistance of a tool from the point that contacts the ESDS device to the handle (Figure 11.44).

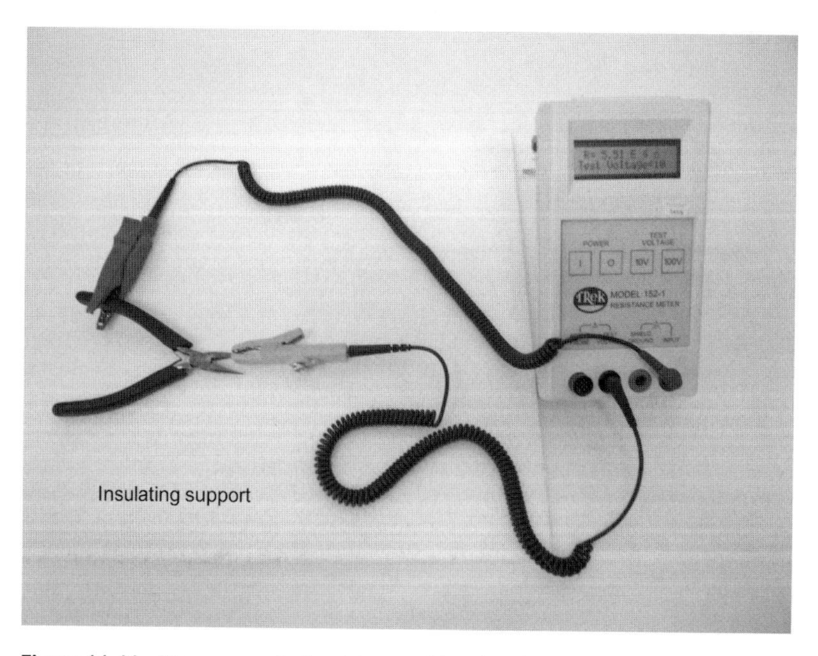

Figure 11.44 Measurement of resistance of hand tool.

Equipment required:

- 10/100 V high-resistance meter
- Test leads and clips
- Tool to be tested

Procedure:

- Connect the test leads and clips to the resistance meter.
- Connect one test lead to the tool bit.
- Connect the second test lead to the tool handle.
- Measure result at 10 V.
- If the resistance is >1 MΩ, measure result again at 100 V.

Common Problems

One problem with this test is that it can be difficult to make good electrical contact with a handle made from hard high-resistance material. Contact can be improved by wrapping a conductive self-adhesive metal tape around the handle to provide a larger area contact. Measuring the tool as held in the hand gets around this problem.

11.9.5.2 Resistance to Ground of Handheld Tool

A more convenient tool resistance to ground system test can be done by a grounded user holding the tool in their hand. A resistance meter is connected to ground, with its second terminal connected to a touch plate (Figure 11.45). To test the tool system resistance to ground, the user holds the tool handle and touches the bit to the touch plate and reads the resistance to ground from the meter. Holding the tool in the hand makes a good connection to the handle and a realistic test. A test electrode can be made from any convenient metal item isolated from ground by an insulating support. Avoid touching the tool tip, or the resistance of the handle is short circuited.

This test is a convenient system test that can easily be done by a grounded user at a workstation. It directly confirms the tool is grounded within required resistance limits. It is important to remember that the resistance result includes the resistance of the operator's ground path as well as the tool resistance.

Equipment required:

- 10/100 V resistance meter
- Wrist strap or footwear-flooring grounding for the test person
- Tool test electrode

Procedure:

- Connect one terminal of the resistance meter to ground.
- Connect the second terminal of the resistance meter to the tool test electrode.
- Wear the grounded wrist strap and hold the tool.

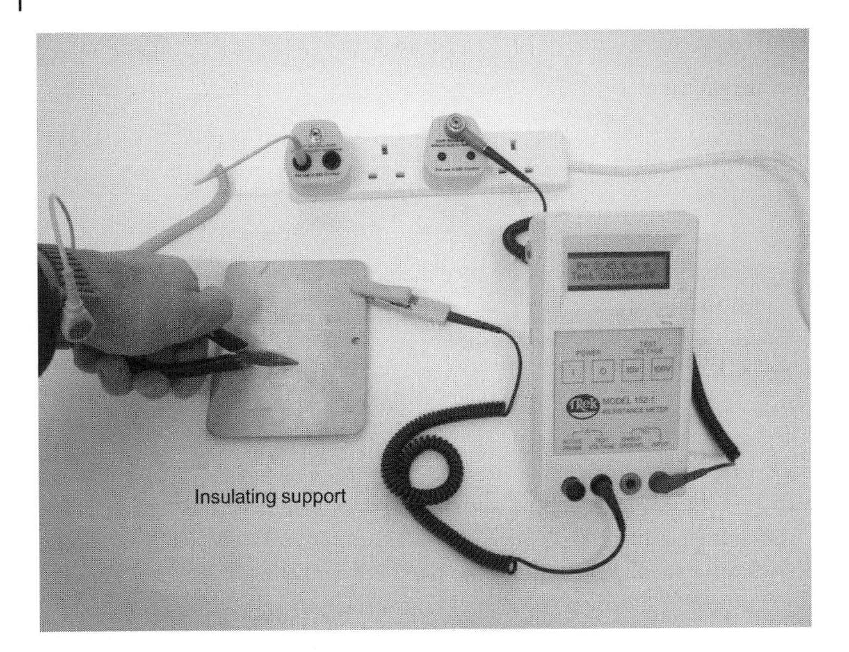

Figure 11.45 Test of resistance to ground of hand-held tool. The person holding the tool can be grounded in their usual manner.

- Touch the tool to the test electrode.
- Measure the result at 10 V.
- If the resistance is >1 MΩ, measure result again at 100 V.

Common Problems

If the user's hand touches the metal tool bit, the resistance of the handle is bypassed by the resistance of the user's hand, and a false low resistance reading is obtained.

11.9.6 Resistance of Soldering Irons

11.9.6.1 Resistance to Groundable Point of Soldering Iron Tip

The resistance from a soldering iron tip to the earthing point (e.g. earth pin on the mains plug) can be measured while the iron is not in use (Figure 11.46). The resistance meter used must be capable of low-resistance measurements.

Equipment required:

- Low-resistance meter
- Test leads and clips

Procedure:

- Connect one terminal of the resistance meter to the soldering iron groundable point.
- Connect the second terminal of the resistance meter to the soldering iron tip.
- Measure the resistance result.

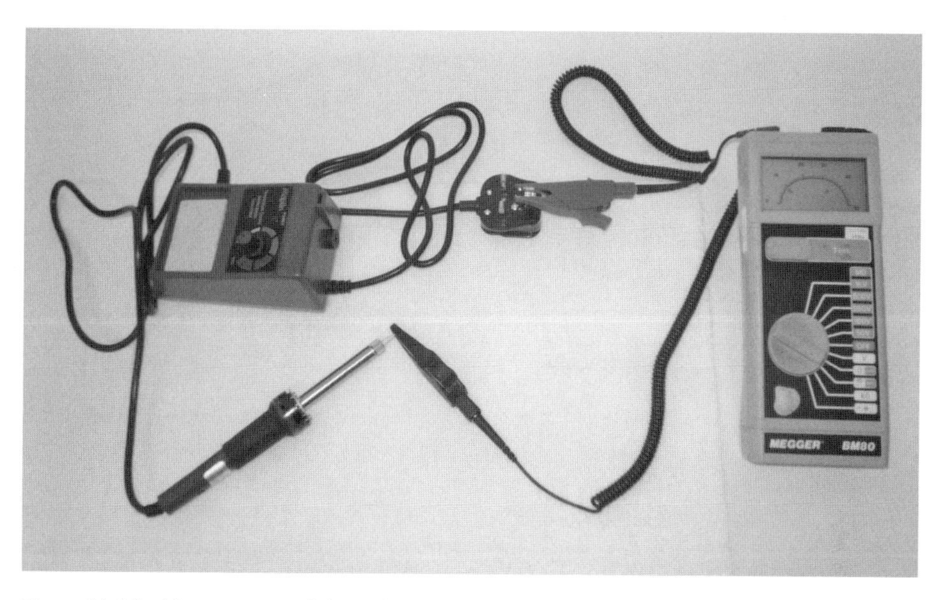

Figure 11.46 Measurement of the resistance to groundable point of a soldering iron bit.

Common Problems

Corrosion of the bit surface contact with the iron can give variable results in this test. But, detection of this is one of the reasons for doing the test!

11.9.6.2 Resistance to Ground of a Soldering Iron Tip

The resistance from soldering iron tip to ground can be easily measured while in use, using a tool test electrode (Figure 11.47). The resistance meter used must be capable of low-resistance measurements. A 0 Ω earthing connector must be used to connect to the EPA ground.

Equipment required:

- Low-resistance meter
- Test leads and 0 Ω earthing connector
- Tool test electrode

Procedure:

- Connect one terminal of the low-resistance meter to EPA ground.
- Connect the second terminal of the resistance meter to the tool test electrode.
- Touch the soldering iron bit to the tool test electrode.
- Measure the result.

Common Problems

Corrosion of the bit contact with the iron can give variable results in this test – but then detecting this part of the reason for doing the test!

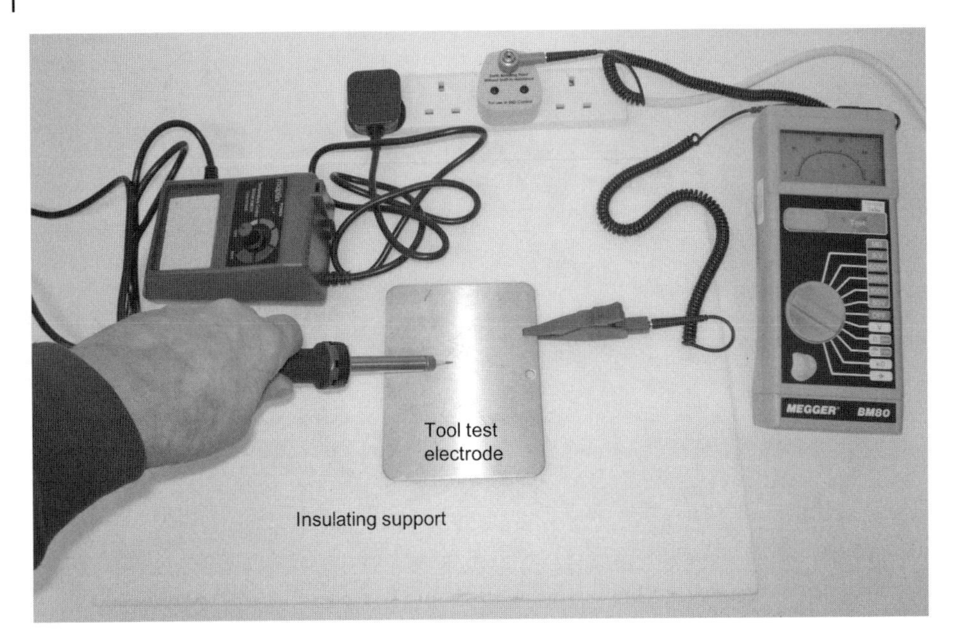

Figure 11.47 Measurement of resistance to ground of a soldering iron tip while in use.

11.9.7 Resistance of Gloves or Finger Cots

It is often useful to measure the resistance through a glove while it is worn to determine whether handheld objects will be isolated from the grounded hand. There are several convenient ways of doing this. A standardized way of doing so is specified in ANSI/ESD SP15.1 using a constant area and force electrode (CAFE) and measuring the resistance between the CAFE and a handheld electrode (Figures 11.48 and 11.49). Gloves can also be evaluated using a charge plate monitor (see Section 11.9.8.2).

11.9.7.1 Measurement of Resistance Through a Glove Using a Handheld Electrode

A person wearing a wrist strap and glove can measure the system resistance through the glove and wrist strap using a resistance meter, as shown in Figure 11.50. This is the same measurement method as given for measurement of wrist straps as worn (see Section 11.8.3.2), except that in this case the test subject wears the glove under test.

11.9.7.2 Testing Resistance Through a Glove Using a Wrist Strap Tester

If the glove has low enough resistance such that it falls within the pass range of a wrist strap tester, this can be used to check the entire wrist strap – hand – glove system (Figure 11.51). This will give a simple pass-fail result rather than a system resistance result. This is the same measurement method as given for measurement of wrist straps as worn (see Section 11.8.3.2), except that in this case the test subject wears the glove under test.

Common Problems

Typical wrist strap testers have a fixed upper "fail" limit, often 35 MΩ. This upper limit must correspond to the upper limit specified in the ESD control program for the resistance

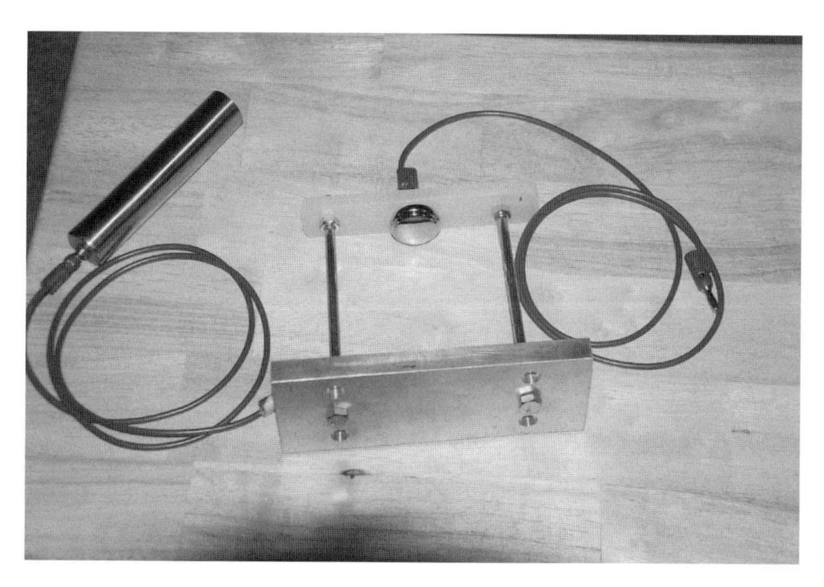

Figure 11.48 Handheld (left) and CAFE (right) electrodes. Source: D E Swenson.

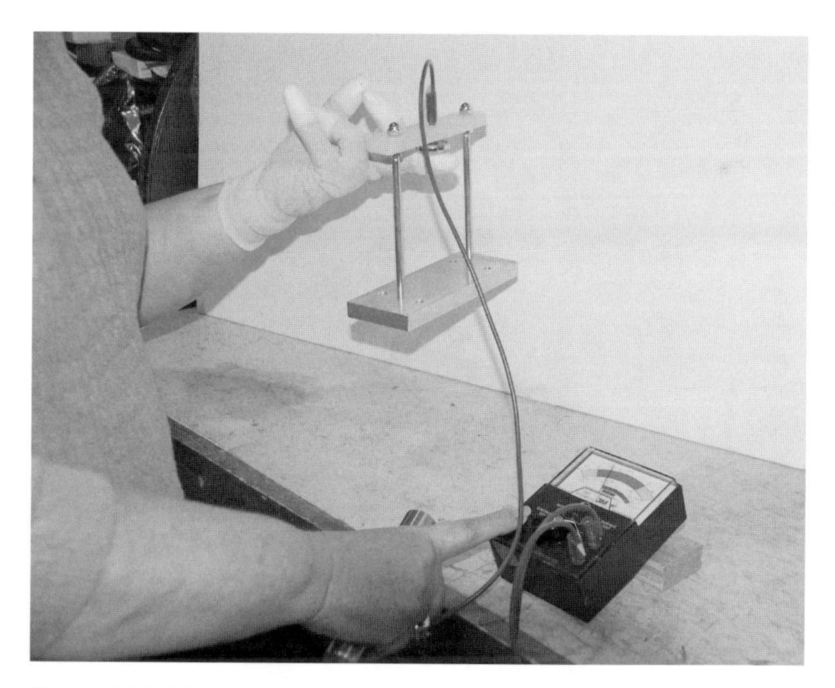

Figure 11.49 Measurement of resistance through glove to ground using CAFE electrode. Source: D E Swenson.

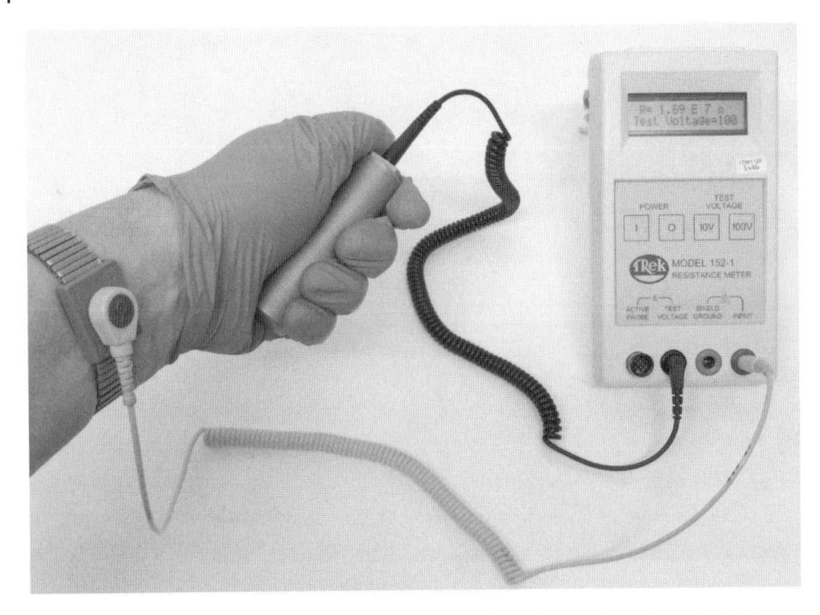

Figure 11.50 Measurement of resistance through a glove using a handheld electrode.

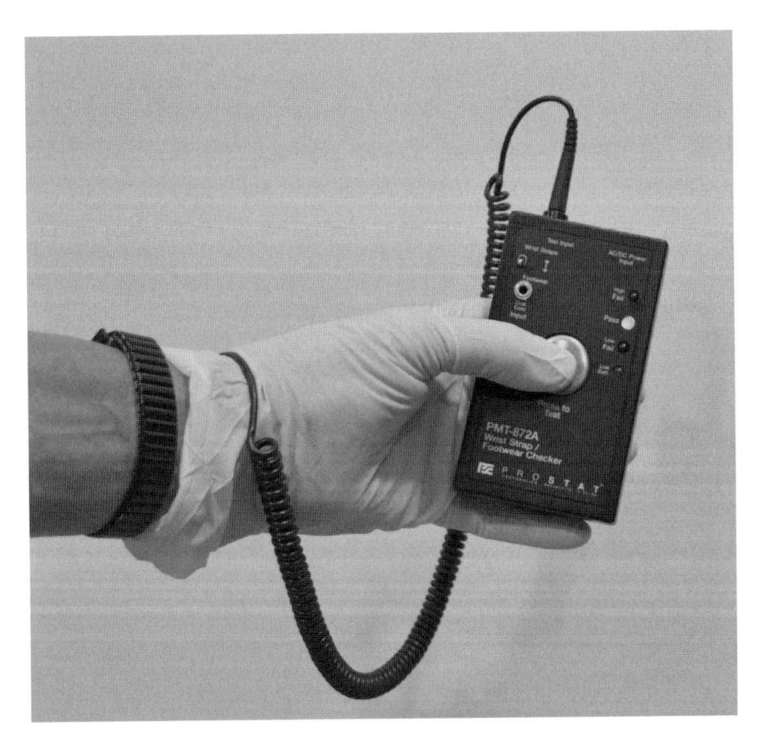

Figure 11.51 Testing of system resistance through a glove and wrist strap using a wrist strap tester. Source: D E Swenson.

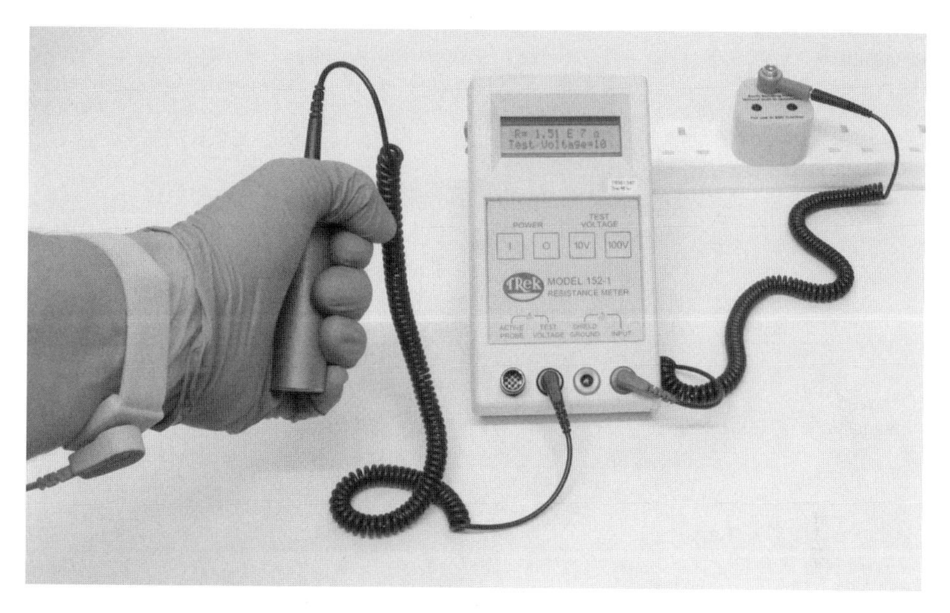

Figure 11.52 Measurement of resistance to ground through a glove using a handheld electrode.

of glove when worn. If the upper limit accepted by the ESD program for the glove is higher than the "fail" threshold of the tester, gloves having resistance above the fail threshold are "failed" incorrectly. If the upper limit accepted by the ESD program for the gloves is lower than the "fail" threshold of the tester, gloves having resistance above the acceptable limit of the ESD program, but below the "fail" threshold of the tester, are "passed" incorrectly.

11.9.7.3 Measurement of Resistance to Ground Through a Glove Using a Handheld Electrode

A system resistance through the glove to ground can be measured by an operator who is already grounded by either a wrist strap or footwear and flooring (Figure 11.52). The resistance measurement is made between a handheld electrode and ESD ground. This is the same measurement method as given for measurement of wrist straps as worn (see Section 11.7.8.4).

11.9.8 Charge Decay Measurements

We are often interested in knowing whether charge can dissipate quickly enough to prevent static build-up. There are many types of charge decay measurements that have been devised for different purposes. They can use widely different equipment and principles of operation.

Charge decay test methods are not, at the time of writing, specified as standard test methods used by the ESD control standards 61340-5-1 and ESD S20.20. The IEC 61340-2-1 (International Electrotechnical Commission 2015a) Section 4.4 gives a contact test method for charge decay measurement using a CPM. This technique gives useful tests for tools, gloves, or finger cots, as shown in Sections 11.9.8.1–11.9.8.3.

In these tests, the charged pate monitor plate is charged to a defined starting level V_i, often 1000 V. The voltage on the plate is observed to drop (decay) during the test toward

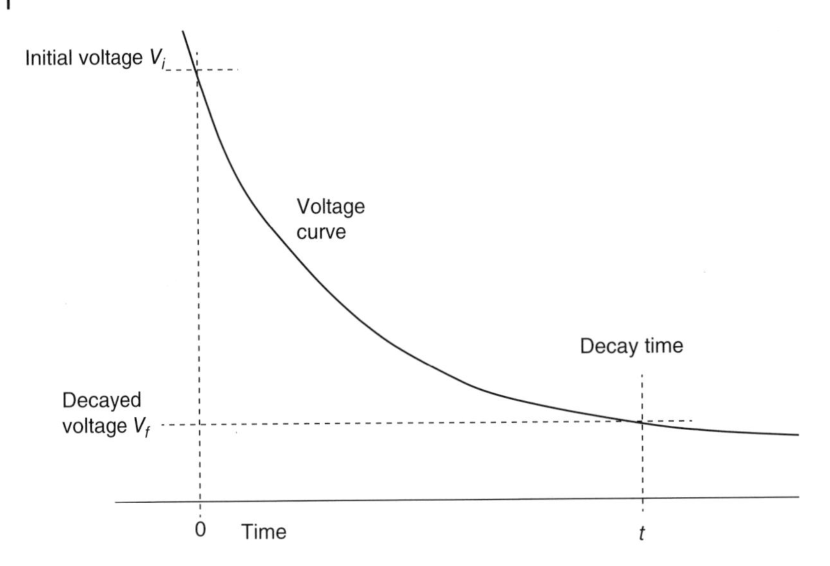

Figure 11.53 Charge decay time.

zero. This typically gives a quasi-exponential voltage curve (Figure 11.53). The time taken for the voltage to reach a defined level V_f, often chosen to be 100 V, is measured.

Slow decay curves can sometimes be monitored manually, but faster decay curves may need a digitizing oscilloscope or chart recorder to aid measurement. The time taken for the voltage to decay from the chosen initial voltage (V_i in Figure 11.53) to the chosen final voltage V_f, is taken as the decay time.

The user should select the final voltage as appropriate for their purpose. Often a fraction of the initial voltage (e.g. 0.1 V_i) or a "hazard threshold voltage" (e.g. 100 V) is specified.

Typically, the charge decay time result will depend on the initial test voltage, perhaps varying in some cases by orders of magnitude. Longer decay times are usually found for lower test voltages.

A commercial CPM can be conveniently used for charge decay tests. This is often limited in test voltage to the 1000 V and final voltage of 100 V designed for use with ionizer testing.

Many charge decay test methods are in effect indirect ways of comparing the resistance to ground of the item measured. The resistance to ground of the item bleeds the charge from the plate capacitance C_p. When touched by a handheld tool, the resistance to ground of the tool R_t is connected in parallel with the plate capacitance C_p. The voltage decays with a time constant R_t C_p. The decay time measured from 1000 to 100 V is approximately 2 R_t C_p. Where the resistance is very high, this technique can be more reproducible than using a direct resistance measurement method.

11.9.8.1 CPM Charge Decay for Tools

A handheld tool held by a grounded operator can be tested using the charge decay method (Figure 11.54) using a CPM. This method works well with high-resistance tool handles.

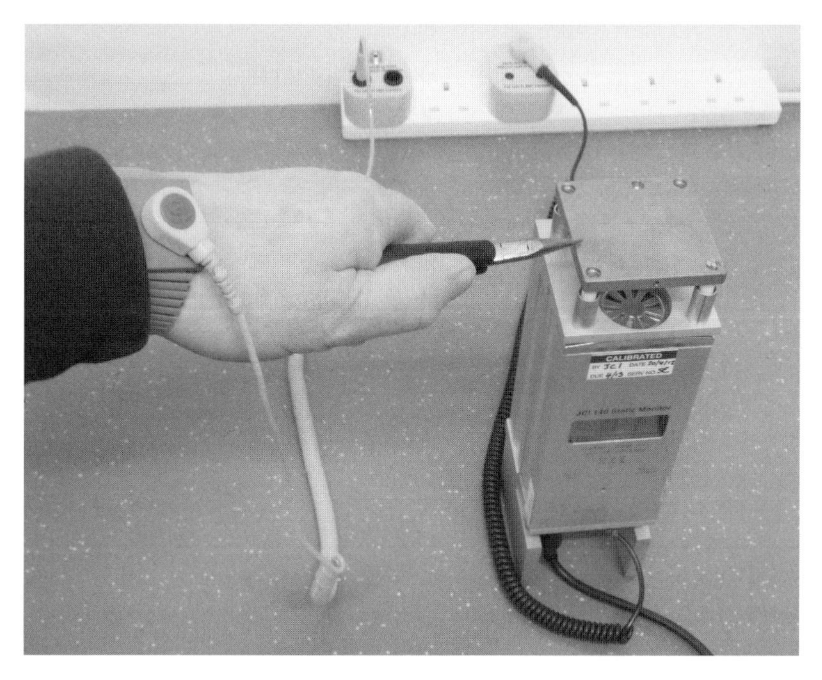

Figure 11.54 Measurement of charge decay time of tools. The tool is touched to the CPM plate and the voltage decay time is measured. Both the operator and CPM are grounded.

Equipment required:

- Charge plate monitor
- Wrist strap or footwear – flooring grounding of operator
- Sample tools for test

Procedure:

- The operator must be grounded.
- Charge the CPM to 1000 V.
- The operator holds the tool in their hand but does not touch the tool blade. The tool blade is then touched to the CPM.
- Observe and time the decay to 100 V.

It is helpful if the decay curve is recorded, as this can reveal the true charge decay characteristics of the tool and be used to document the test for product qualification records. Some examples of typical waveforms are given in Figures 11.55–11.57. A tool that has a low resistance will give a fast drop in voltage to zero (Figure 11.55). A tool with high-resistance handle will show a longer, slower decay time (Figure 11.56). An initial fast drop may be seen in this case due to increasing capacitance and charge sharing as the tool bit contacts the CPM plate.

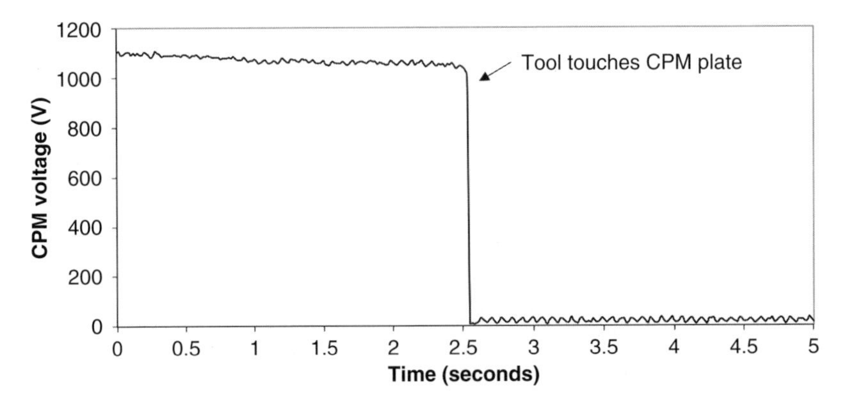

Figure 11.55 Decay curve from tool that has low-resistance handle.

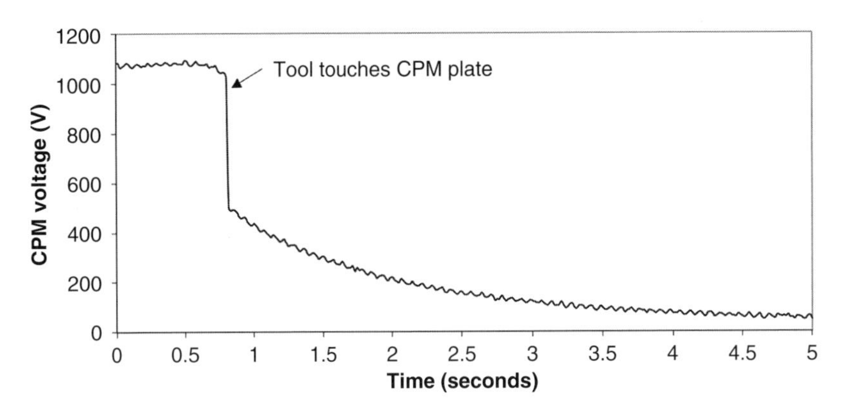

Figure 11.56 Decay curve from tool that has intermediate-resistance handle.

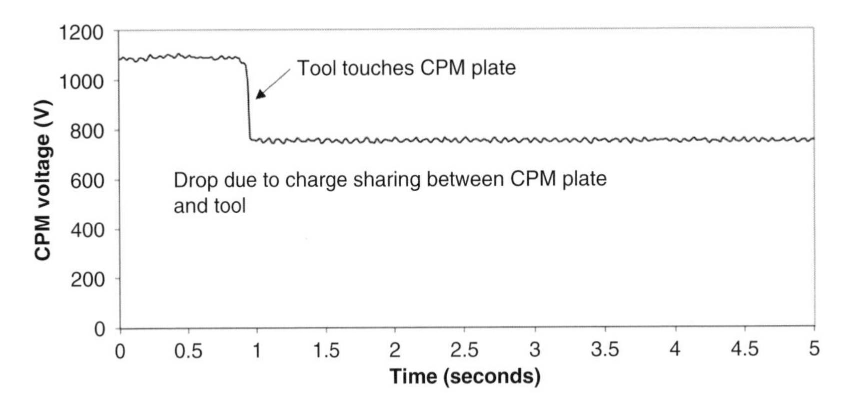

Figure 11.57 Decay curve from tool that has insulative handle and shows capacitive voltage reduction.

If the tool handle is insulative, the plate voltage does not reduce to zero in an acceptable time. Again, there may be a step in voltage as the tool bit contacts the plate as the charge stored on the plate is shared with the tool capacitance (Figure 11.57).

Common Problems

The metal tool blade should not be touched with the fingers, as the resistance of the handle will be bypassed.

Typical digital displays on CPM instruments may update only every half-second or so. Short decay times of a second or less are difficult to measure without monitoring the plate voltage on an oscilloscope type display.

11.9.8.2 Charge Decay of Gloves and Finger Cots

A charge decay test like that for tools using a CPM can also be used for gloves and finger cots (Figure 11.58).

Equipment required:

- Charge plate monitor
- Grounded person for test
- Sample glove or finger cot for test

Procedure:

- Charge the CPM to 1000 V.
- The grounded person wears the glove/finger cot under evaluation and touches the CPM through the glove/finger cot.
- Observe the decay time to 100 V.
- Remove the finger from the plate. Observe whether the plate charges due to contact with the glove material.

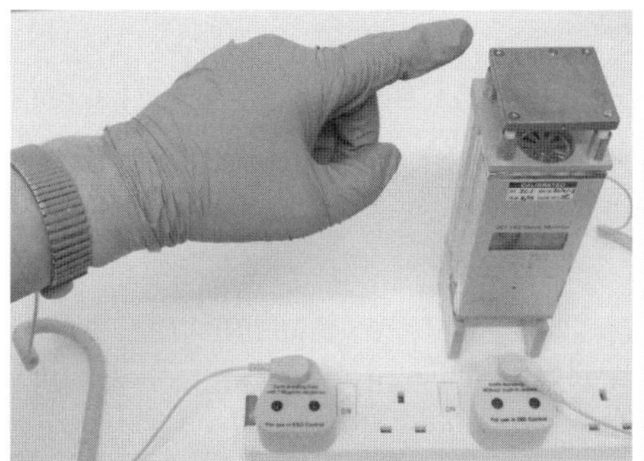
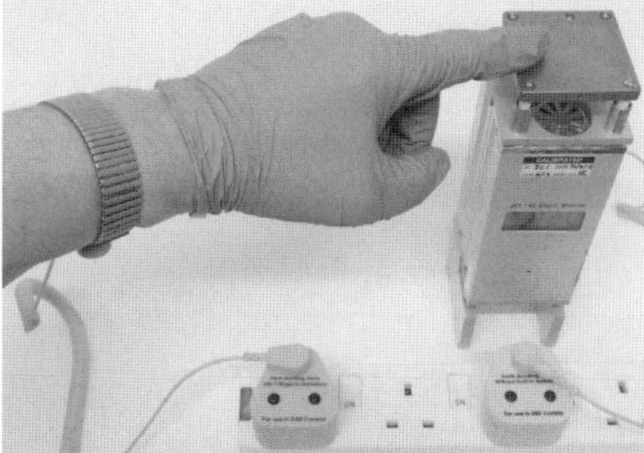

Figure 11.58 Measuring the charge decay of a glove when worn. The gloved finger is brought into contact with the charged plate and the decay observed.

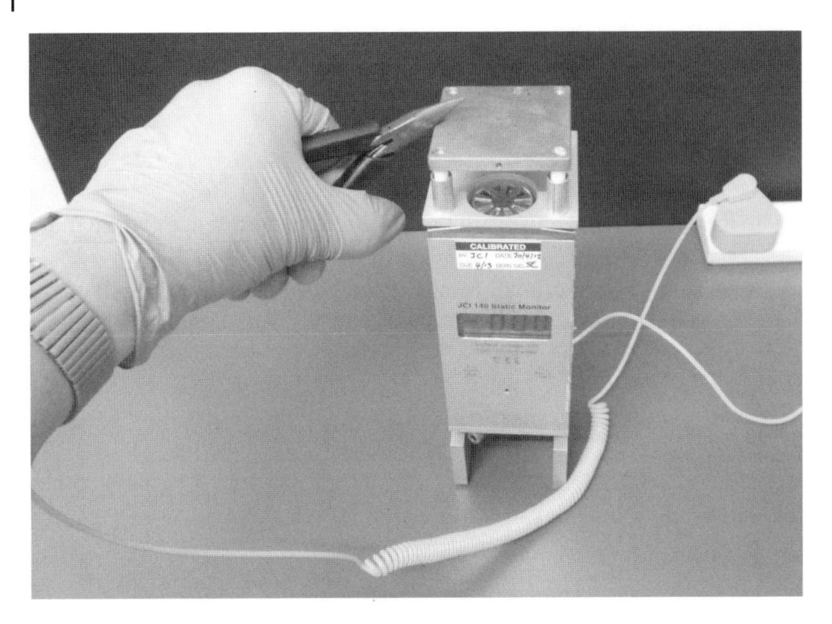

Figure 11.59 Charge decay system test of tool held in gloved hand.

Common Problems

Typical digital displays on CPM instruments may update only every half-second or so. Short decay times of a second or less are difficult to measure without monitoring the plate voltage on an oscilloscope type display.

When the gloved finger is removed from the CPM plate, the plate may become charged. This may indicate unacceptable charging of the plate by the glove material.

11.9.8.3 System Test of Glove and Handheld Tool

A system test can be done of grounding of a handheld tool via a gloved hand (Figure 11.59), using the tool charge decay test method of Section 11.9.8.1. A decay curve like Figures 11.55 or 11.56, with decay time within accepted values, indicates acceptable performance.

Common Problems

Make sure the fingers do not accidentally contact the tool bit, short-circuiting the handle.

11.9.9 Faraday Pail Measurement of Charge on an Object

11.9.9.1 The Faraday Pail

In its simplest form, the Faraday pail is a metal pail isolated from ground that can be used with a coulombmeter or electrostatic voltmeter to measure charge on an item placed within the pail. The pail must be sufficiently large to fully contain the item. A charge is induced on the pail that is equal to the total net charge on the item placed within the pail. Simple unshielded Faraday pails can be improvised from common metal storage containers (Figure 11.60).

Figure 11.60 Unshielded Faraday pail on CPM plate. This pail was purchased in the kitchen department of a local store!

Simple unshielded Faraday pails are prone to errors from induced charge due to fields from nearby charged insulators, personnel, or voltage sources. This can be reduced by containing the measurement pail isolated within an earthed metal screen (Figure 11.61).

Where a Faraday pail is used with a coulombmeter, the charge is measured directly by the coulombmeter.

Where a Faraday pail is used with a CPM or electrostatic voltmeter, the charge gives a voltage rise on the pail that is measured using the voltmeter or CPM. For comparative measurements (e.g. comparing the charging of an item when handled by two different glove types), it may be sufficient to compare the voltages produced on the pail. If the actual charge values Q is required, it can be calculated, knowing the capacitance C_p of the pail in its measurement arrangement and the voltage result V, from the relation.

$$Q = C_p V$$

When used with a voltmeter or CPM, a Faraday pail must usually be zeroed by momentary connection to ground to bring the pail to zero volts. When used with a coulombmeter, the coulombmeter must be zeroed before each measurement.

11.9.9.2 Measurement of Electrostatic Charging of Items Handled with Gloves or Finger Cots Using a Faraday Pail

It can be useful to measure the charging of ESDS items handled while wearing gloves or finger cots. IEC/TR 61340-2-2 (International Electrotechnical Commission 2000) gives general guidance on chargeability testing.

Figure 11.61 Shielded Faraday pail used with a coulombmeter.

The principle of these tests is that a grounded person, wearing the glove or finger cots under test, handles the item and then places them into a Faraday pail (see Section 11.9.9.1) or on to a CPM (see Section 11.9.9.3).

If an unscreened Faraday pail is used, electrostatic fields from charged clothing, gloves, or other insulating items in the vicinity can strongly affect the measurement result. In this case, make sure that all potentially charged items are moved well away from the measurement area and that the operator moves away before the measurement result is noted. If a screened Faraday pail is used, this is much less affected by nearby electrostatic fields.

Equipment required:

- Faraday pail connected to a voltmeter or charge measurement instrument
- Wrist strap grounder
- Sample glove or finger cot for test

Procedure:

- Wear the wrist strap, and ground it.
- The subject holds the product in their hand and handles it in a representative way.
- If necessary, zero the Faraday pail and measurement instrument.
- Place the product in the Faraday pail.
- Move the hand well away from the Faraday pail.
- Note the charge reading.

Common Problems

Touching the Faraday pail can cause errors due to charging or discharging of the pail by contact.

An unshielded Faraday pail is prone to errors due to induced charge from nearby electrostatic field sources such as charged clothing, the user's body, or insulators. Proximity to the user's hand, body, or other conductors can increase capacitance and reduce the readings obtained.

Placing a charged ESDS in a metallic Faraday pail or on a metal CPM plate can cause charged device ESD that could damage the ESDS. This can be prevented by lining the Faraday pail or CPM plate with static dissipative material having surface resistance >10 kΩ.

11.9.9.3 Evaluation of Charging of an Item Using a CPM Plate

Indicative and comparative evaluation of charging of an item can be made using a CPM. In this case, the charged item is placed on the CPM plate instead of into the Faraday pail (Figure 11.62). The CPM plate is first discharged by grounding it before placing the charged

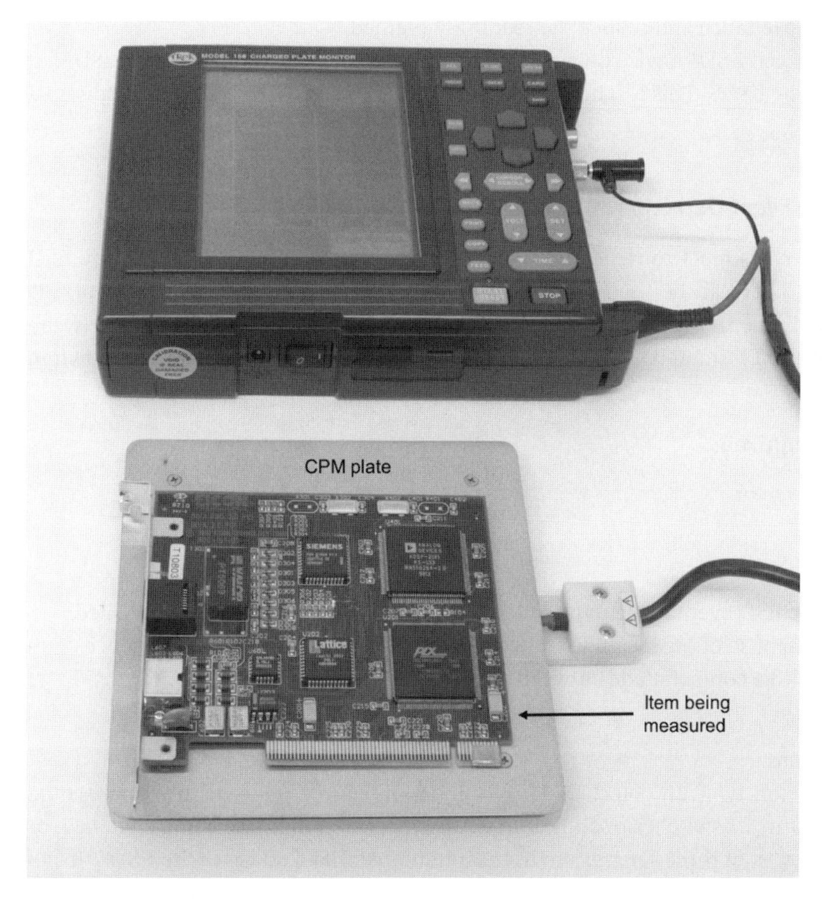

Figure 11.62 Measurement of product charging using a CPM.

product on it. The voltage on the CPM after placing the product on it is indicative of the charge on the product. Different gloves can be compared using repeated tests handling the same item.

The charge induced on the CPM plate is an unknown fraction K of the charge on the item measured. If the CPM plate capacitance C_p is known, the charge Q on the item measured can be estimated from the CPM plate voltage V from the relation.

$$Q = KC_pV$$

The unknown factor K must usually be assumed to be $K = 1$ unless it can be somehow evaluated.

Common Problems

Touching the CPM plate can cause errors due to charging or discharging of the pail by contact.

The CPM plate is prone to errors due to induced charge from nearby electrostatic field sources such as charged insulators. Proximity to the user's hand, body, or other conductors can increase capacitance and reduce the readings obtained.

Placing a charged ESDS device on a metal CPM plate can cause charged device ESD that could damage the ESDS. This can be prevented by lining the CPM plate with static dissipative material.

11.9.10 ESD Event Detection

ESD event detectors react to radio frequency electromagnetic radiation emitted by ESD in the vicinity. They can be used to detect ESD occurring nearby, including in working automated equipment. Some detectors have switched settings to distinguish between charged device and human body ESD. Examples of handheld ESD detectors are shown in Figure 11.63.

Equipment required:

- ESD event detector

Procedure:

- Switch on the ESD event detector and place it near the operation being investigated.
- Watch the response of the event detector during the process. Look for coincidence of an ESD event with a contact made between a conductor and the ESDS.

Common Problems

ESD event detectors typically detect ESD from almost any source in the vicinity. These could be from contactors or switches operating in equipment, room lighting being switched, or other sources. Most of these sources are irrelevant to potential damage to the ESDS. It can be difficult to distinguish between potentially damaging ESD events and irrelevant ESD events. Where the ESDS can be observed through the process, look for coincidence between an ESD event and a contact between the ESDS device and a conductor. If the ESDS cannot

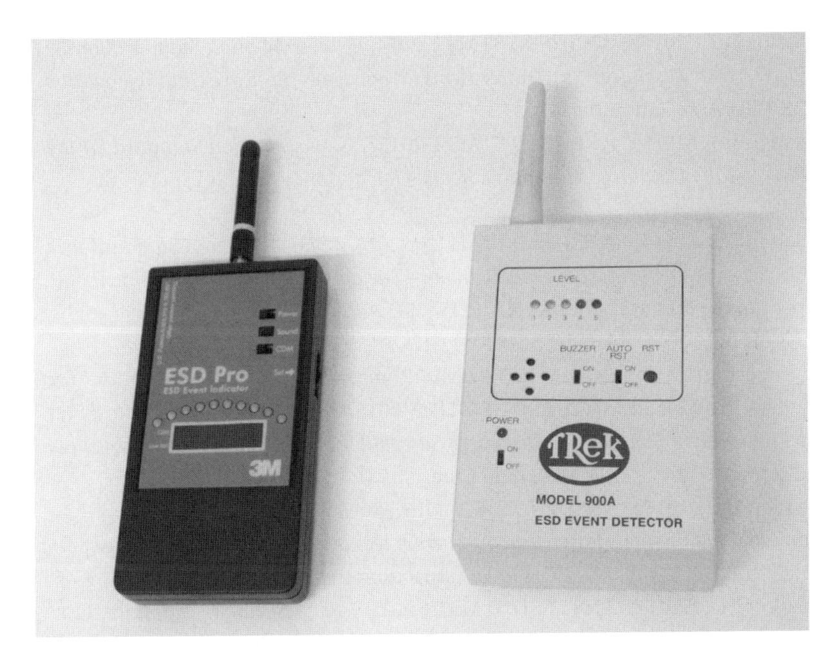

Figure 11.63 Examples of handheld ESD event detectors.

be seen during the part of the process where ESD was noted, it may be necessary to review the process steps while not in operation.

References

British Standards Institute. (2016) BS EN 61340-5-1. *Electrostatics - Part 5-1: Protection of electronic devices from electrostatic phenomena - General requirements.*

EOS/ESD Association Inc. (2013a) ANSI/ESD S1.1-2013. *ESD Association Standard for the Protection of Electrostatic Discharge Susceptible Items – Wrist Straps*. Rome, NY, EOS/ESD Association Inc.

EOS/ESD Association Inc. (2013b) ANSI/ESD STM2.1-2013. *ESD Association Standard for the Protection of Electrostatic Discharge Susceptible Items – Garments*. Rome, NY, EOS/ESD Association Inc.

EOS/ESD Association Inc. (2013c) ANSI/ESD STM7.1-2013. *ESD Association Standard for the Protection of Electrostatic Discharge Susceptible Items – Floor Materials – Resistive Characterization of Materials*. Rome, NY, EOS/ESD Association Inc.

EOS/ESD Association Inc. (2013d) ANSI/ESD STM12.1-2013. *ESD Association Standard Test Method for the Protection of Electrostatic Discharge Susceptible Items – Seating – Resistance Measurement*. Rome, NY, EOS/ESD Association Inc.

EOS/ESD Association Inc. (2014a) ANSI/ESD S20.20-2014. *ESD Association Standard for the Development of an Electrostatic Discharge Control Program for – Protection of Electrical and Electronic Parts, Assemblies and Equipment (excluding Electrically Initiated Explosive Devices)*. Rome, NY, EOS/ESD Association Inc.

EOS/ESD Association Inc. (2014b) ANSI/ESD STM9.1-2014. *ESD Association Standard for the Protection of Electrostatic Discharge Susceptible Items – Footwear – Resistive Characterization*. Rome, NY, EOS/ESD Association Inc.

EOS/ESD Association Inc. (2015a) ANSI/ESD STM3.1-2015. *ESD Association Standard for the Protection of Electrostatic Discharge Susceptible Items – Ionization*. Rome, NY, EOS/ESD Association Inc.

EOS/ESD Association Inc. (2015b) ANSI/ESD STM11.11-2015. *ESD Association Standard for Protection of Electrostatic Discharge Susceptible Items – Surface Resistance Measurement of Static Dissipative Planar Materials*. Rome, NY, EOS/ESD Association Inc.

EOS/ESD Association Inc. (2015c) ANSI/ESD STM11.12-2015. *ESD Association Standard for Protection of Electrostatic Discharge Susceptible Items*. Rome, NY, EOS/ESD Association Inc.

EOS/ESD Association Inc. (2015d) ANSI/ESD STM11.13-2015. *ESD Association Standard Test Method for the Protection of Electrostatic Discharge Susceptible Items – Two-Point Resistance Measurement*. Rome, NY, EOS/ESD Association Inc.

EOS/ESD Association Inc. (2015e) ANSI/ESD S13.1-2015. *Provides electrical soldering/ desoldering hand tool test methods for measuring current leakage, tip to ground reference point resistance, and tip voltage*. Rome, NY, EOS/ESD Association Inc.

EOS/ESD Association Inc. (2015f). ANSI/ESD STM97.1-2015. *ESD Association Standard Test Method for the Protection of Electrostatic Discharge Susceptible Items – Floor Materials and Footwear – Resistance Measurement in Combination with a Person*. Rome, NY, EOS/ESD Association Inc.

EOS/ESD Association Inc. (2016a) ANSI/ESD SP3.3-2016. *Standard Practice for the Protection of Electrostatic Discharge Susceptible Items – Periodic Verification of Air Ionizers*. Rome, NY, EOS/ESD Association Inc.

EOS/ESD Association Inc. (2016b) ANSI/ESD SP3.4-2016. *Standard Practice for the Protection of Electrostatic Discharge Susceptible Items – Periodic Verification of Air Ionizer Performance Using a Small Test Fixture*. Rome, NY, EOS/ESD Association Inc.

EOS/ESD Association Inc. (2016c) ANSI/ESD STM97.2-2016. *Standard Test Method for the Protection of Electrostatic Discharge Susceptible Items – Floor Materials and Footwear – Voltage Measurement in Combination with a Person*. Rome, NY, EOS/ESD Association Inc.

EOS/ESD Association Inc. (2017) ANSI/ESD STM4.1-2017. *ESD Association Standard for the Protection of Electrostatic Discharge Susceptible Items – Worksurfaces – Resistance Measurements*. Rome, NY, EOS/ESD Association Inc.

EOS/ESD Association Inc. (2018a) ESD TR53-01-18. *Technical Report for the Protection of Electrostatic Discharge Susceptible Items – Compliance Verification of ESD Protective Equipment and Materials*. Rome, NY, EOS/ESD Association Inc.

EOS/ESD Association Inc. (2018b) ANSI/ESD STM11.31-2018. *ESD Association Standard Test Method for Evaluating the Performance of Electrostatic Discharge Shielding Materials – Bags*. Rome, NY, EOS/ESD Association Inc.

EOS/ESD Association Inc. (2018c) ANSI/ESD S541-2018. *Packaging Materials for ESD Sensitive Items*. Rome, NY, EOS/ESD Association Inc.

European Committee for Electrotechnical Standardisation (CENELEC). (1992) EN 100015-1. *Basic specification. Protection of electrostatic sensitive devices. Harmonized system of quality assessment for electronic components. Basic specification: protection of electrostatic sensitive devices. General requirements*. Brussels, CENELEC.

International Electrotechnical Commission. (2001) IEC 61340-4-3:2001. *Electrostatics – Part 4-3: Standard test methods for specific applications – Footwear*. Geneva, IEC.

International Electrotechnical Commission. (2000) IEC 61340-2-2:2000. *Electrostatics – Part 2-2: Measurement methods - Measurement of chargeability*. Geneva, IEC.

International Electrotechnical Commission. (2004) IEC 61340-4-5:2004. *Electrostatics – Part 4-5: Standard test methods for specific applications – Methods for characterizing the electrostatic protection of footwear and flooring in combination with a person*. Geneva, IEC.

International Electrotechnical Commission. (2014) IEC 61340-4-8:2014. *Electrostatics – Part 4-8: Standard test methods for specific applications – Electrostatic discharge shielding – Bags*. Geneva, IEC.

International Electrotechnical Commission. (2015a) IEC 61340-2-1:2015. *Electrostatics - Part 2-1: Measurement methods - Ability of materials and products to dissipate static electric charge*. Geneva, IEC.

International Electrotechnical Commission. (2003 and 2015b) IEC 61340-4-1:2003+AMD1:2015 CSV. *Electrostatics - Part 4-1: Standard test methods for specific applications - Electrical resistance of floor coverings and installed floors*. Geneva, IEC.

International Electrotechnical Commission. (2015c) IEC 61340-4-6:2015. *Electrostatics - Part 4-6: Standard test methods for specific applications - Wrist straps*. Geneva, IEC.

International Electrotechnical Commission. (2015d) IEC 61340-5-3:2015. *Electrostatics - Part 5-3: Protection of electronic devices from electrostatic phenomena - Properties and requirements classification for packaging intended for electrostatic discharge sensitive devices*. Geneva, IEC.

International Electrotechnical Commission. (2016a) IEC 61340-2-3:2016. *Electrostatics. Part 2-3: Methods of test for determining the resistance and resistivity of solid materials used to avoid electrostatic charging*. Geneva, IEC.

International Electrotechnical Commission. (2016b) IEC 61340-4-9:2016. *Electrostatics - Part 4-9: Standard test methods for specific applications – Garments*. Geneva, IEC.

International Electrotechnical Commission. (2016c) IEC 61340-5-1:2016. *Electrostatics - Part 5-1: Protection of electronic devices from electrostatic phenomena - General requirements*. Geneva, IEC.

International Electrotechnical Commission. (2017) IEC 61340-4-7:2017. *Electrostatics - Part 4-7: Standard test methods for specific applications – Ionization*. Geneva, IEC.

International Electrotechnical Commission. (2019) IEC TR 61340-5-4:2019. *Electrostatics - Part 5-4: Protection of electronic devices from electrostatic phenomena – Compliance Verification*. Geneva, IEC.

Smallwood, J. (2017). A practical comparison of surface resistance test electrodes. *J. Electrostat.* 88: 127–133.

Smallwood J. (2018) Paper 4B3. Comparison of surface and volume resistance measurements made with standard and nonstandard electrodes. In: *Proc. EOS/ESD Symp. EOS-40* Rome, NY, EOS/ESD Association Inc.

Further Reading

EOS/ESD Association Inc. (2012) ANSI/ESD STM4.2-2012. *ESD Association Standard for the Protection of Electrostatic Discharge Susceptible Items – ESD Protective Worksurfaces – Charge Dissipation Characteristics.* Rome, NY, EOS/ESD Association Inc.

EOS/ESD Association Inc. (2016) ESD TR20.20-2016. *ESD Association Technical Report - Handbook for the Development of an Electrostatic Discharge Control Program for the Protection of Electronic Parts, Assemblies and Equipment.* Rome, NY, EOS/ESD Association Inc.

EOS/ESD Association Inc. (1999) ESD TR15.0-01-99. *Standard Technical Report for the Protection of Electrostatic Discharge Susceptible Items-ESD Glove and Finger Cots.* Rome, NY, EOS/ESD Association Inc.

EOS/ESD Association Inc. (2019) ANSI/ESD SP15.1-2019. *ESD Association Standard Practice for the Protection of Electrostatic Discharge Susceptible Items – In-Use Resistance Measurement of Gloves and Finger Cots.* Rome, NY, EOS/ESD Association Inc.

EOS/ESD Association Inc. (2019) ANSI/ESD STM9.2-2019. *ESD Association Standard for the Protection of Electrostatic Discharge Susceptible Items – Footwear – Foot Grounders Resistive Characterization.* Rome, NY, EOS/ESD Association Inc.

International Electrotechnical Commission. (2018) IEC TR 61340-5-2:2018. *Electrostatics – Part 5-2: Protection of electronic devices from electrostatic phenomena - User guide.* Geneva, IEC.

Smallwood, J. (2005). Standardisation of electrostatic test methods and electrostatic discharge prevention measures for the world market. *J. Electrostat.* 63 (6–10): 501–508.

Vermillion R. (2016) Testing methods for ESD control packaging products. Controlled Environments. Available from: https://www.cemag.us/article/2016/02/testing-methods-esd-control-packaging-products [Accessed 6th June 2019].

12

ESD Training

12.1 Why Do We Need ESD Training?

Training is needed so that all personnel who have some role in electrostatic discharge (ESD) control know enough to fulfill their role successfully and reliably. For effective ESD control, unaccompanied entry into active electrostatic discharge protected areas (EPAs) should be restricted to personnel who have received sufficient ESD training to make sure that they will maintain ESD control procedures according to their role.

According to Snow and Dangelmeyer (1994), untrained personnel account for most deviations from compliance, while trained personnel account for few noncompliances. When employees received training, the number of deviations in those work areas reduced dramatically.

Personnel may need some form of ESD training for many reasons such as

- To recognize ESD-sensitive (ESDS) devices and EPAs, ESD packaging, and ESD control equipment
- To know what ESD control equipment, materials, and procedures to use
- To understand why they need to use them
- To understand where and when to use them
- To know how to test personal wrist straps and footwear and other ESD control equipment according to their role
- To recognize and, if possible, prevent common noncompliances
- To know how to avoid creating noncompliances
- Any safety issues (e.g. high voltage precautions) or use of correct personal protective equipment
- To know what to do when the equipment fails
- To know who to go to for advice
- New techniques, processes, and equipment when they arise
- Specific ESD control requirements or unusual practices used in exceptional EPAs or situations
- Basic electrostatics and ESD knowledge appropriate to their needs

Few organizations have proven documented ESD failure data, and even fewer can trace failure to specific ESD control issues. Where loss of ESD control happens, ESD damage is a risk rather than a certainty. Damage occurs only when ESD occurs to an ESDS device and

The ESD Control Program Handbook, First Edition. Jeremy M Smallwood.

has sufficient strength and is from a potentially damaging source. Only a small proportion of the ESDS device handled will fail, and the failures may be identified during test at a later stage.

Personnel rarely then get the firsthand feedback that loss of ESD control causes ESD failures. Static electricity and ESD are rarely noticeable by sight or sound, and neither can they be felt unless at very high levels. Furthermore, static electricity is not always present in significant levels – it appears and disappears as materials are handled and moved, and even in response to the weather! There are many ways in which static charge is harmlessly dissipated before it can cause damage – ESD may occur only when the many factors involved conspire to prevent static charge dissipation.

If ESD damage occurs, it is usually detected at a test stage long after the damage was caused. Even if it is recognized as an ESD failure, the action or control failure that caused it is usually not obvious.

These factors tend to promote skepticism that ESD damage is a real issue and that ESD control measures are necessary and can make a real difference.

ESD training often seems to be trying to convince trainees of a scarcely believable scenario, and overselling this can be self-defeating (McAteer 1980). Nevertheless, the challenge is to give personnel the understanding they need to know when, where, and how to use ESD control equipment and procedures effectively. They understand why they are doing this and believe that it is important. An ungrounded person is probably the greatest ESD source in manual handling processes. A trained person correctly using ESD control equipment and procedures and preventing noncompliances is arguably a most effective first line of defense against ESD risks.

12.2 Training Planning

The International Electrotechnical Commission (IEC) 61340-5-1 and American National Standards (ANSI)/ESD S20.20 standards require that an ESD Training Plan is written to document the ESD program's training needs. Regardless of whether compliance with a standard is required, it is advisable to have a documented ESD Training Plan covering all aspects of ESD training.

Snow and Dangelmeyer (1994) stated that the goal of training is to reduce deviations to a minimal or zero level. They found that most employees are willing to comply with procedures when offered the right training at the right time. They applied three principles of the psychology of training and learning.

- Train only to affect a measurable change in work behavior
- Motivate students to improve learning
- Consider that students tend to forget information and skills that are not used regularly

Personnel who have different roles often need different training levels or content to fulfill those roles. The Training Plan should document the following:

- The personnel who need ESD training
- The types and content of training required by different personnel

- The need for training before handling ESDS
- The frequency of refresher training
- Policy for maintaining training records and where the records are kept
- Tests and methods used to ensure trainees have understood and can apply their training

12.3 Who Needs Training?

Any person who has a role in ESD control or needs to enter an EPA probably needs some form of ESD training. This may be simple (e.g. do's and don'ts list for visitors) or may need considerable depth (e.g. principles and practice of ESD control, and compliance with standards, for the ESD coordinator).

As a minimum, all personnel who handle ESDS must have sufficient training to ensure they understand and use the equipment and procedures required to prevent ESD damage from occurring. They should receive this training before they start work that involves entering the EPA and handling ESDS devices.

Personnel also usually need periodic refresher training for various reasons. This is in part to remind them of the ESD control practices they must use. Refresher training can also help increase understanding and improve reliability of applying ESD control in practice. Refresher training courses can also be an opportunity to update, explain, and communicate any changes to the ESD control practice that may have been made since the previous training was given.

Personnel who have particular job roles may also need specific training according to their role. The specific training may be simpler than the basic ESD training course (as in the case of a visitor, who may need simple instructions such as what to do, what not to do, and how to wear and test personal grounding equipment), it may be more complex (e.g. for an ESD coordinator or auditor), or it may be just different (e.g. for a cleaner). There may be many reasons why personnel other than those who enter an EPA or handle ESDS devices may also need some form of ESD training specific to their role. Some examples of role specific ESD training are given next.

The ESD coordinator and other personnel involved in developing the ESD control program may benefit from in-depth training in the principle and practice of ESD control and compliance with ESD control standards. Update training may include review of standards updates, trends in ESDS threats and ESD control techniques and practices, or attending conferences and seminars.

Personnel who check, test, and audit an ESD control program may need specific training in ESD measurements as well as auditing practices.

Cleaners who go into an EPA need specific instruction regarding things they must not do, as well as on use of any specific cleaning techniques and materials they must use with floors, bench surfaces, or other ESD control equipment.

Managers and supervisors who have budget responsibility for ESD control or who may need to go into an EPA may need training specific to their role. For successful ESD control implementation, management must be convinced that ESD is a real issue and the control measures are necessary (McAteer 1980). Sadly, nearly 40 years after McAteer's paper, this can still be a challenge.

Managers who need to enter EPAs may also need brief instructions on the use of foot straps, garments, or other ESD control equipment they will use in the EPA, as well as on any activities or actions they must avoid (e.g. touching ESDS). If they accompany visitors into the EPA, they may need instruction on supervision of these visitors.

Trainers who need to develop and present effective ESD training may themselves need training on the specific ESD control practices current in their organization. They may also need sufficient understanding of ESD control principles and practice to answer trainee questions during training sessions. They may need training on how to present effective demonstrations relevant to electrostatics and the ESD control program.

Personnel who purchase ESD-related items may need an overview of ESD control practices and their impact on product specification and procurement. If the ESD control program complies with a standard, they may need a working knowledge of the standard requirements for sourcing ESD control equipment.

Subcontractors who go into an EPA may need either specific training according to their activities or instructions on any areas and activities they must avoid.

Visitors who go into an EPA should normally be accompanied by trained personnel to ensure they do not do anything that might compromise ESD control. Nevertheless, they may need brief instructions on the use of foot straps, garments, or other ESD control equipment they will use in the EPA, as well as on any activities or actions they must avoid (e.g. touching ESDS).

Skeptical engineers who do not believe in ESD control can be among the most resistant and unreliable in using ESD control measures. They can also have a negative effect on others around them, reducing their confidence in the need for ESD control. Conversely, engineers who have good understanding of ESD issues and control measures can be a great asset in implementing an ESD control program, helping to develop an effective ESD control program and helping others understand the importance and how to use ESD control equipment.

One way of summarizing the roles of personnel that need training, and the training they will need, is to present them as a matrix (Table 12.1). This can help plan training courses and their content, as well as who will receive them.

Table 12.1 An example of a matrix of ESD training and personnel roles.

	Training type				
Personnel role	**ESD awareness**	**Use of EPA equipment**	**Testing EPA equipment**	**Principles and practice of ESD control**	**ESD control for managers**
Operators	√	√			
Supervisors	√	√	√		
Managers					√
Audit and test personnel	√	√	√		
ESD coordinators	√	√	√	√	√

12.4 Training Form and Content

12.4.1 Training Goals

Training should address an observable outcome that will help the attendee do their job better. An example is to wear a wrist strap or ESD control foot straps correctly and test them (Snow and Dangelmeyer 1994). Other training goals may be to explain how ESD damages devices or recognize what materials might carry charge. The success of training may be measured as how well the attendee does a task or answers questions.

Training course attendees are better motivated if they understand the purpose of the training and see the relevance to their work. Motivations can include

- Building a better or more reliable product
- Getting credit for doing a better job
- Avoiding frustration and possible blame when devices fail without known cause
- Reducing rework
- Avoiding the cost of failures

Skepticism and disbelief in ESD can be a real demotivating factor. Conversely, demonstration of real evidence of ESD issues in various forms can be highly motivating for attendees, stimulating interest and a desire to find solutions to the issues. The demonstrations should, where possible, be appropriate to the attendee's job role. Various demonstrations are further discussed in Sections 12.6 and 12.7, although it is best if the trainer devises demonstrations tailored to the course and attendee's roles and workplaces.

Snow and Dangelmeyer (1994) found that telling attendees how the learning goals will help each person do a better job and build a better product gave positive motivation. Telling when goals are met also helped to motivate.

Training content should as far as possible be designed to address the needs of the personnel being trained and the practices and processes of the facility. Some training topics may be generally of interest to most personnel. Other topics may be of interest only to personnel from specific job roles.

Topics that may be of more general interest might include

- Recognition of ESDS
- Recognition of ESD protected areas
- Recognition of ESD packaging that might contain ESDS devices
- Awareness of the need to control ESD
- Knowledge of who the ESD coordinator is
- ESD control procedures used in the facility
- The need for and use of ESD control equipment
- How to test personal wrist straps and footwear
- Common noncompliances and how to avoid creating them
- What to do (or who to notify) if failures or noncompliances are found
- New techniques, processes, and equipment when they arise
- Basic electrostatics and ESD knowledge (appropriate to their job role)

Role or area-specific ESD training might include topics such as

- ESD awareness for manager, including the possible financial cost/benefit and other impact of ESD damage and ESD control
- ESD measurement test procedures, for personnel who check ESD control equipment and EPAs
- ESD control, safe working, and special procedures for personnel who work in high-voltage or other areas with special safety issues
- Audit techniques and audit of ESD control procedures
- ESD coordinator training and knowledge development and standards update
- Cleaning regime cleaning materials and practice for personnel who clean within the EPA
- Do's and don'ts for visitors, and guidance for personnel who accompany visitors
- Instructions for contractors working in the EPA
- Instructions for facility maintenance activities in the EPA

12.4.2 Initial Training

Initial ESD training should typically cover topics appropriate to the ESD control program requirements, EPA discipline, and person's job role. For a trainee handling ESDS and working in the EPA, this might include topics such as the following:

- What is static electricity?
- How and when does ESD occur?
- What problems come from static electricity and ESD, and why do we need to avoid them?
- Recognition and correct use of ESD control packaging
- What is secondary packaging, and how do we recognize it?
- What is an EPA, and how do we recognize one?
- Don't bring secondary packaging into an EPA.
- Don't open ESD control packaging outside an EPA.
- Charged people are a major source of ESD damage in manual handling, and personal grounding is an extremely important ESD control measure.
- Use and test of wrist straps and ESD footwear.
- Understanding what unnecessary insulators are, their recognition and why we don't want them in an EPA.
- Don't take unnecessary insulators into an EPA, and remove them if you find them in the EPA.
- ESD control equipment, its recognition, and how to use it.
- For portable equipment such as tools and gloves, recognition of ESD control versions and the importance of not bringing ordinary versions into the EPA.
- The importance of not bringing personal items, which might bring undefined risks, into an EPA.
- Don't remove or store clothing near unprotected ESDS devices.
- Any safety precautions necessary in the EPA.

12.4.3 Refresher Training

Refresher training gives an opportunity to reinforce key aspects of ESD control, especially anything that has been noted to regularly go wrong. It also provides an opportunity to give more depth or detail to topics and updates or changes that may have occurred in the ESD control program. Trainees can be encouraged to give feedback or discuss issues and make suggestions for improvements.

Refresher training can be a good way to review things that have arisen as regular noncompliances. However, persistent noncompliances may not always be best addressed by more or improved training. It may be an indication that an existing ESD control is inconvenient or difficult to use, or at least may not be the best solution. As an example, Snow and Dangelmeyer (1994) reported that a persistent noncompliance related to use of foot straps was eliminated by moving to the use of ESD control shoes.

12.4.4 Training Methods

Any form of ESD training can be used, varying from group watching of videos to one-to-one hands-on practical training. The main consideration is that it should be effective in communicating the topics to be covered.

12.4.4.1 Video, Computer or Internet-Based Training

A video or film or computer or internet-based courses can be effective and cost-effective for training large or small numbers of personnel or a single person, for example on induction into their place of work.

Online resources such as YouTube videos can provide some valuable material for training purposes, but they can be very variable in their quality, veracity, and relevance to the organization's facility and processes.

There is a risk that commercially available training may be too general and not appear to the trainee to relate to the activities and processes they work with in practice. A generic training course may be ineffective if it appears to be irrelevant and difficult to relate to their work. The course should appear authentic and realistic with no exaggeration of the issues.

As the detailed ESD control measures often vary with the organization's ESD control program, it is often better that discussion of detailed control measures is avoided in generic training. Universal and common measures such as wrist strap and ESD control footwear use can be usefully covered.

Short videos can be a good way of teaching basic skills such as putting on and testing a wrist strap or ESD control footwear, with the trainee then practicing the skill (Snow and Dangelmeyer 1994).

12.4.4.2 Instructor-Led Training

Perhaps the most effective means of training is instructor led, either in a group or on a one-to-one basis. Classroom or training facility–based practical training can be very effective for a group, and on-the-job training may be productive for individuals or small numbers of trainees.

The instructor should have a good knowledge ESD control theory and practice and the organization's ESD control procedures and practices, to a level above the level they are to teach. If the necessary expertise does not exist in-house, it should be sourced outside the organization. This might include using an external trainer with appropriate expertise to run a suitable course.

A commercial course prepared and presented by a third-party instructor can be a good way of providing broad-based awareness training or general expertise such as ESD measurements or ESD coordinator level training. A disadvantage of generic commercial training is that it may not align well to the organization's specific ESD control program, processes, and facilities.

An in-house course prepared and presented by an expert instructor (in-house or external) can be an excellent way of providing training that is closely aligned to the organization's ESD program and facilities and include details specific to them.

High-level training, e.g. for the ESD coordinator, may be available only as external courses. Attendance of seminars, conferences, or symposia should be considered as a means of updating knowledge of current standards, trends and ESD control techniques, available equipment, know-how, and expertise.

12.4.5 Supporting Information

It is likely that a substantial amount of information on ESD, standards, and training materials will be collected, both for ESD program development and for training purposes. This should be stored and kept available to personnel as appropriate. These could include

- ESD course handout booklets, CDs, videos, and computer-based presentations
- Written procedures and instructions
- Copies of the organization's ESD control procedures
- Copies of standards related to ESD control
- Books, conference proceedings or papers, and other white papers
- Magazine or other articles
- Reports of studies made in-house or by third parties to investigate ESD issues and questions or in support of ESD program development
- ESD control product data sheets
- ESDS susceptibility data of components

The location and availability of this information and resources should be publicized so that personnel can refer to it.

12.4.6 Training Considerations

According to Snow and Dangelmeyer (1994), five stages should be considered in the training process.

- Preparation
- Delivery
- Instructor-led demonstrations
- Hands-on learning
- Follow-up assessment and training

Many books have been written on the subject of creating and delivering successful presentations. The best presentations use a mixture of verbal and visual elements. Points should be presented in a logical sequence. Each point should lead to following points or build on previous points. I have found that short videos, demonstrations, or audience participation (discussions or activities) are all useful ways of keeping interest in a presentation high.

12.4.6.1 Preparing a Presentation

In preparing a presentation, two fundamental questions must be asked. Who is the audience, and what type and level of background knowledge can they be assumed to have? What are the learning points that they must take from the presentation? From this, the presentation material, including demonstrations, can be developed. The presentation should be made appropriate to the audience in terms such as topics of interest, technical content and level, and depth.

Many presenters find that about one slide per minute of presentation time gives a reasonable estimate of a presentation duration. For some, more time per slide is needed. Don't forget, however, to make time allocation for demonstrations, videos, discussion, or another attendee participation.

If the presentation is to attendees who are unfamiliar with the venue and instructor, it may be necessary to start with some introductions and introductory information. This could include introduction of the presenter, location of the facilities, emergency fire exits, and whether a test alarm is expected. It is a worthwhile precaution to ask attendees to switch off their mobile phones (When presenting, this is a useful reminder to switch mine off also). For a course with a small number of attendees, it can be useful to ask the attendees to give a short sentence of introduction to themselves also.

A long course can be split into specific topics or themes. It is good to vary the presentation and engage the attendees by using demonstration and participation where possible. Questions and discussion can be encouraged. Discussing issues and situations arising in the attendee's workplace can be valuable in demonstrating relevance, providing interesting practical learning points and helping their understanding. It can, however, be a mixed blessing, leading to over-run if not carefully controlled!

If a training session is long, decide if breaks and refreshments will be necessary. Breaks can also be useful for giving the presenter the opportunity to talk to attendees on a one-to-one basis. This can be useful for evaluating whether the attendees have adequately received and understood the material up to that point. Attendees often take to opportunity to ask questions at breaks on topics that they have not had courage to ask in the more public setting of the course.

While it is good to be well prepared and familiar with the presentation material, it can be counterproductive to be over rehearsed or appear to be presenting a memorized speech.

12.4.6.2 Presentation and Attendee Participation

The delivery skills of the presenter make a large contribution to the engagement of the audience. These days, there are many good books on presentation skills that can give guidance and ideas for this. Dressing appropriately so that both the presenter and the audience feel comfortable is important. (An audience of managers might expect the presenter to be in a suit, but "smart casual" may be better for an audience of EPA workers). Try to avoid

Figure 12.1 The author presenting a course.

anything that might be a distraction or a barrier to communication. Try to make the presentation as interesting and enjoyable as possible for the audience and yourself (Figure 12.1).

Eye contact, or an illusion of eye contact, is important, but maintaining unnatural eye contact can make the recipient uncomfortable. In a small group, it may be possible and important to make occasional eye contact with each attendee in a relaxed and random order. In a larger group, there are usually a few attendees around the room who are obviously more receptive and engaged than others. Keeping eye contact with some of these can give the illusion of keeping eye contact with the whole audience. Keep eye contact moving around the room to avoid making anyone feel uncomfortable! If the presenter feels self-conscious about eye contact, looking at different points behind the audience can help give the illusion of it.

The presenter should face the audience whenever possible and try to avoid turning their back. This can be difficult when the display screen is behind the presenter, and it may be necessary to point to parts of it. If possible, the presenter should avoid blocking the view of the screen or demonstrations. Achieving this may require some thought into the arrangement of the room and presenting area when setting up for the presentation. Before starting, walk around the room and look at the presenter's area from the point of view of the entire audience, sitting at selected points, to look from attendee eye height.

Classes where learners can participate, discuss, and engage are more likely to be found interesting and remembered by the audience. Demonstrations can be very valuable, especially if the audience participates in them. As learners best remember what they practice frequently, it is valuable to have attendees practice the skills they will need to use. It is best to immediately follow up this learning by practice in the workplace. Surprisingly, even demonstrations that go wrong can entertain and make the course more memorable. As electrostatic demonstrations can be fickle, I explain early on that at least one will probably go wrong, but I don't know which one. This creates a sense of interested anticipation! When one does go wrong, its failure can often be used to reinforce some other learning point!

Learning is often facilitated by working in a small group and encouraging questions and discussion. I find that attendee questions often bring up very interesting points specifically relevant to the questioner. Questions indicate that the attendee is thinking about the course material presented and trying, in their mind, to apply it in their work.

Unless a course is very short or the audience is large, I prefer to start by briefly introducing myself and then asking each attendee to introduce themselves and their job role in a couple of sentences. This helps "break the ice" and starts the attendees speaking and contributing. It also helps me understand my audience for later formulating answers to questions in appropriate terms and level.

Early on in ESD awareness courses I ask the question, "Who here has experienced electrostatic shocks in everyday life?" I then go on to ask typical circumstances they have felt shocks and to explain that the shock is a form of electrostatic discharge. I encountered very few people who claim not to have experienced an electrostatic shock. Getting the audience to respond to this serves several useful purposes.

- Normally, everyone has experienced a static shock at some time, and everyone can respond – this again helps establish the habit of audience participation.
- It generates interest.
- It shows they have themselves experienced ESD, and it is a matter of everyday experience.
- It gives me the opportunity to explain that to feel a shock the body voltage must be over 2000 V for most people, but many components can be damaged by ESD at lower body voltages we do not feel shocks (this is later substantiated by demonstrating body voltage).

Specific points can often be illustrated by relevant anecdotes from real life. If an attendee has an illustrative anecdote from their experience, this can be even better, bringing the point into relief and immediate relevance to other attendees. Occasionally, attendees have brought to the discussion some evidence of ESD damage in their experience – this can be particularly valuable.

When preparing electronic slide presentations, the visual style, type face, color selection, etc. should be selected for clarity. Transition effects should be used sparingly as they can be distracting. Establishing a consistent slide style helps give a professional presentation image. Be careful to make sure that text and information low down on a slide can be seen be all the audience.

12.4.6.3 Hands-On Learning

For some types of learning, there is no substitute for hands-on exercises for the attendees. Examples are wearing and testing of wrist straps or making ESD measurements and tests for compliance verification. The exercise normally consists of demonstration of the actions to be learned, followed by the learner trying the actions for themselves, and practicing until sufficiently familiar and proficient. The demonstrator can also evaluate the proficiency of the learner during the exercise.

In some cases (e.g. testing of wrist straps and footwear), the learning exercises may be best done in the workplace. In other cases, the exercise can be done in a training facility as part of a course.

12.4.6.4 Follow-Up Assessment and Training

It is necessary to evaluate the attendee's understanding or ability to perform tasks as taught on the course, during or after completion of the course. This evaluation is a requirement of modern ESD control standards. This can be any of the following, for example:

- A practical evaluation (e.g. observation of wearing and testing a wrist strap correctly, or testing a workstation and recording results)
- Question and answer (this can be an interactive discussion or written questions)
- Multiple-choice questionnaire
- Some other form of evaluation

12.4.7 Public Tutorials and Courses

Public tutorials and courses are available from a wide variety of sources such as the ESD Association, IPC, industry groups and organizations, and specialist ESD control consultants, trainers, and equipment suppliers. Some of these, such as the ESD Association tutorials provided in association with its symposia, can lead to certification qualifications.

The level of these courses ranges from basic ESD awareness and control practice to in-depth ESD practitioner level courses. Many of the industry provider courses are at the operator basic ESD awareness and control level. Some providers offer courses at the user's site as well as open public events. Some providers offer courses up to ESD coordinator (program manager) level, including measurements and auditing.

12.4.8 Qualifications and Certification

There are currently a few ESD-related qualifications and certification providers. At the time of writing, the ESD Association is probably the foremost provider of ESD control personnel certification at various levels.

Many independent training companies give ESD training that includes a certificate of attendance of the course. This should not be confused with certification – a certificate of attendance merely shows that a person has attended a course. The certificate of attendance is usually issued without an examination of the attendee's knowledge and understanding.

Certification provides confirmation that a person meets a level of knowledge and problem-solving ability (Newberg 2017). Achieving certification will usually require some extensive training and testing. For the certified professional, it provides credibility in the industry, demonstrating knowledge, experience, and competency. It is a form of professional development and can improve performance and confidence. It can differentiate the individual from those who are not certified in terms of technical skill. It can create opportunities for career advancement and increased earnings. In some organizations, it may give an advantage over noncertified personnel in competition in recruitment applications.

At the time of writing (2017), the ESD Association offers the following:

- TR53 Technician Certification
- ESD Certified Professional Program Manager Certification

- Device Stress Testing Certification
- ESD Certified Professional – Device Design

Each certification course requires attendance to a number of day or half-day courses, which are normally given in association with ESD Association Symposia around the world. As an example, the ESD Certified Professional Program Manager course covers the following topics:

- ESD basics
- How-tos of in-plant ESD auditing and evaluation measurements
- Ionization issues and answers
- Packaging principles
- ESD standards overview
- Device technology and failure analysis overview
- Electrostatic calculations
- Cleanroom considerations
- System-level ESD/electromagnetic interference (EMI), including principles, design, troubleshooting, and demonstrations
- ESD program development and assessment (ANSI/ESD S20.20 Seminar)

After attending the courses, the candidates can take an examination that is also provided at the symposiums and, if successful, leads to award of the certificate.

Alternatively, certification can be gained through National Association for Radio and Telecommunications Engineers (iNARTE) examination. iNARTE provides certifications for qualified engineers and technicians in the fields of telecommunications, electromagnetic compatibility/interference (EMC/EMI), product safety (PS), ESD control, and wireless systems installation.

The iNARTE ESD certification program was implemented in conjunction with the ESD Association in the 1990s. Certification can be obtained at the engineer or technician level (iNARTE 2017). Assessment includes providing a record of the candidate's relevant experience and passing of an iNARTE examination.

IPC is a trade association connecting electronics industries. Among other activities, it provides standards and training for the electronics industries and is widely used across the world.

IPC provides ESD control certification courses via the online IPC learning management system, IPC Edge (IPC 2017). IPC developed these courses with the EOS/ESD Association to train operators and trainers on ESD controls and best practices. This online ESD Certification Program allows users to validate their knowledge and skills by passing an exam on ESD principles.

12.4.9 National and International ESD Groups and Electrostatics Interest Organizations

Many countries have national ESD or electrostatics interest organizations. There are also ESD Association chapters around the world e.g. in Texas, Philippines, India, and Korea (https://www.esda.org/membership/local-chapters/Accessed:Dec.2017). Some national specialist ESD interest organizations are given in Table 12.2.

Table 12.2 Some regional or national ESD interest organizations.

Organization	Region	Activities (languages)	Web site
ESD Association	North America International	Corporate or individual membership, Standards, symposia, conferences, tutorials, white papers (English)	www.esda.org https://www.facebook.com/EOSESDAssociationInc
ESD Association Korea Chapter	Korea	Membership, symposia, conferences, tutorials (Korean)	http://www.esd.or.kr https://www.facebook.com/KOREA-Chapter-of-ESD-Association-186248624800065
STAHA	Scandinavia	Membership, symposia, conferences, tutorials (Finnish and English)	http://www.staha.fi
ESD Forum	Germany	Membership, symposia, conferences, tutorials (German and English)	http://www.esdforum.de
Italian ESD Assoc.	Italy	Membership, symposia, conferences, tutorials (Italian and English)	http://www.esditaly.com/esditaly.html
Dutch ESD-EMC Assoc.	Netherlands	Membership, symposia, conferences, tutorials (Dutch and English)	http://www.emc-esd.nl
ESD Association China	China	Membership, training. (Chinese)	http://chinaesd.org.cn
Japan ESD Association	Japan	Seminars, exhibitions, publications (Japanese)	http://www.jesda.org
Industry Council on ESD Target Levels	International	Invited industry membership, white papers(English)	http://www.esdindustrycouncil.org

Some organizations are interested more generally in electrostatics in industrial processes, and others in academic research (Table 12.3). These organizations often also have an interest in ESD control in electronics manufacture as well as measurements and many other areas of electrostatics. They often include ESD control in their conferences or publications.

12.4.10 Conferences

Many of the organizations listed in Tables 12.2 and 12.3 organize conferences worldwide that may include some ESD-related papers. Their current activities may be found via their web sites.

The organizations listed in Table 12.2 are most likely to arrange specialist ESD control–related conferences and publish papers in their proceedings.

12.4.11 Books, Articles, and Online Resources

There are a range of books available on ESD-related topics. Many of these are concerned mainly with design of on-chip ESD protection, e.g. Amerasekera and Duvvury (2002),

Table 12.3 Some general electrostatics interest organizations.

Organization	Region	Activities	Contact
Electrostatics Society of America	N. America International	Membership, conferences, newsletter	http://www.electrostatics.org
Working Party Static Electricity in Industry	Europe International	Industrial and academic committee. Conference every four years.	
Institute of Physics	UK International	Membership, Dielectrics, and Electrostatics Group Conference every four years	http://www.iop.org/activity/groups/subject/de/page_65558.html

Table 12.4 Some magazines, journals, and online resources that publish ESD-related articles.

Publication	Activities	Contact
InCompliance	Online magazine	https://incompliancemag.com
Interference Technology	Online magazine	https://interferencetechnology.com
Controlled Environments	Online magazine	www.cemag.us
Evaluation Engineering	Online magazine	https://www.evaluationengineering.com
Microelectronics Reliability	Peer reviewed Journal	https://www.journals.elsevier.com/microelectronics-reliability
IEEE Transactions on Device and Materials Reliability	Peer reviewed research Journal	http://ieeexplore.ieee.org/xpl/RecentIssue.jsp?punumber=7298
IEEE Transactions on Electron Devices	Peer reviewed research Journal	http://ieeexplore.ieee.org/xpl/RecentIssue.jsp?punumber=16
IEEE Transactions on device and materials reliability	Peer reviewed research Journal	http://ieeexplore.ieee.org/xpl/RecentIssue.jsp?punumber=7298
Journal of Electrostatics	Peer reviewed research Journal	https://www.sciencedirect.com/science/journal/03043886

Wang (2002), and Voldman (2004). There are several books on ESD control in the manufacturing environment, e.g. Dangelmeyer (1999), McAteer (1990), Welker et al. (2006). Unfortunately, few of these are up to date although their content may still be highly relevant. Because of their age, they are usually not well aligned with current ESD control standards. Welker et al. (2006) is notable in that it focuses on ESD control in the clean room environment.

There are several online magazines that occasionally publish good-quality articles on the subject (Table 12.4). Some academic journals that from time to time publish papers on ESD-related topics are also listed here.

12.5 Electrostatic and ESD Theory

12.5.1 The Pros and Cons of Theory

For those who understand it, inclusion of some level of electrostatics and ESD theory and simple circuit diagrams can be helpful and informative. Conversely, for attendees who do not understand it, theory can be daunting and off-putting and can reduce their confidence that they can understand the course. So, theory and circuit diagrams must be used sparingly and with caution.

In many courses, the attendees are from many backgrounds and job roles. While many in my courses do not understand even simple electronic circuits, there are normally a few who do and appreciate this type of explanation. Normally I include some level of theory in simple nonelectronic explanation terms. Except in basic awareness courses, I often also include a simple circuit explaining electrostatic charging and charge dissipation, appropriate to the level and audience of a course. I may spend more, or less, time on theory depending on the level of interest expressed by attendees and my understanding of their background gleaned from introductions given at the beginning of the course. If necessary, use of the whiteboard and discussion can supplement the core presentation material.

12.5.2 A Technical and Nontechnical Explanation of Electrostatic Charging

In electronic terms, the simple circuit of Figure 2.1 and Figure 12.2 is easily understood by attendees with some electrical circuit knowledge including Ohm's law. Charge generation is represented by a current generator I and flows through a resistance to ground R. Ohm's law gives the voltage V produced as $V = IR$. It can easily be seen that the higher the resistance, the higher the voltage produced. In a higher-level course, this is used to explain why we determine an upper limit of resistance to ground in order to determine a maximum voltage produced by electrostatic charging. The capacitance C represents charge storage. With no current generation, the charge stored in the capacitance drains through the resistance, and the voltage drops away with a characteristic charge decay time given by the product RC.

If the resistance R is very high (as with insulating materials), the voltage generated by even a small current I can be very high. This explains why we specify an upper limit of resistance to limit voltage build-up. For very high resistance, the decay time constant RC also becomes long, showing that charge and voltages can remain for long periods.

A water analogy can be used to explain electrostatic charging to nontechnical people (Figure 12.2). In this explanation, charge is analogous to water, and a basin represents charge storage. The capacity of the wash basin is analogous to the capacitance of the circuit. The level of the water in the basin is like the voltage due to the charge. The water flow into, and out of, the basin is analogous to electrostatic charging and charge dissipation current flow. The basin has a drain that allows water to escape and a tap that provides water input. The size of the drain is analogous to the electronic resistance to ground. A small drain allows only slow water escape, like a high resistance in a circuit.

Clearly with no water input and the drain open, there is no water level in the basin. Most people will accept that for a given water flow into the basin, the water level will depend on the size of the drain hole and rate at which water can escape. If the plug is in the drain hole

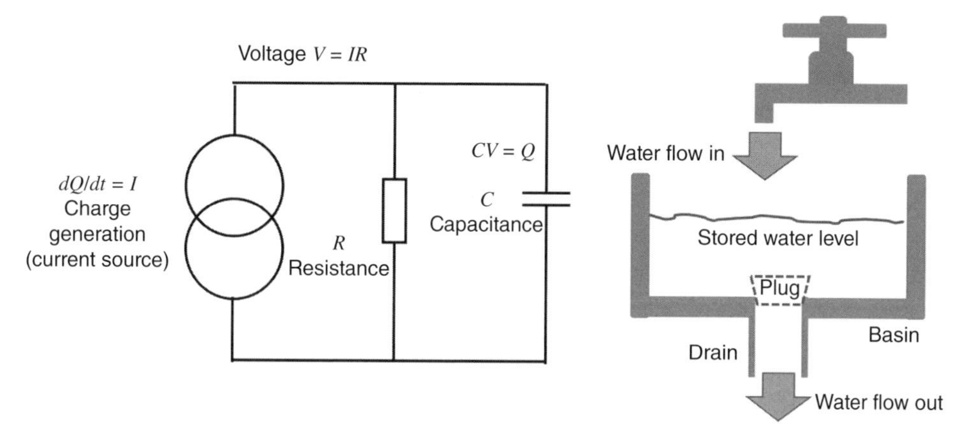

Figure 12.2 Water analogy of a simple model of electrostatic charging that is easily understood by many nontechnical people.

(analogous to an insulator blocking charge dissipation), even a small water flow in (such as a dripping tap) produces a significant water level that can remain for a long time.

Finally, the height of water in a basin for a given charge depends on the shape of the basin. A basin that has a smaller footprint will give a higher water level, and a basin with a larger footprint will give a lower voltage level for a given amount of water. This is analogous to the same charge giving higher voltage for smaller storage capacitance.

12.6 Demonstrations of ESD Control-Related Issues

12.6.1 The Role of Demonstrations

Demonstrations can be very effective for showing the reality of ESD-related issues and bringing into focus the need for ESD control. To be most effective, demonstrations should emphasize the facts that the audience are most interested in – for example, for managers, the cost of ESD damage and rework, possible return on investment, and impact on product quality, yield, reliability, or reputation may be of most interest. For production personnel, the impact on need for rework may be more relevant.

I have found that most people respond well to demonstrations of real static electricity and how ESD can arise. These usually work best if they use materials that are commonly found in the uncontrolled workplace, such as packaging materials, engineering plastics and tapes, and isolated conductors. This shows directly why these materials should be excluded from EPAs where possible.

12.6.2 Demonstrating Real ESD Damage

McAteer (1980) found that to help people believe in ESD damage, they needed to see a device being damaged. A convincing authenticity is provided by showing real ESD damage examples relevant to the organization.

McAteer used samples of different types of ESDS to demonstrate damage, including a metal packaged operational amplifier, metal-oxide-semiconductor (MOS) device, and a small assembly containing a MOS device. He used a curve tracer to show changes to the device parameters resulting from ESD produced during the demonstration. He commented that this type of demonstration persuaded management that the ESD problem was real. They were then anxious to discuss seriously the costs of static failures versus the costs of prevention.

An indirect way of showing the effect of ESD control on failures is if changes in product yield can be correlated with ESD control measure compliance. Snow and Dangelmeyer (1994) found that yield charts showing a decline in failures correlated with increased ESD control are convincing evidence. Sometimes, increases in failure rates can be correlated with a drop in atmospheric humidity, especially below 30% rh. Where this occurs, it is a strong indicator that ESD damage is occurring.

Unfortunately, many organizations are not able to do failure analysis to a level that could reliably identify ESD damaged devices. In this situation, it is difficult if not impossible to provide convincing firsthand evidence of ESD damage to the organization's components, assemblies, or product.

There are, however, many documented accounts of ESD damage given in literature such as the EOS/ESD Symposium Proceedings, journals such as *Microelectronics Reliability* or *IEEE Transactions* (Table 12.4), and books such as Dangelmeyer (1999) and McAteer (1990). Selected examples of these can be used to support ESD training and help present a case for the importance of ESD control.

Unfortunately, it is likely that some people will remain unconvinced of the risk and reality of ESD damage, and importance of ESD control, without direct evidence of ESD damage occurring in their processes.

Sometimes in an ESD control course, some attendees have been able to relate their own company experience of ESD damage. When this happens, their contribution can be very valuable, as a "real-life" anecdote from a fellow attendee's experience can be much more convincing than second-hand and possibly old accounts from an instructor, other organizations, or research publications.

12.6.3 The Cost of ESD Damage

Unfortunately, few seem to have a real understanding of the failure levels and cost of ESD damage in their organization. This is partly because the cost of failure analysis to a level that could confirm ESD damage can be high. In my experience, most organizations do not do failure analysis to this level and so do not have the basic information required to estimate the cost of ESD damage. There are a few published studies of this topic, and these can be quite instructive. These can give useful information for inclusion in a course.

Helling (1996) gave an interesting and useful account of estimating cost of failures and the return on investment in ESD control. His data and examples can be useful to quote to help make points on this topic. His data confirms the following useful general points:

- The cost of failures tends to increase as the product goes through the production process stages.

- The most expensive failures are usually those that occur at the customer's site.
- Cost/benefit indications of over 1:10 can be found for an effective ESD control program.

In a course including relevant attendees such as managers or QA personnel, I often ask the hypothetical question – what is the likely approximate cost of a failure or unreliability at the customer's site for your product? This usually provokes some interesting discussion. In many cases, the answer to this question alone makes it worth taking ESD control seriously.

Many people will not be convinced by accounts of the cost of ESD damage found in other factories, especially if the sources are old. There is no doubt that if it is available or can be determined, real current data on the cost of ESD damage at the attendee's own facility will be far more convincing. In the absence of this data, it can be as valuable to demonstrate cases where component failure levels, although not analyzed and proven due to ESD, nevertheless fell when ESD controls were improved.

12.7 Electrostatic Demonstrations

12.7.1 The Value of Electrostatic Demonstrations

Static electricity cannot be sensed unless it is at very high levels, so it is difficult for people to really understand that it is at work around us much of the time. Using an instrument to reveal the static electricity makes it real to the audience in a way that talking about it and explaining it cannot.

Electrostatic voltages and fields can be revealed with an electrostatic voltmeter or field meter. These can be used as the basis of many simple demonstrations that reveal how static electricity works and can bring real understanding to attendees.

12.7.2 The Pros and Cons of Demonstrations

There are several ways in which demonstrations can be effective as a training method. This is perhaps particularly true of teaching how static electricity works and the effectiveness of some ESD control products. I have found static electricity demonstrations to be most effective if they use materials and objects representative of items that are normally found in a poorly controlled working area. A successful demonstration then shows not only how static electricity works but helps recognition of what sort of materials cause problems. Even experienced people can find that a good demonstration helps them really understand electrostatic action and ESD issues. Developing reliable electrostatic demonstrations can also be a great way of educating the trainer! Some of my most reliable demonstrations have arisen out of adversity and failure of a less reliable demonstration. The demonstration of Figure 12.5 was invented out of need when other demonstrations failed under high humidity conditions in Shanghai!

One problem with live static electricity demonstrations is that they are rather fickle and can unexpectedly fail or give surprising results. This is especially true under high humidity conditions, and demonstrations are best done (where possible) under moderate or dry air conditions. Nevertheless, with practice and selection of materials, demonstrations can be found that are reasonably reliable even at 70% rh. As a precaution, I normally tell my course attendees that it's likely that at least one demonstration will not work, and I don't know up front which it will be. This helps to create a sense of interested attention and anticipation!

The most convincing and relevant demonstrations are often those made using materials that are commonly found in a poorly controlled workplace. For this reason, I generally avoid unusual props like balloons (I once did see balloons in an EPA workstation; I was told it was the operator's birthday!) or Van der Graaf generators. The following examples give some demonstrations that I have used in my electrostatics training.

12.7.3 Useful Equipment for Demonstrations

The heart of good electrostatic demonstrations is a good electrostatic field meter or electrostatic voltmeter to show the presence and magnitude of electrostatic voltages and fields. Field mill instruments are more useful than induction field meters as they avoid problems due drift and do not need zeroing. Beware that nearby charged insulating materials can cause unexpected effects.

While a stand-alone voltmeter or field meter works well for one-to-one interactions or small groups, for a larger class it is better to use an instrument that has an output that can be fed into a suitable display (Figure 12.3). PC-based oscilloscopes such as Picoscope can be used to display the changes in field or voltage over time and project the display onto a large screen display or via a data projector.

A metal plate mounted on a good-quality insulating handle or stand can be used for demonstration of triboelectrification and induced voltages on conductors, equipotential bonding, and many other effects.

A selection of insulating materials and known reliable static generators should be collected, especially those that tend to be found as noncompliant items in the workplace.

Figure 12.3 Demonstrating a charged insulator, in this case a slab of expanded polystyrene foam.

Materials may need to be replaced from time to time as their charging and charge-holding properties tend to diminish with handling. This is due to contamination with oils, salts, and perspiration from the hands.

A selection of ESD control packaging can be useful. Low-charging materials such as pink polythene should be kept in a package separately from the other demonstration materials. This is because they tend to contaminate other materials with their antistat and can make demonstrations of electrostatic charging rather unreliable.

It can be useful to have an electric hairdryer or hot air gun in the demonstration kit to dry out materials under high-humidity conditions. This can then be used as a teaching point, demonstrating how humidity affects electrostatic charging and activity.

An ESD detector can be useful for demonstrating that ESD occurs during experiments. This should be selected to make a loud enough sound when ESD occurs, to be easily heard by all trainees.

12.7.4 Showing How Easy It is to Generate Electrostatic Charge

The main learning point in this demonstration is that many insulating materials charge easily and give strong electrostatic fields. Highly insulating materials tend to charge during handling and hold their charge for long periods. This can be shown using an electrostatic field meter held in a stand (Figure 12.3).

I have often used a plastic document file divider for this demonstration. This just rests in, or on, a pile of demonstration materials until the moment of demonstration. Lifting the material and holding it in front of the electrostatic field meter shows that it is already highly charged. Many plastic packaging materials can also be used for this demonstration. With suitably chosen materials, this demonstration is reliable under most except the most humid conditions.

Another reliable demonstration is the high voltage produced by ordinary tapes such as packing tape on stripping from the reel. This is revealed by stripping a length of tape from the reel and holding it in front of the field meter.

12.7.5 Understanding Electrostatic Fields

The experiment of Figure 12.3 can also be used to show the variation of electrostatic field with distance from the field source. The field meter reading increases as the charged item is move toward the field meter and then drops dramatically as it is moved away.

Depending on the level of course, the instructor can then explain some important learning points, such as the following:

- The field meter is calibrated to read surface voltage but correctly does so only when taking readings from a large flat surface a defined distance away.
- Electrostatic field drops rapidly with increasing distance from the source.
- The closer a field source is to the ESDS, the more concerned we will be about it.
- If an insulator is found to have greater voltage level than a defined risk level, the risk can be reduced by keeping it a sufficient distance away from any ESDS. This is the basis of the field and voltage limits given in the standards.

12.7.6 Understanding Charge and Voltage

People tend to think that voltages are at the root of electrostatics and stay constant unless the item becomes more charged or charge leaks away. This demonstration can be used to show that it is charge that is at the root of electrostatics and that voltages change, appear, and disappear as materials are moved around – even though the charge present is not changed.

I have often used a training booklet with an insulating plastic cover in this demonstration (Figure 12.4). When the booklet is held in front of the field meter with the cover closed, only low or zero electrostatic voltage is shown. The instructor can explain that the cover is already charged by contact with the cover, but no voltage is visible as the charges on the cover and paper are in balance and are close together and so cancel in their effect.

When the cover is opened, a high voltage appears on the cover material. (A high voltage does not appear on the paper as it is a sufficiently conductive material for the charge to escape somewhere.) After closing the cover again, the voltage has normally largely disappeared again as the charge has been rebalanced.

The fact that the cover material is highly charged can also be used to demonstrate electrostatic attraction. Usually the first page of the booklet adheres to the cover due to attractive charge forces.

Another rather interesting demonstration can be made with a document in an insulating plastic document holder (Figure 12.5). With the document in the holder, little voltage or field can be seen by the field meter. If the document is then partially withdrawn from the holder, a voltage appears on the part from which the paper is withdrawn. If the document is reinserted, the voltage disappears.

These demonstrations support the following teaching points:

- Charge is already separated between the plastic and the paper, but no voltages show until it is separated by moving the paper and plastic and charges apart.
- The voltages disappear when the paper and plastic are moved back together.

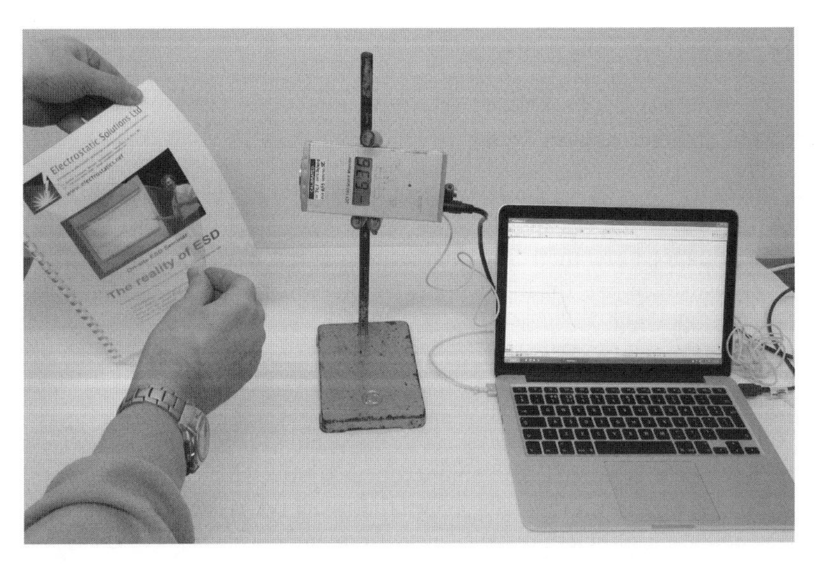

Figure 12.4 Using a clear plastic covered booklet to show how voltages appear and disappear when the cover is opened and closed.

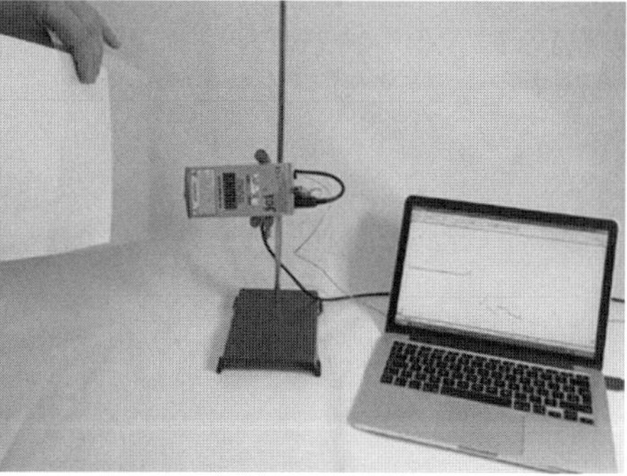

Figure 12.5 Using a document holder to show how voltage appears when the document is pulled out and disappears when it is returned.

I usually neglect two other factors that are present in these experiments, for the sake of reducing possible confusion. First, presence of a conducting material nearby can reduce the apparent voltage produced by a charge. This is a second reason the voltage disappears when the paper is reinserted into the holder. Second, the paper is normally sufficiently conducting that any charge on it can move to and from the body of the demonstrator while they are performing the demonstration.

I have occasionally done these experiments while wearing highly insulating rubber gloves. If this is done, it can often be shown that the paper is positively charged on separation while the plastic is negatively charged.

12.7.7 Tribocharging

Tribocharging can be demonstrated with conductors or insulating materials. As many people, even those more familiar with electrostatics, believe that conductors cannot generate charge, it can be useful to demonstrate tribocharging of a conductor. One way of doing this is to use a metal plate mounted on a highly insulating handle (Figure 12.6). The field meter is positioned to show the voltage on the plate and display it.

A suitable charged plate monitor can also be used to do these experiments. The metal plate, however, can be easier for trainees to relate to metal items and printed circuit board (PCB) used in the workplace. The instructor can make the point that what happens to the plate can be expected to happen to the conductors on a PCB isolated from ground.

In this arrangement, the plate can be tribocharged by rubbing with various materials or other actions. I have typically used the following:

- Negative charging of the plate by flicking it with a wool cloth
- Positive charging of the plate by flicking it with a rubber glove
- Charging of the plate by stripping tape previously applied to it

I have also used this arrangement to demonstrate the charging effect of processes, for example use of a cooling spray can.

Under humid air conditions, the moisture layer condensing on the insulating handle can be sufficient to allow slow discharge of the plate. In these conditions, a hair dryer or hot air gun can be used to temporarily dry the insulator and prevent discharge. This can itself be a

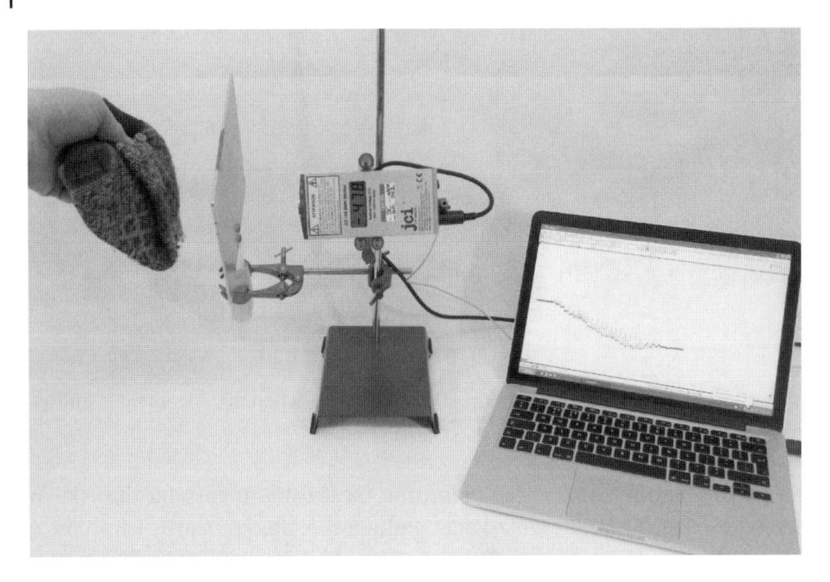

Figure 12.6 Tribocharging a metal plate.

useful teaching point, showing that the discharge of the plate across the insulator is due to moisture on the surface that is removed by drying.

12.7.8 Production of ESD

Once the metal plate is charged, it is easy to demonstrate ESD using an EMI-based ESD detector (Figure 12.7). These instruments detect ESD from the EMI radiated from the ESD. Some emit an audible signal as well as updating a count on each ESD detected.

With the plate charged, a discharge can be initiated by touching it with a ground wire. At the same time as ESD occurs, the voltage on the metal plate can be seen to drop instantaneously to zero.

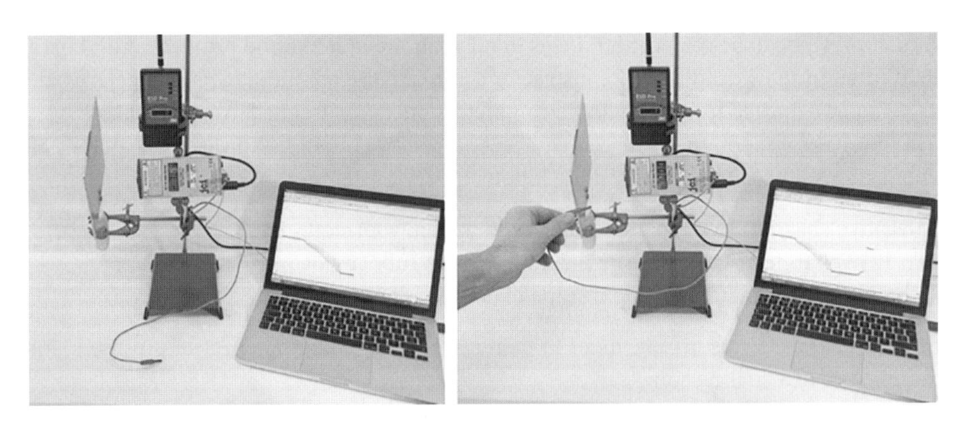

Figure 12.7 Demonstration of ESD generation.

Possible teaching points in this experiment include the following:

- ESD occurs when a charged conductor is touched by another conductor.
- This is true even if the conductor that touches it is not grounded.
- The metal plate acts similarly to the conductors of a PCB or other ESDS device. So, ESD can occur if a charged ESDS device contacts another conductor or is grounded.

The conductor used to touch the plate and generate ESD does not have to be a ground wire and does not have to be grounded. ESD can also be demonstrated by touching with a tool, or even a grounded ESD tool held in the hand of a grounded instructor. Even experienced trainees can sometimes be surprised that ESD can occur when a correctly grounded person handling or working on ESDS using ESD tools. The explanation is of course that the ESD can occur if the ESDS is itself charged. ESD can occur when a person touches an ESDS if either the person is charged or the ESDS is charged, or both.

12.7.9 Equipotential Bonding and Grounding

Following the demonstration of production of ESD on touching in the experiment of Section 12.7.8, it can be demonstrated that ESD is prevented by equipotential bonding (Figure 12.8). First it is shown that if a person touches the metal plate with an ESD tool, ESD occurs because the person and the plate are at different voltage. ESD is registered on the ESD detector.

The demonstrator can then connect themselves to the metal plate using a wrist strap. No connection is made to ground. Usually, some voltage is indicated on the metal plate as the demonstrator moves around and generates body voltage. Nevertheless, no ESD is

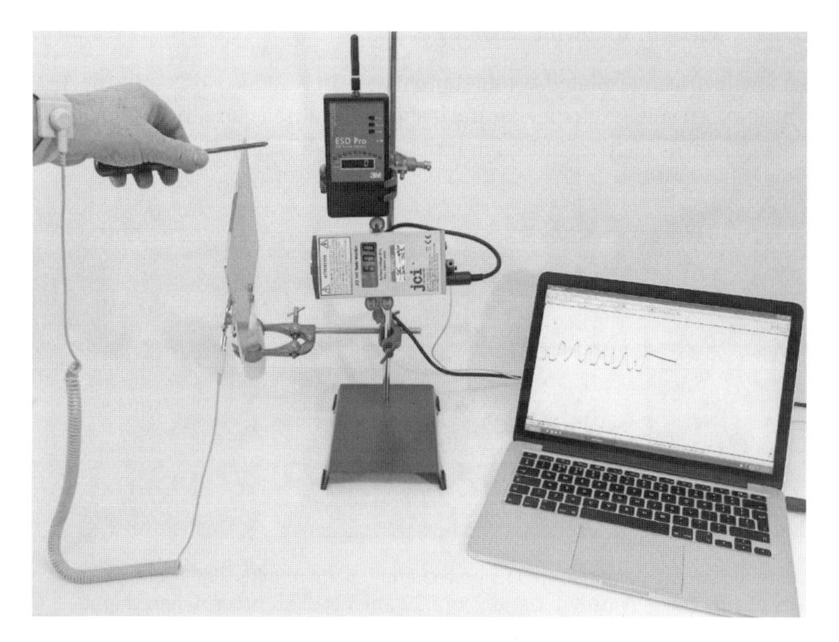

Figure 12.8 Demonstration of equipotential bonding.

produced when the demonstrator touches the plate with the tool. The explanation is that the demonstrator and the plate are at the same voltage and so ESD does not occur.

This experiment supports the following teaching points:

- ESD occurs when two conductors touch and there is a voltage difference between them.
- If there is no voltage difference between the conductors, no ESD occurs when they touch.
- Once an equipotential bonding connection is made, no ESD can occur between the connected conductors. However, at the point that the connection is made, ESD is likely to occur as the conductors will probably be at different voltages.

This experiment can lead into a discussion explaining grounding as a form of equipotential bonding where all conductors are bonded to earth. It can also be explained that in situations where grounding is not possible equipotential bonding can be used to control ESD sources. Current ESD control standards often consider equipotential bonding and grounding to earth both to be acceptable forms of "grounding."

12.7.10 Induction Charging

The metal plate mounted on an insulating handle can be used to show induced voltages on conductors due to electrostatic fields. The metal plate is initially discharged to zero volts in the absence of any electrostatic fields. When a charged insulator is moved closer, the voltage on the conductor changes and can reach surprisingly high levels (Figure 12.9).

ESD can then be made by touching the metal plate with a ground wire. An audible ESD detector can be used to demonstrate the occurrence of ESD. With the ground wire removed, the insulator is moved away, and the plate voltage and the plate can be seen to go to an opposite polarity voltage.

This experiment supports the following teaching points:

- The voltage on an electrically isolated conductor changes in response to nearby electro static fields. The induced voltage changes as the insulator is moved closer or farther away. The plate is not necessarily charged although it can be at high voltage.

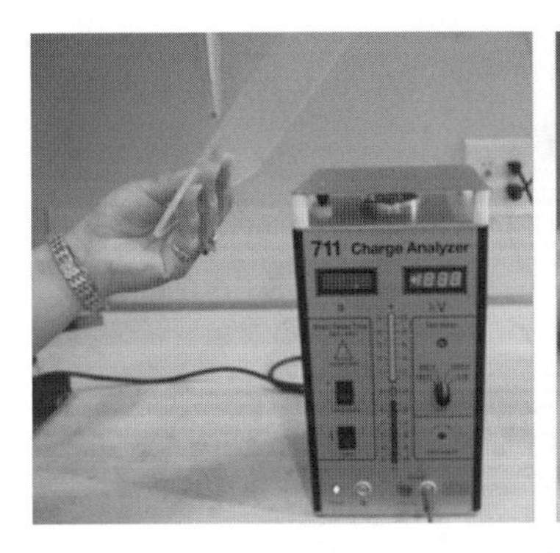

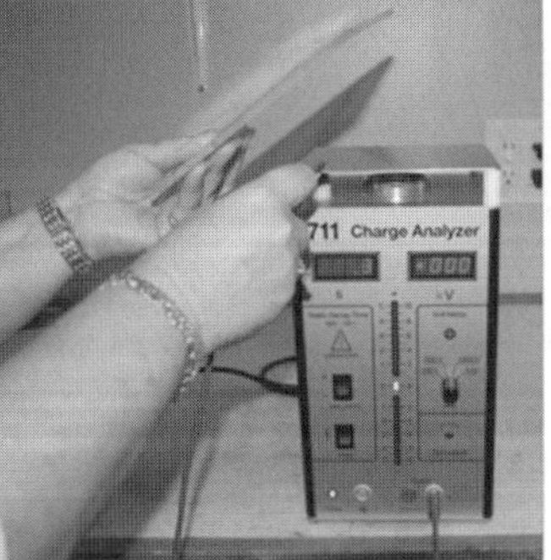

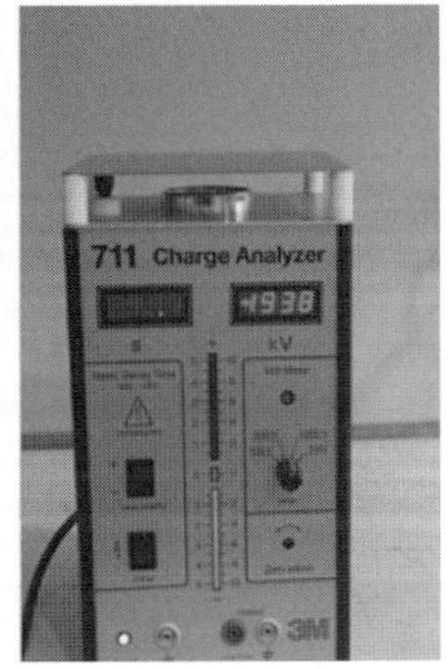

Figure 12.9 Induction charging demonstrated using a charged plate monitor (left) voltage induced on plate by nearby charged insulator, ready for ESD (middle). ESD occurs when charge moves from plate on grounding; (right) plate is left charged to opposite polarity when insulator is removed, ready for another discharge. Source: D. E. Swenson.

- The charged insulator does not need to touch the conductor for the voltage to be induced.
- The induced voltage is negligible if the insulator is kept sufficiently far away.
- ESD occurs if the conductor is touched by another conductor or ground wire when at a different voltage. After touching with the ground wire, the plate is now charged, although at this point it has no voltage, until the field source is moved away.
- After the ground wire is removed and the insulator moved away, the plate can be seen to be charged as it achieves a high opposite polarity voltage.

12.7.11 ESD on Demand – The "Perpetual ESD Generator"

Once it has been shown that an electrostatic field induces a voltage on a conductor, it can be demonstrated that repeated ESD can be made just by moving the conductor in the field and touching it with another conductor.

A charged insulator is used to induce a voltage on the isolated metal plate (Figure 12.10). An ESD detector is used to detect the ESD generated by touching the metal plate with a metal item such as a metal tool. The field meter shows the voltage changes occurring on the metal plate.

After each ESD is made, the insulator is moved in position, and the voltage on the metal plate changes. It is now ready to provide another ESD on touching with the tool. ESD occurs even if the demonstrator is grounded and using an ESD tool to touch the plate.

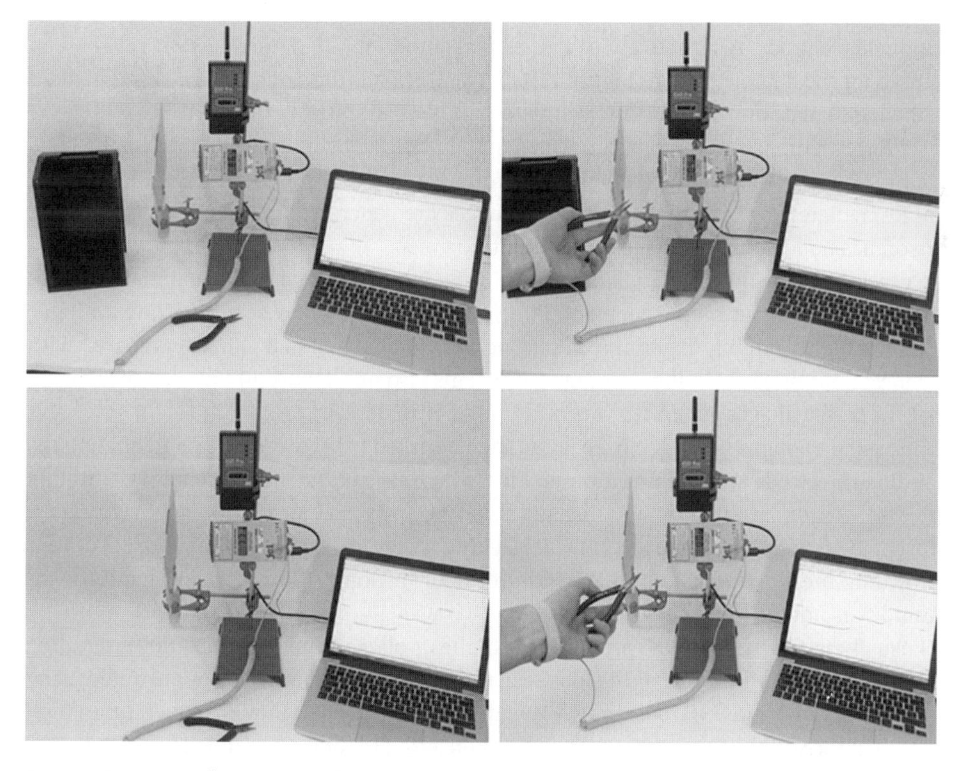

Figure 12.10 The "perpetual ESD generator." The metal plate rises in voltage due to electrostatic field changes. A conductor contacting the plate initiates ESD.

This experiment supports the following teaching points:

- The voltage on an isolated conductor changes in response to nearby electrostatic fields. The voltage changes as the insulator is moved around.
- ESD can be generated on touching the conductor with another conductor at different voltage.
- If the field strength is changed, e.g. by moving the relative positions of the plate and the insulator, the voltage again changes, and ESD can again occur.
- This situation could occur in the workplace if a conductor on an ESDS device or PCB is touched by a metal tool or other item in the presence of an electrostatic field.

The effects shown in Sections 12.7.10 and 12.7.11 are often the main reason for control of electrostatic fields and insulators in an EPA. Electrostatic fields set up the conditions under which ESD is more likely to occur when the ESDS touches another conductor. The stronger the field, the greater the concern and risk of ESD damage.

12.7.12 Body Voltage and Personal Grounding

The voltage developed on the body while walking can be demonstrated using the same arrangement of metal plate with the plate voltage measured by an electrostatic voltmeter as above. The demonstrator connects themselves to the metal plate via a long wire and handheld electrode. Alternatively, a charged plate monitor (CPM) (see Section 11.6.13) can be used for the demonstration (Figure 12.11). The CPM has the same physical arrangement as the metal plate and electrostatic voltmeter but has calibrated performance.

With the demonstrator's body connected to the plate, any voltage developed on the body is now shown on the field meter. For most ordinary footwear and flooring combinations, it can immediately be seen that significant voltages are generated on the body although the person has no idea these voltages are present.

Occasionally the floor material in the training facility is a low-charging material that does not give much voltage on the people walking on it. It is therefore wise to also have available a known high-charging material (e.g. sample of nylon carpet) that can be compared with the local floor material in the training facility. This also has the useful function of showing that the body voltages generated vary with the floor material.

In a small enough class, each trainee can then be asked to try the experiment, holding the handheld electrode and walking around. I try to do this experiment about halfway through the presentation, as relief from sitting and listening and to get all attendees moving about. Most of them will show some level of voltage, even if they are wearing ESD control footwear (providing the flooring is not an ESD control floor or noninsulative material and the air humidity is not too high!).

I usually ask the trainee who produces the highest voltage to help demonstrate grounding using a wrist strap. (Grounding via footwear and flooring could also be demonstrated using an ESD floor mat and ESD foot straps.) The trainee confirms that they are generating a high voltage in the walking experiment. They are then asked to wear a grounded wrist strap and move around in the same way. The body voltage is shown to be zero while wearing the wrist strap. The effect of wrist strap failure can be demonstrated by disconnecting the wrist strap

Figure 12.11 Demonstrating body voltage while walking, using a charged plate monitor. The computer displays the body voltage waveform.

while the trainee is moving around. The body voltage is seen to reappear when the wrist strap is disconnected.

This experiment supports the following teaching points:

- Body voltage of hundreds of volts are usually generated by ordinary actions such as walking in daily life in unprotected areas. The person has no idea these voltages are present unless they are high enough to cause electrostatic shocks. Different footwear and flooring combinations give different body voltage results.
- The body voltage is reduced to less than 100 V by wearing a wrist strap.
- If the wrist strap connection fails, the body voltage reappears. A continuous connection is required to control body voltage.

If a similar experiment is done using ESD footwear and flooring, the body voltage is often greater than when grounded by wrist strap. Using different combinations of footwear and flooring, it can be demonstrated that body voltage typically is different for each combination. In a higher-level course, this can lead to a discussion of qualification of footwear and flooring combinations with regard to body voltage generation, and the necessity of a walk test. It can be demonstrated that body voltage is not usually well controlled if ESD control footwear is used without ESD control flooring or the flooring used without ESD control footwear.

12.7.13 Charge Generation and Electrostatic Field Shielding of Bags

ESD packaging is a poorly understood area of ESD control. Many people do not realize that different types of packaging materials can have very different combinations of properties. They often incorrectly believe all ESD packaging gives adequate ESD protection for ESDS devices.

A simple demonstration can show differences between electrostatic charging and electrostatic field shielding properties of ESD bags. Unfortunately, it is not so simple to demonstrate differences in ESD shielding properties in a course. I normally compare samples of ordinary polythene, pink polythene, black polythene, and metallized ESD shielding bags. I normally introduce the test of ESD bags by doing the test with an ordinary polythene bag (previously selected to give a good strong electrostatic charging response!).

The differences can be demonstrated by inserting an electrostatic field meter into an ESD bag as if it were an ESDS device. The response of the field meter shows electrostatic charge generation and fields experienced within the bag (Figure 12.12). This experiment works best with a "field mill" type of instrument.

The field meter is placed within the bag. With a polythene bag, the field meter is normally already showing high electrostatic fields at this point.

With the field meter within the bag, the bag can be handled, and the bag material moved about. Any tendency to generate electrostatic fields and charge is indicated by the field meter.

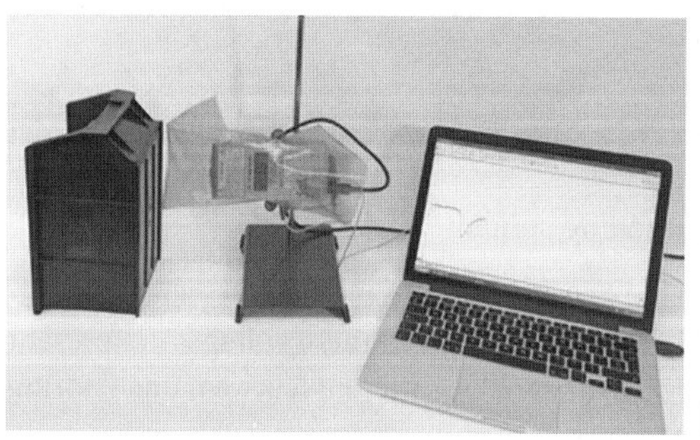

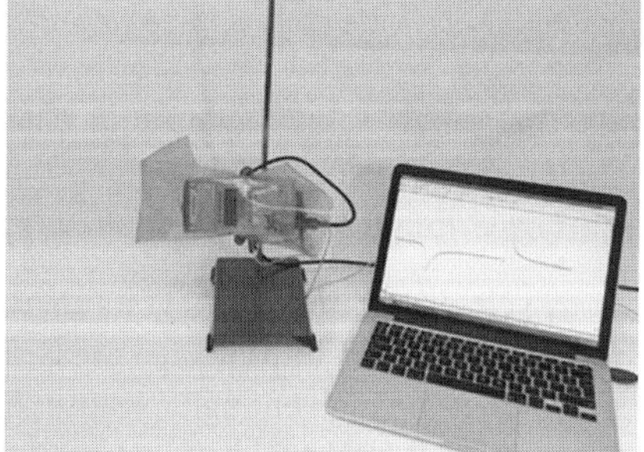

Figure 12.12 Placing a field meter inside an ESD bag gives a means of showing differences in electrostatic charge generation and electrostatic field shielding.

A highly charged insulator can then be brought nearby. If the electrostatic field shielding provided by the bag type is poor, then strong indication of field is shown by the field meter. If the electrostatic field shielding is good, then little indication of field is shown.

It is more convincing if the field meter is then taken out of the bag, and the charged insulator brought nearby to show that it is indeed charged and gives a strong electrostatic field.

When the demonstration is done with pink polythene bags under dry (<30% rh) air conditions, the bag often shows strong electrostatic charging and little electrostatic field shielding. It can be explained to the trainees that this material relies heavily on atmospheric moisture for its low-charging properties. Under dry conditions, its performance can be little different from an ordinary polythene bag. Aging and loss of antistat can also give this loss of performance.

Under humid air conditions (>50% rh) pink polythene gives little static charging and can appear to give good electrostatic field shielding. After demonstrating this, the bag can be dried using a hair drier or hot air gun. After drying, the experiment is repeated, and the bag will often show strong electrostatic charging and little electrostatic field shielding.

Black polythene bags give little electrostatic charging and show good electrostatic field shielding in this test and can be shown to be independent of humidity.

Metallized ESD shielding bags usually show little electrostatic charging in this demonstration. They often show a small electrostatic field appearing within the bag in the electrostatic field shielding test. Heavily crumpled bags may show reduced electrostatic field shielding.

This experiment supports the following teaching points:

- Different types of ESD packaging have different electrostatic charging and electrostatic field shielding properties.
- Ordinary polythene bags generate strong charges and are transparent to electrostatic fields.
- Pink polythene bags have variable properties. With sufficient humidity, they give low electrostatic charging and may also give good field shielding. Under low humidity, or when old, they may appear little different from ordinary polythene.
- Black polythene bags show low electrostatic charging and good electrostatic field shielding.
- Metallized ESD shielding bags show low electrostatic charging and good electrostatic field shielding, although a small residual electrostatic field often appears within the bag.

12.7.14 Insulators Cannot Be Grounded

People often think that insulators can be grounded. It is easy to demonstrate that grounding does not work for an insulator (Figure 12.13).

A rigid charged insulator such as a plastic tray is placed in front of the electrostatic field meter so that the field due to its charge is shown. A ground wire is then clipped to the tray. The electrostatic field does not disappear. The demonstrator explains to the trainees that the charge cannot move on the insulator; therefore, it cannot move to the ground wire to go to earth.

Figure 12.13 Demonstrating that an insulator cannot be successfully grounded.

This experiment supports the following teaching points:

- Charge does not move around quickly on insulating materials and cannot move to the ground wire.
- Insulators cannot be successfully grounded.

12.7.15 Neutralizing Charge – Charge Decay and Voltage Offset of Ionizers

A means is often needed of reducing electrostatic fields and voltages from charged essential insulators. Ionizers provide one means of doing this by neutralizing the charge that creates the field.

This demonstration can follow naturally from demonstration that an insulator cannot be grounded (see Section 12.7.14). A charged insulator is placed in front of the field meter, showing a strong electrostatic field or high voltage (Figure 12.14). A bench fan ionizer is then directed toward the insulator. The electrostatic field can then be seen to gradually reduce toward zero. After some time, the surface voltage has reached a near constant level.

If a chart recorder type oscilloscope or similar display is used in this experiment, the time taken to reduce the voltage to a low level can be measured. The final voltage remaining on the insulator surface can also be measured. It can be explained to the trainees that this offset voltage is due to imbalance in the quantity of negative and positive ions produced by the ionizer.

In this experiment, it can also be shown that the time taken for charge decay can vary greatly with factors such as distance from the ionizer and orientation of the ionizers.

The demonstration can be repeated using a metal plate on insulating handle as the target. The metal plate can be charged by induction or from a voltage source. Using a metal plate, two further demonstrations can then be done.

First, with the ionizer neutralizing the voltage on the charged plate, a residual offset voltage is shown. Temporarily grounding the plate with a wire demonstrates the offset voltage and gives detectable ESD. On removal of the ground wire, the voltage on the metal plate

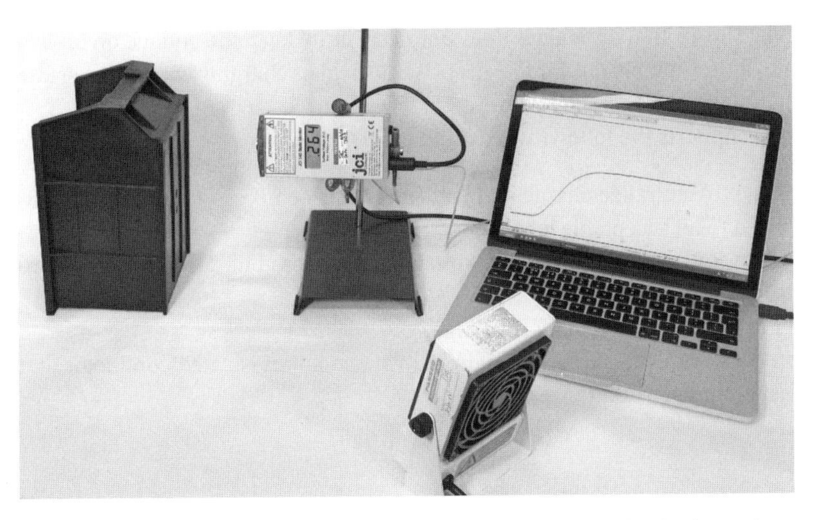

Figure 12.14 Demonstrating charge decay and offset voltage of an ionizer using an insulating material.

can be seen to rise again to the offset voltage. It can then be explained any item within the region around the ionizer becomes charged to the offset voltage. This could, if too high, create ESD risks. The offset voltage may vary with age and condition of the ionizer, and regular test and maintenance is necessary to be sure this is kept within acceptable levels.

In a second experiment a charged insulator can be moved around near the metal plate (Figure 12.15). The metal plate can be seen to vary in induced voltage in response to this

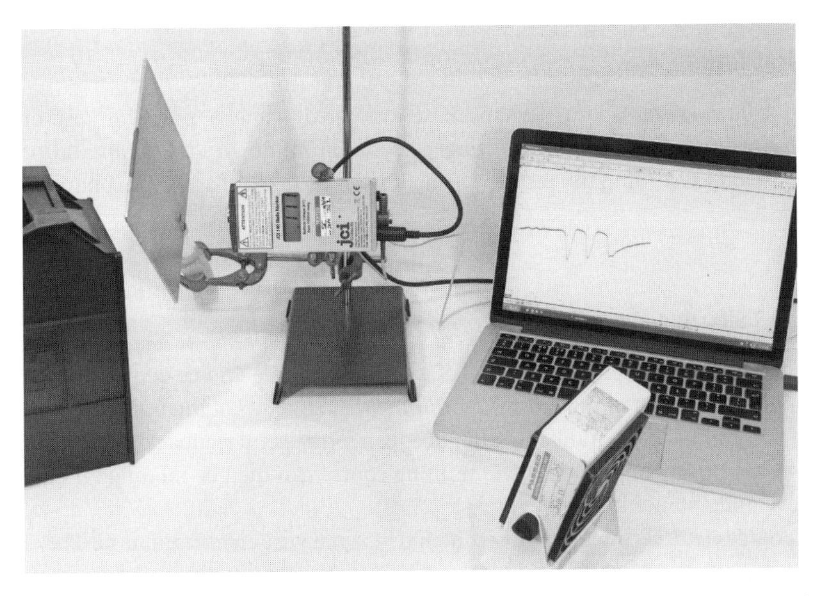

Figure 12.15 Demonstrating inability of an ionizer to keep a metal plate at low voltage in the presence of fast-changing electrostatic fields or charging.

insulator. It can be shown that an ionizer does not necessarily keep the voltage on an isolated conductor to an acceptable level if the electrostatic field or charging conditions create voltage faster than the ionizer can neutralize them.

This experiment supports the following teaching points:

- Ionizers take time to neutralize charge and voltage. It may be necessary to wait for voltages to be reduced to an acceptable level.
- For many types of ionizer, the neutralized voltage is near zero but does not reach zero due to the offset voltage.
- Ionizer offset charges items to the offset voltage.
- Neutralization does not necessarily maintain the voltage on an isolated conductor at a low level in fast-changing conditions.

12.8 Evaluation

12.8.1 The Need for Evaluation

It is important to make sure that trainees have understood and remember key training points. It is also important to ensure they can use key equipment correctly and, in some cases such as personal grounding, test them correctly. This means that some form of test (paper or practical) will be needed to cover key points. The IEC 61340-5-1 and ANSI/ESD S20.20 standards require this as part of their ESD Training Plan (International Electrotechnical Commission 2016; EOS/ESD Association Inc 2014).

12.8.2 Practical Test

Competence in some activities are perhaps best evaluated using a practical test or "on-the-job" observation. An example is testing personal grounding (wrist strap and/or footwear) and correctly recording the result. The actions to be taken if a "fail" is obtained should also be tested.

12.8.3 Written Tests

Written tests could be, for example, in the form of a simple multiple-choice questionnaire. Questions should address each of the most important teaching points of the training.

Evaluation of performance of a number of attendees can form useful feedback as to which points are commonly poorly understood. The training course can then be modified if necessary, to improve understanding in these areas.

Higher-level courses (e.g. certification) often include formal written examination. These can be "open book" or "closed book."

Computer or internet-based courses often have built-in tests that evaluate the understanding of the trainee.

12.8.4 Pass Criteria

Whatever the testing method, a pass level must be established. This could be achievement of a certain mark in a paper test or successful demonstration of an action or procedure in a practical test. Records should be kept of all tests and stored in a convenient place for future reference and evidence.

References

Amerasekera, A. and Duvvury, C. (2002). *ESD in Silicon Integrated Circuits*, 2e. Wiley. ISBN: 0 470 49871 8.

Dangelmeyer, T. (1999). *ESD Program Management*, 2e. Clewer. ISBN: 0-412-13671-6.

EOS/ESD Association Inc. (2014) ANSI/ESD S20.20-2014. *ESD Association Standard for the Development of an Electrostatic Discharge Control Program for – Protection of Electrical and Electronic Parts, Assemblies and Equipment (excluding Electrically Initiated Explosive Devices).* Rome, NY, EOS/ESD Association Inc.

Helling, K. (1996). ESD protection measures – return on investment calculation and case study. In: *Proc. EOS/ESD Symp. EOS-18*, 130–144. Rome, NY: EOS/ESD Association Inc.

INARTE (2017) Electrostatic Discharge Control Certification. Available from: https://inarte.org/certifications/inarte-electrostatic-discharge-control-esd-certification [Accessed 7th Dec. 2017]

International Electrotechnical Commission (2016) IEC 61340-5-1: 2016. *Electrostatics – Part 5-1: Protection of electronic devices from electrostatic phenomena - General requirements.* Geneva, IEC.

IPC (2017) IPC Announces New ESD Control Certification Courses on IPC EDGE Courses provide professional credentials for Certified ESD Trainers (CETs) and ESD Certified Operators (ECOs). Available from: http://www.ipc.org/ContentPage.aspx?Pageid=IPC-Announces-New-ESD-Control-Certification-Courses-on-EDGE [Accessed 7th Dec. 2017]

McAteer, O.J. (1980). An Effective ESD awareness training program. In: *Proc. EOS/ESD Symp*, 189–191. Rome, NY: EOS/ESD Association Inc.

McAteer, O. (1990). *Electrostatic Discharge Control.* San Francisco, CA: McGraw-Hill. ISBN: 0-07-044838-8.

Newberg C (2017) Certification. Available from: https://www.esda.org/certification [Accessed 5th Dec. 2017].

Snow, L. and Dangelmeyer, G.T. (1994). EOS-16 94-1 – 94-12. A successful ESD training program. In: *Proc. EOS/ESD Symp.* Rome, NY: EOS/ESD Association Inc.

STAHA Association. Available from: http://www.staha.fi/index_files/STAHA_leaflet_2013.pdf [Accessed 5th Dec. 2017].

Voldman, S.H. (2004). *ESD Physics and Devices.* Wiley. ISBN: 0-470-84753-0.

Wang, A.Z.H. (2002). *On-Chip ESD Protection for Integrated Circuits.* Klewer Academic Press.

Welker, R.W., Nagarajan, R., and Newberg, C. (2006). *Contamination and ESD Control in High-Technology Manufacturing.* Wiley-Interscience, IEEE Press. ISBN-10: 0 471 41452 2, ISBN-13: 978 0 471 41452 0.

Further Reading

Baumgartner, G. ESD demonstrations to increase engineering and manufacturing awareness. In: *Proc. EOS/ESD Symp. EOS-18*, 156–166. Rome, NY: EOS/ESD Association Inc.

13

The Future

13.1 General Trends

The future, particularly in a fast-moving world like the electronics industry, has even more capacity to surprise than the workings of electrostatics and electrostatic discharge (ESD). Attempting to predict it might seem to be doomed to failure. To the engineers who first discovered ESD damage, it must have been unexpected. Many of those who followed showed disbelief, and skepticism remains rife in the industry to this day. Although the future can never be accurately predicted, some general short- and medium-term trends seem likely to continue. New technologies can of course disrupt the picture at any time. Nevertheless, it's likely that the future reader of this chapter may be surprised and even amused by the difference between my predictions and the intervening reality!

Electronic components will continue to become smaller in their internal feature dimensions, and this will tend to make them more susceptible to ESD damage. New device technologies will continue to be developed. Some of these will be more susceptible, and some less susceptible to ESD.

New device packages and circuit assembly technologies will become available and more widely used. Some may even replace current technologies, as through-hole components have been widely replaced by surface-mounted technology today.

ESD control programs will continue to evolve, as will the standards that support them. As low ESD withstand (very sensitive) components become more widely handled, ESD control programs will need to develop to handle them, both in manual and automated processes.

I hope that ESD withstand data will become more widely published on component data sheets or available to the ESD control professional on demand. Unfortunately, many component manufacturers as yet seem unwilling to make this data easily available.

Knowledge and understanding will be more important – leading to greater need for ESD training at high level and audit, as well as specific aspects such as ESD-related measurements.

Some of these topics are discussed more fully in the following paragraphs.

The ESD Control Program Handbook, First Edition. Jeremy M Smallwood.

13.2 ESD Withstand Voltage Trends

13.2.1 Integrated Circuit ESD Withstand Voltage Trends

The ESD Association publishes from time to time an "ESD Technology Roadmap" (ESD Association 2010, 2013, 2016c). This reviews and predicts the trends in the integrated circuit (IC) industry for the forthcoming few years and their impact on ESD control practices and trends in device testing. The roadmap focuses on the IC manufacturing industry and does not cover devices such as magnetoresistive heads, LEDs, lasers and photodiodes, and thin film transistor flat-panel displays. The 2016 roadmap does look at the impact of novel technology trends such as 2.5d and 3d ICs, micro bumps, and on-package I/O. These trends are highly dependent on technology changes as time goes on.

Typically, the roadmap gives graphs showing the human body model (HBM) and charged device model (CDM) ESD withstand ranges from the mid-1970s and projected over a few years into the future (Figures 13.1 and 13.2). The ranges are a projection by engineers from leading semiconductor manufacturers. The maximum represents what is possible due to technology scaling, and the minimum arises from meeting circuit performance demands, which often require reduced ESD protection.

The graphs show that between the late 1970s when ESD became a widespread issue and the mid-1990s, ESD withstand voltages increased in general. This is because with awareness of ESD issues devices were designed to be more ESD robust, in many cases with built-in on-chip ESD protection networks.

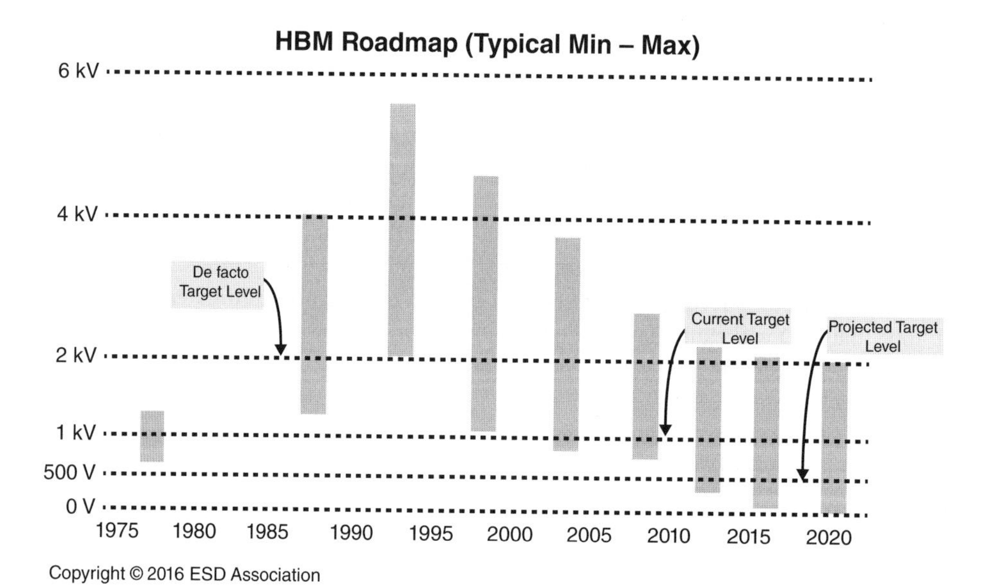

Reproduced Figure 1, from the Electrostatic Discharge (ESD) Technology Roadmap, May 2016, by permission of The EOS/ESD Association, Inc.

Figure 13.1 HBM ESD withstand projections. Source: ESDA (2016c).

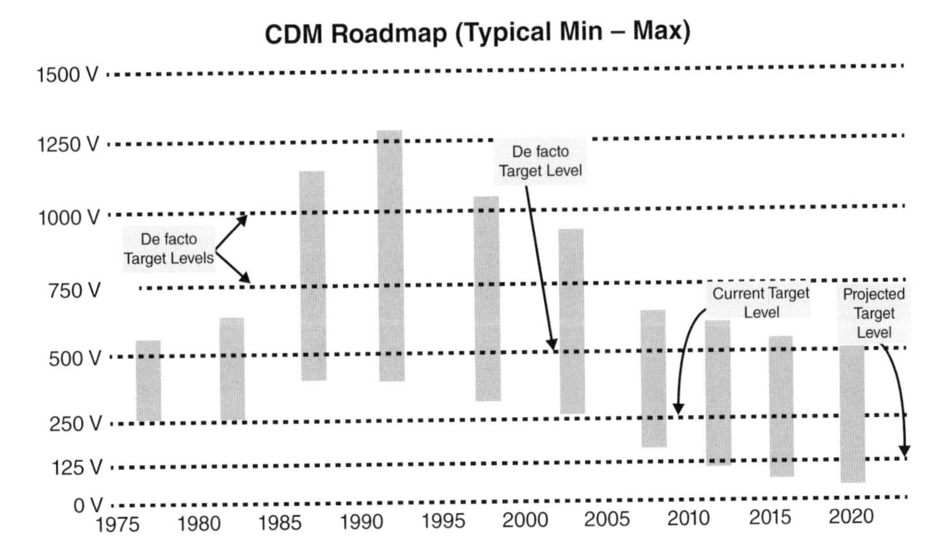

Reproduced Figure 1, from the Electrostatic Discharge (ESD) Technology Roadmap, May 2016, by permission of The EOS/ESD Association, Inc.

Figure 13.2 CDM ESD withstand projections. Source: ESDA (2016c).

From the mid-1990s, withstand voltages fell again as technology pressures made achieving high ESD withstand voltage increasingly difficult.

In the 2000s a group of semiconductor manufacturers joined together to recommend new ESD target levels (Industry Council 2010a,b, 2011). The motivation was to set ESD protection requirements for the safe handling of ICs within electrostatic discharge protected areas (EPAs), while responding to the increasing difficulty in achieving high ESD withstand voltage in the face of technology scaling and other technology changes in IC design. The result was to recommend reducing target HBM ESD withstand voltage to 1000 V and machine model (MM) ESD withstand voltage to 30 V (Industry Council 2011). This was followed by a recommendation to reduce CDM ESD withstand voltage target level to 250 V (Industry Council 2010a). They concluded that "basic" ESD controls were adequate for protection of devices with ESD withstand voltage down to 500 V HBM. "Detailed" ESD controls are required for handling devices with ESD withstand voltages less than 500 V HBM.

The Industry Council for ESD Target Levels recommended HBM ESD withstand target levels are reduced to 1 kV. This is probably only a first step – it seems likely that the HBM ESD withstand target level will be further reduced to 500 V, with a corresponding reduction in CDM ESD withstand target level. CDM target levels have already been reduced to 250 V, and the roadmap suggests that a target of 125 V CDM will be needed in the future. Some high-performance devices already have ESD withstand voltages as low as 100 V HBM.

The 2016 roadmap also gives an expanded view of HBM and CDM ESD withstand trends from 2010 until beyond 2020 (Figures 13.3–13.6). These show that as time goes on, a greater proportion of ESD-sensitive (ESDS) devices will have ESD withstand voltages between 500 and 100 V HBM, and <200 V CDM (Figures 13.5 and 13.6). In Figures 13.3 and 13.4,

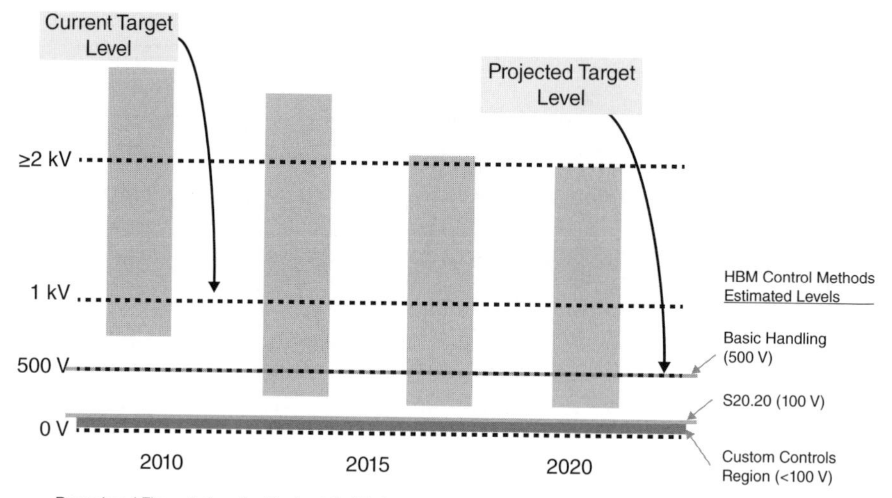

Reproduced Figure 2, from the Electrostatic Discharge (ESD) Technology Roadmap, May 2016, by permission of The EOS/ESD Association, Inc.

Figure 13.3 Technology Roadmap expanded 2010 and beyond HBM ESD withstand projections. Source: ESDA (2016c).

CDM Forward Looking Roadmap (Typical Min – Max)

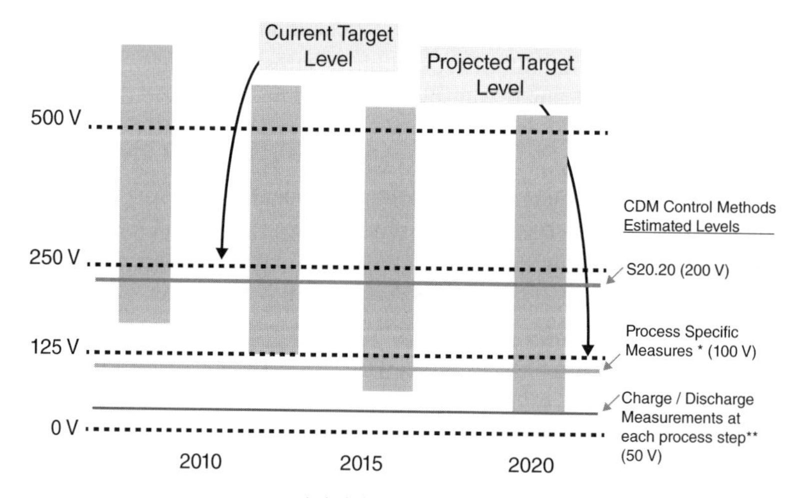

* - Include process specific measures to avoid charging *or* discharge
** - Include process specific measures to avoid charging *and* discharge

Figure 13.4 Technology Roadmap expanded 2010 and beyond CDM ESD withstand projections. Source: ESDA (2016c).

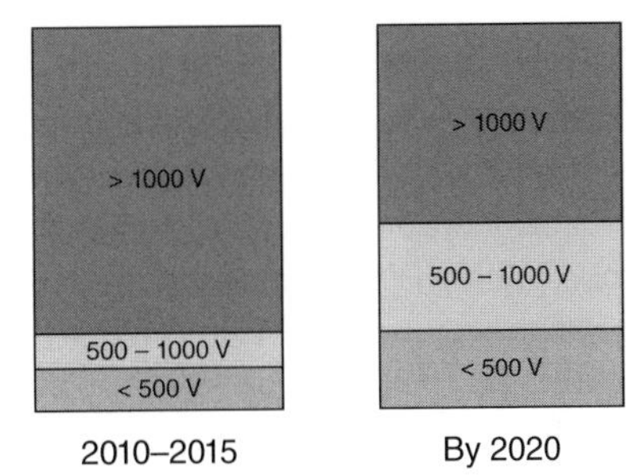

Figure 13.5 Projected change in distribution of HBM withstand voltage among ICs. Source: ESDA (2016c).

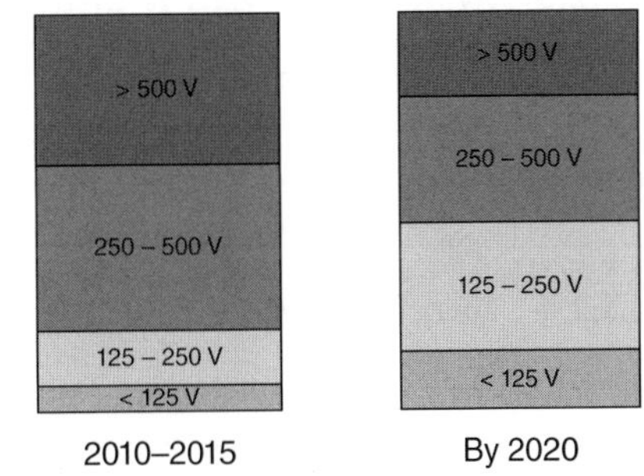

Figure 13.6 Projected change in distribution of CDM withstand voltage among ICs. Source: ESDA (2016ac).

the control levels achieved by an ESD control program compliant with ANSI/ESD S20.20 are also shown. Below these levels (100 V HBM and 200 V CDM), special "custom" and process-specific ESD control measures are usually required.

Looking further than 2020, the roadmap suggests that the range of HBM ESD withstand voltages may not change so much, but the distribution of devices within the range may change, with a greater proportion falling in the 500–1000 V and 100–500 V withstand voltage ranges. Similarly, a greater proportion is likely to fall within the 250–500 V, 125–250 V, and <125 V CDM withstand voltage ranges.

The drivers for these reductions in ESD withstand are rooted in technology changes. The internal device dimensions are reducing into the 22 and 18 nm range, and this tends to lead to inherently more sensitive internal circuitry. Increasing numbers of devices include high-speed input/output (I/O) pins that will need to operate into the 10–30 Gbit range of data rate. Radio frequency circuits are becoming ever more widespread. These device pins often tolerate only minute capacitance added by ESD protection networks, before performance is impaired. Increased device performance often comes at the expense of reducing ESD withstand voltage.

The MM ESD withstand test is already being phased out, and future ESD withstand characterizations will be normally in terms of HBM and CDM withstand voltages.

The CDM ESD susceptibility is also reduced with increasing package size. The roadmap states that current packages of 3000 pins in land grid array (LGA) or ball grid array (BGA) with high-speed I/O at 22 nm barely meet a CDM target of 125 V.

13.2.2 Other Component ESD Withstand Voltage Trends

While the ESDA roadmap deals with the ESD withstand voltage trends for integrated circuits, there are many types of transistor, diode, surface acoustic wave device (SAW), microelectromechanical system (MEMS), magnetoresistive (MR) sensor, and other devices that are susceptible to ESD damage. Some of these are among the most sensitive devices handled in the manufacture of electronic systems. These devices often have little or no ESD protection built into the device. It is likely that these will continue to be among the most ESDS devices handled. Their ESD withstand voltages will continue to depend on the particular device and technology.

13.2.3 Availability of ESD Withstand Voltage Data

There is currently scope for the improvement of the availability and publication of device ESD withstand data. I hope that device manufacturers will eventually publish this data, for example on device data sheets, as a matter of course.

13.2.4 Device ESD Withstand Test

MM ESD withstand voltage test is already being dropped from routine device characterization. The MM test is seen as giving little useful additional information over HBM ESD characterization (Industry Council 2011).

HBM and CDM ESD withstand voltage test practices may continue to be developed to more easily test high pin count and small pin spacing devices and counter increasing test times and costs (ESD Assoc. 2016c). Statistical sampling methodologies are likely to be used for high pin count devices. These changes will entail regular update of the relevant device ESD test standards and test equipment. As devices become more sensitive, introduction of lower voltage test levels may be needed.

The human metal model (HMM) test will continue to be developed for testing device pins that can be exposed to system-level ESD. One area of research is improving the repeatability of this test. Transient latch up (TLU) and other tests may also be further developed. Use of transmission line pulsing as a technique for characterizing ESD protection capability will also be further developed, with standards emerging for this type of testing.

13.3 ESD Control Programs and Process Control

13.3.1 ESD Control Program Development Strategies

While some organizations will be content to apply standard ESD control measures in ESD control programs according to standards such as IEC 61340-5-1 and ANSI/ESD S20.20, others will feel the need to tailor their ESD program to a greater extent. Reasons for this may include, for example, the following:

- Handling of low ESD withstand voltage ESDS devices
- Use of processes having less common ESD risks

- A desire to optimize the ESD control program for best effectiveness and return on investment
- A desire to achieve very low risk of ESD damage (e.g. for high value or high reliability product)

Even in the absence of these motivations, the continuing trends in reduction of device ESD withstand voltages confirm a need for continuing improvement of ESD control procedures and compliance in processes handling these devices.

The Industry Council (2011) commented that ESD Control Programs range across three levels.

- Little or no ESD control
- Basic ESD control programs
- Detailed ESD control programs

It is likely that organizations will need to move toward better ESD controls as time goes on, and more organizations will move into the "detailed ESD controls" category.

Organizations that have little or no ESD control are likely to experience ESD losses even with devices of 2 kV HBM or greater ESD withstand voltage. Without adequate personal grounding, body voltage can easily exceed 2000 V. Even if simple ESD control measures may be present, ESD training and compliance verification are often minimal or absent. It is no coincidence that the key requirements of modern ESD control standards include documentation of the ESD program, an ESD Control Product Qualification Plan, an ESD Training Plan, and a Compliance Verification Plan. Omission of any of these means that the ESD program is likely to be ineffective.

- Lack of an ESD control program document means that ESD control measures will be poorly reproduced and key measures may become omitted.
- Lack of an ESD Control Product Qualification Plan can lead to purchase of equipment that does not effectively fulfill its purpose in the long term.
- Lack of a Compliance Verification Plan means that ESD control equipment failures are unlikely to be discovered and remedied.
- Lack of an ESD Training Plan means that personnel are likely to remain insufficiently aware of the procedures and equipment required to be used in ESD control.

With the increasing proportion of components that have low ESD withstand voltage, organizations that currently pay minimal attention to ESD control are likely to find that increasing losses will be experienced. This will demonstrate the need for establishment of an effective ESD control program having at least good basic control measures.

13.3.2 A Basic ESD Control Program

A basic ESD control program implements the standard ESD control measures with little or no redundancy (Industry Council (2011). The principles of ESD control are that

- ESDS devices are handled only within an EPA in which ESD risks are controlled.
- Outside the EPA, components are protected within ESD protective packaging.

The techniques used to control ESD risks within the EPA are that

- All conductors especially personnel who might contact ESDS devices are equipotential bonded, preferably to earth.
- Unnecessary insulators are removed from the vicinity of ESDS devices.
- Insulators that are necessary to the process or product are evaluated for ESD risk. Unacceptable risks are reduced by some means to an acceptable level. Risks are normally evaluated in terms of electrostatic fields and voltages.

For long-term success, it must include documentation and implementation of the ESD program, an ESD Control Product Qualification Plan, an ESD Training Plan, and a Compliance Verification Plan.

The current ESD control program standards IEC 61340-5-1:2016 and ANSI/ESD S20.20-2014 are designed to protect devices down to 100 V HBM and 200 V CDM. ESD from charged conductors is limited to 35 V.

A basic ESD control program well implemented can be very effective, providing any failures of ESD control equipment are quickly detected and remedied.

13.3.3 Detailed ESD Control Program

Detailed ESD control programs add redundancy to the ESD control program and may use constant monitoring systems to make sure that failures are quickly detected (Industry Council 2011). Redundancy helps prevent single equipment failures causing ESD failures, as a second control measure may also help control the ESD risk.

Handling devices with ESD withstand voltages below 500 V HBM typically requires a more careful level of ESD control, as control equipment failures will often quickly lead to ESD failures occurring. Handling devices with ESD withstand voltages below 100 V HBM takes an ESD control program out of the range of standards such as IEC 61340-5-1 and ANSI/ESD S20.20 and into a region where more specific controls are needed.

The ESD roadmap predicts that tighter limits will be required for ESD control, and possibly more frequent compliance verification. A reduction in CDM ESD withstand voltage is expected due to increasing speed, package sizes, and complexity with multichip packages such as 2.5D and 3D technology.

As device ESD withstand voltages reduce, increased care, advanced, and nonstandard ESD controls will be increasingly necessary. Processes will need to be evaluated with care to identify specific ESD threats. These may need to be countered using specifically designed control measures. ESD risk will be determined less by checklist with fixed limits and more by application an engineering approach by skilled ESD coordinators (Jacob et al. 2012).

The current standard design limits of 100 V HBM, 200 V CDM ESD withstand, and ±35 V on isolated conductors may need to be reduced as lower ESD withstand voltage devices become more common.

13.3.4 Human Body ESD

Control of human body ESD using a wrist strap or ESD control footwear and flooring is a mature and well-understood topic. These solutions work well, providing users correctly implement standard ESD control requirements. This must include

- Use of wrist straps when seated
- Qualification of footwear and flooring by resistance from person to ground, for each combination of footwear and flooring used
- Qualification of footwear and flooring by a body voltage walk test, for each combination of footwear and flooring used
- Compliance verification of footwear and flooring by resistance from person to ground, for each combination of footwear and flooring used

At the time of writing, many users have yet to adequately adopt adequate footwear and flooring compliance verification procedures as per the standards. Use of walk tests is particularly important when handling low HBM ESD withstand voltage devices, as the objective is to keep body voltage below a low level related to the HBM withstand voltage of the devices handled. Studies have shown that body voltage generated is, for some combinations of footwear and flooring, badly predicted by the resistance of the footwear and flooring individually and even in combination.

Working with low ESD withstand devices may require some users adopt lower body voltage limits to reflect the ESD withstand voltage of the devices they handle. For example, an upper limit for body voltage of 50 V might be selected when working with 50 V HBM ESD withstand devices.

13.3.5 ESD Between ESDS and Conductive Items

ESD between ESDS and conductive items include "two-pin" ESD from charged ungrounded conductive items and "one-pin" ESD between the ESDS and another conductive part at a different voltage. A "two-pin" ESD event is one in which the ESD current enters the device through one pin and passes out through another pin. Susceptibility to "two-pin" ESD from charged ungrounded conductive items was formally represented by the MM ESD withstand test. This test is now considered redundant and is falling out of use, in part because it typically gives failures similar to the HBM test but with about 1/30th the withstand voltage or more. A part that shows 1 kV HBM ESD withstand would typically be expected to give >30 V MM ESD withstand (Industry Council 2011). This can be justified by theoretical arguments as well as practical experience.

A "one-pin" ESD event is one in which the ESD is directly between one pin of the ESDS device and another conductive object, with no other ESDS pin being involved in the discharge. Susceptibility to a "one-pin" ESD event is represented by the CDM ESD susceptibility test.

As automated assembly increases, the threat from human body ESD is reduced except in manual parts of processes. The threat from ESD between the ESDS devices and conductive parts correspondingly represents a greater proportion of ESD risks.

At present in the ESD control standards, the threat of ESD from charged conductors is controlled by two requirements.

- Conductors that make contact with ESDS devices must be grounded if possible.
- Conductors that make contact with ESDS devices but cannot be grounded must have the voltage difference between the conductor and ESDS limited to ±35 V.

These requirements address both one-pin and two-pin ESD threats. The charged device ESD threat is also addressed by limitation of electrostatic field in the vicinity of the ESDS device. Both types of ESD damage can also be prevented by preventing contact between the ESDS devices and low-resistivity conductive items.

In practice, it is difficult to measure and verify the voltage difference between the ESDS device and an ungrounded conductor. To do so normally requires measurement of the voltage on the ESDS, measurement of the voltage on the conductor, and then calculating the difference. These can be difficult measurements to make, especially in an operating machine process, as the ESDS device and conductor may be small low-capacitance items. Special electrostatic voltmeters may be required to make these measurements. So, it remains to be seen how practical these requirements are in practice. The standards also do not yet define in a measurable way what is meant by an ungrounded conductor.

13.3.6 "Two-Pin" ESD From Charged Ungrounded Conductive Items

The requirement that the voltage difference between the conductor and ESDS device limited to ±35 V can be viewed as addressing two-pin ESD risk from conductors up to 200 pF capacitance (the MM source capacitance). This protects devices with MM ESD withstand as low as to 35 V, corresponding to HBM ESD withstand around 1000 V. It could be argued that as lower voltage HBM withstand parts become more commonly used, this voltage value might have to reduced.

In practice, the real threat from ESD from conductors is dependent on the capacitance of the conductor. Many conductors in real situations may have lower capacitance than the 200 pF MM value. Some may have higher capacitance. For a lower-capacitance item, a higher voltage could be tolerated, and for higher capacitance a lower voltage may be required, based on the lowest ESD withstand device handled. The threat may also depend on whether the device is susceptible to ESD current, charge, or energy transferred in the discharge (Smallwood and Paasi 2003; Paasi et al. 2003).

13.3.7 "One-Pin" ESD Between the ESDS and Another Conductive Part

Over the last decade or so, there has been considerable interest in developing a methodology for relating process capability to device CDM ESD withstand data. This is particularly of interest in automated processes but also in understanding ESD risks from contact between devices and metal objects (whether grounded or not). Some researchers have used existing CDM withstand voltage as the parameter (e.g. Steinman 2010, 2012), but other researchers have proposed that it is necessary to record peak current or other CDM waveform data, rather than the source voltage, for comparison with real-world conditions (e.g. Tamminen and Viheriäkoski 2007; Tamminen et al. 2017a,b). At the same time, it has proven difficult to make in-process measurements for comparison with CDM waveform parameters. It's likely that this area will continue to be researched over the forthcoming decade especially as device CDM withstand voltages are reduced. New measurement techniques or methodology may be needed before success is achieved in this area (Tamminen et al. 2017a,b). Device ESD withstand measurement practices may need to be modified to record the required data.

13.3.8 Charged Board, Module, and Cable Discharge Events

Charged board and cable discharge events happen when a PCB contacts a conductive item or a cable at a different voltage is connected. The PCB, or the cable, or other conductive object may be the charged item. These events can be highly energetic compared to charged device ESD, because the PCB or cable have high capacitance compared to devices. The resulting damage can look more like electrical overstress than ESD (Olney et al. 2003).

Modules and subassemblies are often potted or enclosed by plastic housings that can easily become charged by triboelectrification by rubbing during handling or contact with packaging. These assemblies are often assumed to be immune to ESD due to the protection afforded by the enclosure as a barrier against direct ESD. Polymer potting compounds and enclosures are, however, "transparent" to electrostatic fields from nearby charged materials. In an uncontrolled area, charged personnel can be an external source of electrostatic fields when holding handheld items. Internal PCBs and other conductors can rise to very high voltages by induction and then discharge by contact of a terminal with another conductor.

Cables can become charged by triboelectrification through contact and rubbing while on a reel, within packaging or being moved over surfaces. They can also become inductively charged by the electrostatic field from nearby charged insulators such as packaging materials. Charged cables can have high and variable capacitance of hundreds of pF. So, a charged cable or wiring loom is a very energetic ESD source. This is often discharged into a PCB connector during assembly of a product or installation of a system. The circuitry connected to the PCB connector may not be designed to survive high levels of ESD, particularly if they are internal system connections rather than those exposed to the outside world (Stadler et al. 2017; Marathe et al. 2017).

These ESD risks are not specifically covered by current ESD standards, but it's possible that they may become so in the future.

13.3.9 Optimization

Optimization of an ESD program in terms of return on investment has long been an aim for some ESD control practitioners. Nevertheless, compliance with standards has often been an overriding goal, particularly in organizations where understanding of ESD controls has been at a lower level. As greater expertise is applied in successful ESD control, that expertise will be also applicable to optimization. More organizations will be able to achieve improved optimization of ESD control alongside compliance with the standards.

13.4 Standards

The core ESD control standards IEC 61340-5-1 and ANSI/ESD S20.20 will continue to be refined and developed for the foreseeable future. At their current stage of maturity, the changes are expected to be incremental rather than major. But, experience with using the standards, the increasing use of low ESD withstand voltage devices, and developing ESD control approaches will be reflected in rewording and changes to the detailed requirements of the standards. As an example, the 61340-5-1:2007 standard set a limit of electrostatic fields at 10000 V m^{-1}, and this was reduced to 5000 V m^{-1} in the 61340-5-1:2016 update. The latter also added some requirements for dealing with ungrounded conductors that make contact with ESDS devices. The ANSI/ESD S20.20 often leads the way in this development, with a new version often appearing a year or so before the 61340-5-1. New versions of the User Guides IEC 61340-5-2 and ESD TR20.20 usually accompany the new standard versions (International Electrotechnical Commission 2018a; EOS/ESD Association Inc 2016b).

Alongside this, some of the related test methods and other standards undergo development and change with experience of their usage. New test methods may emerge for specific applications. Methodologies for process evaluation are likely to be improved and developed and may emerge as standard practice or full standards.

13.4.1 Impact on Future Standards

It is likely that changes in approach to ESD program development and process control (see Section 13.3) may have to be addressed in future versions of the standards. What constitutes an ungrounded conductor will need better definition in terms of measurable parameters, e.g. resistance to ground.

It may be necessary to move from simplified "coverall" requirements to measures more carefully tailored to the real component susceptibility and failure mode. For example, voltage limits on conductors could be linked to the capacitance of the conductor and ESD HBM and CDM withstand voltage of devices handled. It may be possible to establish that small ungrounded conductors below a capacitance limit do not pose a threat to devices having withstand voltage above a given level. Conversely, very sensitive devices, especially voltage-sensitive devices (e.g. low gate capacitance low voltage MOSFETs) or those that might be damaged by high transient peak current levels may require very careful control or elimination of contact with ungrounded conductors, especially those of low resistivity.

13.4.2 ESD Control in Automated Handling

So far, ESD control in automated handling equipment has escaped standardization, although the ESD Association has produced the Standard Practice document ANSI/ESD SP10.1. Part of the problem has been difficulty in providing appropriate standardized ESD controls for automated equipment and processes. Standards apply requirements that then are necessary for compliance, whether they are needed for successful ESD control in a particular situation. An inappropriately specified requirement can represent an unwelcome cost increase or may be difficult to achieve in a machine for technical reasons.

In principle, the controls given in 61340-5-1 and ESD S20.20 (grounding of conductors, treatment of ungrounded conductors, and control of insulators) are also applicable within the critical path of ESDS within a machine. So, it is possible that standardization could progress in one of three ways.

- The 61340-5-1 and S20.20 documents could be developed to directly address automated systems. Guidance for application to automated systems could be incorporated in the 61340-5-2 and ESD TR20.20 User Guide documents.

- Separate standards could be developed covering requirements applicable to automated equipment.
- Separate documents (Technical Report (TR), Technical Specification (TS) in the IEC system, or Standard Practice (SP) in the ESDA system) could be further developed that give guidance rather than requirements for compliance.

13.5 ESD Control Equipment and Materials

13.5.1 ESD Control Materials

New materials will continue to be developed for ESD control as process control requirements become more demanding. Techniques for evaluating material characteristics will also evolve as understanding of ESD risks improves (Viheriäkoski et al. 2017).

13.5.2 ESD Protective Packaging

The development of new component packages and automated system handling techniques will require new ESD protective packaging types and materials. The IEC has produced the IEC TR 61340-5-5 technical report to review current and future requirements for ESD packaging and related test methods (International Electrotechnical Commission 2018b). The diversity of general ESD protective packaging types, and the materials they are made from, will increase with time. This will generate a need for new and better ways of evaluating packaging materials and systems, and a general improvement in understanding of ESD packaging effectiveness. This may lead to new requirements for ESD packaging. For example, the performance of an ESD barrier layer could be defined in terms of volume resistance and breakdown voltage of the package material. For some types of components, transferred charge and induced current on the ESDS within the package, as well as energy attenuation, could be found to be important.

Electrostatic attraction has already become an issue where small components are packaged for automated handling (D E Swenson private communication 2017). This trend is likely to continue as the number of very small components proliferates.

13.6 ESD-Related Measurements

13.6.1 ESD Protective Packaging Measurements

Development of packaging forms and materials will require development of improved test methods for ESD protective packaging materials and systems. At the time of writing, there is already a need to develop resistance test electrodes for use with small feature and non-planar ESD packaging materials. Improved understanding of the necessary requirements for ESD protective packaging for different ESDS devices and applications may also lead to development of new test methods.

13.6.2 Voltage Measurement on ESDS Devices and Ungrounded Conductors

Voltage measurements on small ESDS devices and ungrounded conductors are challenging, especially in operating automated equipment. The different types of voltage measurement equipment, their capabilities, and their limitations are currently poorly understood within industry except among the more expert practitioners. There will be a need for more widespread understanding and use of these techniques.

It is likely that these measurements will remain a central part of ESD risk evaluation in the medium term, despite the realization that voltage is often not the parameter best correlated to ESD risk. The lack of viable alternatives and established industry mind-set continues to frustrate adoption of other parameters as measures of ESD risk.

13.6.3 Measurements Related to ESD Risk in Automated Handling Equipment

Measurements related to ESD risk in automated handling equipment under working conditions is currently one of the most pressing problems for ESD control. Consequently, we can expect to see continuing developments in this area over the next few years. It is difficult to predict whether solutions will be found in direct or indirect measurement of ESD current, electromagnetic radiations, or some other parameter related to the ESD.

13.7 System ESD Immunity

While it has been demonstrated that in general system ESD immunity is not related to component ESD susceptibility (Industry Council 2010a,b, 2012), a system-efficient electrostatic discharge design (SEED) approach has been suggested. This addresses failure related to high-level ESD applied to device pins that are connected to external interfaces such as universal serial bus (USB) connectors, as well as "soft" failures due to conducted or radiated electromagnetic interference from ESD (Stadler et al. 2017; Marathe et al. 2017). This approach may become more widely used and further developed in the future.

13.8 Education and Training

The industry is moving out of an era in which ESD control can be specified effectively by standard measures and recipes and toward increased specification through engineering principles, knowledge, and understanding. This will require a greater level of specialist education and training to be available to industry practitioners. This will require a greater availability of high-level industry courses including continual professional development. Ideally, I hope that more universities and colleges will see value in teaching ESD control at different levels as part of their electronics-related courses.

References

EOS/ESD Association Inc (2010). ESD Association Electrostatic Discharge (ESD) Technology roadmap. Revised March 2010 Available from: https://www.esda.org/assets/Uploads/docs/2013ElectrostaticDischargeRoadmap.pdf (Accessed: 15th May 2017).

EOS/ESD Association Inc. (2013). ESD Association Electrostatic Discharge (ESD) Technology roadmap. Revised March 2013 Available from: https://www.esda.org/assets/Uploads/docs/2013ElectrostaticDischargeRoadmap.pdf (Accessed: 10th May 2017).

EOS/ESD Association Inc. (2014). ANSI/ESD S20.20-2014. *ESD Association Standard for the Development of an Electrostatic Discharge Control Program for – Protection of Electrical and Electronic Parts, Assemblies and Equipment (excluding Electrically Initiated Explosive Devices).* Rome, NY, EOS/ESD Association Inc.

EOS/ESD Association Inc. (2016a) ANSI/ESD SP10.1-2016. *Standard practice for protection of Electrostatic Discharge Susceptible Items – Automated handling Equipment (AHE).* Rome, NY, EOS/ESD Association Inc.

EOS/ESD Association Inc. (2016b). ESD TR20.20-2016. *ESD Association Technical Report - Handbook for the Development of an Electrostatic Discharge Control Program for the Protection of Electronic Parts, Assemblies and Equipment*. Rome, NY, EOS/ESD Association Inc.

EOS/ESD Association Inc (2016c) ESD Association Electrostatic Discharge (ESD) Technology roadmap. revised 2016. Available from: https://www.esda.org/assets/Uploads/docs/2016ESDATechnologyRoadmap.pdf (Accessed: 10th May 2017).

Industry Council on ESD Target Levels (2010a) White paper 2: A case for lowering component level CDM ESD specifications and requirements. Rev. 2.0. http://www.esdindustrycouncil.org/ic/en/documents/6-white-paper-2-a-case-for-lowering-component-level-cdm-esd-specifications-and-requirements (Accessed: 10th May 2017)

Industry Council on ESD Target Levels (2010b) White paper 3: System Level ESD Part I: Common Misconceptions and Recommended Basic Approaches. Rev. 1.0 http://www.esdindustrycouncil.org/ic/en/documents/7-white-paper-3-system-level-esd-part-i-common-misconceptions-and-recommended-basic-approaches (Accessed: 10th May 2017).

Industry Council on ESD Target Levels (2011) White paper 1: A case for lowering component level HBM/MM ESD specifications and requirements. Rev. 3.0. Available from: http://www.esdindustrycouncil.org/ic/en/documents/37-white-paper-1-a-case-for-lowering-component-level-hbm-mm-esd-specifications-and-requirements-pdf (Accessed: 10th May 2017).

Industry Council on ESD Target Levels (2012) White paper 3: System Level ESD Part II: Implementation of Effective ESD Robust Designs. Rev. 1.0 http://www.esdindustrycouncil.org/ic/en/documents/36-white-paper-3-system-level-esd-part-ii-effective-esd-robust-designs (Accessed: 10th May 2017).

International Electrotechnical Commission (2016) IEC 61340-5-1: 2016. *Electrostatics – Part 5–1: Protection of electronic devices from electrostatic phenomena - General requirements.* Geneva, IEC.

International Electrotechnical Commission. (2018a) IEC TR 61340-5-2:2018. *Electrostatics – Part 5-2: Protection of electronic devices from electrostatic phenomena - User guide*. ISBN 978-2-8322-5445-5 Geneva, IEC.

International Electrotechnical Commission. (2018b) IEC TR 61340-5-5:2018. *Electrostatics - Part 5-5: Protection of electronic devices from electrostatic phenomena – Packaging systems used in electronic manufacturing*. Geneva, IEC.

Jacob, P., Gärtner, R., Gieser, H. et al. (2012). Paper 3B.8. ESD risk evaluation of automated semiconductor process equipment – A new guideline of the German ESD Forum e.V. In: *Proc. EOS/ESD Symp. EOS-34*. Rome, NY: EOS/ESD Association Inc.

Marathe, S., Wei, P., Ze, S. et al. (2017). Paper 3A.4. Scenarios of ESD Discharges to USB Connectors. In: *Proc. EOS/ESD Symp. EOS-39*. Rome, NY: EOS/ESD Association Inc.

Olney, A., Gifford, B., Guravage, J., and Righter, A. (2003). Real-world charged board model (CBM) failures. In: *Proc. EOS/ESD Symp. EOS-25*, 34–43. Rome, NY: EOS/ESD Association Inc.

Paasi, J., Salmela, H., and Smallwood, J.M. (2003). New methods of assessment of ESD threats to electronic components. In: *Proc EOS/ESD Symp. EOS-25*, 151–160. Rome, NY: EOS/ESD Association Inc.

Smallwood, J. and Paasi, J. (2003). Assessment of ESD threats to electronic components and ESD control requirements. In: *Proc. Electrostatics 2003. Inst. Phys. Conf. Ser. No. 178 Section 6*, 247–252.

Stadler, W., Niemesheim, J., and Stadler, A. (2017). Paper 3A.1. Risk assessment of cable discharge events. In: *Proc. EOS/ESD Symp. EOS-39*. Rome, NY: EOS/ESD Association Inc.

Steinman, A. (2010). Paper 3B3. Measurements to Establish Process ESD Compatibility. In: *Proc. EOS/ESD Symp. EOS-32*. Rome, NY: EOS/ESD Association Inc.

Steinman, A. (2012). Paper 2B.4. Process ESD Capability Measurements. In: *Proc. EOS/ESD Symp. EOS-34*. Rome, NY: EOS/ESD Association Inc.

Tamminen, P. and Viheriäkoski, T. (2007). Paper 3B.3. Characterization of ESD risks in an assembly process by using component-level CDM withstand voltage. In: *Proc. EOS/ESD Symp. EOS-29*, 202–211. Rome, NY: EOS/ESD Association Inc.

Tamminen, P., Smallwood, J., and Stadler, W. (2017a). Paper 1B.4. Charged device discharge measurement methods in electronics manufacturing. In: *Proc. EOS/ESD Symp. EOS-39*. Rome, NY: EOS/ESD Association Inc.

Tamminen, P., Smallwood, J., and Stadler, W. (2017b). Paper 4B.2. The main parameters affecting charged device discharge waveforms in a CDM qualification and manufacturing. In: *Proc. EOS/ESD Symp. EOS-39*. Rome, NY: EOS/ESD Association Inc.

Viheriäkoski, T., Kärjä, E., Gärtner, R., and Tamminen, T. (2017). Paper 4B.3. Electrostatic discharge characteristics of conductive polymers. In: *Proc. EOS/ESD Symp. EOS-39*. Rome, NY: EOS/ESD Association Inc.

Further Reading

Fung, R., Wong, R., Tsan, J., and Batra, J. (2017). Paper 1B.3. An ESD case study with high speed interface in electronics manufacturing and its future challenge. In: *Proc. EOS/ESD Symp. EOS-39*. Rome, NY: EOS/ESD Association Inc.

Koh, L.H., Goh, Y.H., and Wong, W.F. (2017). Paper 1B.2. ESD Risk Assessment Considerations for Automated Handling Equipment. In: *Proc. EOS/ESD Symp. EOS-39*. Rome, NY: EOS/ESD Association Inc.

A

Appendix A: An Example Draft ESD Control Program

A.1 About This Plan

This appendix shows an example electrostatic discharge (ESD) program derived using the procedure of Chapter 10.

It should not be assumed that this example is a "model" ESD program or that the suggested procedures in writing an ESD control program must be followed. It is merely an example of how things might be approached. Each organization should develop their own ESD Control Program Plan according to their process and facility needs.

This ESD program has been derived as suggested in Chapter 10. Headings from the standard have been used to create a structure for the plans and make sure the required items are covered. The needs of the facility and anticipated operations are also considered to make sure that necessary equipment and convenient control measures are included. The order of the heading may have been changed, and in some cases, headings were deleted because they were unnecessary in an ESD control program plan.

A.2 Description of the Example Facility

The example considers a simple facility in which ESD-sensitive (ESDS) printed circuit boards (PCBs) and assemblies are received at Goods In, removed from their packaging, and individually repacked in bags for dispatch. In some workstations, limited test by connection to electrical mains voltages may be done.

In this example, most ESDS devices are contained within ESD protective packaging that is specified in customer contracts and marked as specified in the contracts. Nevertheless, some ESD protective packaging used within electrostatic discharge protected areas (EPAs) that is not specified in this way.

The ESD Control Program Handbook, First Edition. Jeremy M Smallwood.

In some cases, operators will enter the EPA and work for a short time handling unprotected ESDS devices while standing. In this case, personal grounding will be done most conveniently by wearing ESD control footwear and standing on an ESD control floor. In other cases, operators will sit at the workstation for some time handling unprotected ESDS devices, and a wrist strap will be required. ESD control seats will also be grounded using an ESD control floor.

A typical EPA in the imagined facility includes

- An ESD workstation with a surface on which unprotected ESDS devices may be placed
- An ESD control floor mat for grounding standing personnel (through footwear) and ESD control chair
- An ESD control chair
- Wrist straps for grounding seated personnel, and earth bonding points of connection of straps
- Containers such as tote boxes
- ESD control garments (coats)

A.3 Test and Qualification Procedures

The ESD control Program Plan contains an ESD Control Product Qualification Plan and a Compliance Verification Plan. These require several qualification procedures (QPs) and test procedures (TPs). These are not given here but would be defined and documented including the test equipment to be used, details of the test method, and a simple test procedure.

A.4 ESD Control Program Plan at XXX Ltd

A.4.1 Introduction

This document provides an ESD control program for XXX Ltd. in compliance with IEC 61340-5-1:2016 and ANSI/ESD S20.20-2014.

A.4.2 Scope

This ESD program covers all activities that manufacture, process, assemble, install, package, label, service, test, inspect, transport, or otherwise handle ESDS devices at XXX Ltd. Specifically, these activities include receipt of ESD-susceptible (ESDS) PCBs and assemblies at Goods In, removal from their ESD protective packaging for inspection, and individual repacking in bags for dispatch.

The ESD withstand voltages of the individual types of ESDS devices are not known. The ESD program is designed to be compliant with the current standards and capable of handling electrical or electronic parts, assemblies, and equipment with ESD withstand voltages greater than or equal to 100 V human body model (HBM), 200 V charged device model (CDM) and ±35 V for isolated conductors.

A.4.3 Terms and Definitions

The following definitions and terms are used in this ESD control program:

CDM	Charged device model. ESD stress model that approximates the discharge event that occurs when a charged component is quickly discharged to another object at a different electrostatic potential.
EPA	ESD protected area. Area in which an ESDS can be handled with accepted risk of damage as a result of electrostatic discharge or fields.
ESDS	ESD-susceptible device. A sensitive device, integrated circuit, or assembly that may be damaged by electrostatic fields or electrostatic discharge.
HBM	Human body model. ESD stress model that approximates the discharge from the fingertip of a typical human being onto a pin of a device with another pin grounded.
Shall	"Shall" indicates a requirement of the ESD control program. If it is stated that some aspect shall be done, and if it is not, this constitutes a noncompliance.
Tailoring	Modification of the requirements of the standard after evaluation of the applicability of each requirement. Requirements may be added, modified, or deleted. Tailoring decisions, including rationale, and technical justification, shall be documented.
Static dissipative	A packaging material or item that has surface resistance <100 GΩ and >10 kΩ as per IEC 61340-5-3.
Conductive	A packaging material or item that has surface resistance <10 kΩ as per IEC 61340-5-3.
Insulator	A materials or item that has surface resistance >100 GΩ as per IEC 61340-5-3.
Conductor	Item or material that has surface resistance <100 GΩ.
ESD shielding packaging	ESD protective packaging compliant with IEC 61340-5-3 that has static dissipative inner and outer surfaces and a barrier layer or air gap to prevent ESD attenuate ESD current passing from the outside to the inside of the package.

A.5 Personal Safety

Compliance with local safety laws, regulations, and requirements shall take precedence over ESD control requirements.

Where use of personal protective equipment (PPE) is required, this shall be specified in line with safety requirements and, where possible, also in compliance with ESD control requirements. PPE specified for use within this ESD control program shall be specified within this ESD Control Program Plan.

To avoid electric shock risk, personnel who might come into accidental contact with high voltages shall be protected by control measures that limit the current flow under fault conditions to less than 0.5 mA. To limit current and provide protection against electric shock in case of accidental contact of the wearer with 250 VAC electric power systems, the minimum resistance from the wearer's body tested to groundable point (wrist strap) or metal plate (footwear as worn) shall be 750 kΩ (see Section A.7.2 of this plan).

A.6 ESD Control Program

A.6.1 ESD Control Program Requirements

This ESD Control Program Plan shall cover the following:

- The requirements of the ESD program
- ESD training
- ESD control product qualification
- ESD control product compliance verification
- EPA ground and grounding and bonding systems
- Personal grounding systems
- EPA requirements
- ESD protective packaging
- Marking for ESD control and protection purposes

A.6.2 ESD Coordinator

An ESD coordinator shall be appointed. The ESD coordinator shall have responsibility for implementing the ESD control program including establishing, documenting, maintaining, and verifying the compliance of the program as well as updating the program in response to changes in practices, operations, or facilities. Other personnel may be appointed as necessary to assist the ESD coordinator in these tasks.

A.6.3 Tailoring ESD Control Requirements

Where necessary the requirements of the standard may be omitted or modified providing the rationale for these is documented and referenced by this ESD control program. The rationale for requirements added to those of the standard shall also be documented and referenced.

Aspects of this ESD control program that are subject to tailoring include use and handling of essential papers (see Section A.7.3.2).

A.7 ESD Control Program Technical Requirements

A.7.1 ESD Ground

Mains electrical safety protective earth is used as ESD control earth (ground).

Table A.1 Personal grounding equipment requirements and test criteria.

ESD control item	Test procedure	Pass criteria	Test frequency
Wrist strap and cord when worn, wearer's body to cord end groundable point	TP1	>750 kΩ <35 MΩ	Each day before first use
Footwear when worn, tested between wearer's body and one foot placed on metal plate	TP2	>750 kΩ <100 MΩ	Each day before first use
Temporary wrist strap bonding point	Visual inspection	Correctly connected	Each day before first use
Wrist strap bonding point	TP3	<1 MΩ	Monthly

A.7.2 Personal Grounding

All personnel shall be grounded when handling ESDS, either by wearing

- A wrist strap connected to an wrist strap bonding point
- ESD control footwear on both feet and standing on an ESD control floor

Personnel handling ESDS when seated shall be grounded using a wrist strap. Wrist straps shall make good direct contact with the wearer's skin.

The personal grounding equipment requirements and test criteria are listed in Table A.1.

A.7.3 ESD Protected Areas (EPA)

A.7.3.1 General EPA Requirements

ESDS devices shall be handled out of ESD protective packaging only when within an EPA. ESDS devices shall be within designated packaging at all times when taken outside an EPA.

EPA boundaries shall be clearly defined and identified with suitable boundary markings and signage at the entrances (For example Figure A.1).

EPA workstations shall not be used for purposes other than operations and processes in which ESDS are handled.

Personnel who have completed and passed ESD awareness training may enter the EPA unaccompanied. Personnel needing to enter the EPA who have not completed this training shall be accompanied by a trained person at all times while within the EPA.

Figure A.1 Example of signage showing the entrance to an EPA.

A.7.3.2 Insulators

The following items are considered likely to be made from insulators unless tested and approved for use within an EPA:

- Secondary (non-ESD protective) packaging materials
- Items made of plastics
- Personal items
- Clothing
- Papers
- Furniture

All items made of insulators shall be excluded from EPAs unless they have been designated essential to the process or operations.

Tailored Requirement for Essential Papers

It is deemed essential to the inspection process that records are kept and signed off on in paper form as inspection of ESDS devices is completed. These papers and associated pens shall be kept and used in a designated area at least 30 cm from the position where unprotected ESDS devices are handled. When not in use, papers shall be kept in static dissipative document holders.

Essential papers shall never be brought into the EPA directly from photocopying as they may be highly charged. Photocopied papers should be left for at least an hour under moderate humidity atmosphere (>30% rh) after copying for charge to dissipate before bringing them into the EPA.

Electrostatic Fields and Voltages

Electrostatic fields at the workstation where ESDS devices are handled shall be maintained $<5000\ \mathrm{V\ m^{-1}}$ measured using TP6. Any insulators or electrostatic field sources that can have surface voltage >2 kV measured using TP7 shall be kept >30 cm from positions where ESDS devices are handled. Any insulators or electrostatic field sources that can have surface voltage >125 V measured using TP7 shall be kept >2.5 cm from positions where ESDS devices are handled.

A.7.3.3 Isolated Conductors

Any conductor that has resistance to ground >100 GΩ shall be considered isolated. Any conductor that has resistance to ground <1 GΩ can be considered adequately grounded.

All conductors that make contact with ESDS devices shall be grounded unless covered by ESD risk evaluation and verified voltage requirements below. Small isolated conductors that do not make contact with ESDS devices (e.g. hand tool tips) are permitted. Where isolated conductors make contact with ESDS devices, the ESD risk should be evaluated and suitable ESD controls devised and documented in the ESD Control Program Plan.

Isolated conductors shall not make contact with ESDS devices unless the voltage can be demonstrated to be controlled to $< \pm 35$ V.

A.7.3.4 ESD Control Equipment

All equipment used in EPAs shall be qualified and approved for EPA use.

A.7.4 ESD Protective Packaging

Where ESD protective packaging is specified in customer contracts or other related documentation, the specified packaging shall be used to contain ESDS devices whenever they are to be taken out of the EPA. In all other cases, ESD shielding packaging shall be used to contain ESDS devices whenever they are to be taken out of the EPA.

ESD protective packaging used within the EPA may be static dissipative, conductive, or ESD shielding. Packaging materials that have surfaces that are insulators shall not be brought into the EPA.

Pink polythene shall not be brought into EPAs. The properties of this material are highly dependent on age and atmospheric humidity.

Marking of ESD protective packaging shall be as specified in Section A.7.5.

A.7.5 Marking of ESD-Related Items

A.7.5.1 General

Where marking of items for ESD control purposes is specified in customer contracts or other related documentation, the specified marking shall be used.

In other cases, where marking ESD control equipment or items for ESD control purposes is deemed necessary, it shall be defined in this plan.

A.7.5.2 Marking of ESD Protective Packaging

Where marking of ESD protective packaging is specified in customer contracts or other documentation, the specified marking shall be used.

In all other cases, ESD shielding packaging shall be used to contain ESDS whenever they are to be taken out of the EPA.

The outer surface of ESD protective packaging of ESDS devices taken out of the EPA shall be marked with a symbol identifying it as ESD protective packaging, for example the symbol given in Figure A.2. The main packaging function should be indicated by a letter code as follows:

- *S* to mean electrostatic discharge shielding
- *F* to mean electrostatic field shielding
- *C* to mean electrostatic conductive
- *D* to mean electrostatic dissipative

Figure A.2 ESD shielding packaging symbol. (Reproduced by permission of the EOS/ESD Association Inc.)

A.7.5.3 Marking of ESD Control Equipment Used in EPAs

ESD control equipment other than packaging that can be used inside or outside the EPA shall be identified by markings showing their ESD control properties. The preferred marking is the ESD protective symbol shown in Figure A.2 without the letter code.

Table A.2 EPA equipment requirements and test criteria.

ESD control item	Test method	Pass criteria	Test frequency
Work surface or storage rack	TP4. Measurement of resistance to ground from surface	< 1 GΩ	6 monthly
Floor	TP4. Measurement of resistance to ground from surface	< 1 GΩ	6 monthly
Seat	TP4. Measurement of resistance to ground from. surface	< 1 GΩ	6 monthly
ESD control garments	TP5. Point-to-point resistance of ESD garments	< 1 GΩ	6 monthly
ESD Protective packaging	TP8 Surface resistance	< 100 GΩ	6 monthly

A.8 Compliance Verification Plan

All items required for ESD control must be tested on a regular basis. Suitable test methods and equipment used shall be defined in company test procedures based on IEC TR 61340-5-4:2019.

The test procedures used, pass criteria, and frequency of testing are given in Tables A.1 and A.2. The ambient temperature and humidity at the time of test shall be recorded. Test records shall be kept by the ESD coordinator for a minimum of two years.

When possible, occasional checks should be made under low ambient humidity conditions (< 30% rh).

A.9 ESD Training Plan

A.9.1 General Requirements of the ESD Training Plan

The personnel who need ESD training and the type of training given are defined in Table A.3. Suitable tests shall be defined for evaluation of trainee comprehension and ability to execute required ESD control procedures.

Table A.3 ESD training matrix.

Personnel	ESD awareness	ESD audit and measurements	EPA cleaning	Principles and practice of ESD control
Managers and supervisors responsible for EPAs	yes			
All personnel who handle ESDS or enter the EPA	yes			
ESD coordinator				yes
Cleaners	yes		yes	
ESD auditing personnel	yes	yes		

A.9.2 Training Records

Current ESD training records shall be kept by the human resources department.

A.9.3 Training Content and Frequency

A.9.3.1 ESD Awareness Training

ESD awareness training shall be given to any person who is to enter and work in the EPA, before they first enter the EPA. Exceptions to this requirement are the ESD coordinator and visitors accompanied by trained personnel.

Refresher ESD awareness training shall be given on an annual basis.

The ESD awareness course shall cover the following:

- Awareness of the need for ESD control
- Demonstration of basic static electricity
- Recognition of EPAs and EPA boundaries
- Recognition of EPA compliant and noncompliant materials and equipment
- Identification of ESDS
- Recognition and correct handling of essential insulators such as papers
- What to do, and not to do, in an EPA
- Use and testing of wrist straps and ESD control footwear
- Introduction to the ESD coordinator and their role

Trainees shall be tested, for example, by:

- Multiple choice questionnaire covering key ESD awareness points
- Practical use and testing of wrist straps and ESD control footwear

A pass mark of 80% shall be achieved and correct use and testing of wrist straps and ESD control footwear demonstrated by the trainee before they are allowed to work in the EPA.

A.9.3.2 ESD Audit and Measurements Training

ESD audit and measurements training shall be given to personnel who are required to check and test ESD control equipment, before they start these duties. Refresher training may be required at the discretion of the ESD coordinator.

The ESD audit and measurement training shall cover the following:

- Hands-on experience of compliance verification measurement of EPA equipment
- Basic EPA auditing techniques

Trainees shall be tested by an observed practical measurement and audit practice. Correct use of the organization's equipment and procedures by the trainee shall be observed before they commence audit and measurement work.

A.9.3.3 EPA Cleaning Training

EPA cleaning training shall be given to personnel who are required to clean within an EPA facility, before they start these duties. Refresher training may be required at the discretion of the ESD coordinator.

The EPA cleaning training shall cover the following:

- What to do and what not to do as a cleaner working in an EPA

- Cleaning materials and techniques for the EPA floor and bench mats

Trainees shall be tested by questionnaire and observation of practice.

A.9.3.4 Principles and Practice of ESD Control Training

Training in the principles and practice of ESD control shall be given to the ESD coordinator, before they commence responsibility for the ESD control program. Continuous professional development in the subject shall be undertaken.

Training in the principle and practice of ESD control will be conducted by techniques such as the following:

- External courses according to availability
- Reading of books, articles, and other resources
- Reading of standards and associated user guides and other standardization documents
- Familiarisation with existing ESD Control Program Plans and facilities

Where available, study for relevant ESD control qualifications should be encouraged.

A.10 ESD Control Product Qualification

Unless otherwise specified, all ESD control materials, equipment used in EPAs, and ESD protective packaging selected shall as a minimum have a data sheet or other document showing compliance with IEC 61340-5-1 or ANSI/ESD S20.20 and requirements of this ESD Program Plan.

Table A.4 ESD control product qualification tests and pass criteria.

ESD control item	Qualification procedure	Pass criterion
ESD control footwear, while worn	QP1. Resistance from person to ground while standing on ESD floor	Resistance from the wearer's body to ground >100 kΩ and <1 GΩ
	QP2. Body voltage walk test of footwear and flooring system	No peaks in body voltage greater than 100 V
Wrist strap	TP1	>750 kΩ <35 MΩ
Floor mat	QP3. Resistance to groundable point of work surfaces or mats	<1 GΩ
Chair or seat	TP4. Measurement of resistance to ground from surface	<1 GΩ
Work surface on which ESDS may be placed	QP3. Resistance to groundable point of work surfaces or mats	<1 GΩ
	QP4. Point-to-point resistance of work surface	<1 GΩ
ESD control garment	TP5. Point-to-point resistance of ESD garments	< 1 GΩ
ESD Protective packaging	TP8 Surface resistance	< 100 GΩ

The ESD coordinator shall maintain a list of items and equipment approved for use in EPAs. Listed items shall be identified by manufacturer, type, or other specific information.

Further product qualification requirements and procedures for specific ESD control items are given in Table A.4.

Proprietary wrist strap and footwear checkers used for daily checking of wrist straps and footwear shall be verified to have pass criteria as specified in Table A.1.

References

EOS/ESD Association Inc. (2014) ANSI/ESD S20.20-2014. *ESD Association Standard for the Development of an Electrostatic Discharge Control Program for – Protection of Electrical and Electronic Parts, Assemblies and Equipment (excluding Electrically Initiated Explosive Devices).* Rome, NY, EOS/ESD Association Inc.

International Electrotechnical Commission. (2015) IEC 61340-5-3:2015. *Electrostatics. Protection of electronic devices from electrostatic phenomena. Properties and requirements classifications for packaging intended for electrostatic discharge sensitive devices.* Geneva, IEC.

International Electrotechnical Commission (2016) IEC 61340-5-1: 2016. *Electrostatics – Part 5-1: Protection of electronic devices from electrostatic phenomena - General requirements.* Geneva, IEC.

International Electrotechnical Commission. (2019) IEC TR 61340-5-4:2019. *Electrostatics - Part 5-4: Protection of electronic devices from electrostatic phenomena – Compliance Verification.* Geneva, IEC.

Index

e